GRASSES AND GRASSLAND ECOLOGY

Grasses and Grassland Ecology

David J. Gibson
Southern Illinois University, Carbondale

OXFORD
UNIVERSITY PRESS

Great Clarendon Street, Oxford, OX2 6DP,
United Kingdom

Oxford University Press is a department of the University of Oxford.
It furthers the University's objective of excellence in research, scholarship,
and education by publishing worldwide. Oxford is a registered trade mark of
Oxford University Press in the UK and in certain other countries

© Oxford Univeristy Press 2009

The moral rights of the authors have been asserted

First Edition published in 2009

All rights reserved. No part of this publication may be reproduced, stored in
a retrieval system, or transmitted, in any form or by any means, without the
prior permission in writing of Oxford University Press, or as expressly permitted
by law, by licence or under terms agreed with the appropriate reprographics
rights organization. Enquiries concerning reproduction outside the scope of the
above should be sent to the Rights Department, Oxford University Press, at the
address above

You must not circulate this work in any other form
and you must impose this same condition on any acquirer

Published in the United States of America by Oxford University Press
198 Madison Avenue, New York, NY 10016, United States of America

British Library Cataloguing in Publication Data

Data available

Library of Congress Cataloging in Publication Data

Data available

ISBN 978–0–19–852919–4

Links to third party websites are provided by Oxford in good faith and
for information only. Oxford disclaims any responsibility for the materials
contained in any third party website referenced in this work.

Preface

>...a vast expanse of level ground unbroken save by one thin line of trees which scarcely amounted to a scratch upon the great blank; until it met the glowing sky wherein it seemed to dip...There it lay, a tranquil sea or lake without water...with the day going down upon it...it was lovely and wild, but oppressive in its barren monotony.
>
> Charles Dickens (*American Notes for General Circulation*, 1842) describing Looking Glass Prairie near Lebanon, Illinois

Grasslands evoke emotion, they are the largest biome on Earth, they represent a tremendous source of biodiversity, they provide important goods and services, and they are the place where as a species, we first stood up and walked. As a result, grasses and grasslands are widely studied. However, after teaching an upper level/postgraduate course on grassland ecology for several years, I realized that a suitable textbook was needed. Hence, I have written this book in the hope that it will be useful not just for the students taking my grassland course, but researchers, land managers, and anyone who has an interest in grasslands anywhere in the world.

The book brings together a huge literature from ecological, natural history, and agricultural disciplines. The nomenclature for plant names is synonymized according to the USDA Plants National Database (http://plants.usda.gov/index.html) as of May 2008, or, for species not in the database, the name used in the original source. As a result, names are changed from the original source when a more up-to-date name is provided in the database. Some familiar plants have new names: for example, where I talk about tall fescue I use *Schedonorus phoenix* in place of the older *Festuca arundinacea*. Older names are included in the index with a reference to the new name.

My own experience in grasslands results from an early and abiding love of the chalk grasslands of southern England. The dune grasslands of Newborough Warren, Anglesey and the montane grasslands of Snowdonia, Wales have also been important to me ever since my days as a doctoral student at the University College of North Wales (now the University of Bangor). My research in grasslands has been inspired by many mentors and colleagues, but I must mention in particular Paul Risser (formerly of the University of Oklahoma), the late Lloyd Hulbert (Kansas State University), and the Long Term Ecological Research group at Konza Prairie during the late 1980s as having a large and enduring influence on my understanding of the North American prairie. The writing of this book follows from the start that they gave me.

It takes a long time to write a book, but I am grateful to Southern Illinois University Carbondale for providing a sabbatical leave in Spring 2006 that allowed time to write several chapters. Many, many people helped me write this book. In particular, I thank colleagues who answered my questions and read drafts of various chapters including Roger Anderson (Chapter 10), Elizabeth Bach (Chapter 1), Sara Baer (Chapters 7 and 10), Ray Callaway (Chapter 6), Ryan Campbell (Index), Gregg Cheplick (Chapter 5), Keith Clay (Chapter 5), Jim Detling (Chapter 9), Stephen Ebbs (Chapter 4), Don Faber-Langendoen (Chapter 8), Richard Groves (Chapter 8), Trevor Hodkinson (Chapter 2), Allison Lambert (Chapters 1, 2, 3, 4, and 8), Susana Perelman (Chapter 8), Wayne Polley (Chapter 4), David Pyke (Chapter 10), Steve Renvoize (Chapter 3), Paul Risser (Chapter 1), Tim Seastedt (Chapter 7), Rob Soreng (Chapter 2), and Dale Vitt (Chapter 8). Thanks to Daniel Nickrent, Hongyan Liu, Gervasio Piñeiro, Sam McNaughton, Dale Vitt, Steve Wilson, and Zicheng Yu for allowing me to reproduce their excellent photographs. John Briggs kindly supplied the satellite image for Plate 13 and Howard Epstein an original for Fig 4.3. Cheryl Broadie and Steve Mueller of the SIUC IMAGE facility were a tremendous help in preparing several of the photographs and figures. Helen Eaton, Ian Sherman, and the staff at Oxford University Press have always been tirelessly helpful in ensuring that this book was actually published. Finally, grateful thanks to my wife Lisa and our children Lacey and Dylan for providing the love and emotional support necessary for the marathon of writing a book.

Carbondale, Illinois D.J.G
May 2008

Contents

Preface   v

1 Introduction   1
1.1 Grasslands: a tautological problem of definition   1
1.2 Extent of the world's grasslands   2
1.3 Grassland loss   4
1.4 Grassland goods and services   12
1.5 Early grassland ecologists   18

2 Systematics and evolution   21
2.1 Characteristics of the Poaceae   21
2.2 Traditional vs modern views of grass classification   22
2.3 Subfamily characteristics   24
2.4 Fossil history and evolution   29

3 Ecological morphology and anatomy   35
3.1 Developmental morphology—the phytomer   35
3.2 Structure of the common oat *Avena sativa*   36
3.3 Culms   36
3.4 Leaves   43
3.5 Roots   46
3.6 Inflorescence and the spikelet   49
3.7 The grass seed and seedling development   51
3.8 Anatomy   54

4 Physiology   58
4.1 C_3 and C_4 photosynthesis   58
4.2 Forage quality   68
4.3 Secondary compounds: anti-herbivore defences and allelochemicals   73
4.4 Silicon   78
4.5 Physiological integration of clonal grasses and mechanisms of ramet regulation   79

5 Population ecology   81
5.1 Reproduction and population dynamics   81
5.2 Fungal relationships   94
5.3 Genecology   102

6 Community ecology — 110
6.1 Vegetation–environment relationships — 110
6.2 Succession — 113
6.3 Species interactions — 115
6.4 Models of grassland community structure — 121
6.5 Summary: an issue of scale — 128

7 Ecosystem ecology — 129
7.1 Energy and productivity — 129
7.2 Nutrient cycling — 141
7.3 Decomposition — 149
7.4 Grassland soils — 153

8 World grasslands — 160
8.1 Ways of describing vegetation — 160
8.2 General description of world grasslands — 162
8.3 Examples of regional grassland classifications — 179

9 Disturbance — 184
9.1 The concept of disturbance — 185
9.2 Fire — 187
9.3 Herbivory — 194
9.4 Drought — 207

10 Management and restoration — 211
10.1 Management techniques and goals — 211
10.2 Range assessment — 221
10.3 Restoration — 235

References — 243
Plant index — 289
Animal index — 297
Subject index — 299

CHAPTER 1

Introduction

Each grass-covered hillside is an open book for those who care to read. Upon its pages are written the conditions of the present, the events of the past, and a forecast of those of the future. Some see without understanding; but let us look closely and understandingly, and act wisely, and in time bring our methods of land use and conservation activities into close harmony with the dictates of Nature.

John E. Weaver (1954)

The purpose of this first chapter is to introduce the grassland biome. Grasslands are the most extensive, arguably the most useful to human society, yet the most threatened biome on the planet. Nevertheless, it is surprisingly difficult to unambiguously define grassland. Here we seek such a definition (§1.1); discuss where grasslands occur (§1.2); how they are being lost, fragmented, and degraded (§1.3); and summarize their outstanding and immense value (§1.4). The chapter finishes with a brief biography of John Bews and John Weaver, two of the early pioneers of grassland ecology (§1.5).

1.1 Grasslands: a tautological problem of definition

At its simplest, grassland can be defined as 'a habitat dominated by grasses', however a more useful, strict, but all-encompassing definition is elusive (Table 1.1). Indeed, many authorities do not provide a definition, assuming merely that we know a grassland when we see one. Others prefer to define a grassland by the absence of specific features; for example, Milner and Hughes (1968) offered a floristic definition (Table 1.1), but added that a more useful approach is to consider grassland physiognomically or structurally as 'a plant community with a low-growing plant cover of non-woody species.'

The main points that these varied definitions have in common are the prevalence of grasses (members of the Poaceae), an infrequent or a low abundance of woody vegetation, and a generally arid climate. Risser's (1988) definition is perhaps the most comprehensive as it encompasses these ideas. Other factors that are important and which together help characterize natural grasslands in many parts of the world include deep, fertile, organic-rich soils (frequently Chernozems—see Chapter 7), frequent natural fire (Chapter 9), and large herds of grazing mammals (Chapter 9). Semi-natural or seeded grasslands (e.g. amenity grasslands: Chapter 8) may lack some of these features, especially natural disturbance.

As with most disciplines there is a series of terms, almost a language, relating specifically to grasslands (Table 1.2). A number of these terms have their origin in a long history of grassland management for grazing, i.e. range management (Chapter 10). Many of these terms have become formalized by the appropriate professional organizations, e.g. the Forage and Grazing Terminology Committee representing more than half a dozen agencies and societies from the USA, Australia, and New Zealand. Several of the terms listed in Table 1.2 are often used synonymously (e.g. prairie and grassland in the North American Midwest), but others refer specifically to a particular type of grassland (e.g. savannah) or geographic location (e.g. veld, spinifex). The distinction between **meadow** and **pasture**, reflecting their use and management through mowing and grazing respectively, goes back hundreds of years and is reflected in several European languages; e.g. French *pré* and *prairie*, German *Wiese* and *Weide*, and the Latin *pratum* and *pascuum* (Rackham 1986). Terms such as **forage** and **herbage** refer specifically to the fraction

Table 1.1 Grassland definitions

Definition	Source
'…the great empty middle of our continent…is indivisible. It endures…aridity is the first defining and implacable factor…a place of journeys…treeless plains…'	Manning (1995)
'…a plant community in which the Gramineae are dominants and trees absent.'	Milner and Hughes (1968)
'…a vegetation type dominated by grasses but containing many broadleaf herbs (forbs).'	Bazzaz and Parrish (1982)
'Land on which the vegetation is dominated by grasses.'	The Forage and Grazing Terminology Committee (1992)
'A general lack of woody vegetation helps define grasslands…'	Knapp and Seastedt (1998)
'[grasslands]…are dominated primarily by grasses (Gramineae) and grass-like plants (mostly Cyperaceae)…climates generally have distinct wet and dry seasons and are noted for temperature and precipitation extremes.'	Sims (1988)
'…terrestrial ecosystems dominated by herbaceous and shrub vegetation, and maintained by fire, grazing, drought and/or freezing temperatures.'	Pilot Assessment of Global Ecosystems; White et al. (2000).
'One-fourth or more of the total vegetation consists of primarily herbaceous communities in which the Gramineae are the dominant life form…the grasses give character and unity of vegetal structure to the landscape….an overstory of scattered trees and shrubs may be present.'	Kucera (1981)
'…any plant community, including harvested forages, in which grasses and/or legumes make up the dominant vegetation.'	Barnes and Nelson (2003)
'A region with sufficient average annual precipitation (25–75 cm [10–30 inches]) to support grass but not trees.'	Stiling (1999)
'<1 tree per 5 acres…on slopes of 2–4%'	Anderson (1991)
'…types of vegetation that are subject to periodic drought, that have a canopy dominated by grass and grasslike species, and that grow where there are fewer than 10 to 15 trees per hectare.'	Risser (1988)

of grassland useful to the range manager. Forage is a particularly relevant and important component since it defines the portion used as feed for domestic herbivores. Forage science is a specialized agricultural discipline in its own right (e.g. Barnes et al. 2003). Overall, the terms and language associated with grasslands reflect both what these ecosystems have in common and their diversity related to an extensive, worldwide distribution.

1.2 Extent of the world's grasslands

Globally, grasslands occur on every continent (excluding Antarctica) occupying 41–56 × 10^6 km^2, covering 31–43% of the earth's surface (World Resources 2000–2001, and references therein) (Plate 1). This range in estimates reflects differences in defining grassland by different authorities, particularly in the extent to which cropland, tundra, or shrublands are included. The Pilot Analysis of Global Ecosystems (PAGE) Classification, the most widely accepted account, estimates grasslands as covering 52 544 000 km^2 or 40.5% of the total land area (White et al. 2000; World Resources 2000–2001). The PAGE Classification of grasslands excludes urban areas as defined by the Nighttime Lights of the World Database (estimated at 1010 km^2) but is nevertheless broad, including savannahs (17.9 × 10^6 km^2), open and closed shrublands (16.5 × 10^6 km^2), and tundra (7.4 × 10^6 km^2), as well as non-woody grasslands (10.7 × 10^6 km^2). The PAGE classification is based on satellite imagery of land cover from the International Geosphere-Biosphere Programme Data and Information System (IGBP-DIS) DISCover 1 km resolution land cover maps obtained using the Advanced Very High Resolution Radiometer (AVHRR) data (Loveland et al. 2000). Under the PAGE classification, grasslands occupy more of the earth's surface than the other major cover types, i.e. forests (28.97 × 10^6 km^2) or agriculture (36.23 × 10^6 km^2) (White et al. 2000). Nearly 800 × 10^6 people live throughout this large

Table 1.2 Grassland terminology (source indicated where appropriate). The definitions reproduced here from Thomas (1980) are derived and modified from those also provided by Hodgson (1979)

Term	Definition	Source
Forage	Any plant material, including herbage but excluding concentrates, used as feed for domestic herbivore	(Thomas 1980)
Herbage	The above-ground parts of a sward viewed as an accumulation of plant material with characteristics of mass and nutritive value	(Thomas 1980)
Meadow	A tract of grassland where productivity of indigenous or introduced forage is modified due to characteristics of the landscape position or hydrology, e.g. hay meadow, wet meadow	(Forage and Grazing Terminology Committee 1992)
Pasture	A type of grazing management unit enclosed and separated from other areas by fencing or other barriers and devoted to the production of forage for harvest primarily by grazing	(Forage and Grazing Terminology Committee 1992)
Pastureland	Land devoted to the production of indigenous or introduced forage for harvest primarily by grazing.	(Forage and Grazing Terminology Committee 1992)
Permanent pasture	Pastureland composed of perennial or self-seeding annual plants that are grazed annually, generally for 10 or more successive years	(Barnes and Nelson 2003)
Prairie	French term for grassland, used now to describe the grasslands of the North American Great Plains. Defined in the USA as nearly level or rolling grassland, originally treeless, and usually characterized by fertile soil	(Forage and Grazing Terminology Committee 1992)
Rangeland	American term for land on which the indigenous vegetation is predominantly grasses, grass-like plants, forbs, or shrubs and is managed as a natural ecosystem	(Forage and Grazing Terminology Committee 1992)
Sod	A piece of turf dug up, or pulled up by grazing animals	(ThomasF 1980)
Savannah	Grassland with scattered trees or shrubs; often a transitional type between true grassland and forestland and accompanied by a climate with alternating wet and dry seasons	(Forage and Grazing Terminology Committee 1992)
Spinifex	Australian steppe area dominated by the grass spinifex (*Triodia* spp.) with occasional shrubs (*Acacia* spp.) and low trees (*Eucalyptus* spp.)	(Skerman and Riveros 1990)
Steppe	Semi-arid grassland characterized by short grasses occurring in scattered bunches with other herbaceous vegetation and occasional woody species.	(Forage and Grazing Terminology Committee 1992)
Sward	An area of grassland with a short (say <1 m tall) continuous foliage cover, including both above-ground and below-ground plant parts but excluding any woody plants.	(Thomas 1980)
Veld	Afrikaans term for South African grassland with scattered trees of shrubs	Bews (1918)

area, more than the number in forests ($c.450 \times 10^6$ people), but less than those in agricultural areas (2.8×10^9 people: 1995 estimates). Most of the 800×10^6 grassland inhabitants live in savannah regions (413×10^6), and within the savannah region the majority (266×10^6) are in sub-Saharan Africa (White *et al.* 2000). By contrast, the least populated grassland region is the tundra (11×10^6) and within that only 104 000 people live in North American tundra grasslands.

Across the globe, grasslands are most extensive in sub-Saharan Africa (14.46×10^6 km²) followed by Asia (excluding the Middle East) at 8.89×10^6 km², and the grasslands of Europe, North America, and Oceania (including New Zealand and Australia) at $6000–7000 \times 10^6$ km² each (Table 1.3). On an area basis, 11 countries across the globe have each in excess of 1×10^6 km² of grassland area, with the top country being Australia with 6.6×10^6 km² (Table 1.4). It is notable that each of the major regions of the world is represented in this list. On a percentage area basis, 11 countries have >80% of their area under grassland (Table 1.5), with Benin having the largest percentage at 93.1%. Of these countries, 9 are in sub-Saharan Africa, and all are quite small with less than 1×10^6 km² land area (Mozambique is the largest at 825 606 km²). Australia, with the largest expanse of grassland of

Table 1.3 Grassland area (km² × 10⁶) and population in world regions, excluding Greenland and Antarctica. Adapted from White et al. (2000)

Region	Savannah	Shrubland	Non-woody grassland	Tundra	Global grassland	Population (000)
Asia[a]	0.90	3.76	4.03	0.21	8.89	249 771
Europe	1.83	0.49	0.70	3.93	6.96	20 821
Middle East and North Africa	0.17	2.11	0.57	0.02	2.87	110 725
Sub-Saharan Africa	10.33	2.35	1.79	0.00	14.46	312 170
North America	0.32	2.02	1.22	3.02	6.58	6 125
Central America and Caribbean	0.30	0.44	0.30	0.00	1.05	30 347
South America	1.57	1.40	1.63	0.26	4.87	56 347
Oceania	2.45	3.91	0.50	0.00	6.86	3 761
World	17.87	16.48	10.74	7.44	52.53	789 992

[a] Excluding Middle Eastern countries.

Table 1.4 Top countries for grassland area (countries with >1 000 000 km² of grassland)

Country	Region[a]	Total land area (km²)	Total grassland area (km²)
Australia	Oceania	7 704 716	6 576 417
Russian Federation	Europe	16 851 600	6 256 518
China	Asia	9 336 856	3 919 452
United States	North America	9 453 224	3 384 086
Canada	North America	9 908 913	3 167 559
Kazakhstan	Asia	2 715 317	1 670 581
Brazil	South America	8 506 268	1 528 305
Argentina	South America	2 781 237	1 462 884
Mongolia	Asia	1 558 853	1 307 746
Sudan	Sub-Saharan Africa	2 490 706	1 292 163
Angola	Sub-Saharan Africa	1 252 365	1 000 087

[a] Asia excludes Middle Eastern countries.

From White et al. (2000).

any country, has the sixth highest percentage land area under grassland (85.4%).

Grasslands contribute ecosystem functions to many of the world's major watersheds; i.e. water catchment areas that provide a functional unit of the landscape. A survey of 145 mapped watersheds representing 55% of the world's land area (excluding Greenland and Antarctica) showed 25 watersheds possessing >50% grassland cover. These watersheds include the Senegal, Niger, Volta, Nile, Turkana, Shaballe, Jubba, Zambezi, Okavango, Orange, Limpopo, Mangoky, and Mania, of which 13 are in Africa, 5 in Asia, 3 in South America, 2 in North America, 1 shared between North and Central America, and 1 in Oceania (Fig. 1.1). None of the mapped watersheds in Europe had more than 25% grassland.

1.3 Grassland loss

Covering such a large area of the earth's land surface, it is no surprise that grasslands have been

Table 1.5 Top countries for percentage of grassland area (countries with >80% grassland area)

Country	Region[a]	Total land area (km²)	Grassland area (%)	International tourists (000 yr⁻¹) (±% 10 yr change)[b]	International tourist receipts (million US$ yr⁻¹) (±10 yr change)[b]
Benin	Sub-Saharan Africa	116 689	93.1	145 (+150)	28 (−7)
Central African Republic	Sub-Saharan Africa	621 192	89.2	23 (+331)	5 (−6)
Botswana	Sub-Saharan Africa	579 948	87.8	693 (+160)	174 (+361)
Togo	Sub-Saharan Africa	57 386	87.2	ND	ND
Somalia	Sub-Saharan Africa	639 004	86.7	10 (−74)	ND
Australia	Oceania	7 704 716	85.4	4059 (+180)	8503 (+471)
Burkina Faso	Sub-Saharan Africa	273 320	84.7	693 (+160)	32 (+459)
Mongolia	Asia	1 558 853	83.9	87 (−57)	21 (ND)
Guinea	Sub-Saharan Africa	246 104	83.5	96 (ND)	4 (ND)
Mozambique	Sub-Saharan Africa	788 938	81.6	ND	ND
Namibia	Sub-Saharan Africa	825 606	80.6	405 (ND)	214 (ND)

[a] Asia excludes Middle East countries.
[b] Change from 1985–1987 to 1995–1997.
ND, no available data.
From White et al. (2000).

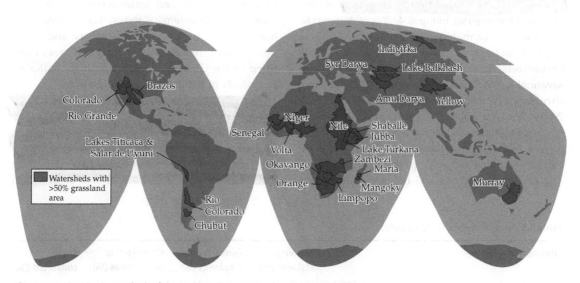

Figure 1.1 Grassland watersheds of the world. With permission from White et al. (2000).

heavily used throughout human history. For example, in Australia, a close interaction between humans and grassland fire is though to have existed for 40 000 years or more (Gillison 1992). In Papua New Guinea, a similar relationship with slash-and-burn agriculture is thought to have existed for at least 9000 years (Gillison 1992). Indeed, it is widely held that *Homo sapiens* may have arisen in the savannahs of Africa (Stringer 2003). Further back, grasslands are associated with the co-evolution of major plant and animal groups—the grasses and grazing mammals, respectively (Chapter 2).

Grasslands have been lived in and used by people throughout human history. Inevitably this has led to tremendous changes, and most recently the loss of much of this biome. The major modifications to grassland cover are due to:

- agriculture
- fragmentation
- invasive non-native species
- fire (lack of)
- desertification
- urbanization/human settlements
- domestic livestock.

Of these, the first three—agriculture, fragmentation, and non-native species—pose perhaps the greatest threat to native grasslands and are discussed below. The extent of urbanization is reported in Table 1.6, desertification is discussed in Chapter 8, and the effects of fire and domestic grazers are discussed in Chapters 9 and 10.

Globally, there has been large-scale conversion of grassland to human-dominated uses; of the world's 13 terrestrial biomes, 45.8% of temperate grasslands, savannahs and shrublands, 23.6% of tropical/subtropical grasslands, savannahs, and shrublands, 26.6% of flooded grasslands and savannahs, and 12.7% of montane grasslands and shrublands have been converted (Hoekstra *et al.* 2005). With only 4.6% of the habitat protected, temperate grasslands, savannahs, and shrublands have a higher Conservation Risk Index (ratio of habitat converted to habitat protected = 10:1) than any of the other terrestrial biomes. This means that habitat conversion exceeds protection in grasslands more than in any other terrestrial biome; only 1 ha of grassland is protected for every 10 ha lost. The Conservation Science Program–World Wildlife Fund–US Global 200 programme identified 17 grassland ecoregions worldwide that are 'critically endangered' (Table 1.7), with an additional 13 that are considered 'vulnerable' (Olson and Dinerstein 2002). Ecoregions are fine-scale regional ecological areas within a biome that are characterized by local geography and climate and a unique assemblages of species. The critically endangered and vulnerable grassland ecoregions comprise some of the world's most diverse and spectacular grasslands, including, for example, the Terai–Duar savannahs and grasslands in southern Asia, which are alluvial grasslands dominated by 7 m high *Saccharum* spp. (elephant grass) and Asia's highest density of tigers, rhinos, and ungulates. Other critically endangered areas include the South African fynbos and south-west Australian forests and scrub ecoregions, both of which include extensive grassy components and harbour very high levels of diversity and endemism. All of the critically endangered areas identified in the Global 200 programme are suffering the effects of habitat loss and alteration.

The greatest alteration to grasslands worldwide has been through transformation to agricultural land creating in many places a grassland/agricultural mosaic, or in other areas wholesale conversion to agriculture (Fig. 1.2). The greatest loss of grassland area is in North

Table 1.6 Estimated remaining and converted grassland (%)

Continent and region	Remaining in grasslands (%)	Converted to croplands (%)	Converted to urban areas (%)	Total converted (%)
N. America: Tallgrass prairie in the United States	9.4	71.2	18.7	89.9
S. America: Cerrado woodland and savannah in Brazil, Paraguay, and Bolivia	21.0	71.0	5.0	76.0
Asia: Daurian steppe in Mongolia, Russia, and China	71.7	19.9	1.5	21.4
Africa: Central and eastern Mopan and Miombo in Tanzania, Rwanda, Burundi, Dem. Rep. Congo, Zambia, Botswana, Zimbabwe and Mozambique	73.3	19.1	0.4	19.5
Oceania: South-west Australian shrublands and woodlands	56.7	37.2	1.8	39.0

From White *et al.* (2000).

America where only 9.4% of the original tallgrass prairie remains (Table 1.6). Locally the loss can be even greater, as in the state of Illinois, where only 0.01% (9.5 km²) of high-quality native prairie remained by 1978 (Illinois Department of Energy and Natural Resources 1994). Ten other states have also reported declines of >90% in the extent of tallgrass prairie. Globally, large areas of grassland

Table 1.7 Critically endangered grassland ecoregions

Major habitat	Biogeographic realm	Ecoregion
Tropical and subtropical grasslands, savannahs, and shrublands	Afrotropical	Sudanian savannahs
	Indo-Malayan	Terai-Duar savannahs and grasslands
Temperate grasslands, savannahs, and shrublands	Nearctic	Northern prairie
	Neotropical	Patagonian steppe
Flooded grasslands and savannahs	Afrotropical	Sudd-Sahelian flooded grasslands and savannahs
	Indo-Malayan	Rann of Kutch flooded grasslands
	Neotropical	Pantanal flooded savannahs
Montane grasslands and shrublands	Afrotropical	Ethiopian highlands
		Southern rift montane woodlands
		Drakensberg montane shrublands and woodlands
Mediterranean forests, woodlands, and scrub	Afrotropical	Fynbos
	Australasia	South-western Australia forests and scrub
	Nearctic	California chaparral and woodlands
	Neotropical	Chilean matorral
	Palearctic	Mediterranean forests, woodlands, and scrub
Deserts and xeric shrublands	Afrotropical	Madagascar spiny thicket
	Australasia	Carnavon xeric scrub

From Olson *et al.* (2000).

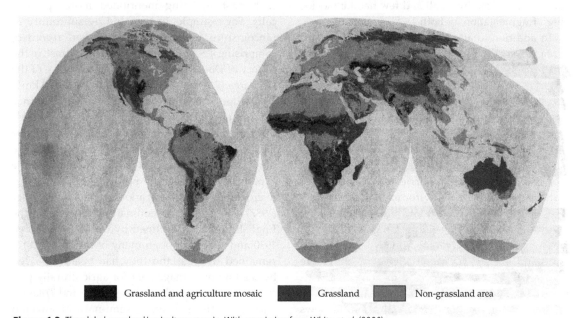

Grassland and agriculture mosaic   Grassland   Non-grassland area

Figure 1.2 The global grassland/agriculture mosaic. With permission from White *et al.* (2000).

have also been converted to agriculture in South America and Oceania (21% and 56.7% remaining, respectively). In all areas, grassland loss is predominantly through conversion to cropland rather than to urban areas. Excluding agricultural mosaics from remaining grasslands reduces global grassland area by $c.7.1 \times 10^6$ km^2, particularly in sub-Saharan Africa (3.5×10^6 km^2) (White et al. 2000). Substantial areas in South America (1.4×10^6 km^2) and Asia (1.2×10^6 km^2) have also been altered by agriculture (Fig. 1.2).

Wholesale loss to cropland leads to immediate loss of grassland, but fragmentation, in which blocks of habitat are dissected into smaller units, can produce a more insidious and gradual decline in ecosystem structure and function. The extent of grassland fragmentation was illustrated by White et al. (2000) who reported that 37% of 90 grassland regions analysed in North America and Latin America occurred as small linear patches. In another analysis, they showed that road networks in the Great Plains of the USA fragmented 70% of the prairie into patches of <1000 km^2. The extent of grassland fragmentation can be deceptive at first glance for without factoring in the road networks, it appeared that 90% of the grassland was composed of blocks of 10 000 km^2 or more. Globally, White et al. (2000) estimated that nearly 37% of grasslands are characterized by small and few habitat blocks, high fragmentation, or both.

In addition to affecting estimates of grassland area, fragmentation increases the edge to area ratio of surviving patches, leading to a smaller effective undisturbed interior. Edge habitats may be structurally and physically different from the interior with compacted or altered soil (including agricultural run-off adjacent to crop fields), and a high density of woody vegetation. Fragmentation can alter the natural disturbance regime; roads, for example, provide a barrier to the spread of landscape fire. These environmental differences can lead to several direct and indirect effects on the flora and fauna including a high diversity of exotic species, low diversity of native species, low nest success, and high predation rates on native birds. Fragmented populations may be genetically isolated, reduced in size, and susceptible to inbreeding, genetic drift, and extinction. Reviews of numerous studies attest to these detrimental affects (Saunders et al. 1991). For example, in Ohio grasslands, the upland sandpiper, bobolink, savannah sparrow, and Henslow's sparrow, were found to be area-sensitive being encountered most frequently only on grassland tracts >0.5 km^2 (Swanson 1996). Habitat fragmentation was found to disrupt plant–pollinator and predator–prey interactions in central European calcareous grasslands (Steffan-Dewenter and Tscharntke 2002). In grasslands of the US Great Plains, intensive land use following grassland fragmentation was shown to provide hotspots for exotic birds (Seig et al. 1999).

Paradoxically, many conservationists in Europe lament the decline of managed semi-natural grassland because many of these grasslands are species-rich. By the end of last Ice Age ($c.10 000$ BC), the British Isles were covered with forests, with grasslands being rare and restricted to a few high-altitude montane areas (Pennington 1974). Extensive forest clearance and farming began around 4000 BC and continued for centuries, into Roman times (Rackham 1986). Some of the earliest written records of grassland use are recorded in the Domesday Book, William the Conqueror's survey of Britain dating from 1086. Mowed meadows were the best-recorded land-use in Domesday, accounting for 1.2% of the land area surveyed ($c.1214$ km^2), with pasture being mentioned more sporadically. For example, in the Lindsey subcounty of Lincolnshire, 178 km^2 of meadow were recorded, comprising 4.5% of the land area. By contrast, in the county of Dorset, pasture accounted for 28% of the land area (716 km^2, Fig. 1.3), meadows 1% (28 km^2), and woodland 13% (328 km^2). However, since this anthropogenic expansion of grassland across the English landscape, in the last 200 years there has been an even more dramatic decline in these now-valued semi-grasslands concomitant with the expansion of grassland 'improvement', conversion to cropland, and urbanization. For example, Fuller (1987) estimated that semi-natural pastures in the English lowlands declined by over 97% between 1930 and 1984. Although many of these areas still remained in grassland they had been improved by seeding or management for agriculturally preferred species, notably *Lolium perenne* and *Trifolium repens*. Of the surviving semi-natural lowland

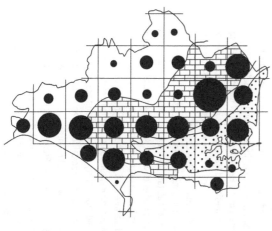

Figure 1.3 Area of pasture in Dorset, England in 1086 as recorded by the Domesday Book. Each black circle represents, at the scale of the map, the total area of pasture within each 10-km square. With permission from Rackham (1986).

▨ Chalk downland region
▨ Heathland region

grasslands, communities of special floristic interest represented only 1–2% (Blackstock et al. 1999).

Other areas around the world where species-poor grassland of low conservation value has expanded include the pastures of the eastern and south-eastern USA in which over 14×10^6 ha of the introduced European *Schedonorus phoenix* (syn. *Festuca arundinacea*) has been planted (Buckner and Bush 1979). Similarly, in California, 10×10^6 ha of native California prairie once characterized by perennial bunchgrasses including *Nassella pulchra* and *N. cernua* is now dominated by exotic annual grasses, including *Avena fatua, A. barbata, Bromus mollis, B. diandrus, B. rubens, Hordeum marinum, H. murinum, H. pusillum,* and *Vulpia myuros* (Heady et al. 1992). Sown pastures cover 436 000 km² (5.6%) in Australia, none of which were present before the arrival of European settlers in 1780 (Gillison 1992). In temperate areas, these sown pastures are dominated by European species including *Lolium perenne* and *Trifolium repens*. To the range manager these planted exotics are beneficial as they can provide nutritious forage year round.

However, to many conservationists, the exotic species listed above are indicative of the larger problem of invasion of grasslands by non-native species. In some cases, as noted above for the California prairie, exotics can completely replace the native species causing a fundamental alteration to the physiognomy and structure of the whole ecosystem. In other cases, a single species may be more insidious, seemingly moving into vacant microhabitats. Nevertheless, any exotic species introduction affects the complex dynamics of grassland to some extent (see §9.2.4). The extent to which grassland is subject to invasion by exotic species depends in part on the severity of any changes in the environment or to the natural disturbance regime. Overgrazing, drought, infrequent fire, excessive trampling, or other stresses can all allow exotics to establish; after which, changes can become self-reinforcing (Chaneton et al. 2002; Weaver et al. 1996). Fragmentation, mentioned above, and increased adjacency of other ecosystems such as agricultural land and forest, increases the influx of exotics through dispersal. Although many exotics are introduced and spread accidentally, others are deliberately introduced. For example, seeding with exotic species is used to rehabilitate degraded pastures in Australia (Noble et al. 1984). *Medicago* spp. have been widely used in winter rainfall, semi-arid zones with success, but the species have subsequently become widely naturalized.

The total extent to which exotic species have altered grasslands is unclear except where wholesale changes are evident as with the California prairie noted earlier. Floristically, the magnitude of exotic species invasion in grassland can be high. For example, 17% (70 of 410) of plant species on the Pawnee National Grasslands in eastern Colorado, USA, and 28% (16 of 56) of grasses in Badlands National Park, South Dakota, USA are exotics (Licht 1997). Of 100 organisms listed as the 'world's worst' exotics, 13 land plants that invade grassland were included (Table 1.8).

In addition to the widespread exotics listed in Table 1.8, numerous other exotics have invaded, become naturalized, and altered native grasslands on a more regional basis; these include the following (Poaceae except where indicated):

• *Centaurea* spp. (knapweeds: Asteraceae) and other thistles including *Cirsium vulgare* and *Carduus*

Table 1.8 "World's worst" invasive land plants of grassland

Name	Common name	Family	Life form	Origin	Where it's invading
Arundo donax (L.)	Giant cane	Poaceae	Perennial grass	Indian subcontinent	Subtropical and temperate principally in riparian zones and wetlands, but also chaparral, oak savannahs, and pastures in South Africa and the United States.
Chromolaena odorata (L.) King and Robinson	Agonoi	Asteraceae	Perennial shrub	Central and South America	Pastures in tropical Africa and Asia
Fallopia japonica var. *japonica* (Houtt.) Ronse Decraene	Donkey rhubarb	Polygonaceae	Herbaceous perennial	Japan	Dense clumps in riparian areas, woodland, and grassland in most of North America, Northern Europe, Australia and New Zealand. Spreads along river banks, and into derelict/development land, amenity areas and roadsides.
Hedychium gardnerianum Ker (Himal.)	Kahila garland-lily	Zingiberaceae	Herbaceous perennial	Eastern Himalayas	Wet habitats in the Federated States of Micronesia (Pohnpei), French Polynesia and Hawaii, New Zealand, South Africa, La Réunion, Jamaica and the Azores islands. Recorded from improved pastures in New Zealand.
Imperata cylindrica (L.) Beauv.	Alang-alang	Poaceae	Perennial grass	SE Asia, Philippines, China and Japan	Throughout the temperate and tropic zones (particularly the SE US) in all habitats from dry flatwoods to the margins of permanent bodies of water. Alters the natural fire regime, producing intense heat when it burns.
Lantana camara L.	Largeleaf lantana	Verbenaceae	Ornamental shrub	SW USA, Central and South America	Weed of pastures and other habitats in >50 countries throughout tropics, subtropics and temperate regions. Decreased productivity of pastures and poisons cattle.
Leucaena leucocephela (Lam.) De Wit	Leucaena	Fabaceae	Tree	Mexico and Central America	Widely promoted for tropical forage and reforestation, now spreading widely in >20 countries across all continents except Europe and Antarctica. Problem in open (often coastal or riverine) habitats, semi-natural, disturbed, degraded habitats and other ruderal sites. Specifically noted invading grasslands in Ghana, Florida, and Hawaii

Species	Common name	Family	Growth form	Native range	Notes
Mikania micrantha (L.) Kunth.	Mile-a-minute weed	Asteraceae	Perennial vine	Central and S. America	A weed of tea, rubber, oil palm and other crops, and forests and pastures in Australia, India, Bangladesh, Sri Lanka, Mauritius, Thailand, the Philippines, Malaysia, Indonesia, Papua New Guinea, and many of the Pacific islands
Mimosa pigra L.	Catclaw mimosa	Fabaceae	Woody shrub	Mexico, Central and South America	Introduced to Australia, and many countries in Africa and Asia. Invades and spreads along watercourses and seasonally flooded wetlands in tropic and subtropical regions. In the Northern Territories of Australia it is a problem in grassland (Mitchell grass) and low Eucalyptus and Melaleuca savannah
Pinus pinaster Aiton.	Cluster pine	Pinaceae	Tree	Mediterranean	Widely planted, now invading shrubland, forests, and grassland in South Africa, Chile, Australia, New Zealand. Regenerates profusely after fire and severely depletes the local water table
Psidium cattleianum Sabine	Strawberry guava	Myrtaceae	Evergreen shrub or small tree	Brazil	Mauritius, Hawaii, Polynesia, the Mascarenes, Seychelles, Norfolk Island, and peninsular Florida. A problem mostly in forests, but also invades grassland in south Florida and Micronesia
Pueraria montana var. lobata (Willd.) Maesen & S. Almeida	Kudzu	Fabaceae	Semi-woody vine	Asia	Widespread across $2-3 \times 10^6$ ha in eastern USA, \$500 million estimated losses per year. Also in Japan and New Zealand. Invades all habitats, including pastures, except periodically flooded areas
Schinus terebinthifolius Raddi	Brazilian holly	Anacardiaceae	Tree	Argentina, Paraguay, Brazil	Southern USA, Hawaii, and Spain. Pioneer of disturbed sites as well as undisturbed natural vegetation including pastures and grassland

From IUCN/SSC Invasive Species Specialist Group (2003).

acanthoides—rangeland of the US Midwest and rolling Pampas of Uruguay (Lejeune and Seastedt 2001; Soriano 1992).
- Annual grasses including *Bromus* spp. (brome, cheat grass), *Avena fatua* (wild oat)—throughout North American and Canadian grasslands, especially in the west (Hulbert 1955; Mark 1981).
- Annual legumes (Fabaceae) *Medicago* spp. (medic), *Stylosanthes* spp. (pencilflower), and *Trifolium* spp. (clover), and perennials including *Lolium perenne* (perennial ryegrass), *Dactylis glomerata* (cocksfoot), and *Trifolium repens* (Fabaceae, white clover)—naturalized following introduction in modified native grasslands and pastures in temperate and tropical Australia (Donald 1970; Moore 1993).
- *Acaena magellanica* (bidibid, Rosaceae, New Zealand native)—forms dense carpets replacing once extensive *Poa cookii* and *Pringlea antiscorbutica* communities in sub-Antarctic islands (Hnatiuk 1993).
- *Sporobolus indicus* (wire grass, South African native)—with the exotic woody *Psidium guayava* Raddi (Myrtaceae) dominates hills on Viti Levu and other Fijian islands (Gillison 1993).
- *Acacia mearnsii* (black wattle, Australian native)—cost of watershed invasion in South Africa estimated at $1.4 billion in water lost to impoundments through transpiration (de Wit *et al.* 2001).

1.4 Grassland goods and services

All grasslands are dominated by members of the Poaceae family, the fifth most speciose family (>7500 species) (Chapter 2), and the most widespread. Grasses are also the most important food crop on earth, with corn, wheat, maize, rice, and millet accounting for most crops grown for food. (The Fabaceae with the legumes including soybean, beans, lentils, pulses, and peas are also important food crops and perhaps come second.)

Efforts at assigning an economic value to ecosystems and ecosystem functions (Costanza *et al.* 1997) are difficult and uncertain at best, and certainly controversial (Pimm 1997; Sagoff 1997). Ecosystem services and functions can be categorized into four groups (Farber *et al.* 2006):

- supportive functions and structures (e.g. nutrient cycling, primary production, pollinator services)
- regulating services (e.g. sequestration of CO_2, prevention of soil loss, maintenance of soil fertility)
- provisioning services (e.g. plant materials and game)
- cultural services (e.g. ecotourism, scenery, religious value).

A problem is that many of the most valuable services are precisely those that have no clear market value (Sala and Paruelo 1997). On the basis of 17 ecosystem services and functions, the estimated the total value of the grass/rangeland biome was $232 ha^{-1} yr^{-1} (1997 $s); considerably less than forests ($969 ha^{-1} yr^{-1}) and wetlands ($14 785 ha^{-1} yr^{-1}), but 2.5 times that of cropland ($92 ha^{-1} yr^{-1}) (Costanza *et al.* 1997). Other, less controversial, economic models which specifically incorporate the value of grasslands include those of Ogelthorpe and Sanderson (1999), and Herendeen and Wildermuth (2002). Of these, Ogelthorpe and Sanderson's (1999) study is unique in combining a ecological-based floristic model based on the composition of grassland vegetation types with a more conventional management model to cost the supply of desired goods (ewes and lambs for market) and derive optimum policy. By contrast, Herendeen and Wildermuth (2002) conducted an economic analysis of an agricultural county in Kansas, USA, developing energy, water, soil, and nitrogen budgets. They showed that grazed rangeland (70% of county land area) was essentially self-sustaining, contributing little to any depletion of resources, independent of outside subsidies, and not disruptive of natural cycles. This contrasted sharply with other activities including row crop agriculture.

Williams and Diebel (1996) considered the economic value of grassland under two categories: use value and non-use value. The former includes services that people obtain from direct interaction with the grassland resource and consist of grazing livestock, harvesting native or cultivated plants, hunting wildlife, recreational activities (e.g. hiking, bird watching, photography), educational activities, erosion control and water quality enhancement, and research activities. By contrast, non-use values are those associated with intangible uses and include: existence and option, aesthetics, cultural-historical and sociological significance, ecological or biological mechanisms, and biological diversity (see §7.1.3

and §9.1). The authors bemoaned the difficulty of pricing these services but concluded that accounting for these resources in economic indices would be critical for prairie conservation. The debate on pricing ecosystem functions, goods, and services continues to consider the efficacy of integrating economic and ecological components (Costanza and Farber 2002; Farber *et al.* 2006, and papers therein).

In a more general sense, occupying such a large area of the earth's surface, grassland ecosystems provide many important goods and services that can be quantified, at least to some extent. These are considered below under four broad headings (White *et al.* 2000): food, forage and livestock, biodiversity, carbon storage, and tourism and recreation. Other important functions of grasslands include the provision of drinking and irrigation water, genetic resources, amelioration of the weather, maintenance of watershed functions, nutrient cycling, human and wildlife habitat, removal of air pollutants and emission of oxygen, employment, soil generation, and a contribution to aesthetic beauty (Sala and Paruelo 1997; World Resources 2000–2001).

1.4.1 Food, forage, livestock, and biofuels

The most widespread use of grassland worldwide is in the production of domestic livestock (principally mammalian herbivores: cattle, sheep, goats, horses, water buffalo, and camels). In addition, large numbers of wild herbivores also depend on grasslands, and in many cases share the land with domestic herds. Wild native herbivores are regarded as beneficial, adaptive, or even critical for grasslands. For example, the American bison is considered a keystone species for the US tallgrass prairie (see §9.3.1). The non-economic benefit to grasslands (e.g. enhanced biodiversity) of introduced domestic livestock is less clear (McIntyre *et al.* 2003), and in the common case of overgrazing, is clearly detrimental. Certainly, however, in subsistence rangeland economies, domestic livestock serve many benefits in addition to food and cash including dung for fuel and flooring (Milton *et al.* 2003).

The densities of livestock in grasslands range from 1 to >100 head/km^2, with the highest densities in the world in the Middle East, Asia, and Australia (White *et al.* 2000). Recent trends (1986–1988) include an average 5.6% increase in livestock numbers in developing countries with extensive grassland. Some of the largest increases include Mongolia, where cattle increased by 40% and sheep and goats by 31.5% over this period. The largest increase was seen in Guinea where livestock increased by 104%. Although these increases may reflect somewhat improved economic conditions in these countries, they are also indicative of overgrazing and a degraded range. White *et al.* (2000) estimate that 49% of grasslands worldwide were lightly to moderately degraded, with at least 5% strongly to extremely degraded.

As a source of forage for livestock, whether grazed directly or harvested for consumption elsewhere (e.g. feed lots), there is a long-standing appreciation and dependence in agriculture for grasslands (Flint 1859). In the USA, forage accounts for about 57% of the total feed of beef cattle (Barnes and Nelson 2003). Dairy cattle use a somewhat lower proportion of forage in their diet (*c.*16%) than do beef cattle, swine (13%), poultry (<5%), horses and mules (3%), and sheep and goats (2%). The value of forages in the USA is $27.8 billion (1998 estimate), and exceeds the value for other crops. Hay accounts for $11.7 billion, exceeded only by the value of corn and soybeans in the market ($19.1 and $14.7 billion, respectively) (Barnes and Nelson 2003).

There is a large variety of forage plants, and for the most part they are grasses or legumes (Moore 2003). In Europe, Australia, and New Zealand, the most important forages are *Lolium perenne* (perennial ryegrass) and *Trifolium repens* (clover). In the north-eastern USA introduced cool-season species, particularly white clover and Kentucky bluegrass *Poa pratensis* are important, with introduced warm-season species including *Cynodon dactylon* (Bermuda grass), *Paspalum dilatatum* (dallisgrass), *Paspalum notatum* (bahiagrass), and *Sorghum halepense* (Johnsongrass) becoming predominant in the south-eastern USA (Barnes and Nelson 2003). The cool-season *Schedonorus phoenix* (tall fescue) is the most widely planted forage in the transition between the north-eastern and south-eastern regions, whereas in the mid-west and western USA several native warm-season prairie grasses remain dominant such as *Andropogon gerardii* (big bluestem) and *Sorghastrum nutans* (Indian grass), which are replaced towards the west with shorter grasses such as *Bouteloua gracilis* (blue grama) and *Bouteloua dactyloides* (syn. *Buchloe dactyloides*, buffalograss) in the south and *Agropyron* spp. and *Pseudoroegneria* spp. (wheatgrasses), and *Elymus* spp. (wildryes) in the north. A wide variety of native and introduced species are used as forage in tropical regions and these include *Chloris gayana* (Rhodes grass), *Cynodon dactylon*, *Digiitaria decumbens* (Pangola grass), *Panicum maximum* (Guinea grass, zaina), and *Pennisetum clandestinum* (Kiyuku grass) (Skerman and Riveros 1990). A wide variety of legumes are planted as forage, with over 30 genera often planted in mixture with grasses in the tropics alone (Skerman *et al.* 1988). Because of their association with *Rhizobium* spp. bacteria, legumes are highly nutritious and can increase levels of nitrogen in the soil (Skerman *et al.* 1988). Important species include *Trifolium repens* and *T. pratense* (white and red clover), *Medicago sativa* (alfalfa), *Lespedeza* spp. (annual and perennial lespedezas), *Vicia sativa* (common vetch), *Lotus corniculatus* (birdsfoot trefoil) in temperate areas (Moore 2003; Rumbaugh 1990), and *Centrosema pubescens* (Centro, butterfly pea), *Pueraria phaseoloides* (tropical kudzu), *Stylosanthes guianensis* (Schofield stylo, Brazlian stylo), and *S. hamata* (Caribbean stylo) in the tropics (Skerman *et al.* 1988; Skerman and Riveros 1990). Of these forage legumes, *M. sativa* is the most widespread as it is planted on $>32 \times 10^6$ ha worldwide, with the most widespread plantings in the USA (13.3×10^6 ha) (Michaud *et al.* 1988).

A number of perennial grasses are used increasingly in Europe and North America as renewable bioenergy sources, as they can be grown with minimal maintenance on marginal soils and harvested to produce large volumes of carbon-rich biomass. These bioenergy crops include a number of high biomass, rhizomatous bunchgrasses such as miscanthus *Miscanthus* spp., switchgrass *Panicum virgatum*, kleingrass *P. coloratum*, buffalograss *Bouteloua dactyloides*, elephant grass *Pennisetum purpureum*, reed canarygrass *Phalaris arundinacea*, big bluestem *Andropogon geradii*, giant reed *Arundo donax*, and tall fescue *Schedonornus phoenix*. Ideal bioenergy crops share the ecological traits of C_4 photosynthesis (Chapter 4), high water-use efficiency, partitioning of nutrients below ground in the dormant season, lack of known pests or diseases, rapid spring growth, long canopy duration, sterility, and perennial life form. The forage from these bioenergy crops allows the production of clean-burning liquid biofuels including ethanol for generating heat and electricity and as a transportation fuel. The energy content of biofuels is 17–21 MJ kg^{-1}, which compares favourably with the energy content of fossil fuels (21–28 MJ kg^{-1}). It has been estimated that if only 20% of the current agricultural land in the state of Illinois were planted to *Miscanthus*, 145 TWh of electricity could be generated which exceeds the 137 TWh of electricity consumed annually by the state including Chicago, the third largest US city (Heaton *et al.* 2004). The potential economic and environmental benefits of using bioenergy crops include near-zero emissions of greenhouse gases (the carbon emitted from burning the crop biomass is equal to or less than the carbon fixed via photosynthesis to produce the biomass), and improved carbon sequestration and soil and water quality. Economic estimates for using *P. virgatum* in the USA predict an annual increase in net farm returns of $6 billion, a decrease in government farm subsidies of $1.86 billion, and a reduction of 44–159 Tg yr^{-1} of greenhouse gas emissions (McLaughlin *et al.* 2002). Bioenergy crops do not require replanting for >15 years and can be harvested annually for biomass. As in native prairie (Chapter 7), planting these perennial grasses leads to large amounts of belowground carbon sequestration with higher levels of organic carbon and nitrogen in the soil than conventional crops like corn (see §1.4.3). The sequestered soil carbon has potential use as a carbon credit. Emissions of CO_2 from burning bioenergy crops are substantially lower than from conventional sources; for example, CO_2 emissions from *P. virgatum* were estimated at 1.9 kg C GJ^{-1} compared with 13.8, 22.3, and 24.6 1.9 kg C GJ^{-1} from gas, petroleum, and coal, respectively (Lemus and Lal 2005). Although planting grasses as biofuels has economic and environmental advantages, the ecological traits of a biofuel species (described above) are also known to contribute to invasiveness, prompting concerns regarding the ecological risks involved in introducing and planting them (Raghu *et al.* 2006). Many of the biofuel species listed above are invasive outside their native range.

Native grassland has the potential to be a valuable bioenergy source. Mixtures of native grassland perennials, referred to as **low-input high-diversity** (LIHD) biofuels, were shown to exhibit bioenergy yields 238% greater than monocultures. Moreover, these LIHDs are carbon negative with more CO_2 sequestered in the soil and roots (4.4 Mg CO_2 ha^{-1} yr^{-1}) than released during conventional biofuel production (0.32 Mg CO_2 ha^{-1} yr^{-1}) (Tilman *et al.* 2006a).

1.4.2 Biodiversity

By any measure, the world's grasslands qualify as important repositories of biodiversity. As noted earlier and discussed in more detail in Chapter 2, the major cereal crops of the world are grasses, and their ancestors arose in grasslands. Modern improvement of cereal and forage cultivars through breeding continues today to draw upon the genetic reserves in grasslands.

The richness of grassland biodiversity is exemplified by the following observations summarized from White *et al.*'s (2000) analysis:

- Forty of the world's 234 Centres of Plant Diversity (CPD) identified by the International Union for Conservation of Nature and Natural Resources (IUCN)-World Conservation Union and World Wildlife Fund-US occur in grasslands, with an additional 70 CPDs containing some grassland habitat. To qualify as a CPD mainland areas must contain >1000 vascular plants with >10% endemism.
- Grassland/savannah/scrub is the main habitat in 23 of 217 Endemic Bird Areas identified by Birdlife International, of which 3 have the highest rank for biological importance (Peruvian High Andes, central Chile, and southern Patagonia).
- Thirty-five of 136 terrestrial ecoregions identified on the basis of outstanding diversity and as priorities for conservation by the World Wildlife Fund-US Global 200 programme are grassland.
- Ten of 32 North American and 9 of 34 Latin American grassland ecoregions are rated as globally outstanding for biological distinctiveness by the World Wildlife Fund-US.

- Perhaps the biologically richest grassland worldwide is that of the Agulhas Plain, part of the Cape Floristic Province of South Africa. This region, considered as one of the world's 25 biodiversity hotspots for conservation (Myers and Mittermeier 2000), contains a flora of 1751 species, including 23.6% regional endemics and 5.7% local endemics (Cowling and Holmes 1992) in a landscape of shrubland, savannah, and fynbos (Rouget 2003).
- By contrast, some grasslands, such as those of the US Great Plains, are of relatively recent origin and so support comparatively few species and low levels of endemism (Axelrod 1985). Nevertheless, species in the Great Plains tend to be characterized by high levels of ecotypic differentiation (e.g. Gustafson *et al.* 1999; Keeler 1990, McMillan 1959b, and see Chapter 5;) adding to the biological richness and the value of the system (Risser 1988).

As with other biomes, there is growing concern over the loss of biodiversity through habitat loss and alteration (see above). White *et al.* (2000) identified 697 areas worldwide that were at least 10 km^2 in size and 50% grassland cover that were afforded IUCN category I, II, or III level of protection (i.e. nature reserve, wilderness area, national park or monument status). These areas totalled 3.9×10^6 km^2, more than the 1.6×10^6 km^2 of similarly protected forest, but only 7.6% of the total 52×10^6 km^2 of grasslands worldwide. The largest area of protected grasslands is 1.3×10^6 km^2 in sub-Saharan Africa. On a percentage basis, the highest level of protection is in North America where 0.8×10^6 km^2 of 6.6×10^6 km^2 (12%) is protected. Nevertheless, within the Great Plains there are large differences in the level of protection afforded different grasslands; in 1994 only 0.07% (91.8 km^2) of Tamaulipus Texas semi-arid plain was protected (at any level), compared with 13.3% (120 000 km^2) of west-central semi-arid prairies (Gaulthier and Wiken 1998). Of the different grasslands worldwide, temperate grasslands are afforded the least protection (0.69%), ranging from 0.08% in the Argentine Pampas to 2.2% in the South African grassveld (Henwood 1998b). Not surprisingly, there is growing concern at this poor level of protection with the IUCN and World Commission on Protected Areas taking a lead role in raising public awareness of the problem and organizing the political will to protect more grassland (e.g. Henwood 1998a; IUCN-WCPA 2000).

1.4.3 Carbon storage

A global service of grasslands and other ecosystems is the storage of carbon. Through photosynthetic fixation of CO_2, producers remove carbon from the atmosphere. Simultaneously, respiration recycles carbon, again as CO_2, back into the atmosphere. As the basis for life, the efficient operation of the global carbon cycle is critical. Its successful operation depends also on three carbon reservoirs, namely the atmosphere, the oceans, and the terrestrial biosphere. This overly simplistic description of the carbon cycle (more details can be found in most general ecology textbooks, and see §7.2) should, however, be sufficient to emphasize the importance of this cycle.

Grasslands provide a significant service towards global carbon storage by virtue of high levels of carbon accrual and sequestration below ground in the soil. Up to 90% of grassland biomass is below ground (see §7.1.2), and levels of soil carbon are higher in grasslands than in forests, agroecosystems, or other ecosystems (Table 1.9). Carbon storage below ground is particularly high at high latitudes where decomposition rates are generally low and soil organic matter and hence carbon stocks can build up over thousands of years (see §7.3). Above ground, low-latitude areas have high production because of the higher temperatures, but below-ground storage is low. The total global carbon store in grasslands is comparable to that in forests because grassland areas are so extensive worldwide, but on a unit area basis it is less, although still comparable to that of agroecosystems. Thus, grasslands are a potentially important sink for carbon in the terrestrial biosphere, a fact of particular importance given the increase in anthropogenic increase in atmospheric carbon since the beginning of the Industrial Revolution (IPCC 2007). The extent and time frame over which terrestrial carbon sinks, including grasslands, will change in response to global climate change is uncertain, but a matter of concern (Grace et al. 2001a). Uncertainty lies in predicting changes in decomposition and photosynthetic rates to changing CO_2, temperature, and nutrient supplies (e.g. altered nitrogen availability).

The high organic matter content of grassland soils has been their Achilles heel in some respects, as it led to widespread cultivation in areas such as the North American Great Plains. Conventional tillage practices oxidize soil organic matter, and, as a result, soil carbon has declined 20–60% in former grasslands that have been cultivated (Burke et al. 1995; Mann 1986). Recovery to steady-state conditions of the active soil carbon pool after cultivation

Table 1.9 Carbon storage of grasslands compared with forests and agroecosytems. Values are in Gt C and show minimum and maximum estimates

Ecosystem	Vegetation	Soils	Total	C stored/area (t C/ha)
Grasslands				
High-latitude	14–48	281	295–329	271–303
Mid-latitude	17–56	140	158–197	79–98
Low-latitude	40–126	158	197–284	91–131
Total	71–231	579	650–810	123–154
Forests	132–457	481	613–938	211–324
Agroecosystems	49–142	264	313–405	122–159
Other[a]	16–72	160	177–232	46–60
Global total	268–901	1484	1752–2385	120–164

[a] Includes wetlands, barren areas, and human settlements.

Adapted from White et al. (2000).

can take 50 years or more after restoration with native perennial grasses under the US Conservation Reserve Program (Baer et al. 2002). In this study, the labile carbon pool resembled native prairie within 12 years of restoration. Other causes of carbon loss from grasslands include fire, grazing, and exotic species. Burning savannahs, for example, releases carbon to the atmosphere, and contributes up to 42% of gross CO_2 to global emissions (White et al. 2000). Sala and Paruelo (1997) estimated carbon sequestration in grasslands of eastern Colorado, USA to be worth c.$200 ha^{-1}, quite a bit more than the average cash return of $47 ha^{-1} yr^{-1} from meat, wool, and milk. The loss of carbon from grassland soils to the atmosphere occurs rapidly following conversion to cropland, but the reverse process of carbon accrual following agricultural abandonment is very slow (60 kg C ha^{-2} yr^{-1}), with the system regaining its original value at a rate of only $1.20 ha^{-1} yr^{-1}. Grasslands take up more methane and emit less nitrous oxide than cropland at an estimated cost of $0.05 ha^{-1} yr^{-1} and $0.60 ha^{-1} yr^{-1} respectively (Sala and Paruelo 1997).

1.4.4 Tourism and recreation

Increasingly, grasslands provide a destination for tourists (ecotourism) and a location for recreational activities including hiking, fishing, viewing of game animals, safaris, cultural and spiritual needs, and aesthetic enjoyment. Ecotourism is defined as 'responsible travel to natural areas that conserves the environment and improves the well-being of the local people' (Honey 1999). It is characterized by its benefits to the visitor as well as conservation and the people of the host country. Honey (1999) identifies seven characteristics of ecotourism: (1) involves travel to natural destinations, (2) minimizes impact, (3) builds environmental awareness, (4) provides direct financial benefits for conservation, (5) provides financial benefits and empowerment for local people, (6) respects local culture, and (7) supports human rights and democratic movements.

The importance of grasslands in meeting these criteria is difficult to gauge accurately, but some of the useful indicators include the number of tourists and the money that they spend. In an assessment of countries with grasslands making up 80% or more of the land area, the numbers of international tourists between 1995 and 1997 ranged from 10000 per year in Somalia to 4.0×10^6 in Australia (Table 1.5). Over the 10-year period from 1985–1987 to 1995–1997 the numbers of international tourists entering these countries increased by 150% (Benin) to 331% (Central African Republic). Somalia was an exception, with tourist numbers decreasing 74%, hardly surprising given the political problems in the country (civil strife and famine). Similarly, where figures are available for these countries, international tourism receipts ranged from US$5 $\times 10^6$ to US$8503 $\times 10^6$ per year during 1995–1997 in the Central African Republic and Australia, respectively, reflecting a change from 1985–1987 of −7% (Benin) to 471% (Australia) (data for Somalia were unavailable) (Table 1.5). In addition to visits and money spent, the numbers of safari hunters and revenues from the hunting industry also increased over this period. Although these data do not prove a positive relationship between the occurrence of grassland and tourism or recreation, they are certainly consistent with the idea that grasslands provide this service. The general consensus from conservation agencies such as the World Resources Institute is that tourism and recreation represent a significant economic grassland service, but that the continued decrease in grassland extent and biodiversity raises serious concern about a potential decline in the capacity of grasslands to maintain these services over the long term (World Resources 2000–2001).

The attractiveness of grasslands and the willingness of tourists to pay to view large herbivores such as elephants provide a powerful incentive for landowners to develop environmental business opportunities. For example, game farms adjacent to South Africa's Kruger National Park generate 15 times the income from tourism and employ 25 times more people than cattle farming (Milton et al. 2003). In some countries with large expanses of grassland habitat, such as Namibia in which 13% of the country is set aside for nature conservation, tourism is one of the most important sectors of the economy (Barnes et al. 1999). The Etosha National Park, an area of desert grassland in Namibia, was the most important specific attraction named by visitors surveyed, following their preference for unique, unspoiled nature/landscape and wildlife/animals (Barnes et al. 1999). Similarly, in Botswana where Kalahari savannah and woodland dominate the landscape, wildlife viewing was estimated to have a clear economic advantage over cattle farming on about one-third of wildlife land (Barnes 2001). Ecotourism is an economically significant and fast-growing component of the economy, especially in rural areas, of an increasing number of developing countries with extensive areas of grassland (Kepe 2001). In developed countries, too, the economic merit of ecotourism in habitats containing grassland is increasingly considered a viable and appropriate land use (Norton and Miller 2000).

Wildlife tourism to the Serengeti grassland in central Africa is providing an economic boost to Tanzania, one of the poorest countries in the world. The Serengeti National Park, with an area of 14 763 km^2 contains some 4×10^6 animals, including migratory zebras *Equus burchelli*, elands *Taurotragus oryx*, and wildebeest *Connocheatas taurinus* along with free-roaming lions *Panthera leo*, elephants *Loxodonta africana* and other grazing ungulates such as Thomson's gazelle *Gazella thompsonii* and buffalo *Syncherus caffer*, spread across a tropical/subtropical bunchgrass savannah dominated by grasses including *Themeda trianda* (McNaughton 1985) (see §6.1 and Plate 4). The Serengeti is one of 12 national parks and 14 game reserves in Tanzania, all of which are at the heart of a tourism boom. Honey (1999) assessed the success of Tanzania in meeting the 7-point criteria for ecotourism concluding that the country ranked high on points 1 and 3 (involves travel and builds environmental awareness) and was making reasonable progress towards attaining points 2, 4, 5, and 7 (minimal impact, financial benefit for conservation and local people, and support of human rights), but was doing poorly with regard to point 6 (respect of local culture). Specifically, the local Masai people, despite being at the forefront of ecosystem preservation, remain the subject of prejudice and are viewed more as a tourist attraction than as a valued and important cultural group in their own right. By contrast, in neighbouring Kenya, which owns the northern part of the Serengeti, the ecotourism scorecard report is somewhat worse. Kenya has been Africa's most popular wildlife tourism destination since the 1960s (point 1). However, recent civil unrest (2007–8) has seriously damaged the tourist industry, and protected areas have suffered from over-exploitation, poaching, and poor management (point 2), resulting in little concern for the needs and human rights of the rural farmers and pastoralist (points 6 and 7) or environmental protection (point 4). On the positive side, Kenya has conducted a number of innovative ecotourism experiments such as community conservation schemes that have had some success (point 5) and raised environmental awareness at least among some segments of the population (point 3).

1.5 Early grassland ecologists

Many early ecologists and botanists were interested in grassland. We can go back to Charles Darwin's comments lamenting the invasion of exotic thistles in the Pampas of Argentina (Darwin 1845). However, two early grassland ecologists come to mind as having significantly influenced the subsequent development of the discipline. This section will briefly outline the life and times, and contributions, of J.E. Weaver (North America) and J.W. Bews (South Africa). Coincidentally, both men were born in the same year, 1884; the year that Mark Twain's *Huckleberry Finn* was published, that Belize became a British colony (until 1981), that the Statue of Liberty was unveiled in New York, and the 17-year old pianist and composer Scott Joplin arrived in St Louis.

1.5.1 John William Bews

J.W. Bews (1884–1938), MA, DSc, Professor of Botany in the Natal University College, Pietermaritzburg, South Africa, was a pioneer in plant ecology in South Africa. He was originally from the Orkney Islands off the north coast of Scotland, and had academic training at the University of Edinburgh. Bews published widely on the floristics, ecology, and systematics of South African vegetation (Table 1.10) (Gale 1955). His highly original ideas on human ecology were influenced by the general, politician, and botanist Jan Christian Smuts (Anker 2000). His contributions were recognized by the award of the South Africa Medal (Gold) in 1932.

Table 1.10 Selected books and monographs by John W. Bews and John E. Weaver (as single author except where indicated)

John W. Bews	
1913	An oecological survey of the midlands of Natal, with special reference to the Pietermaritzburg district. *Annals of the Natal Museum* 2, 485–545
1916	An account of the chief types of vegetation in South Africa, with notes on the plant succession. *Journal of Ecology* 4, 129–159.
1917	The plant ecology of the Drakensberg range. *Annals of the Natal Museum* 3, 511–565.
1918	*The grasses and grasslands of South Africa*. P. David & Sons, Printers, Pietermaritzburg
1920	The plant ecology of the coast belt of Natal. *Annals of the Natal Museum* 4, 367–469.
1921	*An introduction to the flora of Natal and Zululand*. City Printing Works, Pietermaritzburg
1923	(with R.D. Aitken) *Researches on the vegetation of Natal*. Series I. No. 5. Government Printing and Stationery Office, Pretoria
1925	(with R.D. Aitken) *Researches on the vegetation of Natal*. Series II. No. 8. Government Printing and Stationery Office, Pretoria
1925	*Plant forms and their evolution in South Africa*. Longmans, Green, London
1927	*Studies in the ecological evolution of the angiosperms*. Wheldon & Wesley, London
1929	*The world's grasses; their differentiation, distribution, economics and ecology*. Longmans, Green, London
1935	*Human ecology*. Oxford University Press, London
1937	*Life as a whole*. Longmans, Green, London
John E. Weaver	
1918	(with R.J. Pool and F.C Jean) *Further studies in the ecotone between prairie and woodland*. University of Nebraska, Lincoln, NE
1929	(with W.J. Himmel) *Relation between the development of root system and shoot under long- and short-day illumination*. American Society of Plant Physiologists, Rockville, MD
1930	(with W.J. Himmel) *Relation of increased water content and decreased aeration to root development in hydrophytes*. American Society of Plant Physiologists, Rockville, MD
1932	(with T.J. Fitzpatrick) *Ecology and relative importance of the dominants of tallgrass prairie*. s.n., Hanover, IN
1934	(with T.J. Fitzpatrick) *The prairie*. Prairie/Plains Resource Institute, Aurora, NE (reprinted 1980)
1938	(with F.E. Clements) *Plant ecology*. McGraw-Hill, New York
1954	*North American prairie*. Johnsen Publishing, Lincoln, NE
1956	(with F.W. Albertson) *Grasslands of the Great Plains: their nature and use*. Johnsen Publishing, Lincoln, NE
1968	*Prairie plants and their environment; a fifty-year study in the Midwest*. University of Nebraska Press, Lincoln, NE

This prestigious award recognizes the exceptional contribution to the advancement of science, on a broad front or in a specialized field, by an eminent South African scientist. He was the first principal of the University of Natal, where his legacy lives on in the John Bews Building that houses the faculties of science and agriculture and their library.

Bews conducted extensive floristic work, some of the first on the grassland plant communities of South Africa. A strong proponent of Clementsian succession, the second stage of Bews's grassland work dealt with the developmental history of the various plant communities. His grassland work is important because he not only summarized what was then known about the floristics and ecology of grasslands in South Africa (Bews 1918), but in a longer treatise produced a general worldwide classification of grassland vegetation (Bews 1929). His 'phylogenetic arrangement' of world grasslands was derived from Schimper's early classification which divided the vegetation of the world into woodland, grassland, and desert. Bews incorporated the then current understanding of grass evolution with Clementsian ideas on the successional relationships between forests and grasslands. As a result he produced a vegetation classification which allows accommodation of grasslands from around the world. The classification was essentially a functional type approach that can be used today (see Chapter 8). He also placed the systematics of grasses into the context of angiosperm evolution in a series of papers (Bews 1927, collected together in book form in the same year).

1.5.2 John Earnest Weaver

J.E. Weaver (1884–1966), Professor of Plant Ecology at the University of Nebraska, was a pioneer ecologist of the North American tallgrass prairie. He studied the prairie for over 50 years, leaving a legacy of >100 publications including 17 books (Table 1.10). His work covered every aspect of prairie ecology, and he is remembered today for his detailed study of plant root systems (Weaver 1958, 1961; Weaver and Darland 1949a), competition between plants (Weaver 1942), and community composition (Weaver 1954; Weaver and Albertson 1956). His drawings and photographs of excavated root systems have never been equalled, and are superseded only perhaps through the use of more modern techniques for tracking root system development (Chapter 3). He was particularly concerned with the effects of grazing (Weaver and Tomanek 1951) and the 'great drought' of 1934 (Weaver and Albertson 1936, see §9.4). Much of his work was published as substantial, lengthy articles and monographs. Fondly remembered by over 100 master's and doctoral students, he is said to have remarked of the prairie 'look carefully and look often' (Voigt 1980). He was honoured as Research Associate with the Carnegie Institute of Washington, President of the Nebraska Academy of Sciences, President of the Ecological Society of America, and honorary President of the International Botanical Congress.

CHAPTER 2

Systematics and evolution

...the study of grass classification or taxonomy does more than satisfy our curiosity about the diversity of living things and the way in which they have evolved.

Stebbins (1956)

The rapid development as far as we can judge of all the higher plants within recent geological times is an abominable mystery.

Charles Darwin, letter to Sir Joseph Hooker (1879).

Grasses are the dominant plants of grasslands (Chapter 1) and of agriculture. The systematics and evolution of this large, diverse, and phylogenetically advanced family are described in this chapter. **Agrostology** is the science of grass classification, and, as noted by Gould (1955), is essential to the study of grassland. However, the taxonomic treatment of the grasses has a fascinating history in itself as investigators moved from the use of primary morphological and anatomical characters to include cytogenetic, physiological, and molecular characters allowing the development of increasingly evolutionarily informative classifications (Stebbins 1956). The evolution of this family, although still incompletely understood, represents an interesting example of adaptation and co-evolution with environmental and biotic factors, especially aridity and grazing.

2.1 Characteristics of the Poaceae

Grasses are members of the family Poaceae, alternatively known as the Gramineae, within the Class Liliopsida (the **monocotyledons**) (Table 2.1). Within the Liliopsida, the grasses are placed in the order Poales which includes 17 families including the closely related Joinvilleaceae, Ecdeiocoleaceae, and Flagellariaceae (see §2.4.1) as well as the familiar Juncaceae, Cyperaceae, and Bromeliaceae (Stevens 2001 onwards). The systematic groups within the family are still in a state of flux, but 7500 to 11 000 species are recognized depending on the authority, divided among 600–700 genera within 25 (sometimes >50) tribes and 12 subfamilies (3–12) (Flora of China Editorial Committee 2006). The largest genera are *Panicum* (panic grass, *c*.500 spp.), *Poa* (bluegrass, *c*.500 spp.), *Festuca* (fescue, *c*.450 spp.), *Eragrostis* (lovegrass, *c*.350 spp.), *Paspalum* (paspalum, *c*.330 spp.), and *Aristida* (threeawn, *c*.300 spp.); however, the monophyly of some of the genera is doubtful. Economically all the important cereal crops are grasses, including the wheats (*Triticum* spp.), rice (*Oryza sativa*), maize (*Zea mays*), oats (*Avena* spp.), and barley (*Hordeum vulgare*), sorghum (*Sorghum* spp.), millets (*Panicum* spp., *Pennisetum* spp.), and sugar cane (*Saccharum officinarum*). Furthermore, grasses represent major sources of forage (see §4.2).

A succinct description of the Poaceae provided by the Grass Phylogeny Working Group (GPWG 2001) is as follows:

A monophyletic family recognizable by the following synapomorphic characters: inflorescence highly bracteate. Perianth reduced or lacking. Pollen lacking scrobiculi, but with intraexinous channels. Seed coat fused to inner ovary wall at maturity, forming a caryopsis. Embryo highly differentiated with obvious leaves, shoot and root meristems, and lateral in position.

A more extensive characterization of the Poaceae is provided by Mabberley (1987), describing the family as:

Usually perennial and often rhizomatous herbs or (bamboos) ± woody and tree-like but without secondary thickening; cell-walls, especially epidermis ± strongly silicified, vessel-elements usually in all vegetative organs; stems usually terete and with hollow internodes; roots often with root-hairs but often with endomycorrhizae also. Leaves distichous (spirally arranged in e.g. *Micraira*), never 3-ranked, with usually open sheath and elongate lamina usually with basal meristem and pair of basal auricles (narrowed to a petiolar base above sheath in many bamboos); ligule usually adaxial at junction of lamina and sheath, rarely 0. Flowers usually wind-pollinated, usually bisexual, in 1–∞-flowered spikelets, spike-like to panicle-like secondary inflorescences; spikelets usually with a pair of subopposite bracts (glumes) and 1-several distichous florets often on zig-zag rhachilla, the florets usually comprising a pair of sub-opposite subtending scale-like bracts (lemma and palea), 2 or 3 small lodicules (up to 6 in Bambusoideae). Anthers (1–)3 or 6 (especially Bambusoideae, where up to >100 in *Ochlandra*), anthers elongate, basifixed but deeply sagittate so as to appear versatile, with longitudinal slits and nearly smooth pollen grains. Gynecium (2 (3 in Bambusoideae)), 1-locular with 2(3) stigmas, often large and feathery, ovule 1, orthotopous to almost anatropous, (1)2-tegmic. Fruit (caryopsis) usually enclosed in persistent lemma and palea, usually dry-indehiscent, integuments adnate to pericarp, the seed rarely falling free of these accessory structures such as when pericarp becomes mucilaginous when wet and expelling the seed on drying out; embryo straight with well developed plumule covered by a closed cylindrical coleoptile, radicle with a similar coleorhiza, and enlarged lateral cotyledon (scutellum), all peripheral to copious starchy endosperm usually with proteinaceous tissue and sometimes also oily, rarely (*Melocanna*) absent. X = 2–23.

Additional details of some of these features are discussed in this chapter and Chapters 3 and 4.

Members of the grass family are frequently confused with the superficially similar sedges (Cyperaceae) and rushes (Juncaceae). However, a number of important features allow these families to be distinguished unambiguously even by the non-expert (Table 2.2).

2.2 Traditional vs modern views of grass classification

Because of the large size and great diversity of the family, it has proved difficult to produce a systematic treatment of the grasses that is widely

Table 2.1 Classification of grasses

Taxonomic level	Name
Class	Liliopsida (Monocotyledons)
Subclass	Commelinidae
Order	Poales
Family	Poaceae (Gramineae)
Type genus	Poa (e.g., *Poa pratensis* L., Kentucky Bluegrass)

Table 2.2 Features allowing the grasses (Poaceae) to be distinguished from the sedges (Cyperaceae) and rushes (Juncaceae). Note, exceptions occur for most character states listed. Also see Table 20.1 of Campbell and Kellogg (1986)

Characteristic	Poaceae	Cyperaceae	Juncaceae
Leaves	2-(rarely 3-) ranked, flat, non-channelled	3-(rarely 2-) ranked, flat, channelled	2-many ranked, flat or terete
Ligule	Usually present	Absent	Absent
Stem cross-section	Terete (round), rarely compressed	Triangular	Terete
Internodes	Hollow or solid	Solid	Solid, with septae
Inflorescence	Spikelet (1 or more florets above the glumes) in panicles, racemes, or spikes	'Spikelet' in racemes, panicles, etc.	Panicles, heads, corymbs, solitary
Flowers/florets	Floret (lemma, palea). Usually tiny scales = lodicules = perianth	3-merous, chaffy, scales, or bristles; inflated in *Carex* (perigynium)	3-merous, 6 scale-like perianth parts ('drab lilies')
Stigma	2 (in frequently 3)	2–3	3
Anthers	Attachment flexible (attached to filament above the base)	Attached at the base, not flexible	Attached at the base
Fruits	Caryopsis (grain), thin pericarp fused to seed coat (pericarp loose in *Sporobolus*)	Achene or nutlet, often lenticular or trigonous; style sometimes persistent	Loculicidal capsule
Seeds	1 seeded	1 seeded	Multiple seeded
Habitat	Mostly terrestrial	Terrestrial and aquatic (emergent)	Terrestrial and aquatic (emergent)

agreed upon or has remained stable for very long. As in other areas of systematics, molecular methods have revolutionized the field (e.g. GPWG 2001; Hodkinson et al. 2007b; Soreng and Davis 1998; Zhang 2000). Understanding the systematics of grasses continues to be an active area of research. A brief history is given below. More detailed treatments can be found in Stebbins (1987), Watson (1990), Chapman (1996, Chapter 6), Clark et al. (1995) and Soreng et al. (2007). Web-based electronic resources for the family include:

- The World Grass Species Database (http://www.rbgkew.org.uk/data/grasses-db.html)
- Grass Genera of the World Database (http://delta-intkey.com/grass)
- Grass Manual on the Web (http://www.herbarium.usu.edu/webmanual/)
- Catalogue of New World Grasses (CNWG: http://mobot.mobot.org/W3T/search/nwgc.html)

The use of grasses in agriculture and the naming of grasses extends back at least 2000 years. In ancient Greece, Theophrastus (370–287 BC) recognized in his *Enquiry into Plants* at least 19 different grasses including what we now know as 2 bamboos (*Bambusa* and *Dendrocalamus*) and 3 species of wheat (*Triticum aestivum*, *T. dicoccum*, and *T. monococcum*) (Chapman 1996). Nevertheless, until the mid-eighteenth century the names of grasses were just compiled in lists without any taxonomic order. For example, in 1708 Johann Scheuchzer published *Agrostographiae Helvetica Prodromus*, one of the first papers dealing just with grasses. However, in 1753 Carl Linnaeus in *Species Plantarum* provided the starting point of a binomial nomenclature for flowering plants. In *Genera Plantarum* (1767) he included more than 40 grass genera including many that are well known and currently recognized such as *Andropogon* (bluestems), *Panicum* (panic grasses), *Hordeum* (barley), and *Poa*. Of these, *Panicum* was the most diverse genus with 23 binomials described in *Species Plantarum*. However, his classification was a sexual system based on numbers of floral parts, and hence highly artificial. Later classifications of the flowering plants, including those of the grasses, endeavoured to be natural, based on assessments of adaptive radiations or character homologies.

Robert Brown (1810) was the first to understand the grass spikelet by recognizing it as a reduced inflorescence branch. He also recognized the two great subdivisions of the Poaceae—the Panicoideae and Pooideae subfamilies—describing the spikelets of each group, and their tropical–subtropical vs cool-climate distribution and adaptations, respectively.

In 1878 the English botanist George Bentham published a widely accepted natural classification based on morphological characteristics of the inflorescence and fruit (Bentham 1878). He recognized 13 tribes within the Panicoideae and Festucoideae (roughly equivalent to the Pooideae). His classification scheme formed the basis for numerous later treatments of the grasses including that of Bentham and Hooker (1883) and Bews's (1929, see Chapter 1) synopsis of the world's grasses. Hitchcock and Chase used Bentham's scheme in their 1935 and 1950 classification of US grasses in which 14 tribes in 2 subfamilies were recognized (Hitchcock and Chase 1950). Hitchcock and Chase's classification was used as a standard for treatments of North American grasses and grasslands and was followed by most US floras up until the 1980s.

Up to this point, grass classification was based on the use of readily observable morphological features. In the 1920s and 30s the use of additional morphological, anatomical, cytological, and physiological characters, often recognizable only from microscopic examination, heralded development of the 'new taxonomy'. The taxonomic arrangement of genera based on these characters often differed sharply from that based on traditional accepted homologies of characteristics of the inflorescence, and caused considerable changes in the focus of the classical system (Stebbins 1956).

The Russian cytologist N.P. Avdulov used chromosome studies, leaf anatomy, characteristics of the first seedling leaf, organization of the resting nucleus, and starch grain characters to recognize the two 'great' subfamilies of the Panicoideae and Pooideae (Avdulov 1931). In 1932 and 1936 the Frenchman H. Prat used Avdulov's characters, traditional characters, and characters based on the leaf epidermis to recognize three subfamilies, extended in 1960 to six (Prat 1936).

Both Avdulov's and Prat's systems were phylogenetic, meaning that the groups recognized were intended to be hierarchical and reflect the prevailing knowledge of genetic and evolutionary history within the family. The earlier morphologically based 'natural' classifications of Bentham were natural in the sense that the groups reflected similarities among the taxa, but not necessarily their evolutionary history. The even earlier sexual classification of Linnaeus, which was highly artificial, placed often distantly related taxa in close proximity because of the presence of shared number of parts. Construction of phylogenetic systems, although intellectually satisfying, may not be the most convenient systems for purposes of identification (Stebbins and Crampton 1961). Nevertheless, all modern systems are evolutionary, intended to reflect phylogeny.

One of the first to use the phylogenetic systems of Avdulov (1931) and Prat (1936) was English botanist C.E. Hubbard in his treatments of British grasses (Hubbard 1954). Stebbins and Crampton (1961), and later Gould and Shaw (1983) followed suit with their treatment of the grasses of North America. Both of these classifications recognized six subfamilies, including the long-recognized Panicoideae and Pooideae.

Clayton and Renvoize (1986, 1992) published a phylogenetic treatment of grass genera arranged as 40 tribes in 6 subfamilies (Bambusoideae, Arundinoideae, Centothecoideae, Pooideae, Chloridoideae, and Panicoideae). Their treatment has been highly influential because, although traditional in the sense of presenting generic descriptions and conventional keys, it was the first, detailed, worldwide, revision of grass genera and classification since Bews and Bentham. As such, it forms the basis of the most modern classification of the grasses. At about the same time, Watson and Dallwitz (1988, 1992 onwards, http://delta-intkey.com) published a computer database in DELTA (DEscription Language for TAxonomy) describing 785 genera based on 496 characters in which initially 5 and then later 7 subfamilies were recognized. The search facility of the database allows character descriptions to be matched to unknown specimens and thus represented a major new development in the storage and retrieval of taxonomic data. None of these classifications up to this point were significantly influenced by DNA sequence data.

Among the recent floristic accounts of the grasses is that of the Flora North America project (Barkworth *et al.* 2003, 2007, http://hua.huh.harvard.edu/FNA/;) with the subfamilial classification based on analysis of molecular and morphological data by the Grass Phylogeny Working Group (GPWG 2000, 2001) and tribal treatment of Clayton and Renvoize (1986, 1992, the latter with only a couple of exceptions suggested by the GPWG). The Flora of Australia and Flora of China projects similarly base their classifications of the Poaceae on the GPWG scheme (Flora of Australia 2002; Flora of China Editorial Committee 2006). The GPWG classification was based on a representative set of 62 grasses (0.6% of all grass species and *c.*8% of the genera) plus 4 outgroup taxa. Six molecular sequence data sets (*ndhF, rbcL, rpoC2, phyB, ITS-II*, and GBSSI or *waxy*), chloroplast restriction site data, and morphological data were used in a phylogenetic analysis that allowed a classification based on recognition of 11 previously published subfamilies (Anomochlooideae, Pharoideae, Puelioideae, Bambusoideae, Ehrhartoideae, Pooideae, Aristidoideae, Arundinoideae, Chloridoideae, Centothecoideae, and Panicoideae) and the proposal of one new subfamily (Danthonioideae) (GPWG 2001). Several subsequent analyses of molecular data (e.g. Davis and Soreng 2007) have continued to improve the systematic understanding of grass phylogeny (Hodkinson *et al.* 2007b).

2.3 Subfamily characteristics

Throughout the historical development of grass systematics, two large divisions of the family have consistently been recognized: the subfamilies Panicoideae and Pooideae. These two subfamilies were first recognized by Robert Brown in 1810 as the two primary subgroups of the Gramineae (as the Poaceae was then called). Bentham's (1881) treatment also recognized these subgroups as the tribes Panicaceae and Poaceae, respectively, as did Hitchcock and Chase (1950). As noted above (§2.2), the 'new taxonomy' of the 1960s onwards expanded the number of subfamilies as newly recognized groups were split out from the two larger groups. As described below (§2.4), the modern circumscription of 12 subfamilies (GPWG 2000, 2001) is explicitly phylogenetic and reflects division of the family into 2 main clades (monophyletic groups), the BEP and PACCAD clades, with 2 subfamilies sister to these main clades (Anonochlooideae, Pharoideae, and Puelioideae) (Fig. 2.1). All the subfamilies recognized are well supported as monophyletic, except for Centothecoideae (GPWG 2001). The latter is accommodated within the Panicoideae.

2.3.1 Descriptions of subfamilies

The following subfamily descriptions are based on the more extensive account provided by the GPWG (2001) and Kellogg (2002). Additional, detailed descriptions of the subfamilies are provided in Chapman (1996) and Chapman and Peat (1992), except that these sources use Clayton and Renvoize's (1986) scheme in which five subfamilies were recognized. The ordering of the subfamily descriptions provided below follows that depicted in the cladogram proposed by the GPWG shown in Fig. 2.1.

The first three subfamilies—the Anomochlooideae, the Pharoideae, and the Puelioideae—are basal (early-diverging lineages) in the overall Poaceae cladogram (Fig. 2.1), with the Anomochlooideae being the earliest lineage to diverge among extant grasses. Two major clades are recognized, the BEP clade which includes the Bambusoideae, Ehrhartoideae, and the Pooideae,

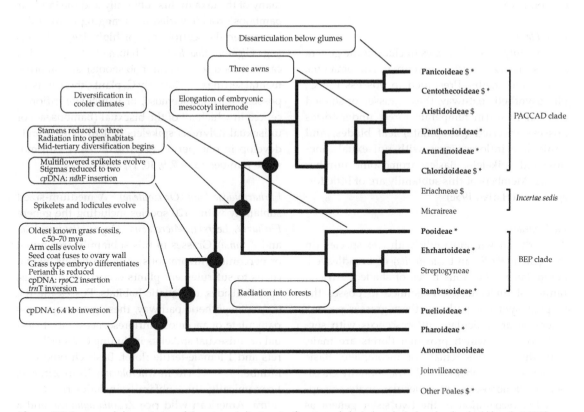

Figure 2.1 Summary phylogenetic tree of the grasses indicating significant morphological, ecological, and molecular (chloroplast DNA) events in the evolution of the family. The 12 subfamilies described in the text appear in boldface. Marked taxa: *, at least some included species have unisexual flowers/florets; $, at least some included species have a C_4 carbon fixation pathway, Kranz anatomy, or both. Dark circles indicate nodes strongly supported by all data combined (bootstrap > 99). Reproduced with permission of the Missouri Botanical Garden Press (GPWG 2001).

and the PACCAD clade which includes the Panicoideae, Arundinoideae, Centothecoideae, Chloridoideae, Aristidoideae, and Danthonioideae.

Anomochlooideae
A small subfamily (four species among two genera: *Anomochloa marantoidea* and *Streptochaeta*—three spp.) of perennial, rhizomatous herbs of shaded tropical forest understories. Presumed to possess the C_3 photosynthetic pathway, these grasses lack grass-type spikelets; instead, the inflorescences ('spikelet equivalents') have complicated branching patterns and are comprised of bracts and are one-flowered and bisexual. Basic chromosome numbers: $x = 11$ or 18. The presence of an adaxial ligule as a fringe of hairs is the single morphological character supporting a monophyletic origin of the subfamily. These grasses have no apparent economic value.

Pharoideae
A subfamily with 12 species (including the genera *Pharus* and *Leptaspis*) of perennial, rhizomatous, monoecious herbs. Presumed to possess the C_3 photosynthetic pathway, these grasses of shaded tropical to warm temperate forest understories possess inverted (resupinate) leaf blades, and paniculate inflorescences with unisexual, one-flowered spikelets. Basic chromosome number $x = 12$. Members of this subfamily are of little forage value (Harlan 1956).

Puelioideae
A poorly known subfamily with *c*.14 species (in the genera *Puelia* and *Guaduella*) of broadleaved, perennial, rhizomatous herbs of shaded, African rainforest understories. Presumed to posses the C_3 photosynthetic pathway, these grasses possess racemose or paniculate inflorescences with several florets of which proximal florets are male, with distal florets female or incomplete. Basic chromosome number $x = 12$. Traditionally classified as bamboos, recent molecular analyses support the recognition of the two sister genera as an early-diverging branch within the Poaceae and sister to the BEP and PACCAD clades (Clark *et al.* 2000). These grasses have no apparent economic value.

BEP clade
BEP is an acronym for the three subfamilies included in this clade: the Bambusoideae, the Ehrhartoideae, and the Pooideae. All are C_3 grasses, but the Bambusoideae and Ehrhartoideae are generally most abundant in warm tropical and subtropical regions whereas the Pooideae (the 'cool-season' grasses) are best represented in cool and cold regions.

Bambusoideae A large, ancient subfamily with *c*.1400 species of perennial (rarely annual), rhizomatous herbaceous or woody plants in 88 genera (including the genera *Arundinaria*, *Bambusa*, *Ochlandra*, and *Pariana*). Members of this subfamily are found in temperate and tropical forests, tropical high montane grasslands, riverbanks, and sometimes savannahs. The woody culms characterize many of the taxa in this subfamily and the familiar bamboos provide varied uses ranging from building materials, scaffolding for high-rise buildings (e.g. *Gigantochloa laevis*: Chapman 1996), garden canes, a food source (bamboo shoots), and in ornamental settings. Exclusively C_3 plants, these grasses possess spicate, racemose, or paniculate inflorescences in which all of the bisexual (Bambuseae) or unisexual (Olyreae) spikelets of 1 to many florets develop in one period of growth. Basic chromosome numbers: $x = 7, 9, 10, 11$, and 12.

Ehrhartoideae (syn. Oryzoideae) A medium-sized subfamily with *c*.120 species including the genera *Ehrharta*, *Leersia*, *Microlaena*, *Oryza*, *Potamophila*, and *Zizania*). Grasses in this subfamily are annual or perennial, rhizomatous or stoloniferous, herbaceous to suffrutescent plants occurring in forests, open hillsides, or aquatic habitats. Possessing the C_3 photosynthetic pathway, these grasses possess paniculate or racemose inflorescences with bisexual or unisexual spikelets including 0–2 sterile florets and 1 female-fertile floret. Basic chromosome number $x = 12$ (10 in *Microlaena*; 15 in *Zizania*). Economically, this subfamily includes rice *Oryza sativa*, American wild rice *Zizania aquatica*, and a problematic weed, the perennial *Leersia hexandra*.

Pooideae The largest subfamily with *c*.3560 species (including the genera *Agrostis*, *Bromus*, *Diarrhena*,

Elymus, Festuca, Lolium, Nardus, Poa, Stipa, and *Sesleria*) of annual or perennial, herbaceous plants of cool temperate and boreal regions, and high mountain regions of the tropics. Exclusively C_3 plants, these grasses possess spicate, racemose, or paniculate inflorescences in which the spikelets are predominantely bisexual, infrequently unisexual or mixed, including 1–many female-fertile florets, compressed laterally. Basic chromosome numbers: $x = 7$ (Bromeae, Triticeae, Poeae generally, few Brachypodieae), 2, 4, 5, 6, 8, 9, 10, 11, 12, 13. Economically, this large subfamily includes many important grasses (e.g. *Lolium perenne, Poa pratensis,* and *Schedonorus phoenix*), ornamental and amenity grasses (e.g. varieties of *Briza, Deschampsia,* and *Festuca*), and cereals (e.g. wheat, barley, oats, and rye).

PACCAD clade
A monophyletic clade including the Panicoideae, Arundinoideae, Centothecoideae, Chloridoideae, Aristidoideae, and Danthonioideae subfamilies. The clade is strongly supported by molecular analysis; nevertheless, the only morphological character linking all the species is the presence of a long mesocotyl internode in the embryo (Kellogg 2002). For the most part, members of this clade grow in warm climates and/or flower late in the growing season; hence they are often referred to as warm season grasses. Taxa of uncertain phylogenetic affinity (tribes Eriachneae, Micraireae, and genus *Cyperochloa*) were all left *Incertae Sedis* (GPWG 2001) but were nevertheless placed in the PACCAD clade on the basis of limited molecular evidence (Fig. 2.1). In support of earlier work by Pilger (1954, 1956 in Lazarides 1979), the Flora of North America and the Flora of Australia projects both recently recognized the Micrairoideae as a subfamily (containing the genera *Micraira, Eriachne, Isachne, Pheidiochloa*) thereby expanding the PACCAD clade into the PACMCAD clade (Barkworth *et al.* 2007; Flora of Australia 2005). However, these genera are often represented in analyses with incomplete data and their true phylogenetic relationships with other groups at the base of the PACCAD clade remain uncertain (R. Soreng, personal communication).

Panicoideae One of the largest and best-recognized subfamilies with *c.*3550 species (including *Andropogon, Panicum,* and *Saccharum*) of annual or perennial, primarily herbaceous grasses of the tropics and subtropics, but also in temperate regions. All the different photosynthetic pathways are represented (C_3, C_4 including PCK, NAD-ME, and NADP-ME: see Chapter 4 for explanation of C_4 pathways) including some C_3/C_4 intermediates. Inflorescences are panicles, racemes of spikes, or a complex combination of these, with bisexual (unisexual in monoecious or dioecious members) spikelets frequently paired in long-short combinations usually with 2 glumes, 1 sterile lemma, and 1 often compressed female-fertile floret. Basic chromosome numbers: $x = 5$, (7), 9, 10, (12), (14). Economically important plants include forage grasses, e.g. *Panicum maximum* (Guinea grass), *Paspalum notatum* (Bahai grass), and *Pennisetum purpureum* (elephant grass), and some cereals (e.g. *Echinochloa crus-galli* (Japanese millet), *Panicum miliaceum* (proso millet), *Pennisetum glaucum* (pearl millet), *Sorghum bicolor,* and *Zea mays* (maize). A number of major weeds are in this subfamily, including *Digitaria sanguinalis, Imperatra cylindrica,* and *Echinochloa crus-galli* (when not planted as a crop).

Arundinoideae A small subfamily, often treated as a 'dustbin' group for taxa of uncertain affinity, of 33–38 species in 15 genera including *Amphipogon, Arundo, Dregeochloa, Hakonechloa, Molinia, Moliniopsis,* and *Phragmites*). Eight of the genera from the crinipoid group (*Crinipes, Dichaeteria, Elytrophorus, Letagrostis, Nematopoa, Piptophyllum, Styppeiochloa,* and *Zenkeria*) are only provisionally placed in this group by the GPWG (2001). Recent analysis indicates the presence of a monophyletic arundinoid core group, although this subfamily has often been regarded as polyphyletic in a broad sense. Members of this subfamily are perennial (rarely annual), herbaceous to somewhat woody plants of temperate and tropical areas. The reeds (*Phragmites*) are found in marshy habitats. Photosynthetic pathway is C_3. Inflorescences are usually paniculate with bisexual florets of 2 glumes ± a sterile lemma and 1-several female-fertile florets. Basic chromosome numbers: $x = 6, 9, 12$. A number of economically useful grasses are included in this subfamily such as *Phragmites australis*, common 'reed' used for thatching and screens (Chapman 1996) and *Molinia caerulea* (purple moor grass), which is bred as an ornamental. *Arundo donax* (giant cane), native in Asia, is a problematic invasive (Table 1.8) following its escape from cultivation for use as an ornamental and screening plant.

Centothecoideae A small subfamily of 10–16 genera with *c.*45 species (including *Calderonella*, *Centotheca*, *Chasmanthium,* and *Thysanolaena*) of annual or perennial, herbaceous or reedlike, warm temperate woodlands and tropical forest. Exclusively C_3 plants, these grasses are, for the most part, characterized by unusual leaf anatomy (e.g. palisade mesophyll and laterally extended bundle sheath cells). Inflorescences are racemose or paniculate with bisexual or unisexual (1–) 2–many-flowered spikelets often compressed laterally. Basic chromosome number: $x = 12$. Members of this small subfamily have been historically placed in the Bambusoideae because of the superficial similarity to members of this subfamily. There is no morphological synapomorphy that supports the subfamily as monophyletic. Consequently, the circumscription of the centothecoid clade itself remains uncertain and based largely on molecular data (Sánchez-Ken and Clark 2000). Unusual leaf anatomy characterizes most members of the subfamily, including the presence of palisade mesophyll and laterally extended bundle sheath cells. Economically members of this subfamily are of limited value except *Centhtotheca lappacea* which provides excellent fodder (Chapman 1996). *Chasmanthium latifolium* (Indian wood oats, northern sea oats) is cultivated as an ornamental in the USA where it is also native.

Chloridoideae A large subfamily of *c.*1400 species including the widespread *Chloris* (*c.*55 spp.) and *Eragrostis* (*c.*350 spp.), of herbaceous (rarely woody) annuals and perennials of the dry tropics and subtropics (some in temperate zones). Photosynthetic pathway predominantly C_4 (PCK, NAD-ME, except NAD-ME in *Pappophorum*), although C_3 in *Eragrostis walteri* and *Merxmuellera rangei*. Inflorescences are paniculate with spicate branches and usually bisexual spikelets of two glumes and 1–many female-fertile florets, usually laterally compressed. Basic chromosome numbers: $x = (7), (8), 9, 10$. Many members of this family exhibit tolerance to drought and saline conditions, or high pH (Chapman 1996). A number of range grasses of dry habitats are in this subfamily including *Astrebla* spp. (Mitchel grasses, Australia: Plate 8), *Bouteloua dactyloides* (buffalograss, North America), and *Chloris gayana* (Rhodesgrass, central Africa). Two members of the subfamily are cereal grasses, *Eragrostis tef* (t'ef, Central Ethiopia), and *Eleusine coracana* (finger millet, India, China, and Africa), and *Astrebla lappacea* is cultivated as fodder. Several *Eragrostis* spp. are undesirable weeds (Watson and Dallwitz 1992 onwards).

Aristidoideae A small subfamily of *c.*350 species (including three genera: *Aristida*, *Sartidia*, and *Stipagrostis*) of annual or perennial, ceaspitose, herbaceous grasses from mostly xerophytic temperate to tropical zones, often in open habitats. Photosynthetic pathways include C_3 (*Sartidia*) and C_4 (*Aristida* NADP-ME; *Stipagrostis* NAD-ME). Inflorescences are paniculate of spikelets with bisexual florets including lemmas with three awns, disarticulating above the glumes. Basic chromosome number: $x = 11, 12$. In this subfamily *Aristida* spp. and *Sartidia* spp. are of limited economic value and are frequently significant weeds of poor nutritional value and minor grazing value in dry regions (e.g. *Aristida dichotoma*, *A. longiseta*, *A. oligantha* which have awns that can damage livestock). *Stipagrostis* species are of value as cultivated fodder (*S. ciliata*, *S. uniplumis*) and as important native pasture species: e.g. *S. ciliata*, *S. obtusa*, *S. plumose* (Watson and Dallwitz 1992 onwards).

Danthonioideae A small subfamily of *c.*300 species in 18–25 genera of perennial, occasionally annual, herbaceous or rarely suffrutescent grasses. Members of this subfamily occur in mesic to xeric open habitats, mostly in the southern hemisphere (*Danthonia* and *Schismus* are native in the northern hemisphere). Photosynthetic pathway is C_3. Inflorescences are paniculate or less commonly racemose or spicate with bisexual or unisexual laterally compressed spikelets with 1–6 (–20) female-fertile florets. The lemma includes a single awn. Basic chromosome numbers: $x = 6, 7, 9$. The presence of haustorial synergids in the ovule, distant styles, a ciliate ligule, a several flowered spikelet, an embryo mesocotyl, and the absence of Kranz anatomy and chloridoid microhairs are diagnostic features allowing the Danthonioideae to be distinguished from other subfamilies. Members of this subfamily have limited economic value although *Danthonia spicata* is used as a poor-quality forage grass in North America, *Pentaschistis borussica* is an important native pasture species in Africa, as are *Schimus arabicus* and *S. barbatus* in Euroasia.

2.4 Fossil history and evolution

The phylogenetic picture that emerges from the recent *rbcl* sequence analyses is that the Poaceae are monophyletic with a Cretaceous origin 83–89 Ma (million years ago) (Janssen and Bremer 2004; Michelangeli *et al.* 2003). The Anomochlooideae represents the earliest-diverging extant lineage (GPWG 2001). The next diverging lineage is the Pharoideae, followed by the Puelioideae. The rest of the grasses are believed to form a clade, with the BEP and PACCAD clades forming two major monophyletic groups (Fig. 2.1). The origin of the BEP and PACCAD clades is unclear and may have been as early as 55 Ma, certainly no later than the Late Eocene (34 Ma) (Prasad *et al.* 2005; Strömberg 2005). Viewed as a whole, the BEP + PACCAD clade includes the majority of grass species and is supported as monophyletic by six morphological synapomorphies (shared unique ancestral traits): loss of the pseudopetiole, reduction to two lodicules, loss of the inner whorl of stamens, loss of arm and fusoid cells, loss of lamina on the first seedling leaf (Bambusoideae and Orzyeae only), and evolution of unisexual florets (most lineages) (GPWG 2001). The phylogenetic patterns within and between the BEP and PACCAD clades are uncertain and unclear. Part of the problem lies in a bias of sampled taxa from the northern hemisphere, and the lack of sampling in general (Hodkinson *et al.* 2007a). Much of the evolutionary action may have taken place across the Gondwanan continent. The extensive radiation and diversification may have been too rapid to allow clear resolution of phylogenetic pattern with the data currently available (Kellogg 2000) although phytolith assemblages indicate diversification of the chloridoids within the PACCAD clade by at least 19 Ma (Strömberg 2005).

Some of the morphological changes associated with evolution of the grasses and the major groups are summarized in Fig. 2.1. Some traits appear to have evolved once, such as the early and accelerated development of the embryo relative to seed and fruit maturation. Several leaves, a vascular system, and organized shoot and root apical meristems are produced in the grass embryo (Kellogg 2000). This feature separates the grasses from the Joinvilleaceae and Ecdeiocoleaceae (their nearest relatives) and all other monocots (see §2.4.1). Other traits have evolved more than once, such as C_4 photosynthesis which is exclusive, in the grasses, to several closely related subfamilies of the PACCAD clade (Kellogg 2000) (Chapter 4). Complicating the issue are a large number of apparent reversals, such as the apparent regaining of pseudopetioles in the Bambusoideae and some members of the PACCAD clade, and the regaining, three or four times, of the inner whorl of stamens within the bambusoid/ehrhartoid clade (GPWG 2001). In the absence of knowledge of the underlying genetics of these traits, it is also possible that these changes really represent retained primitive characters, or the evolution of superficially similar, but novel, characters.

2.4.1 The Joinvilleaceae and Ecdeiocoleaceae connection

Early workers (e.g. Engler 1892, and see review in Cronquist 1981) assumed the grasses and sedges (Cyperaceae) to be closely related and placed them into the Glumiflorae or Cyperales, on the basis of floral reduction and biochemistry. Such a close relationship between these two groups is not now believed. In 1956, Stebbins suggested that the grasses were evolutionarily close to the most primitive Liliaceae such as members of the Flagellariaceae and Restionaceae. The Flagellariaceae (one genus, *Flagellaria*, of four species) are small old-world tropical herbs with grass-like leaves with tendril-like tips, whereas the Restionaceae are a mostly southern hemisphere family (38 genera of 400 species) that mostly lack leaf blades, but appear to take the place of grasses in some areas of South Africa and Australia. In the Cape region alone there are 10 endemic genera and 180 endemic species

(Mabberley 1987). Later, Dahlgren *et al.* (1985) and Stebbins (1987) suggested that the grasses may, in fact, be closest to the Joinvilleaceae, a small family that was only recognized in 1970, consisting of a single genus (*Joinvillea*) comprising two species (*J. ascendens* and *J. bryanii*) limited in their distribution to western Malaya and the Pacific islands. With long, narrow leaves with open sheathing bases, unbranched and hollow stems, and bisexual flowers of six scale- or bract-like perianth segments each, the plants bear a superficial similarity to some grasses (Heywood 1978). Nevertheless, the Joinvilleaceae is an isolated group of obscure origins. Recent analyses (GPWG 2001; Hodkinson *et al.* 2007a and references therein; Michelangeli *et al.* 2003) place the Joinvilleaceae and Ecdeiocoleaceae as sister groups and the closest living relatives to the grasses, with an obscure and uncertain relationship to the Restionales and Cyperales (Fig. 2.1). The gain of multicellular microhairs is a structural synapomorphy supporting the sister relationship between the Joinvilliaceae and the grasses (Michelangeli *et al.* 2003). The presence of a 6.4 kb inversion in the chloroplast DNA genome, and the occurrence of long–short cell alternations in files of cells adjacent to stomatal files in the leaf epidermis, are synapomorphies previously used to join this clade (Kellogg 2000), which have recently been shown to join the Joinvilliaceae–grass clade with the Ecdeiocoleaceae (a family of two genera, *Ecdeiocolea* and *Georgeantha*, often included in the Restoniaceae) (Michelangeli *et al.* 2003). The short cells go on to make stomata or silica bodies. The common ancestor of these groups presumably evolved both of these characters.

2.4.2 Biogeographic origin

It is unclear where the grasses evolved. The current distribution of the early-diverging lineages of grasses presents a fragmented picture. The basal Anomochlooideae are restricted to South and Central America, the Pharoideae are pantropical, and the Puelioideae are found only in tropical Africa. Furthermore, of the sister families, Joinvilleaceae occur on Borneo, New Caledonia, and Pacific islands (e.g. Hawaii) and Ecdeiocoleaceae are limited to south-west Australia. Two possibilities that would account for these distributions are (1) long-distance dispersal across the Atlantic and Indian Oceans, or (2) radiation across a continuous Gondwanan equatorial continent followed by subsequent isolation and further evolution. Phytolith evidence (see §2.4.3) is consistent with the view that the divergence and spread of the BEP and PACCAD clades occurred across Gondwana before the biogeographic separation of the Indian subcontinent from the rest of Asia by c.80 Ma (Prasad *et al.* 2005). However, the paucity of grasses in the early Tertiary or late Cretaceous fossil record (see below) does not help in resolving this issue (Hodkinson *et al.* 2007b).

2.4.3 Fossil history

The early fossil record of the grasses is poor. Grass-like leaves and putative floral structures have been observed in late Cretaceous sediments. For example, Cornet and colleagues (Cornet 2002) recovered fossils of grass-like leaves with paniculate inflorescences and grass-like flowers from the Turonian–Raritan Upper Cretaceous (90 Ma) clays in New Jersey, USA. Included with these fossils were platanaceous leaves (plane tree family), ericaceous leaves (heather family), and possible lauraceous leaves (laurel family). Many of these plants may have grown on delta levees and overbank deposits, close to the site of deposition. Other sediments from the site indicate that the depositional environment was a coastal plain forest of pines (*Prepinus* sp., *Pinus* sp.) and angiosperm trees (*Dewalquea* spp., Platanaceae) along with the achlorophyllous saprophytic *Mabelia connatifila*—the oldest unequivocal fossil monocot from the Upper Cretaceous 90 Ma (Gandolfo *et al.* 2002).

Some of the earliest fossil grass remains are of pollen grains, which are of limited value because of their ultrastructural uniformity. Monoporate pollen assumed to be fossil Poaceae is placed into one of several form genera including *Graminidites*, *Monoporoites*, and *Monoporopollenites* (Macphail and Hill 2002). However, confident assignment of these fossils to the Poaceae requires the observation of minute channels or holes penetrating the outer pollen wall (Kellogg 2001). The oldest confirmed records of grass pollen are from the Paleocene (55–65 Ma) of South America and Africa. Reports of older Late Cretaceous Masstriichtian-age grass or grass-relative fossil pollen grains cannot be attributed unambiguously to the Poaceae. The earliest recorded grass pollen from North America is from the uppermost Eocene (*Graminidites gramineoides*). Grass pollen becomes abundant thereafter in the succeeding Oligocene and Miocene.

The earliest accepted fossil records of the grasses are phytoliths (silicified plant tissues: see §4.4) preserved in Late Cretaceous (Maastrichtian) titanosaurid dinosaur coprolites from central India (Prasad *et al.* 2005). These fossils include morphotypes attributable to at least five taxa of grasses within the [Bambusoideae + Ehrhartoideae] and PACCAD or Pooideae clades. The taxonomic diversity of these phytolith morphotypes suggests an evolution, diversification, and spread of basal Poaceae 65–71 Ma in Gondwana before India became geographically isolated.

The earliest evidence of mega (macro)-fossils confirmed as grasses is from the Eocene Wilcox formation (*c.*54 Ma) of western Tennessee, USA (Crepet and Feldman 1991). These sediments contain macrofossils of spikelets, inflorescence fragments, leaves, and whole plants. The fossils reveal spikelets with two florets subtended by two alternate, slightly unequal, keeled glumes. Three exserted stamens per floret are present, with dorsifixed anthers (Fig. 2.2). However, the bracts are poorly preserved and it is possible that there is only one floret with six stamens (Soreng and Davis 1998). Vegetative remains reveal small, perennial plants with leaves arising from a rhizome. These and other features preserved in the pollen support the diagnosis that these plants fit clearly within the Poaceae *sensu lato*. In the absence of additional diagnostic characters, the subfamily affinity of these fossils is uncertain although the authors suggested similarities with the Pooideae or Arundinoideae—both C_3 subfamilies. Nevertheless, these fossils represent the earliest clear evidence of grasses, and of wind-pollinated herbaceous monocots. Furthermore, the well-defined nature of these plants (their affinities, although uncertain, do not suggest a primitive grass) is consistent with an Upper Cretaceous origin of the family.

The second-oldest grass macrofossil is a female spikelet of the extant genus *Pharus* (Bambusoideae)

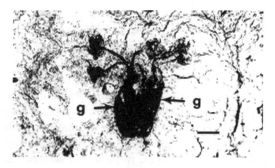

Figure 2.2 The oldest macrofossil of a grass. A spikelet with two florets with glumes (g) is visible. Bar = 1 mm. Reproduced with permission from Crepet and Feldman (1991).

found in association with mammalian hair preserved in amber from the late Early Miocene–early Middle Miocene (15–20 Ma) (Iturralde-Vinent and MacPhee 1996; Poinar and Columbus 1992). Hooked hairs on the lemma provide the earliest evidence for dispersal via attachment to animal fur (**epizoochory**). Assignment of the fossil to a modern-day extant genus is further support for the idea that diversification of the grass family had occurred quite a bit earlier than indicated in the fossil record. By the time the following Oligocene epoch (34–23 Ma) ended, it is quite clear from the fossil record that grasses representing several separate extant tribes and genera were present. By the late Miocene (7–5 Ma) there is evidence that grasses possessing the C_4 photosynthetic pathway had evolved (§2.4.4 and Chapter 4). Thomasson *et al.* (1986) described fossil leaves with 'chloridoid' Kranz anatomy, typical of C_4 grasses. How much earlier than this the C_4 pathway evolved is unknown, although there is speculation that biochemical precursors of C_4 photosynthesis may have existed as far back as the Permo-Carboniferous (Osborne and Beerling 2006).

2.4.4 Ecological origin of grasses and grasslands: relationship to arid conditions and rise of grazing mammals

The original grasses evolved in deep shade or forest margins, characteristics retained by the extant *Anomochloa, Streptochaeta, Pharus, Puelia, Guaduella,* the bamboos, and the basal pooid *Brachyelytrum*. Little diversification of the grasses occurred in these habitats for millions of years. Major diversification is associated with spread into open habitats in the mid-Miocene (Kellogg 2001).

The spread of grasses and their subsequent evolution is believed to reflect climatic factors to a large extent. For example, current distribution of the Andropogoneae tribe (in Panicoideae) shows a close relationship with tropical areas of high midsummer rainfall (Hartley 1958). The C_4 Panicoideae and Chloridoideae dominate tropical and subtropical grasslands. On a narrower taxonomic basis, differentiation of the genus *Poa* is closely tied to regions of high latitude and high altitude. In the USA, *Poa* spp. form more than 5% of the grass flora in areas of cool summer temperatures, i.e. areas below the 24 °C midsummer (July) isotherm (Hartley 1961).

Climatically, the diversification of grasses and the spread of grasslands coincides with increased aridity, especially during the Oligocene (34–23 Ma). In North America, for example, the Rocky Mountain uplift led to aridity of the Great Plains causing a retreat of the forests in the late Oligocene and Miocene (23–5.3 Ma) (Coughenour 1985). In Africa, increased continental elevation led to increased aridity, also during the Oligocene, allowing the spread of grasslands.

Thus, grasses were able to spread during periods of increased aridity because of the possession of a number of traits adapted for drought; these include basal meristems, small stature, high shoot density, deciduous shoots, below-ground nutrient reserves, and rapid transpiration and growth (Coughenour 1985). The basic scenario postulated by Stebbins (1987) included the following selection pressures and associated adaptations by grasses driving their evolution:

- trampling by large herbivores → heavy sympodial rhizome system
- grazing → basal leaf meristems
- grazing by hypsodont ungulates and/or phytophagous insects → silicification of epidermal cells
- open conditions of savannahs → wind pollination
- wind pollination → condensation of compound racemes → racemes of spikelets, more intense pollen clouds, increased target for pollen
- diurnal shifts in temperature and moisture → scale-like perianth, and fleshy lodicules that open and close easily.

The extent to which any of the changes described above reflects true adaptation to aridity or a co-evolutionary relationship with grazers, one in which both partners reciprocally adapt to each other over time, is uncertain and contentious. It is uncertain whether traits presumed to relate to drought and grazing tolerance are truly beneficial (**aptations**), incidentally beneficial (**exaptations**), or the result of selection to confer the present benefit (**adaptation**). Although climatic factors have probably played a large part in the evolution of the grasses and the spread of grasslands, the potentially co-evolutionary relationship with grazing mammals has been the subject of much conjecture and speculation (Coughenour 1985). Certainly, evolutionary change in grazing animals, specifically ungulates (hoofed mammals) appears to track the climatically driven development of extensive savannahs and grasslands from the mid-Miocene onwards.

The grasses arose in the late Cretaceous–early Tertiary, but they diversified and spread into large vegetation formations in the mid-late Miocene (12–5.3 Ma). The first appearance of extensive, open grasslands varied worldwide. In North America it was towards the end of the Miocene (8–5 Ma) (Axelrod 1985) (Fig. 2.3), although savannah or short-sod grasslands appear to have arisen in the early Miocene (c.19.2 Ma) (Strömberg 2004). In South America, grass-dominated ecosystems may have arisen as early as the Eocene–Oligocene boundary (34 Ma) (Jacobs *et al.* 1999). By contrast, in central and western Europe the grasslands today are secondarily anthropogenic and arose from farming and pastoralism following the spread of Neolithic husbandary in the Holocene (Bredenkamp *et al.* 2002). The modern North American tallgrass prairie is also of recent origin following the retreat of woodland during the warm, dry hypsothermal period c.4000 years ago.

The rise of grass-dominated ecosystems was probably preceded by the evolution of the C_4 photosynthetic pathway in the middle Miocene (or perhaps even earlier in the Oligocene) under

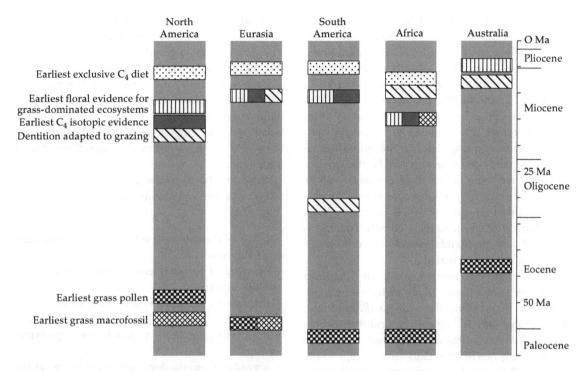

Figure 2.3 Generalized summary of the establishment of grass-dominated ecosystems worldwide. Reproduced with permission of the Missouri Botanical Garden Press from Jacobs et al. (1999).

conditions of low atmospheric partial pressure of CO_2 (the CO_2-starvation hypothesis; see §4.1). At an earlier date, dentition adapted to grazing appeared (i.e. hypsodont, high-crowned, cheek teeth) (Jacobs et al. 1999). The finding of grass phytoliths in late Cretaceous dinosaur coprolites (see §2.4.3) is evidence that grasses were part of their diet. Late Cretaceous Gondwanatherian mammals possessed hypsodont cheek teeth and could have eaten grasses too, suggesting the possibility of an even earlier co-evolutionary interaction between grasses and vertebrate herbivores than previously thought (Piperno and Hans-Dieter 2005).

The rapid and global spread of C_4-dominated grassland in the late Miocene (5–8 Ma) was at the expense of woodland and reflects changes in the climate conducive to frequent fire (Keeley and Rundel 2005) including increased seasonality of rainfall (Osborne 2008). Increased seasonality would have allowed high production (i.e. fuel) during warm, monsoon-like, moist conditions alternating with a fire-susceptible dry season. Fire in the dry season inhibits and kills trees and creates a high-light environment also favouring C_4 grasses. Moreover, increased frequency and severity of drought conditions again favours C_4 grasses at the expense of woody vegetation. This combination of climatic conditions created a novel environment allowing the expansion of C_4-dominated grassland. A predictable and frequent fire regime allowed the grasslands to be maintained.

The extent to which grass evolution is a consequence of adaptation to herbivory is unclear. For example, siliceous phytoliths in grass epidermal cells can be viewed as an adaptation of grasses to herbivory (the abrasive silica wearing down the teeth, thus providing a grazing deterrent). Conversely, the evolution of hypsodonty may be a response to increased siliceous grass in the diet, or, evolution of both herbivores and grasses may have been reciprocal. Phytoliths recovered from sediments indicate a C_3-dominated grassland in the

North American Great Plains in the early Miocene (25 Ma), at least 7×10^6 years before adaptations of horses in the Great Plains to grasslands (Strömberg 2002, 2004). Conversely, other hypsodont ungulates, such as oreodonts and camels, occurred in the Great Plains during the Eocene (55–34 Ma), well before the Miocene. Tying the story even closer together, anomalously high species richness of browsing ungulates in the mid-Miocene (c.18–12 Ma) woodland savannah of North America is postulated to be the result of elevated levels of primary productivity, itself a consequence of perhaps higher-than-present levels of atmospheric CO_2 (Janis et al. 2004). As these climatic conditions changed and grasslands spread, the numbers of browsing species decreased.

The evolution of horses is pertinent (Chapman 1996). The genus *Equus*, which includes the modern horse (*E. caballlus*), the wild ass (*E. hemionus*), the African ass (*E. africanus*), and the zebras (*E. burchellii, E. zebra,* and *E. grevyi*), is the sole remaining genus of an evolutionary lineage in the Equidae that included several now-extinct members. Derived from five-toed mammals, the modern horse is single-toed; an adaptation well suited to running in open country such as grassland. As the grasslands expanded through the Pliocene, the now-extinct single-toed ancestor of the modern horse, *Pliohippus*, became more widespread. At the same time, the three-toed *Hiparion* and *Neohipparium* diminished in abundance as woodland areas decreased. A hoof with three toes is adapted to dodging sideways as well as running forwards: more necessary in a woodland than a grassland.

In summary, the major phases in the evolution of grasses and grasslands are (Jacobs et al. 1999) (Fig. 2.3):

- origin of Poaceae in forest margins or shade during Cretaceous
- opening of forested environments in early to middle Tertiary
- increase in abundance of C_3 grasses in middle Tertiary
- origin of C_4 grasses in middle Miocene (probably earlier)
- spread of C_4-grass-dominated ecosystems at the expense of C_3 grassland and/or woodland in the late Miocene.

CHAPTER 3

Ecological morphology and anatomy

Perhaps the season when the sight of the green meadows most delights us, is early spring. How beautiful are they, as the sunlight comes down upon their gleaming blades, and the blue heavens are hanging over them!

Anne Pratt, The Flowering Plants, Grasses, Sedges, and Ferns of Great Britain, and their allies the Club Mosses, Pepperworts and Horsetails (1873)

Since, in general, increase in xerophytism has everywhere meant evolutionary advance, all these vegetative features are a very useful guide towards an understanding of the main evolutionary trends.

(Bews 1929)

The common possession of a number of characteristic morphological features makes grasses readily recognizable to even the lay person—although, as noted in Chapter 2, there is often confusion with other similar families such as the rushes (Juncaceae) and sedges (Cyperaceae). In this chapter the morphology and anatomy of a grass plant is described, with mention where appropriate of the ecological relevance and importance of the different features. A number of other sources can be consulted for additional details such as Metcalf (1960), Langer (1972), Chapman (1996), Gould and Shaw (1983), or Renvoize (2002) as well as electronic databases (see §2.2) including Watson and Dallwitz (1992 onwards). There is considerable diversity in form and function that relates to the evolutionary adaptation of grasses to their environment. In general, the trends in grass evolution reflect a series of adaptive reductions in size, complexity, and number—albeit with notable exceptions and a number of reversals (Stebbins 1982).

3.1 Developmental morphology— the phytomer

Central to the understanding of grass morphology and growth is the **phytomer** concept (Moore and Moser 1995). This posits that the phytomer, consisting of an internode and node together with the leaf blade and sheath at the upper end, and an axillary bud at the lower end, is the fundamental growth unit of a grass (Briske and Derner 1998; Gould and Shaw 1983). A single phytomer matures from top to bottom (blade, sheath, internode). A stack of phytomers (i.e. a shoot) matures from bottom to top. The oldest phytomer is at the bottom of the stack, with the longest leaf to protect the growing point; the youngest phytomer at the top of the stack has only a short leaf that affords less protection. A grass plant thus consists of a collection of **ramets** (clonally produced parts capable of potentially independent existence: Silvertown and Lovett Doust 1993), each of which is a repeated series of phytomers successively differentiated from individual apical meristems (Briske and Derner 1998). A **tiller** (a secondary stem) is a collection of phytomers differentiated from a single apical meristem (Moore and Moser 1995). Morphological variation within and among grasses is a consequence of variation in the number and size of ramets that themselves comprise a number of phytomers. This variation encompasses diminutive grasses such as *Poa annua* through arboreal giants such as *Dendrocalamus* spp. amongst the bamboos that can grow up to 40 m (Metcalf 1960). Quantification of the variation among phytomers in a single grass plant can provide a basis for understanding morphogenetic and environmental constraints on plant development (Boe *et al.* 2000). For example, in the perennial prairie grass *Andropogon* *gerardii*, blade length, blade weight, and sheath weight decrease among phytomers of the main culm in acropetal fashion, whereas sheath length can remain constant.

Meristematic, or growth, tissue in grasses is located immediately above the nodes in the stems, and at the base of the leaf blade. Because the basal nodes of perennial grasses tend to be short, and the apical meristem is borne close to the ground enveloped by protective leaf sheaths, grasses are well adapted for rapid regrowth following grazing.

3.2 Structure of the common oat *Avena sativa*

The common oat provides a useful starting point for understanding the morphology of a typical grass plant (Fig. 3.1). The main parts to note above ground are the culm, the leaves, and the flower head. The **culm** is the grass stem and consists of several cylindrical tubes of unequal length closed at the joints by solid tissue. The joints, or **nodes**, are generally a darker colour. The hollow portions of the stem between the nodes are the **internodes**.

The **leaves** of grasses are borne in two rows, alternating on opposite sides of the culm. The leaf consists of three portions: the **blade**, **sheath**, and **ligule**. The blade is the upper flattened portion; the sheath is cylindrical embracing the culm, open along one side with one margin overlapping the other (in *A. sativa* at least). The ligule is a thin membrane that occurs at the junction of the sheath and the blade.

The flower head terminates the culm and consists of a main axis plus several spreading branches or pedicels. **Spikelets** are borne at the tips of the pedicels. In *A. sativa*, the flower head is a panicle by virtue of the spreading branches. Spikelets (S on Fig. 3.2) consist of several scales, alternatively borne in two rows on opposite sides of a short stem, the **rachilla**. The spikelet is the basic reproductive unit in grasses, historically forming the basis for classification (Chapter 2). Each spikelet is subtended by two outer scales, the **glumes** (G_1, G_2), which envelop the spikelet. In *A. sativa*, there are 2–3 florets per spikelet. Each floret (FS) consists of an outer scale, the **lemma** (L), and an inner scale, the **palea** (P). The lemma has a short extension at the end, an **awn**. The flower

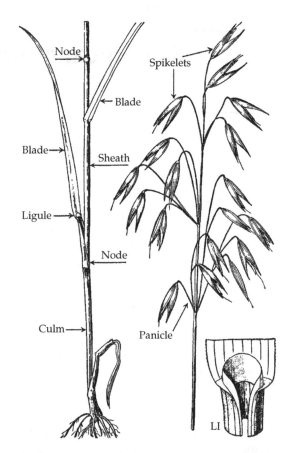

Figure 3.1 Common oat *Avena sativa*, × 0.5. LI: detail showing base of leaf blade and ligule. Reproduced with permission from Hubbard (1984).

(FL) is composed of three parts, a pair of small scales (**lodicules**, LO), three **stamens**, and a **pistil**. The pistil consists of a single ovary, with a single **ovule**, plus two feathery **styles**. When fertilized, the ovule develops into the grain.

It should be emphasized that the structure of *Avena sativa* described above represents a 'typical' grass, and there is extensive variation among species, described below, that both forms the basis of classification (Chapter 2) and reflects adaptation to the environment.

3.3 Culms

The vegetative shoot apex of a grass contains the apical or terminal meristem at its tip and is usually

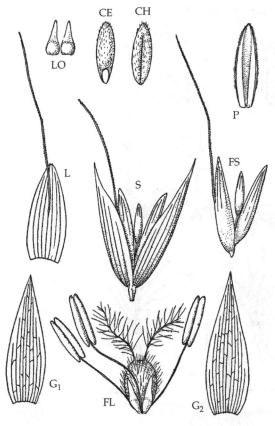

Figure 3.2 Spikelet of common oat *Avena sativa*, FL, LO, × 6; rest × 2. CE, grain (abaxial view showing embryo); CH, grain (adaxial view showing hilum); FL, flower; FS, floret; G_1 and G_2, glumes; L, lemma; LO, lodicules; P, palea; S, spikelet. Reproduced with permission from Hubbard (1984).

located close to the ground protected by enveloping leaf sheaf bases. Shoots originate either from the seed embryo (the primary shoot) or from vegetative buds (axillary buds) in a leaf axil of an older shoot (secondary shoots or tillers). Thus, culms may branch to form a compound shoot system. The location and extent of branching and ramet development largely determines the physical structure of the grass plant (Fig. 3.3). **Stolons** are aboveground branches that tend to run prostrate over the soil surface (e.g. *Agrostis stolonifera*), whereas below-ground branched stems are referred to as **rhizomes** (e.g. *Poa rhizomata*). Both structures possess leaf buds and (at least, vestigial or scale) leaves, can branch, and may give rise to aerial stems, and roots, at the nodes. Rhizomes are common in many perennial grasses and may contain substantial amounts of storage tissue.

Branching of the culm is either sympodial or monopodial. **Sympodial branching** occurs where there is successive development of lateral buds, just behind the apex. The main axis stops growing in this case. Conversely, **monopodial branching** occurs when the culm increases in length by division of the apical meristem, with branching by lateral branches occurring in acropetal succession (Usher 1966). Culms with aerial stems branching extensively at the base produce a **caespitose** (clumped or tufted) habit (e.g. *Schizachyrium scoparium*, *Festuca ovina*, *Koeleria macrantha*), whereas branching in the upper culm produces a shrubby appearance (e.g. *Andropogon glomeratus*). Thus, culms can be erect to decumbent, creeping, shrubby, or even treelike. Most commonly, the shoot tip of lateral branches emerges from the apex of the enclosing leaf sheath (i.e. intravaginally) producing the caespitose habit, although the more diffuse stoloniferous or rhizomatous architecture of sod grasses (e.g. *Festuca rubra*) develops when the shoot tip breaks through the sheath (i.e. extravaginally). Rhizomes and stolons can both branch and produce adventitious roots at the nodes that can be especially important in allowing vegetative reproduction or persistence if the plant is fragmented. *Microstegium vimineum*, for example, is an annual grass, invasive in North America, which produces extensively branched prostrate culms that root freely at the nodes (Gibson *et al.* 2002). Rhizomes and stolons can integrade in some species, e.g. *Cynodon dactylon* (Fig. 3.3).

The production of secondary (or branch) shoots from axillary buds can be important in the growth and population persistence of grasses. For example, *Digitaria californica*, a dominant perennial grass in the semi-desert ranges of the USA, possesses axillary buds on most internodes of the culm, except the panicle internode. The current year's crop of basal culms are produced in the spring or summer from axillary buds (Cable 1971). The importance of axillary buds for population persistence was demonstrated in *Bouteloua curtipendula* and *Hilaria belangeri* in semi-arid oak–juniper savannah, in Texas, USA (Hendrickson and Briske 1997). Dormant axillary buds attached to the base of reproductive parental

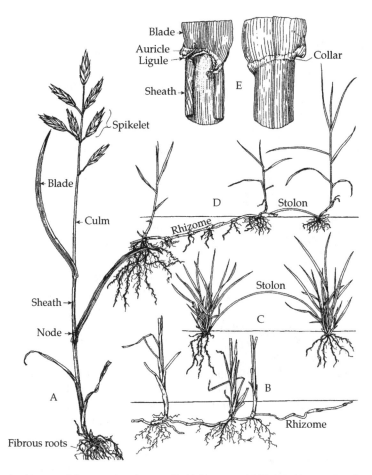

Figure 3.3 Structure and architecture of the grass plant: A, general habit (*Bromus unioloides*); B, rhizomes; C, stolon; D, rhizome and stolon integradation (*Cynodon dactylon*); and E, leaf at junction of sheath and blade, showing adaxial (left) and abaxial (right) surface. Reproduced with permission from Gould and Shaw (1983).

tillers remained viable for 18–24 months, exceeding parental tiller longevity by 12 months. These buds thus provide a meristematic source for tiller recruitment into populations that exceeds the longevity of the seed bank of many perennial grasses (see §5.1.4).

Perennial grasses in which ramets are replaced annually can live for many years, with some estimates of longevity exceeding 1000 years (Briske and Derner 1998). Making age determinations or estimates of genet natality (births) in an organism that is continuously growing and senescing is difficult, but has been aided by the development of DNA-based molecular marker techniques. For example, randomly amplified polymorphic DNA (RAPD) methods (Gibson 2002) were used to estimate genet natality of *Festuca rubra* at 3–7 genets m^{-2} yr^{-1} (Suzuki *et al.* 1999). Clonal development of caespitose grasses depends on ramet production and death rate, with clones showing a developmental sequence of pre-reproductive, reproductive, and post-reproductive stages of 5–10, 15–30, and 15–25 years to complete, respectively (Gatsuk *et al.* 1980). *Deschampsia caespitosa* clones in northern Europe, for example, die back in the centre during the mature reproductive stage, with the whole clone fragmenting in the post-reproductive phase (Fig. 3.4). The clonal fragments of old genets (plants arising as individuals from a seed; Gibson 2002) are free-living and capable of producing new ramets, but they may be short-lived, contributing little to population maintenance (Briske and Derner 1998). The

ECOLOGICAL MORPHOLOGY AND ANATOMY 39

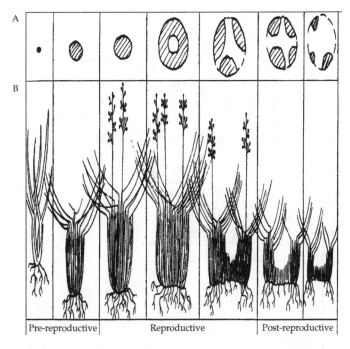

Figure 3.4 Schematic showing architectural development of *Deschampsia caespitosa* clones in northern Europe. A, aerial view; B, side view. 35–60 years is required for clones to grow from seedlings to senescence, note the development of hollow crowns in the reproductive phase followed by clonal fragmentation in the post-reproductive phase. Reproduced with permission from Briske and Derner (1998) redrawn from Gatsuk *et al* (1980).

physiological integration conferred among ramets of clonal grasses is discussed in Chapter 4.

Tillers are aerial axillary shoots, consisting of a culm and its associated leaves. **Tillering** is the production of new tillers. Intravaginal tillers grow up between the leaf sheaths, and extravaginal tillers burst through the enclosing sheath bases or arise from buds not enclosed by existing sheaths. Tufted or casepitose grasses exhibit intravaginal tillers, whereas the production of stolons (e.g. *Poa trivialis*) or rhizomes (e.g. *Elytrigia repens*) is through extravaginal tillers. Aerial tillers can arise from the nodes of extended stems. For example, the annual grass *Microstegium vimineum* produces aerial tillers as it grows through the season (Gibson *et al*. 2002).

Much agricultural and ecological research has been devoted to understanding and quantifying tiller dynamics. Some of the basic calculations made to quantify tiller dynamics include **tiller appearance rate (TAR)**—the rate at which tillers become apparent to the eye (of course, they form earlier). The most frequent forms of TAR that are calculated include:

- **net absolute TAR (NTAR):**

$$\text{NTAR} = \frac{N_2 - N_1}{t_2 - t_1}$$

where N_1 is the number of live tillers (per plant, pot, unit area) at time t_1 and N_2 is the number of live tillers at time t_2 (Thomas 1980). If $N_2 = N_1 +$ (number of new tillers produced between t_1 and t_2) then the expression gives the gross TAR.

- **proportional PAR (PTAR):**

$$\text{PTAR} = \frac{1}{N_1} \times \frac{N_2 - N_1}{t_2 - t_1}$$

where net and gross rates can be calculated depending upon how N_2 is determined, as with absolute TAR.

Net and gross tiller death rates can be estimated in a similar manner by measuring the rate of 'appearance' of dead tillers (e.g. Hendon and Briske 1997).

Tiller appearance rates for the caespitose grass *Arrhenatherum elatius* are illustrated in Fig. 3.5.

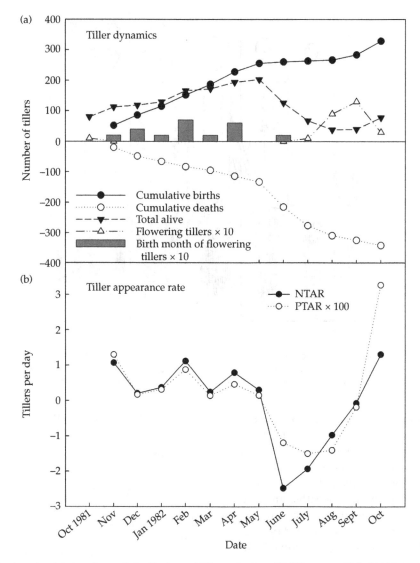

Figure 3.5 Tiller dynamics of *Arrhenatherum elatius*, Newborough Warren, Anglesey, UK (Gibson, unpublished): (a) Seasonal tiller flux; (b) tiller appearance rates. Tiller births were obtained from 25 cm² plots placed in the centre of five tussocks. NTAR, net tiller appearance rate (tillers d⁻¹); PTAR, proportional tiller appearance rate (tiller d⁻¹).

For both NTAR and PTAR, values >0 reflect growth of the grass clumps in terms of tiller numbers, and values <0 indicate a decrease in the tiller population. For *A. elatius*, the figures indicate that clumps were growing from September through May, and experienced negative growth (tiller loss) during the summer. This seasonal flux in tiller numbers is also illustrated in the plot of cumulative tiller births and deaths (Fig. 3.5).

There is continuous tiller death, but it was most pronounced during the summer and the accompanying decrease in tiller births at this time led to a sharp decline in the population numbers of living tillers. In the dune grassland where this grass was growing, it was the dominant species and the seasonal tiller dynamics influenced the rest of the plant community (Gibson 1988a, 1988c).

Similarly, in the cool-season perennial *Schedonorus phoenix* new tillers are produced throughout the year, especially in the late winter to early spring, and after flowering in late summer (Gibson and Newman 2001). Tiller mortality is highest in June because of intraclonal competition among tillers. By contrast, complete winter dieback of tillers is typical for most warm season perennial grasses. For example, across much of its range in the US Great Plains, the warm-season perennial *Schizachyrium scoparium* dies back in the late autumn (i.e. complete tiller mortality), with new tillers emerging in late spring through summer. However, in Texas, in the southern part of its range, a few tillers on the periphery of clumps can persist from one growing season to the next (Butler and Briske 1988).

The tillers form a hierarchy with a few large tillers and lots of small ones, many of which are repressed. Tillering is under genetic control but is highly modified by the environment since it requires intense meristematic activity and cell enlargement; both processes requiring energy and resources. An increase in tillering stimulates leaf production and growth leading to an increase in total leaf area per plant. The following general affects of the environment on tiller production have been observed (Langer 1972):

- Tillering is temperature sensitive, with an inhibition of tillering at high temperatures related to respiration rates and soluble carbohydrate content of the plant. Warm conditions at night appear to be more deleterious than warm days. Temperate species have relatively low optimum temperatures for tillering: 18–24 °C in *Lolium perenne* and 10–25 °C in *Dactylis glomerata* (Langer 1972). By contrast, tillering in subtropical species increases up to 35 °C.
- Light interacts with temperature and high light intensity favours tiller production. The current rather than past light intensity is the most important, as plants repressed in shade resume tillering when brought back out into the light. Light quality can affect tiller dynamics. Experimentally increasing the red:far red ratio enhanced spring tiller production in *Sporobolus indicus* and *Paspalum dilatatum*, and increased tiller death for both species during the autumn (Deregibus et al. 1985). An increase in the red:far red ratio is typically associated with the shade effect of leaf canopies.
- Soil moisture affects tillering, with drought reducing tiller production, reducing tiller size and the percentage of reproductive tillers, and increasing mortality rate. For example, in a lightly grazed *Schizachyrium scoparium* pasture a mean of 51 tillers plant^{-1} during a year with above average precipitation was reduced to a mean of 26 tillers plant^{-1} in a drought year (Butler and Briske 1988). Whereas 20% of the tillers were reproductive in the wet year, there were zero reproductive tillers in the drought year.
- Mineral nutrition status of the soil and hence of the plant is important for tiller production. Tillering increases with increased supply of nitrogen, phosphorus, and potassium. Nitrogen is the most important of these, and interacts with phosphorus and potassium; at low nitrogen, tillering is suppressed even with additional phosphorus or potassium. The response to fertilizer in the field depends on nutrients in the soil and the developmental stage of the plant.
- Grazing, although causing mortality of plant parts directly affected, can stimulate tillering through release of buds in the axis of leaf bases on the stem. For example, continuous grazing of *Schedonorus phoenix* by sheep increased the number of tillers while reducing leaf extension rates (Mazzanti et al. 1994). By contrast, grazing-sensitive species exhibit declines in tiller production and hence total tiller numbers with herbivory. Sensitivity to herbivory can depend upon the timing of defoliation; for example, defoliation of *Eriochloa sericea* during pre-culm and post-culm stages displayed greater cumulative tiller mortality (28% and 11%, respectively) compared to undefoliated plants (39% and 21%, respectively) (Hendon and Briske 1997).

In some sense grass plants can be regarded as a population of tillers, just as trees have been described as a population of buds (Jones 1985). Tiller life history is dependent on the life history of the species and the environmental conditions and season. Tillers may be born and die after only a few weeks remaining vegetative throughout their life, or they can persist vegetatively from one season to the next or longer. Ultimately if a tiller does become reproductive then it dies after flowering is complete. In *Arrhenatherum elatius* we noted earlier the seasonal flux in tiller dynamics (Fig. 3.5). In addition, there is distinct seasonality in tiller survivorship (Fig. 3.6). Tiller production increases rapidly following flowering late in the summer, and most of the tillers born in the autumn overwinter and live into the early summer. By contrast, tillers born in early summer live only one or two months. *Arrhenatherum elatius* flowered from July to October (Fig. 3.6). Tillers that flowered were born from November the year before through June, but 60% of them were born from February to April.

An understanding of culm growth is best related to phytomer production (§3.1). For example, short-shoot growth (e.g. *Bouteloua gracilis*) is a consequence of many phytomers. Late elongation of the inflorescence uses only a few phytomers, leaving many phytomers at the base, each with axillary buds (a stack of phytomers), which then can elongate to produce more growth (tillering). Conversely, long-shoot growth (e.g. *Andropogon gerardii*) can occur with few phytomers; early elongation of the inflorescence uses most of the phytomers, leaving few phytomers at the base. Thus, with few axillary buds there is only minimal tillering.

Specialized rhizome development occurs in the bamboos through the production of extensive jointed, segmented rhizomes. Lignification of the ground tissue makes the rhizomes strong and rigid. Rhizome development in bamboos takes one of two forms. **Pachymorph rhizomes** are short and thick, pear-shaped, with lateral buds giving rise only to rhizomes and culms arising only from the apex of the upturned rhizome; this gives rise to densely clumped bamboos, e.g. *Bambusa* spp. By contrast, **leptomorph rhizomes** of 'running bamboos' are long and slender with every node bearing a shoot bud and roots. The rhizome tip grows continuously forward through the soil allowing the plant to form sometimes massive clonal colonies that can cover entire hillsides, e.g. *Phyllostachys*

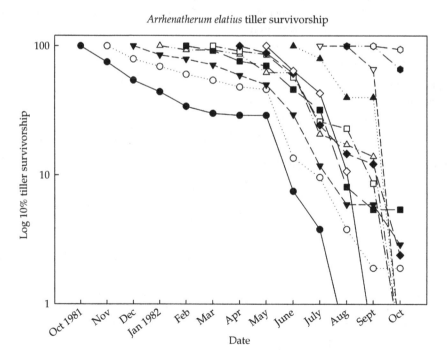

Figure 3.6 Survivorship of *Arrhenatherum elatius* tillers, Newborough Warren, Anglesey, UK (Gibson, unpublished). Newborn tillers were marked the first week of each month from October 1981 to October 1982 in 25 cm^2 plots placed in the centre of five tussocks. Each line represents a different tiller cohort.

nigra which forms dense forests in the rain forest of east Maui, Hawaii. Rhizome cuttings from both pachymorph and leptomorph species allow easy propagation so long as some buds and culms are attached to the cutting.

In addition to the architectural function of branching and vegetative reproduction, rhizomes act as storage organs for grasses. Sugars and starches can occur in large amounts, and support initial growth in the spring and regrowth following grazing. For example, when total non-structural carbohydrate concentration in the rhizomes of *Sorghum halepense* was experimentally reduced by 60%, regrowth from the rhizomes was not possible. Grass rhizomes have been used for human consumption; for example, rhizomes of *Phragmites australis* are sometimes eaten or processed for starch by native people in Tasmania (Australian National Botanic Gardens 1998).

Other storage or vegetative reproduction organs include bulbs and corms which may be developed at the base of culms. True bulbs are rare, but occur in *Poa bulbosa*, whereas corms (or corm-like swellings) occur in several genera, including *Poa, Melica, Molinia, Colpodium, Arrhenatherum, Beckmannia, Hordeum, Phleum, Ehrharta*, and *Panicum* (Clark and Fisher 1986). The corms (haplocorms of Evans 1946) are fleshy swellings at the base of stems, and can be particularly important in vegetative reproduction. In *Arrhenatherum elatius* var. *bulbosum*, the corms occur in a string of four or five basal internodes, each up to 1 cm broad with a regenerative bud that can germinate. As with rhizomes, the food reserves in these structures can be important in the diet of animals. Yellow babbons (*Papio cynocephalus*), for example, will dig up and eat the corms of *Sporobolus rangi* in the Amboseli region of East Africa (Amboseli Baboon Research Project 2001).

Flowering culms arise when the apex of a vegetative culm increases its growth rate and initiates the inflorescence. Internodes elongate and, typically, the flowering stem grows as an erect structure. Leaf primordia are formed rapidly on the elongating stem apex, and buds in the axil of these primordia grow rapidly. Leaf growth is inhibited and bud growth ensues as they become spikelet primordia. With respect to a single stem, this change in function of the apical meristem is permanent. An exception is **vegetative prolifery** (also called proliferation, but only incorrectly, vivipary) that can occur as in *Festuca viviparoidea* (viviperous sheep's fescue) where a branched reproductive inflorescence can revert to vegetative growth (Stace 1991). In this situation the spikelet above the glumes is converted into a leafy shoot (Clark and Fisher 1986). Proliferation has been reported in *Agrostis, Deschampsia, Eragrostis, Festuca, Oryza, Phleum, Poa, Setaria, Sorghum*, and *Zea*.

The switch to flowering is terminal for **monocarpic** plants (dying after one fruiting season), whether annual grasses or **semelparous** (once-flowering) perennials. Many genera have both annual and perennial members, e.g. the brome grasses with the annual *Bromus japonicus* and *B. tectorum*, and the perennial *B. inermis* among others. Many bamboos are long-lived and monocarpic, often flowering simultaneously over large areas at intervals of 100 years or more before dying. For example, there are 35 species of *Sasa* in Japan all of which show these characteristics (Makita 1998). Whether a perennial plant is monocarpic or **polycarpic** (fruiting many times), the individual reproductive shoot is monocarpic dying after reproduction is complete.

The nature of the inflorescence is described in §3.6.

3.4 Leaves

Leaves develop from primordia in the dermatogen and hypodermis (outer cell layer) on the flanks of the conical apical dome of the stem. Leaf primordia are initially entirely **meristematic** (possessing ability for cell division), but meristematic activity becomes limited to an intercalary region which includes the ligule on the inner (adaxial) face and the leaf lamina abaxially. The location of the intercalary meristem means that leaves can continue to grow if the tip of the lamina is removed, e.g. through grazing or mowing. Stems are usually short so leaves arise close together. The leaves are arranged on alternate sides of the apex leading to **distichous phyllotaxy**, i.e. the leaves are arranged in two vertical rows. **Spiral phyllotaxy** in grasses is extremely rare, but occurs in members of the Australian genus *Micraira* (Watson and Dallwitz 1988).

The grass leaf consists structurally of three parts; the blade (lamina), the sheath, and the ligule. The first leaf of a culm branch or lateral shoot consists of a membraneous, modified sheath structure, lacking a blade, known as the **prophyll** or **prophyllum**. This structure protects the immature lateral stem axis by being initially tightly appressed to it. Subsequent leaves include a blade with the lamina emerging from the shoot rolled or folded about the midrib. The lamina is normally linear or lanceolate, with characteristic parallel nerves, narrow relative to its length, elongate, laminate, and flat (capable of being flattened if rolled). Leaves can be stiff, setaceous, needle-like, or acicular (permanently inrolled—e.g. *Miscanthidium teretifolium* which has a blade that is almost cylindrical with the adaxial surface recognized only by a small groove). Leaf margins are entire, smooth, or scabrous. In some species the margins are so strongly scrabrous that the grass has become well known for its ability to cut the skin, e.g. rice cutgrass (*Leersia oryzoides*). In size, grass leaves range from the diminutive *Monanthocloë littoralis* with blades usually <1 cm to the large *Neurolepis nobilis* (bamboo) with blades up to 4.5 m long and 30 cm wide.

Leaf shape is related to environmental conditions. In the humid tropics grass leaves are often large, often with ovate or oblong blades. By contrast, in semi-arid regions they are often narrow and linear, becoming involute (inrolled) under drought conditions. *Cladoraphis spinosa*, a grass of deserts in the western Karroo (South Africa), is an extreme example of xeromorphic adaptation in which the linear-lanceolate to lanceolate leaf blades are hard, woody, and needle-like, up to only 6 mm wide and rolled (Watson and Dallwitz 1988).

Epidermal features of leaves often reflect adaptations to xeric conditions—e.g. bulliform cells at base of furrows (grooves) between the vascular bundles (i.e. in the intercostal zone) on the adaxial surface allow the leaf to inroll in drought (protecting stomata). Bulliform cells are found, for example, in the leaves of the fine-leaved fescues (e.g. *Festuca ovina*), and in marram grass *Ammophila arenaria* that grows on often-droughty coastal sand dunes. Another characteristic feature of the epidermis of grass leaves is the presence of silica cells. These are short cells occurring in rows, in pairs, or singly, and possessing a silica body (**phytolith**). The size and shape of silica bodies varies with taxa, and their presence in sediments is used to reconstruct the evolution of grassland ecosystems (Strömberg 2004). Silica causes the teeth of grazers to wear down (Chapter 4) and its presence has been implicated in the co-evolution of grasses and herbivores (Chapter 2). Stomata occur in the intercostal zone on the adaxial and abaxial surfaces alternating in files with the oblong intercostal cells, and vary in size from 15 to 50 µm with their appearance depending on the shape of the subsidiary cells. Surface features associated with the epidermis include papillae on the long cells (Bambusooideae) or lining grooves, rare glandular hairs (Pappophoreae tribe in the Chloridoideae), and one- or two-celled microhairs and prickle hairs (common and widespread throughout the Poaceae).

The **leaf sheath** is the basal part of the leaf. It is interpreted as a flattened petiole, and affords protection to the shoot and developing branches within. Clasping the stem, leaf sheaths occur either with their free margins overlapping, or, infrequently, their margins fused (connate) to form a tube (e.g. *Glyceria, Festuca, Bromus*). The leaf sheath frequently has a distinct midrib that extends into the blade. At the top of the sheath and at the base of blade on the abaxial surface is a collar which can be a petiole-like contraction, as in most bamboos. Modifications of the leaf sheath include husks around the ear of maize (*Zea mays*). Sometimes (rarely) the sheath and blade are not clearly differentiated, e.g. *Neostapfia colusana*, a tufted Californian annual, and *Orcuttia* spp. (Gould and Shaw 1983).

The **ligule** is a taxonomically diagnostic feature. Occurring on the adaxial surface at the apex of sheath, the ligule varies widely in texture, size, and shape (Fig. 3.7) (additional abaxial ligules occur in, for example, *Bambusa forbesii*). Commonly membraneous (Bambusooideae, Ehrhartoideae, Pharoideae, and Pooideae), white or brownish, the ligule can be a stiffened membrane (e.g. *Sorghastrum nutans*), a ciliate fringe of hairs (Aristidoideae, Arundinoideae [except *Arundo*], and Danthonioideae [exceptions are *Monachather* and *Elytrophorus*]) (Renvoize 2002), or absent (*Neostapfia*,

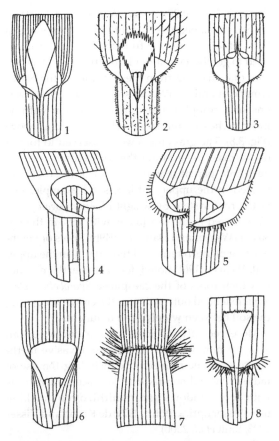

Figure 3.7 Ligules and auricles at junction of leaf sheath and blade (× 5). 1, *Phleum bertolonii*; 2, *Bromus sterilis*; 3, *Melica uniflora*; 4, *Schedonorus pratensis*; 5, *Schedonorus phoenix*; 6, *Alopecurus pratensis*; 7, *Sieglingia decumbens*; 8, *Anthoxanthum odoratum*. Reproduced with permission from Hubbard (1984).

Orcuttia) (Chaffey 2000). The type of ligule is usually uniform within a genus, but in *Panicum* both types occur among the different species. Ear-like projections of tissue called **auricles** can occur at the junction of the sheath and the blade (e.g. *Schedonorus phoenix*) or there can be a tuft of hairs here (several *Eragrostis* spp). The role of the ligule is unclear but it has been viewed as some sort of aerial root cap. The **passive hypothesis** suggests that it acts to prevent water, dust, and harmful spores and insects from reaching the tender parts of the sheath and developing culm (Chaffey 1994; Chaffey 2000). The **active hypothesis** suggests that the ligule has an additional physiological role. Although they lack stomata, the presence of well-developed chloroplasts in membraneous ligules containing starch grains indicates a photosynthetic function. The presence of rough endoplasmic reticulum, numerous mitochondria, hypertrophid dictyosomes and associated vesicles, and paramural bodies in the cells of membraneous ligules suggest a secretory role. The secretion of extracellular products may act as a lubricant to assist the exsertion of the enclosed leaf or culm (Chaffey 2000).

Leaves are usually similar among different shoots. However, **scale leaves** occuring on stolons and rhizomes are generally relatively small, pale green in colour, and non-photosynthetic. In bamboos there is a progressive elaboration of leaf shape on successive nodes up a culm. In these grasses the lower leaves have only a leaf sheath that provides little photosynthetic contribution to the plant. Further up, the lamina becomes well developed, and effectively photosynthesizes. The change from one leaf type to another up the stem can be gradual or clear cut (e.g. *Sasa*) (Chapman 1996).

Leaves are produced continually while the tiller is alive or until a culm switches over to the production of an inflorescence. There is thus a continuous turnover of leaves with new leaves on a tiller expanding as older, lower leaves die. In *Schedonorus phoenix*, for example, there is a flush of new leaf production in the spring, with a new leaf emerging when the lamina of its predecessor on a tiller has fully expanded (Gibson and Newman 2001). Leaf expansion continues until the ligule emerges. Leaf longevity is a cost-benefit issue to a grass, as it is for other plants, in which leaf lifespan is a balance between the costs of construction and maintenance, and the benefits accrued through carbon gain (Kikuzawa and Ackerly 1999). The youngest upper-culm leaves are physiologically the most active; the older, lower leaves, which may be shaded by the upper leaves, photosynthesize at low rates (Skeel and Gibson 1998). Reported leaf lifespan or longevity ranges from 31 ± 1.4 days in *Bouteloua gracilis* (Craine and Reich 2001) to 25 months in *Oryzopsis asperifolia* (McEwen 1962). In the latter report, *O. asperifolia* leaves are recorded as overwintering and remaining green throughout. Leaf longevity also varies depending when in a season a leaf is produced, as early-season leaves may senesce by late summer. Photosynthesis, leaf respiration, leaf nitrogen concentrations, and specific leaf area are all positively correlated among one another and negatively correlated with leaf longevity (Reich *et al.* 1997). Leaf longevity appears to be determined at least in part by nitrogen supply rates and therefore can be altered greatly by plant feedback to the cycle (§7.2). Sodium fertilizer increases leaf longevity in *Lolium perenne*, principally as a result of decreased leaf senescence rather than increased leaf production (Chiy and Phillips 1999). C_3 grasses were observed to show a moderate increase in leaf longevity under elevated CO_2 compared with C_4 grasses which showed no such increase (Craine and Reich 2001). This difference is attributable to the physiological difference among C_3 and C_4 grasses and the relationship between carbon gain and nitrogen cycling (Chapter 4). Significant decreases in leaf lifespan and increases in leaf production were observed along a Mediterranean old-field succession sequence (Navas *et al.* 2003). These changes were partly due to differences in life forms among stages; for example, the increase in leaf production recorded was explained by the replacement of therophytes (annuals, e.g. *Aegilops geniculata, Brachypodium distachyon*), occurring only in the earliest successional stage, by hemicryptophytes (perennials, e.g. *Brachypodium phoenicoides, Bromus erectus*) in the later stages.

3.5 Roots

More than 60% of the living biomass of grasslands is below ground. Estimates of below-ground peak biomass across a range of tallgrass prairie sites ranged from 700 to 2100 g m^{-2} in the top 90 cm of soil (Rice *et al.* 1998). Generally, the extent of the root system is correlated with above-ground growth. Root:shoot ratios range from 0.7:1 to 4:1 with range grasses mostly between 0.8:1 and 1.5:1 (Gould and Shaw 1983). The proportion of biomass allocated below ground varies among species and with environmental conditions. Generally, the proportion of plant biomass below ground decreases with increasing soil moisture (Marshall 1977). Consequently, ecologists need to make a special effort to understand this part of the system. Early researchers laboriously dug pits and trenches in the ground, endeavouring to trace grassland roots (Weaver 1919, 1920). Later, Weaver and Darland (1949a) developed the monolith method in which a block of soil is removed from the field and water is used to gently separate soil from the roots. The picture that emerged from these studies of the North American prairie (Weaver and Darland 1949b) was one of a complex and dynamic system with as much niche differentiation below ground as above (Fig. 3.8). The grass roots were observed to form a dense mass penetrating the soil column entirely, deep into parent materials. Most of the roots are located in the upper soil layers; for example, 43% of the root system by weight of *Andropogon gerardii* occurred in the top 10 cm, with 78% in the top 30 cm (Weaver and Darland 1949b). Roots of neighbouring plants intermixed thoroughly in the upper soil. Weaver (1961) noted, for example, the manner in which roots of the caespitose *Sporobolus heterolepis* extended outward from the crown occupying the soil between widely spaced bunches.

More recently, ecologists have developed an array of non-destructive and less invasive methods to examine root activity including the use of radioactive and non-radioactive tracers, rhizotrons with root windows and minirhizotrons making use of laser optics (Böhm 1979; de Kroon and Visser 2003; Smit *et al.* 2000).

Grass roots at maturity are wholly adventitious and fibrous. **Seminal roots**, which are the first roots to grow in a seedling (see §3.7), function for only a short period of time, but during their short life appear to absorb more nutrients per unit weight than the adventitious roots (Langer 1972). When the seminal roots die they are replaced by **adventitious roots**, which consist of a primary root plus 2–7 first-order branches. The primary root may be indistinguishable from other roots and does not resemble the familiar tap root of many herbaceous dicots. In tufted grasses the adventitious roots arise from basal nodes of the main axis, and from nodes on stolons, rhizomes, and tillers near the ground: i.e. any node in contact with the soil can develop roots. Roots developing from nodes develop from leaf bases and push through the subtending leaf sheath. Whorls of adventitious roots can arise from nodes above elongated internodes up to 1.5 m above ground, e.g. the prop or buttress roots of *Sorghum* spp.

Figure 3.8 Below-ground niche separation of root systems in the North American prairie. 1 and 7, *Schizachyrium scoparium*; 2, *Psoralidium tenuiflorum*; 3 and 5, *Koeleria macrantha*; 4, *Astragalus crassicarpus*; 6, *Brickellia eupatorioides* var *eupatorioides*. Note that the grasses place their roots at different levels, whereas the forbs extend much deeper. Scale is in feet. Reproduced with permission from Weaver and Fitzpatrick (1934).

With little dominance of any main axis (except as a seedling), the topology of grass root systems is diffuse. Whereas other plants exhibit a topological response to limiting resources, grasses appear to show little variation in root diameter (usually <5 mm in diameter) at different branching levels (Robinson *et al.* 2003). Nevertheless, perhaps because of the diffuse nature of grass root systems, the fine root mass is higher and fine root length greater in temperate grasslands than in any other biome (1.5 kg m^{-2}, 112 km m^{-2}, respectively) (Robinson *et al.* 2003). Per unit of root mass, temperate grasslands have far more fine roots than any other system (fine/total root mass ratio = 1.0).

Root hairs cover long lengths of the root epidermis, unlike in dicots in which root hairs are limited to the area behind the root tip. The root hairs are long and persistent, developing from **trichoblasts**—cells which grow less rapidly then other epidermal cells that are not going to become root hair cells. Root hairs are responsible for water and nutrient uptake. The density of root hairs can be related to mycorrhizal status (Chapter 5, Tawaraya 2003), soil pH (Balsberg and Anna 1995), and texture (Bailey 1997).

3.5.1 Distribution of roots in soil and turnover

Root biomass, production, and disappearance fluctuate seasonally. Biomass increases through the growing season with adequate soil water. The top 10 cm of the soil is the most dynamic and sensitive to environmental conditions with respect to root growth (Rice *et al.* 1998). Root growth has been shown to decline towards the end of the growing season, not because of decreasing soil temperature, but, in the tundra grass *Dupontia fisheri*, because of decreasing daylength (Shaver and Billings 1977).

Environmental factors including fire, grazing, temperature, and nutrients affect root production and turnover. Root biomass is generally higher under frequent burning, and this is related to higher rates of root production, not decreased senescence (Rice *et al.* 1998). Grazing produces a variable response of the grass root system (Milchunas and Lauenroth 1993). In a global comparison of 236 grassland sites, root biomass was unrelated to species compositional differences associated with grazing. However, positive effects of grazing occurred in 61% of sites where grazing had a positive effect on annual net primary productivity. At the local scale, root systems of *Schizachyrium scoparium* were observed to deteriorate, decreasing in density, under increased grazing intensity (Weaver 1950). The intensity of clipping, used as an experimental surrogate for grazing, has been shown to be inversely related to root production (Branson 1956). Indeed, in the latter study, which was conducted on five US range grasses (*Pascopyrum smithii*, *Pseudoroegneria spicata*, *Hesperostipa comata*, *Poa pratensis*, and *Bouteloua gracilis*) clipping effects showed a more pronounced effect on the root system than the above-ground system. This suggests that under heavy grazing, effects would be exhibited below ground before they were apparent above ground. Above-ground tissue removal, especially if in excess of 50%, can cause root growth to stop for a period of time, presumably as the plant goes through a period of resource conservation while foliage tissues are restored (Crider 1955; Jameson 1963). The loss of roots after clipping is most rapid near the root tips.

Supplemental nutrients in the short term cause an increase in root biomass, but in the long term the response may be different (Rice *et al.* 1998). The amount and depth of soil moisture is more important than nutrients. Shallow soil moisture leads to shallow root systems, and with deep soil moisture the roots go deeper. The distribution of root weight in the soil column and the total amount of roots vary with soil type, particularly with respect to permeability (Weaver and Darland 1949b). Light intensity incident on above-ground leaves affects root growth. Langer (1972) describes how an 18-fold reduction in light intensity on *Bromus inermis* led to a 30-fold decrease in root weight compared with only a 5-fold decrease in shoot weight.

Root production follows a seasonal pattern. Under *Lolium perenne/Trifolium repens* (ryegrass/clover) swards, new adventitious roots were produced through late winter to early spring, with the rate falling off in April or May to a low rate during the summer (Garwood 1967). Differences among four grasses in these grasslands (two cultivars of *L. perenne*, *Dactylis glomerata*, and *Phleum pratense*) were minimal. More adventitious roots were produced in February–April than at any other time of the year. These roots did not branch immediately and elongated slowly at first, but some of them continued growth in the lower soil horizons until midsummer. The roots produced in the late summer lived only a few days and failed to penetrate to any great depth in the soil. By contrast, some of the September–November roots lived and grew through the winter into the following spring and summer. The general pattern of a summer decline in root production has been observed in several species (Huang and Liu 2003; Murphy *et al.* 1994). The roots of annual grasses do not normally, of course, live for more than a year. However, winter annuals, which germinate in the autumn, rapidly grow numerous roots following germination that overwinter becoming well developed in time for spring growth. The roots of the annual *Bromus tectorum* (cheatgrass), for example, can penetrate to >30 cm deep in the soil by mid-November, eventually reaching 1.2–1.5 m below the surface (Hulbert 1955).

As indicated above, much of the root system of perennial grasses dies and is replaced each year; 50% in *Poa pratensis* and *Agrostis capillaris* (Sprague 1933). Root turnover, i.e. the net result of production of new roots and disappearance of old ones, is high in drought conditions; e.g. root turnover was 564% during a drought year in US tallgrass prairie, compared with 389% the following year when there was adequate rain. Globally, root turnover in grasslands increases exponentially with mean annual temperature (Gill and Jackson 2000). Root turnover rates for temperate grasslands range from 0.83 yr^{-1} in the central Netherlands to no annual turnover in US shortgrass steppe, with an overall mean of 0.47 yr^{-1} (Lauenroth and Gill 2003). In tropical grasslands, by contrast, mean root turnover is 0.87 yr^{-1}, and in high-latitude (mostly tundra) grasslands turnover is 0.29 yr^{-1}. Many individual roots stay alive for more than 2 years. The following percentages of main adventitious root axes were observed to survive for two growing seasons: *Koeleria macrantha*, 30%; *Hesperostipa spartea*, 57%; *Schizachyrium scoparium*, 23%; *Andropogon gerardii*, 45%; *Bouteloua curtipendula*, 36%; and *Sorghastrum nutans*, 37% (Weaver and Zink 1945).

3.6 Inflorescence and the spikelet

The **inflorescence** is the flowering portion of the shoot (Fig. 3.1). It is delimited at the base by a culm node bearing the uppermost leaf. There are normally no leaves within the inflorescence itself. The upper culm leaf is usually reduced in blade length compared with other leaves, but the sheath may be enlarged and envelops the developing inflorescence. In some grasses, the mature inflorescence remains within or partially within the sheath (e.g. *Microstegium vimineum*, *Sporobolus compositus*).

There is tremendous variation in inflorescence morphology among grass genera and species, and this variation forms the basis of classification (Chapter 2). The variation described below is constant mostly within a species. The bamboos have atypical inflorescences, including florets arranged in pseudospikelets, glumes with the potential for psuedospikelet formation or subtended by bracts and/or prophylls, and atypically large numbers of floral parts (e.g. 4–6 glumes in *Nastus*) (Table 4.1 in Chapman 1996).

Inflorescences are generally borne above ground on erect to prostrate flowering culms. However, there are a few grasses in which cleistogamous spikelets are borne on underground culms, e.g. *Amphicarpum purshii*, *A. muhlenbergianum*, and *Enteropogon chlorideus*. The latter is known commonly as buryseed umbrellagrass because of the spikelet-bearing rhizomes; seed set is highest in the cleistogomous spikelets (Barkworth *et al.* 2003) (see Chapter 5).

Flowering can occur within 3 months in short-lived annuals, or it can be delayed many years as in some bamboos. Flowers are borne in sessile or pedicelled spikelets on the inflorescence axis or in simple or compound branching systems. Terminal spikelets mature first and the basal ones last. As in many things, *Zea mays* is an exception; maturation in the pistillate inflorescence ear of corn begins in the middle proceeding up and down the ear.

There are three main types of inflorescence (Plate 2); the panicle, the raceme, and the spike, although this categorization represents extremes of a continuum.

- In a **panicle**, spikelets are borne on stalks (pedicels) on primary, secondary, even sometimes tertiary, branches arising from the main axis. The inflorescence is a panicle in most grasses, e.g. *Panicum*, and the panicle represents the original form of inflorescence in grasses. The branching pattern in a panicle may be loose (e.g. *Avena*, *Bromus*, *Calamagrostis*, *Eragrostis*, *Panicum*, *Poa*), or contracted, narrow, cylindrical, and dense (i.e. contracted as in *Phleum*, *Alopecurus*, *Phalaris*, *Lycurus*). Spikelets can be sessile or short-pedicelled along primary inflorescence branches, and can be digitate at the culm apex or distributed along the main inflorescence axis; e.g. all members of the Chlorideae tribe, and many genera in the Paniceae and Andropogoneae.
- In a **raceme**, spikelets are stalked on pedicels directly on the main axis. The inflorescence can be reduced to a single terminal spikelet as in *Danthonia unispicata*. Some grasses such as *Echinochloa* spp. and *Paspalum* spp. have a paniculate inflorescence with racemose branches.
- In a **spike**, the spikelets are all seated on the main inflorescence axis itself—there are no stalks. Spikelets may be solitary at the nodes, e.g. *Lolium*, *Triticum*, *Agropyron*, or two or more per node, e.g. *Hilaria*, *Elymus*, *Sitanion*. *Hordeum*, however, has a spike with both a single sessile and two short-pedicelled spikelets at same node.

Whatever the type, inflorescences can be uniform or mixed with separate male, hermaphrodite, female, or sterile spikelets. *Tripsacum*, for example has proximal spikelets which have female florets only and distal spikelets which have male florets only. *Zea mays* has separate male and female inflorescences (tassels and ears). *Bouteloua dactyloides* is dioecious with separate male and female plants.

In this case, differences between male and female plants of *B. dactyloides* do not translate into reproductive allocation, or growth and success in the environment (Quinn 1991).

The **spikelet** is the basic unit of grass systematics and consists of a shortened stem axis (rachilla) plus one to several florets, delimited at the base by two floral bracts, the glumes (Fig. 3.2). The spikelet is a determinate structure, although an exception to this occurs in the bamboos where psuedospikelets can occur, e.g. *Bambusa* and other tropical/subtropical bamboos. A pseudo-spikelet occurs when glumes subtend a bud which can become a flowering branch; the flowering branches themselves can have glumes subtended by buds which can develop. This repeating pattern is **iterauctant**. In a spikelet, one or more florets are grouped together, and usually subtended by two glumes. **Glumes** are usually membraneous and are barren in that there is neither floret or bud in their axil, and so they are considered equivalent to a leaf sheath. Glumes assume a protective function in the spikelet and can be awned or unawned. Spikelets can be perfect, staminate, pistillate, or sterile depending upon presence/absence of pistil and stamens in the floret (see subfamily characteristics in Chapter 2).

Florets (consisting of the lemma, palea, lodicules, stamens, and pistil) are borne on a rachilla above the glumes in the spikelet, their number is finite and characteristic of a species, and ranges from 1 per spikelet (e.g. *Stipa*) to > 40 in *Eragrostis oxylepis*. Reduction to a single floret per spikelet has occurred independently in several groups. The Pooideae, Chloridoideae, Bambusoideae, and Arundinoideae mostly have several florets per spikelet. The Panicoideae spikelet is two-flowered (perfect upper, staminate or sterile lower floret), and several pooid and cloridoid grasses have a single perfect floret per spikelet. Florets are either functional or non-functional. In pooids, non-functional floret remnants occur furthest away from the glumes; in panicoids non-functional florets occur nearest the glumes.

There are two membranes at the base of the floret; the **palea** and the **lemma** which is below the palea. The palea, lemma, and floret together form the **anthoecium**. Either or both the palea or lemma may be present, absent, or reduced. The lemma is derived from a leaf sheath and exhibits much variation among taxa, making it very important taxonomically. The lemma is lanceolate or oval in shape, keeled, texture varying from papery to coriaceous, nerved or with 1 to about 15 nerves, firm or hard, dorsally compressed or rounded on the back, apices entire (unawned) or bifid with an awn. When present, awns can occur as a projection from the lemma either terminally (ending as an extension of the midrib) or dorsally (arising from the back of the midrib). Generally a straight, narrow structure, the awn can be bent towards the middle, smooth, or setaceous. In *Aristida* the awn is trifid or three-branched. When awned, the lemma base can be reduced to a membraneous bifid scale or narrow wing. The awn can be hygroscopic twisting on being hydrated, allowing the dehisced grain to move and thus aiding dispersal. Even when not hygroscopic, awns can aid dispersal by enabling the caryopsis to be caught up in animal fur or feathers. The palea is a modified prophyll, usually two-nerved and two-keeled, transparent to glume-like, opposite the lemma, with its dorsal surface against the rachilla. The palea is without an awn except in the Australian *Amphipogon* where two awns are present on the palea (Bews 1929).

Grass flowers (lodicule, stamens, and pistil) are hermaphroditic, male or female only, or one or more of each. The floret consists of a pistil with two stigmas (three in bamboos) and one or two whorls of three stamens each with an anther attached either near the middle (versatile) or at the base (basifixed) on a slender filament. Exceptionally, there are more anthers, e.g. *Ochlandra* (Bambusoideae) with 6–120 stamens. The pistil consists of a one-loculate ovary, with a single ovule, with usually two styles and stigmas. The shape of the ovary varies. Stigmas are two-branched (three-branched in bamboos), feathery, sessile, and elevated on a single style or on separate styles. In *Zea mays* there is a single long, filamentous style per ovary; the 'silk' of the corn ear. The ovary, when fertilized, produces one seed.

The **lodicules** are small narrow scales at the base of the floret. There are normally two inconspicuous lodicules, although some species have one or none, and there are three conspicuous lodicules in bamboos. Evolutionarily the lodicules represent perianth segments. Lodicules are small, pale green or white and cyclically arranged (like the stamens, but unlike the alternate glumes, lemmas, and paleas). They become turgid at anthesis, forcing the floret to open.

The anthers in a grass floret are 1–14 mm in length, yellow or cream through pink and red to purple, and most are glabrous. Dehiscence is by slits or pores. The pollen varies in size but little in morphology, and is more or less spheriodal–ovoid, 14–130 µm. There is a solitary ulcerate germination pore surrounded by a distinct annulus, covered with an operculum. The sexine (outer pollen wall) is either punctate or maculate (i.e. with separate or clustered granules) (Clifford 1986). The limited morphological variation of pollen among grasses revealed under light microscopy has meant that the presence of grass pollen in palynological samples generally indicates little more than the presence or grasses in the vegetation at the time of deposition. This itself is of value in determining the development of grassland ecosystems (Chapter 2), but does not allow discrimination among grassland vegetation types.

In some genera, glumes, lemmas, or sterile branchlets can become indurated and fused to form a protective so-called **involucre** (Bews 1929). For example, in *Cenchrus*, the involucre consists of bristles united at the base to form a spiny cup or bur protecting the spikelets. This structure is easily caught up in animal skin and fur (causing great pain in the soles of human feet!), aiding in dispersal.

Evolutionary trends in the inflorescence have been the subject of debate for >200 years (Bews 1929; Clifford 1986; Kellogg 2001). Condensation (compression and simplification), suppression, multiplication, and fusion of parts from a relatively unspecialized panicle to a derived and reduced inflorescence with correspondingly reduced spikelets has occurred throughout the family numerous times (Table 3.1). The evolution of the floret is believed to be from a 'lily' type ancestor, with fusion of carpels (i.e. to a monocarpellate condition) and sharing of a single cauline ovule, reduction of stamens from six to three, conversion of three petals (possibly) to three lodicules, and loss to two lodicules. The occurrence of petal-identity genes in the lodicules of maize and rice supports the idea that the lodicules are reduced, modified petals. The three ancestral sepals may have become the palea, or the palea and lemma may be derived from associated floral bracts (Chapman 1996). Other features such as the elaboration or loss of awns have been regarded as evolutionarily advanced (Bews 1929).

3.7 The grass seed and seedling development

The fruit of most grasses, known as the **caryopsis**, consists of a single seed fused with an enclosing pericarp. In common terminology, the grass fruit or seed is really a seedlike grain consisting of the caryopsis either with or without enclosing inflorescence structures (e.g. bracts, rachilla). The seed is free from the pericarp in only a few genera,

Table 3.1 Comparison of presumed primitive and advanced grass spikelet characters

	Primitive	Advanced
Spikelet	Large, many-flowered	Small, few- or one-flowered
Glumes	Large, leaflike in texture, several-nerved, awnless or short-awned	Variously modified, reduced or absent, can be highly developed for flower protection or seed dispersal
Lemma	Like the glumes	Conspicuously different from glumes
Palea	Present, two- to many-nerved	Nerveless, reduced or absent
Lodicules	Six or three	Two or one
Stamens	Six, in two whorls	Three, two, or one, in one whorl
Stigmas	Three	Two or one

From Gould and Shaw (1983).

e.g. *Sporobolus, Eleusine*, and is then referred to variously as an **achene** or **utricle**. The caryopsis is either free from the lemma and palea (e.g. wheat) or permanently enclosed (*Aristida, Stipa*, and members of Paniceae). In the Andropogoeae, glumes permanently enclose the caryopsis along with a pedicle and a section of the rachis. The grain may enlarge during development and ripening exceeding the size of the glumes, lemma, and palea. In some bamboos the pericarp is free from the seed and the fruit is a berry that can be as large as an apple (e.g. *Dinochloa, Melocalamus, Melocanna, Ochlandra*), or a nut (*Dendrocalamus, Schizostachyum, Pseudostachyum*) (Bews 1929).

The caryopsis alone is rarely the unit of dispersal in grasses. Disarticulation of grass diaspores occurs below, between, or above the glumes and at all nodes. Panicoideae disarticulate below the glumes, in *Paspalum* and *Panicum* the spikelets fall separately. Spikelets can fall in groups, e.g. chloridoid grasses. In *Schedonnardus paniculatus* and some *Eragrostis* and *Aristida* spp., the entire inflorescence breaks off, tumbleweed like. Additional examples are provided in Table 3.2. Along with other adaptations such as the presence of awns which can be barbed, retorse, or unarmed, the morphology of the grass diaspore has important implications for its dispersal.

Following dispersal, the grass caryopsis may or may not be ripe and ready to germinate. The ecology of seed dispersal is discussed in detail along with seed dormancy and germination cues in Chapter 5. Seed (caryopsis) morphology and seedling growth is discussed below.

The grass embryo is highly specialized and appears as an oval depression on the flat side of the caryopsis next to lemma. On the opposite side of the grain to the embryo is the **hilum**; a line (e.g. *Lolium perenne, Avena fatua*) or dot (*Holcus lanatus, Briza media*) marking the point of attachment of the seed to the pericarp. The embryo includes primordia of one, two, or more foliage leaves and the primary root, and primordia of several adventitious roots. The **scutellum** is an haustorial organ, equivalent to the single cotyledon, and functions in enzyme secretion and absorption of nutrients from the endosperm. Also in the embryo is the **coleoptile** (part of the single cotyledon, an open sheath with a pore at tip, homologue of first foliage leaf), the **epiblast** (an outgrowth of the coleorhiza), and the **coleorhiza** (the first part of the plant to emerge on germination, producing a tuft of anchoring hairs, protecting the radicle; the primary root emerges through it); all are peculiar to the grass embryo. The embryonic shoot is known as the **plumule**, and the embryonic primary root as the **radicle**. The **mesocotyl** is a vascular trace in the scutellum extending down into the coleorhiza and up into the coleoptile. **Endosperm** occurs as a large elliptical structure in the seed formed from the fusion of two polar nuclei of the embryo sac and a sperm nucleus. The endosperm provides nutrition for the embryo

Table 3.2 Diversity of grass diaspores

Dispersal unit	Example	Features
Whole inflorescence	*Scrotochloa*	Inflorescence breaks off as a whole unit
	Spinifex	Plants dioecious. Female inflorescence shed entire, male inflorescence sheds spikelet separately
Groups of spikelets	*Tristachya*	Triad of three spikelets with fused pedicels shed together
Whole spikelet	*Sorghum*	The functional (sessile) spikelet shed together with the remnant of the non-functional (pedicellate) spikelet
Whole anthoecium	*Hordeum*	Palea and lemma adherent to the caryopsis
Caryopsis only	*Triticum*	Cultivated wheat which is free threshing
Seed only	*Sporobolus* (some)	Caryopsis modified to extrude a mucilage-coated seed
Multiple dehiscence	*Catalepsis*	Rachis fragments after which spikelets are then separately deciduous
'Adherent' dispersal	*Aristida* (some)	A tumbleweed-like 'grass ball'

From Chapman (1996).

and developing seedling, and is usually solid and starchy (liquid in some including *Koeleria, Trisetum*, and all Aveneae). Variation in embryo structure is related to the presence or absence of the mesocotyl, the epiblast, the scutellum cleft (i.e. whether or not the scutellum is fused to the coleorhiza), and whether the first leaf is rolled or folded, and allows an evolutionary sequence of embryo types to be recognized (Renvoize 2002). The bambusoid embryo with the epiblast and scutellum present and a folded first leaf is considered to be the most primitive type. The absence of a mesocotyl and presence of an epiblast compared with the presence of a mesocotyl and a scutellum cleft is consistent with the division between the BEP and PACCAD clades, respectively (Chapter 2).

3.7.1 Seedling growth

The first sign of germination is a swelling of the caryopsis as it imbibes water, followed by an enlargement of the coleorhiza and coleoptile, and then elongation of the primary root as it pushes through the coleorhiza. Two pairs of transitory node roots develop within a few hours or days (Fig. 3.9). The primary root and transitory node roots constitute the seminal or primary root system, consisting of 1–7 roots, depending upon the species and the environment at the time. Mesocotyl roots can arise adventitiously from the mesocotyl. The coleoptile is negatively geotropic and positively phototropic, and it grows up enclosing the primary shoot and mesocotyl. The coleorhiza is positively geotropic, and produces the primary seminal roots and transitory node roots. The prophyll emerges first from the coleoptile. Emerging from the coleoptile, the primary shoot consists of a culm and several leaf stages, and produces cotyledonary node roots (crown roots) which appear at the third leaf stage. The mesocotyl and cotyledonary node roots are the only roots of mature plants; the seminal roots remain alive and active as absorbing organs for up to 4 months before dying. In *Bromus inermis*, for example, adventitious roots begin between 5 and 14 days of germination, and by 40 days, adventitious roots constitute 50% of the total roots.

Growth of the grass plant from seedling to inflorescence maturity can be expressed in a series of

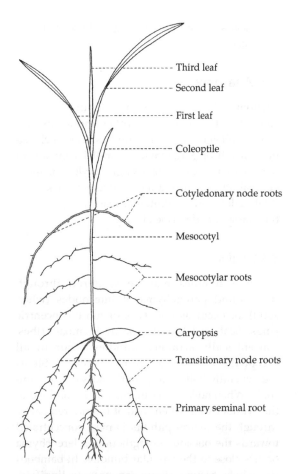

Figure 3.9 Hypothetical grass seedling showing development of adventitious roots (cotyledonary node roots and mesocotylar roots). Reproduced with permission with permission from Gould and Shaw (1983).

developmental growth stages. For example, in rice (*Oryza sativa*), seedling, vegetative, and reproductive stages based on morphological criteria can be recognized (Counce *et al.* 2000). Seedling development consists of unimbibed seed (S0), radicle and coleoptile emergence (S1, S2), and prophyll emergence (S4) stages; vegetative development consists of V1, V2...VN stages where *N* denotes the number of leaves with collars on the main stem; and reproductive development consists of 10 growth stages; panicle emergence (R0) and development (R1, R2, R3), anthesis (R4), and grain development (R5–R9). Recognition of developmental growth stages provides a management aid for growers as well as a means of communicating growth data among researchers.

3.8 Anatomy

Anatomically, the grasses share a number of the characteristic features of other monocots, i.e. closed collateral vascular bundles each enclosed in a sheath of parenchyma, parallel venation in the leaves, possession of a single cotyledon, and a general absence of regular secondary thickening (Mabberley 1987). Additional features specific to the grasses are described below.

3.8.1 Culms

Vascular bundles may be scattered through the ground parenchyma of internodes (when solid), or occur as one, two, or more concentric rings. Scattered bundles occur in most tribes, concentrically arranged bundles occur in all except the Andropogonea and Paniceae. Stems are generally hollow in pooids and solid in panicoids. When hollow, vascularization is confined to the cylinder wall; when solid, it is scattered partly through the central pith but is most concentrated towards the outside. Strengthening sclerenchyma occurs close to the vascular bundles. In bamboos the whole ground tissue can become lignified, providing immense strength.

The vascular bundles of stems and leaves possess two very large metaxylem vessels outside a row of smaller metaxylem, with protoxylem to the inside. The phloem is to the abaxial side of the metaxylem and has clearly differentiated sieve tubes and companion cells. A cavity occurs on the adaxial side of the metaxylem in place of protoxylem; this cavity is due to differences in elongation rates and differentiation of xylem and cells. In leaves, the vascular strand is enclosed by photosynthetic tissue.

Vascular bundles from leaves pass vertically down the culm, remaining unbranched through several internodes before bifurcating and then joining the vascular bundles of the stem at the nodal plexus. At this point **anastomosis** (cross-linking) of vascular strands occurs, allowing assimilate movement across the plant and between leaves. A number of bundles anastomize at each node keeping the total number approximately the same throughout.

Bundles vary in size; the central bundle of leaves is the largest, leading to a well-defined midrib in many species. Larger and smaller bundles occur in alternating series on either side of the midrib. Strengthening sclerenchyma is associated with the larger bundles in leaves and stems.

3.8.2 Roots

Adventitious roots arise in parenchymatous ground tissue at the nodes, just below the intercalary meristem. The vascular system of a grass root is a discrete polyarch. Large roots have 10–14 exarch xylem groups alternating with groups of phloem around a central pith. A 1–2-layered pericycle is usual, but is not always present. Endodermal cell walls thicken as the root matures, casparian strips occur on radial cell walls, parenchyma cells occur between the xylem and phloem, and pericycle and central pith cells lignify as the root matures. The cortex remains parenchymatous and may break down in older roots. In some species the cortex is interspersed with cavities, e.g. the semi-aquatic *Hymenachne amplexicaulis* whish has large air spaces in the cortex (Renvoize 2002). In small roots; the vascular tissues form a protostele with a single large metaxylem vessel.

3.8.3 Leaves

The ground tissue (mesophyll) of a grass leaf consists of short chlorenchyma cells and colourless parenchyma cells. Palisade and spongy layers are rarely differentiated in grasses. The chloroenchyma cells are thin walled and vary from being irregular in shape with air spaces to tightly packed with a moderate to distinctly radiating arrangement around the vascular bundles. The vascular bundles are surrounded by a bundle sheath comprising one or two cell layers. The outer layer (or single layer when there is only one) is a parenchyma sheath comprised of thin-walled cells. The inner cell layer (the endodermis or mestome sheath) is comprised of small cells with thickened inner and radial

Table 3.3 Six types of leaf-blade anatomys

	Pooid (festucoid)	Bambusoid	Arundinoid	Panicoid	Aristidoid (*Aristida* only)	Chloridoid
Inner mesotome sheath/ endodermis	Well developed, thick-walled cells	Present	Poorly defined with thin cell walls	Typically lacking, although present in e.g. *Eriochloa sericea*	Lacking	Present at least around largest bundles
Outer parenchyma sheath	Very indistinct. Small thin-walled cells, with chloroplasts	Present, always thick-walled and round/elliptical with chloroplasts	Large cells lacking chloroplasts	Large cells with or without starch plastids	Present as double layer, inner sheath cells larger than outer. Cell walls thick with specialized plastids	Single cell layer with specialized plastids
Chlorenchyma of mesophyll	Loosely or irregularly arranged. Large air spaces. Cells with chloroplasts.	Arm-cells with large fusoids cells perpendicular to vascular bundles	Cells tight or densely packed. Arm-cells occasional, fusoid-cells lacking.	Irregular or radiate from parenchyma sheath	Long-narrow radially arranged cells around vascular burdles	Long, narrow, cells in one layer radiating from bundles, with few chloroplasts
Specialized cells for starch storage (Kranz cells)	Lacking	Lacking	Present in some genera	Present in some genera	Present	Present

After Brown (1958).

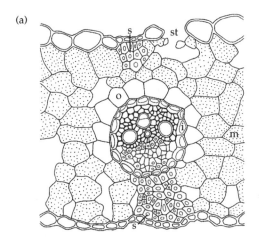

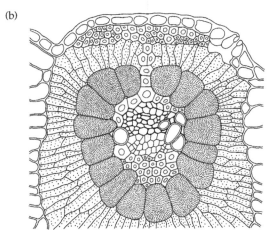

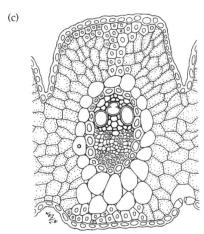

Figure 3.10 Leaf anatomy of grasses. Portions of transverse leaf sections: (a) *Poa* sp., pooid-type leaf; note double vascular sheath of large, thin-walled outer cells (o) lacking chloroplasts and small inner cells (i) with thickened inner and radial walls.

walls. Six types of bundle sheath are recognized on the basis of the variation in anatomy (Brown 1958) (Table 3.3). These six types correspond to environmental conditions, with the aristidoid and chloridoid types being typical of arid land grasses possessing the C_4 photosynthetic pathway (Chapter 4). The pooid, chloridoid, and arundinoid types are illustrated in Fig. 3.10.

Sclerenchyma fibres are associated with most vascular bundles, except the smallest, and/or the fibres occur in clusters. The fibres can be continuous from the vascular bundle to the epidermis forming a girder or I-beam construction (Gould and Shaw 1983; Metcalf 1960) (Fig. 3.10c).

The cells of leaves that are specialized to carry out malic or aspartic acid decarboxylation and the Calvin–Benson cycle part of C_4 photosynthesis (Chapter 4) are known as **Kranz cells** (Brown 1974, 1975). The term was coined by D.G. Haberlandt in 1882, describing the circular leaf sheath in the Cyperaceae. *Kranz* is a German word meaning wreath, ring, or rim. Kranz cells occur in 10 families of angiosperms: in monocots in the Poaceae and Cyperaceae, and in dicots in the Amaranthaceae, Chenopodiaceae, Asteraceae, Euphorbiaceae, Molluginaceae, Nyctaginaceae, Portulacaceae, and Zygophyllaceae. Kranz cells are associated with physiological, anatomical, and cytological specialization. These cells are confined usually to the parenchyma sheath surrounding the vascular bundle of leaves. The cells are elongated, parallel to the main axis of the bundle. In the grasses, Kranz cells and the associated Kranz anatomy is restricted to the Panicoideae (some spp.), Chloridoidea (all members), Aristoideae (except *Sartidia*), and Arundinoideae (*Stipagrostis, Aristida, Asthenatherum, Alloeochete*) subfamilies.

The mesophyll (m) consists of large, loosely and irregularly arranged chlorenchyma cells. Sclerenchyma strands (s) above and below the vascular bundle forms an 'I-beam' girder of supporting tissue. (b) *Bouteloua hirsuta* var *pectinata*, chloridoid-type leaf showing large outer bundle sheath and irregular inner sheath of smaller cells with uniformly thickened walls. Small, tightly packed, radially arranged chlorenchyma cells surround the bundle illustrating typical Kranz anatomy. (c) *Cortaderia selloana*, arundinoid-type showing one large vascular bundle with outer or parenchyma sheath (o) lacking chloroplasts Reproduced with permission from Gould and Shaw (1983).

The characteristics of Kranz cells in the grasses include: walls thicker than mosophyll cells and with numerous pits and plasmodesmata; chloroplasts different from mesophyll chloroplasts (large size, high number per Kranz cell, specific positions within the cell, and many large grana). In the Chloridoidea and Aristoideae the Kranz parenchyma sheath is exterior to the mestome sheath. Kranz Panicoideae has usually one sheath (Kranz sheath) although a mestome sheath is present in some taxa. In the Arundinoideae, *Stipagrostis* of the Aristideae tribe has two sheaths; the outer is Kranz, and the inner comprises thin-walled cells without chlorophyll and is not considered a mestome, and in *Aristida* there are two chlorophyllous Kranz sheaths (i.e. a double parenchyma sheath).

CHAPTER 4

Physiology

The primary form of food is grass. Grass feeds the ox: the ox nourishes man: man dies and goes to grass again; and so the tide of life, with everlasting repetition, in continuous circles, moves endlessly on and upward, and in more senses than one, all flesh is grass.

John James Ingalls (1872), Kansas,
US Senator 1873–1891

Grasses are remarkable in their combination of several unique physiological features. These include the possession in more taxa than any other group of the advanced C_4 photosynthetic pathway (§4.1), highly nutritious forage (§4.2), the production of economically important chemicals including starches, sugars, and other carbohydrates (§4.3), and the possession of large amounts of silica in their tissue (§4.4). All of these physiological features, and others, fundamentally affect the ecology of grasses (§4.5), and are described in this chapter.

4.1 C_3 and C_4 photosynthesis

More so than any other group of plants the grasses have evolved an advanced mechanism of CO_2 assimilation, the C_4 dicarboxylic acid pathway. The C_4 **pathway**, as it is known, allows the plant to concentrate atmospheric CO_2 in such a manner as to avoid photorespiration. As a consequence, plants possessing the C_4 pathway have an advantage over plants in which this pathway is absent (i.e. C_3 plants) in hot and dry environments. The outcome is a fundamental difference in the ecology of C_4 vs C_3 plants that is manifest at scales ranging from a few metres to biomes. It is important to remember that all grasses operate a fully functional C_3 pathway; the C_4 pathway is a supplemental pathway present in some. The light-dependent reactions in which energy is captured from sunlight, oxygen is released from water, and the energy carrier molecules of adenosine triphosphate (ATP) and nicotinamide adenine dinucleotide (NADPH) are produced are also the same in all grasses regardless of whether or not the C_4 pathway is present. The main differences between C_3 and C_4 grasses are presented below in this section and summarized in Table 4.1.

4.1.1 Discovery of the C_4 pathway

The C_4 pathway was discovered in the economically important sugar cane (*Saccharum officinarum*) and maize (*Zea mays*) with details of C_4 photosynthetic carbon metabolism being described by the Australian physiologists M.D. Hatch and C.R. Slack in 1966. Soon after this discovery it was realized that the specialized Kranz anatomy of leaves (Chapter 3) was associated with photosynthesis, that C_4 plants have higher $^{13}C/^{12}C$ ratios in their tissues than C_3 plants, and that C_3 and C_4 plants have different phytogeographic distributions in relation to climate (Hattersley 1986).

4.1.2 Distribution and evolution of the C_4 pathway in grasses

The C_4 pathway occurs in about half of the Poaceae (i.e. in c.4500 species). Of the 12 subfamilies (Chapter 2), the Chloridoideae are predominantly C_4 with only some C_3 members, the Artistidoideae has both C_3 and C_4 members, and the Panicoideae has taxa representing all the types of C_4 pathways along with some C_3/C_4 intermediates. The other nine subfamilies, including all of the BEP clade, are exclusively C_3, or in the case of the primitive

Table 4.1 Summary of differences between C_3 and C_4 grasses. Note: several taxa exhibit intermediate characteristics

Character	C_3 grasses	C_4 grasses
Systematics and anatomy		
Systematic occurrence in grasses	All grasses, BEP clade members exclusively C_3	Most members of Chloridoideae, Aristidioideae, Panicoideae
Leaf anatomy	No Kranz anatomy	Kranz anatomy (bundle sheath cells)
Location of release of CO_2 to Calvin cycle	Chloroplasts in mesophyll	Chloroplasts in bundle sheath cells
Physiology		
Initial CO_2 fixing enzyme	RUBP carboxylase/oxygenase (Rubisco)	PEP carboxylase
Initial product of CO_2 fixation	2 molecules of PGA	PEP
Susceptible to photorespiration?	Yes	No
Isotopic $^{13}C/^{12}C$ ratio	−22 to −35	−9 to −18
Internal partial pressure (P_i) of CO_2 in mesophyll (pattern reversed in bundle sheath cells for C_4 plants).	c.25 Pa	c.10 Pa
CO_2 compensation point	4–5 Pa, increases rapidly with temperature	0–0.5 Pa
Response to increasing atmospheric CO_2	Increased photosynthesis and growth (at least initially)	Minimal to no response
Photosynthetic response to increasing O_2	Negative correlation	Unaffected
Photosynthetic response to increasing temperature	Decreases as photorespiration increases with rising temperature	High rates of net CO_2 assimilation maintained at high temperatures
Photosynthetic response to increasing light intensity	CO_2 assimilation saturates at high light intensity	CO_2 assimilation saturates at higher light intensity than C_3 plants.
Enzymatic affinity (K_m) of Rubisco for CO_2	Low	High
Water use efficiency (WUE)	Low	High
Leaf tissue N concentration	200–260 mmol N m^{-2}	120–180 mmol N m^{-2} (with 3–6 times less Rubisco)
Nitrogen use efficiency (NUE) and photosynthesis per unit of leaf nitrogen (PNUE).	Low	High
Other		
Forage quality/digestibility/crude protein level	Higher	Lower
Mycorrhizal symbiosis in low P soil	Facultative mycotrophs	Obligate mycotrophs
Ecological advantage	Cool, moist environments	Hot, dry, high-light environments

Anomochlooideae, Pharoideae, and Puelioideae, presumed to be so. Genera that include both C_3 and C_4 members include *Alloteropsis, Neurachne*, and *Panicum* in the Panicoideae and *Eragrostis* in the Chloridoideae. A few species possess characteristics intermediate between those of C_3 and C_4 species, including the New World *Steinchisma decipiens, P. hians*, and *P. spathellosa*, and the Australian *Neurachne minor*. C_4 photosynthesis also occurs in a few non-grass angiosperm families, including the Cyperaceae (sedges, e.g. some *Cyperus, Scirpus*) and Hydrocharitaceae (frog's-bit family, e.g. *Vallisneria spiralis* and *Hydrilla verticillata*) in the monocots, and 16 dicot families including some members of the Amaranthaceae (e.g. *Alternanthera, Amaranthus*) and the Chenopodiacea (e.g. some *Atriplex, Bassia, Suaeda*) (Sage 2004).

Evolution of C_4 photosynthesis is believed to be polyphyletic with at least 45 independent origins, and, in grasses, may have evolved up to 11 times (Hattersley 1986; Sage 2004). Atmospheric CO_2 levels were high in the Cretaceous but declined in the mid-Tertiary. Atmospheric CO_2 concentrations lower than they are at present first occurred during the Oligocene epoch, 23–34 Ma. The **CO_2-starvation hypothesis** suggests that these conditions, along with aridity in tropical areas, provided the main selective force favouring evolution of the CO_2-concentrating mechanisms and lack of photorespiration in C_4 species (Cerling et al. 1998; Osborne and Beerling 2006), although not necessarily the spread of C_4-dominated grasslands later in the mid-Miocene (Chapter 2) (Osborne 2008). Since C_4 photosynthesis represents a complex mixture of physiological (e.g. the enzyme PEP carboxylase) and morphological (e.g. Kranz anatomy, Chapter 3) traits, it is probable that C_4-evolution evolved early in the Paleocene, perhaps as much as 50 Ma (Apel 1994). Compared with the evolution of oxygenic photosynthesis as a whole, C_4 photosynthesis would have appeared during the second half of the last hour over the course of a 24 h time frame (Apel 1994). Nevertheless, the oldest undisputed C_4 fossils are panicoid grass leaves with Kranz anatomy from 12.5 Ma. Isotopic carbon ratios from palaeosols (fossil soils) and the teeth of herbivores in East Africa and North America from 14–20 Ma are consistent with the presence of C_4 plants in mid-Miocene landscapes (Sage 2004). Molecular clock evidence on the divergence of maize and sorghum, and maize and *Pennisetum* (all C_4), suggests that C_4 photosynthesis arose in plants even earlier during the early Miocene, 23–34 Ma. As noted in §2.4.4, the evolution of C_4 photosynthesis in grasses allowed the later widespread expansion of C_4-dominated grasslands in low latitudes by the end of the Miocene, and temperate C_4 grasslands by 5 Ma (Cerling et al. 1997; Sage 2004).

4.1.3 Biochemistry of C_4 photosynthesis

The primary feature of the C_4 pathway that distinguishes it from the C_3 pathway is the location and manner in which atmospheric CO_2 is taken up. In C_3 photosynthesis a molecule of CO_2 is taken up and enters the carbon reduction (PCR or Calvin) cycle in the chloroplast. The CO_2 is 'fixed' (bonded covalently) to ribulose 1,5-bisphosphate (RuBP) which is split to form two molecules of 3-phosphoglycerate (PGA), a reaction catalysed by the enzyme RuBP carboxylase/oxygenase (Rubisco) in the mesophyll. PGA is a three-carbon molecule, hence the name

C_3 photosynthesis. A full cycle of the Calvin cycle reduces 6 molecules of CO_2, uses energy from 18 molecules of ATP and 12 molecules of NADPH produced in the accompanying light-dependent reactions of photosynthesis, and produces 2 molecules of glyceraldehyde 3-phosphate (G3P) as product. The cycle begins again with the regeneration of RuBP. G3P is transported from the chloroplast to the ground substance (cytoplasmic matrix) of the cell where it is rapidly converted to glucose 1-phosphate and fructose 6-phosphate, precursors of sucrose. The G3P that remains in the chloroplast is converted to starch and stored temporarily during the day as grains in the stroma. At night the starch can be converted to sucrose and exported from the chloroplast.

A problem with C_3 photosynthesis is that Rubisco has a concentration and temperature-dependent affinity for O_2 as well as CO_2. The solubility of CO_2 declines with increasing temperature more so than does that of O_2. In addition, substrate-saturated enzyme activity (V_{max}) of Rubisco increases at a faster rate with increasing temperature in the presence of O_2 than in the presence of CO_2, and proceeds at a faster rate (Lambers et al. 1998). Whereas carboxylation, as described above, produces two molecules of PGA, oxygenation when it occurs produces only one PGA and a molecule of the two-carbon compound phosphoglycolate (GLL-P). GLL-P is transported form the chloroplast and after a series of transformations one molecule of GLL-P is used to produce a molecule of serine, then glycerate, before regenerating as PGA. In the process a molecule of CO_2 and one molecule of NH_3 are released. This process is referred to as **photorespiration** because it depends on light and is a respiratory process (releasing CO_2). Photorespiration therefore reduces the efficiency of the Calvin cycle, causing a decline in net photosynthesis at high temperatures.

The uptake of CO_2 in C_4 photosynthesis proceeds to attach bicarbonate from CO_2 to phosphoenolpyruvate (PEP) through catalysis by the enzyme PEP carboxylase, creating oxalacetate (OAA), a four-carbon organic acid. This reaction takes place in chloroplasts of the leaf mesophyll where CO_2 is comparatively abundant. PEP carboxylase only catalyses the conversion of PEP to four-carbon organic acids and, unlike Rubisco, is unaffected by O_2 levels. The organic acids are transported to chloroplasts in specialized bundle sheath cells where the CO_2 is released into the normal Calvin cycle. This 'CO_2 pump' enhances CO_2 concentration 10–20-fold at the site of Rubisco in the bundle sheath cells (Apel 1994).

There are three subtypes of the C_4 cycle (Fig. 4.1) related to the nature of the organic acid transported to the bundle sheath cell, the enzyme that catalyses the decarboxylation step, the anatomy of the bundle sheath cells (Chapter 3), and the facility to reduce CO_2 leakage from the bundle sheath back into mesophyll airspaces. In C_3 grasses there are two layers of cells around the vascular bundles, neither of which is photosynthetic. In C_4 plants, there is either a double (XyMS+) or single (XyMS–) layer of bundle sheath cells reflecting the presence or absence, respectively, of a second inner layer, the mestome sheath, adjacent to metaxylem. The outer tangential and radial cell walls of the bundle sheath cells may also be suberized, reducing CO_2 leakage from the bundle sheath back into the mesophyll. The three C_4 subtypes are distinguished according to the decarboxylating enzyme:

- The **PEP-CK (synonyms PCL and PEP) subtype** has Kranz anatomy and is XyMS+ with suberized lamella. The outer bundle sheath layer is photosynthetic with agranal chloroplasts arranged centrifugally (facing towards the outside of the vascular bundle). CO_2 is incorporated in the mesophyll cells first into oxaloacetate then as aspartate before transport to the bundle sheath cells where it is converted back to oxaloacetate by cytosolic phosphoenol pyruvate (PEP carboxykinase). Carbon is returned to the mesophyll as PEP or alanine. Examples include *Brachiaria*, *Chloris*, *Panicum*, *Spartina*, *Zoysia*, genera characteristic of regions with both dry and wet seasons. In Australia, *Eragrostis* species that are PEP-CK-like are most frequent in northern high-rainfall tropical and humid subcoastal and coastal areas, compared with NAD-ME-like species that are more frequent in areas where rainfall is <30 cm yr^{-1} (Prendergast *et al*. 1986).
- The **NADP-ME subtype** is XyMS– with large agranal C_3 chloroplasts (suggesting a poorly developed photosystem II) in a single-layered bundle sheath that has a suberized lamella. PEP is fixed with CO_2 to give four-carbon oxaloacetate, which is converted to malate and transported to the bundle sheath chloroplasts where it is decarboxylated by NADP-malate decarboxylase (aka NADP-malic enzyme) using NADP as co-factor. Carbon is returned to the mesophyll as pyruvate. Toxic secondary compounds (§4.3) are often present in NADP-ME plants, whereas they tend to be absent in NAD-ME C_4 plants (Ehleringer and Monson 1993). These compounds can act as feeding deterrents to insect herbivores. Such protection from herbivory may be necessary to protect the 'exposed' protein contained within the long and rectangular thin-walled bundle sheath cells. By contrast, the bundle sheath cells of NAD-ME grasses tend to be short and cubical, hence with a higher surface to volume ratio they are more difficult to crush. The bundle sheath cells of NAD-ME grasses protect their protein content better than NADP-ME grasses, obviating the need for toxic secondary compounds. Examples of NADP-ME grasses include *Saccharum*, *Zea*, *Sorghum*, and genera common in moist to semi-arid habitats of warm regions. *Neurachne*, *Paractaenum*, *Paraneurachne* and *Plagiosetum*, all NADP-ME panicoids, have distributions entirely restricted to arid climates in Australia (Prendergast *et al*. 1986). Most grasses species in the Panicoideae use the NADP-ME subtype.
- The **NAD-ME subtype** is XyMS+ with granal chloroplasts arranged centripetally (i.e. towards the inner cell wall), which maximizes the diffusion pathway from the site of CO_2 release to the bundle sheath/mesophyll interface. The bundle sheath cells lack a suberized lamella and so may leak CO_2 more than the other types but have the lowest surface area of bundle sheath cells exposed to mesophyll tissue or intercellular spaces. CO_2 is incorporated in the mesophyll and transported as aspartate to the bundle sheath cells where it is converted in the mitochondria to malate and decarboxylated by NAD-malate decarboxylase with NAD as a co-factor and transported to the chloroplasts for incorporation into the Calvin cycle. Mitochondria are particularly abundant in the bundle sheath cells of NAD-ME plants compared with in the NADP-ME subtype. Carbon is regenerated as pyruvate or alanine and returned

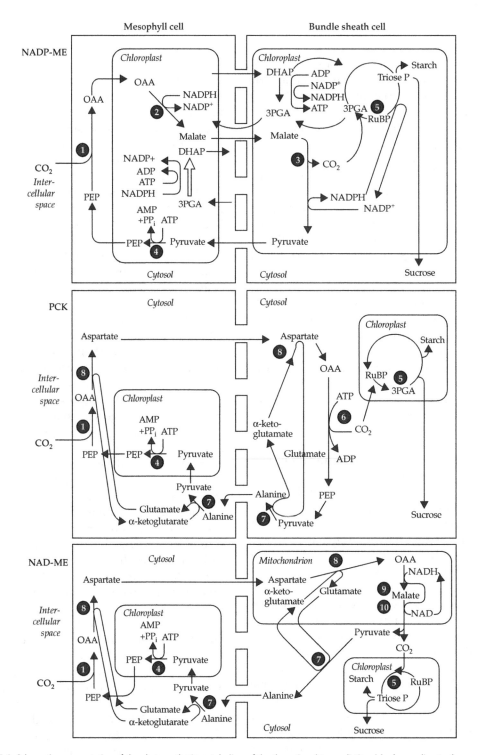

Figure 4.1 Schematic representation of the photosynthetic metabolism of the three C_4 subtypes distinguished according to the decarboxylating enzyme. NADP-ME, NADP-requiring malic enzyme; PCK, PEP carboxykinase; NAD-ME, NAD-requiring malic enzyme. Numbers refer to enzymes: (1) PEP carboxylase, (2) NADP-malate dehydrogenase, (3) NADP-malic enzyme (4) pyruvate-P_i-dikinase, (5) Rubisco, (6) PEP carboxykinase, (7) alanine aminotransferase, (8) aspartate amino transferase, (9) NAD-malate dehydrogenase, (10) NAD-malic enzyme. With kind permission of Springer Science and Business Media from Lambers et al. (1998).

to the mesophyll. This subtype is characteristic of members of the Chloridoideae including grasses of warm regions, especially in dry habitats, and includes *Buchloë*, *Cynodon*, *Eragrostis*, *Panicum*, *Sporobolus*, *Triodia*, and *Triraphis*. NAD-ME species were found to be most frequent in arid climates in Australia (Prendergast 1989).

Ecologically, as described above, there are differences in the distribution of the three C_4 subtypes. However, C_4 subtype is not necessarily an adaptation to a particular climate and is not the sole determinant of distribution. For example, in Australian *Eragrostis* species, distribution was more closely related to Kranz anatomy cell chloroplast position than C_4 subtype: species with centripetal chloroplasts were dominant in arid climates whereas more humid climates were dominated by species with centrifugal/peripheral chloroplasts (Prendergast *et al.* 1986).

At least 20 plant species, including a number of grasses, exhibit anatomical and biochemical features that are intermediate between C_3 and C_4 plants. These C_3/C_4 intermediates have a weakly developed Kranz anatomy similar to the NAD-ME subtype and Rubisco is present in both the mesophyll and the bundle sheath cells. Hence there are reduced rates of photorespiration and low CO_2-compensation points compared with C_3 plants. Two main types of C_3/C_4 intermediates are recognized:

- C_4 enzyme activity is very low so there is no functional C_4-acid cycle, but these plants possess light-dependent recapture of CO_2 by the mesophyll cells following CO_2 release in photorespiration in the bundle sheath cells. Thus, CO_2 is scavenged, lowering the compensation point. Examples of this type of C_3/C_4 intermediate include some species of *Neurachne* and *Panicum*.
- High activity of C_4 enzymes and CO_2 is rapidly fixed into C_4 acids and transferred to the Calvin cycle. However, there is only limited operation of the C_4 cycle, and CO_2 is not effectively concentrated in the bundle sheath cells to allow the low quantum yields characteristic of true C_4 plants. An example is the Australian grass *Neurachne minor*.

C_3 and C_4 plants differ in the $^{13}C/^{12}C$ ratio in tissues with C_3 plants having lower values (–22 to –35) than C_4 plants (–9 to –18) (Chapman and Peat 1992). In the atmosphere $^{12}CO_2$ predominates, with $^{13}CO_2$ accounting for only 1%. Rubisco reacts more readily with $^{12}CO_2$ than it does with the heavier $^{13}CO_2$, so the enzyme discriminates against $^{13}CO_2$. Excess $^{13}CO_2$ diffuses back to the atmosphere, so plant material has less ^{13}C than the atmosphere. However, there is less discrimination against $^{13}CO_2$ in C_4 plants because (1) $^{13}CO_2$ that is discriminated against by Rubisco in C_4 plants is prevented from escaping because of a diffusion barrier between the bundle sheath and mesophyll, and (2) excess $^{13}CO_2$ is scavenged by PEP carboxylase and carbonic anhydrase (Lambers *et al.* 1998). The result is that the $^{13}C/^{12}C$ ratio can be used as a 'signature' of the relative contribution of C_3 and C_4 plants in organic matter such as forage, unidentifiable roots, paleosoils, or animal tooth enamel (e.g. Cerling *et al.* 1993). For example, examination of isotope ratios in the diet of extant elephants (*Loxodonta* in Africa, *Elephas* in Asia) indicated a diet dominated by C_3 browse (Cerling *et al.* 1999). Isotope ratios in soil organic matter in late Holocene paleosoils indicated the occurrence of C_4 plants, suggesting an early expansion of tropical high-altitude grassland (*páramo*) in areas of the Bogota basin in the Columbian Andes that are today dominated by native C_3 vegetation (Mora and Pratt 2002).

4.1.4 Effects of C_4 photosynthesis on light, temperature, and moisture responses

Biochemical differences between C_3 and C_4 photosynthesis lead to a number of profound physiological differences between grasses possessing these two pathways. The most important differences described in detail below relate to the advantage that C_4 plants have over C_3 plants in warm and arid environments. Briefly, C_4 plants are able to maintain high photosynthetic rates at high temperatures (due to lack of photorespiration) and have high **water use efficiency** (WUE) (due to low internal CO_2 levels, allowing low stomatal conductance for the same CO_2 assimilation rates) and **nitrogen use efficiency** (NUE) (due to lower requirements for nitrogen-rich Rubisco).

The CO_2 compensation point (the point at which CO_2 uptake equals CO_2 evolution) is lower in C_4

than in C_3 plants (0–0.5 Pa CO_2 and 4–5 Pa CO_2, respectively). Thus, C_4 plants are better able to fix CO_2 when the stomata are closed, and thus to conserve water in dry conditions. The CO_2 compensation point of C_3 plants is sensitive to temperature, increasing rapidly as temperature increases (Williams and Markley 1973). This gives C_4 plants an additional physiological advantage over C_3 plants at higher temperatures. C_4 photosynthesis is also unaffected by O_2 concentration (because of the lack of photorespiration), whereas C_3 photosynthesis declines as oxygen concentration increases.

The internal partial pressure (P_i) of CO_2 in the mesophyll is c.25 Pa in C_3 plants compared with c.10 Pa in C_4 plants; however, this pattern is reversed in the bundle sheath cells where CO_2 is released to the Calvin cycle in an environment that is not compromised by photorespiration. High P_i of CO_2 in the bundle sheath allows for favourable kinetic properties of Rubisco in the bundle sheath to be taken advantage of. In C_4 plants the enzymatic dissociation constant (the Michaelis–Menten constant, K_m (CO_2)) for Rubisco is high (i.e. the enzyme has a low affinity for its substrate), meaning that CO_2 is not as tightly bound to Rubisco as is necessary for C_3 plants. High K_m for Rubisco in C_4 plants allows for a high (fast) catalytic activity leading to more moles of CO_2 to be fixed per unit of Rubisco and time compared with C_3 plants. In addition, the K_m for CO_2 of PEP carboxylase is lower (hence, high affinity) than the K_m for Rubisco in C_4 plants.

The WUE is high in C_4 plants because stomata of C_4 plants need not be opened and subject to transpirational water loss for as long or as extensively as those of C_3 species to assimilate CO_2 at a given rate. C_4 plants can produce 1 g of biomass per 250–350 g of water transpired, whereas C_3 plants transpire 650–800 g water for every gram of biomass produced (Ehleringer and Monson 1993). Consequently C_4 plants may be advantaged in arid environments. For example, WUE was 40–170% greater for the C_4 grass *Andropogon gerardii* than for co-occurring C_3 forbs, although during a wet year there were no differences in maximum rates of photosynthesis between the two groups of plants (Turner et al. 1995).

Light response curves (plots of photosynthetic rate vs light intensity) for C_3 plants saturate (i.e. flatten) at high light intensity as CO_2 assimilation becomes the limiting factor for accumulation. The slope of this curve is steeper in C_4 than in C_3 plants, independent of O_2 concentration (not the case for C_3 plants), and saturates at a higher light intensity. Thus, C_4 plants have an advantage over C_3 plants in high-light, open environments.

At high temperatures (i.e. <25–30 °C), quantum yield and **light-use efficiency** (LUE) of C_3 plants is low, and declines with increasing temperature as the oxygenating activity of Rubisco increases (i.e. as photorespiration increases). By contrast, the quantum yield of C_4 plants is high and unaffected by temperature. At low temperatures, the quantum yield and LUE of C_4 plants is lower than that of C_3 plants because C_4 plants require two additional ATP molecules derived from the light-dependent reactions to regenerate one molecule of PEP from pyruvate. This gives C_3 plants an advantage over C_4 plants at low temperatures. Hence, because of the lack of photorespiration in C_4 plants, LUE is unaffected by temperature whereas it declines with temperature in C_3 plants.

Tissue nitrogen levels are lower in C_4 plants (120–180 mmol N m^{-2} in leaves) than in C_3 plants (200–260 mmol N^{-2}) (Ehleringer and Monson 1993) because 3–6 times less of the nitrogen-rich Rubisco is needed and because of the low levels of photorespiration enzymes. Consequently, NUE and PNUE (photosynthesis per unit of leaf nitrogen) is high in C_4 plants. However, high PNUE does not appear to translate into an advantage for C_4 plants on low-nitrogen soils (Lambers et al. 1998).

4.1.5 Ecological ramifications

Tropical areas where C_4 photosynthesis evolved continue to be centres of distribution for C_4 plants including grasses. C_4 species dominate tropical and temperate grasslands with abundant warm-season precipitation. The frequency of C_4 grasses correlates positively with minimum growing season temperatures (Terri and Stowe 1976). For example, the geographic distribution of C_4 grass taxa in North America ranges from 12% of the grass taxa in the north to 82% in the desert south-west (Fig. 4.2). Along the Atlantic and Pacific coastal regions, the proportion of C_4 grasses increases as

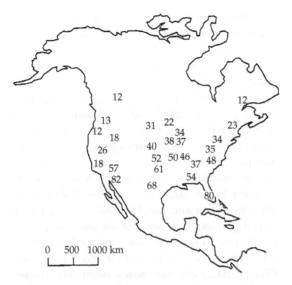

Figure 4.2 Geographic distribution of C_4 grass taxa in North America. Numbers indicate the percentage of grass taxa that are C_4. With kind permission of Springer Science and Business Media from Lambers et al. (1998), redrawn from Terri and Stowe (1976).

latitude decreases on both coasts and as July minimum temperature increases (Wan and Sage 2001). The proportion of NADP-malic enzyme C_4 grasses was positively correlated with mean annual precipitation.

Along a north–south gradient from North Dakota to south Texas the C_3/C_4 ratio of graminoids decreased from 1.9 to 0.8 in a survey of remnant true prairie in North America, and was negatively correlated with annual precipitation and temperature, and positively correlated with soil organic matter (Diamond and Smeins 1988, and see §6.1.2). In the Sinai, Negev, and Judean deserts of the Middle East, the occurrence of C_4 grasses increased with decreasing rainfall (Vogel et al. 1986). C_4 grasses were found to dominate areas of high year-round temperature. The NADP-ME C_4 subtype (27 species in 18 genera) was most frequent where water stress was not a dominating factor (such as in rocky crevices and irrigated fields) whereas the NAD-ME subtype (16 species in 7 genera) was most frequent in xeric, open desert areas. NADP-ME grasses were mostly summer active perennials, whereas the NAD-ME subtype was represented mostly by winter perennials. The PEP-CK subtype was not abundant in this Middle Eastern flora (10 species in 6 genera) and was intermediate between the other 2 subtypes in its distribution, although PEP-CK grasses were also somewhat confined to rock crevices and irrigated fields.

At more local scales, numerous studies document differential distribution of C_3 and C_4 grasses in response to local moisture, irradiance, and temperature gradients (e.g. Archer 1984; Barnes et al. 1983; Gibson and Hulbert 1987). For example, in the Mule Mountains of south-east Arizona, USA, 69 C_4 species representing 6 angiosperm families were encountered accounting for 13.5–22.3% of vascular species (69–96% of the Poaceae encountered) in vegetation ranging from pygmy conifer-oak scrub (*Pinus cembroides, Junierus deppeana, Quercus arizonica, Q. hypoleucoides,* and *Q. emoryi*) in mesic areas to desert grassland (*Elyonurus barbiculmis, Bouteloua radicosa, Heteropogon contortus*) in xeric situations (Wentworth 1983). The proportion and percentage canopy cover of C_4 species increased with increasingly warm and dry environmental conditions in vegetation occurring over granite. In contrast, vegetation occurring over limestone that had been heavily grazed did not show this relationship.

Even on a small spatial scale, local soil moisture conditions can lead to differences in the abundance of C_3 vs C_4 taxa in grasslands. For example, across a 400 m hillside in remnant mixed-grass prairie in Colorado, USA, three C_4 grasses (*Bouteloua gracilis, B. curtipendula,* and *Schizachyrium scoparium*) were found to dominate the rocky, dry mid-slope, whereas C_3 grasses dominated the moister top (*Pascopyrum smithii*) and bottom (*Poa pratensis*) (Archer 1984; Barnes et al. 1983).

Overall, it remains unclear whether it is temperature or moisture that determines the differential distribution of C_3 and C_4 grasses. More than likely it is the interplay between these two factors that is important (Vogel et al. 1986).

4.1.6 Effects of global climate change

Global temperature and atmospheric CO_2 levels have been increasing due to anthropogenic causes for the last 250 years, since the start of the Industrial Revolution. The Intergovernmental Panel on Climate Change predicts that atmospheric

CO_2 will double in the mid–late twenty-first century and cause a 1.4–5.8 °C rise in mean temperature (IPCC 2007). Because CO_2 is a substrate for photosynthesis, increased atmospheric CO_2 may substantially increase photosynthetic rates and growth, decrease leaf transpiration through stomatal closure, and increase WUE of land plants.

For C_3 plants in particular, increased atmospheric CO_2 provides a large 'fertilization' effect by decreasing photorespiration and increasing carbon assimilation rates. Experimental studies of C_3 grasses conducted at elevated (mostly double ambient) CO_2 levels (i.e. $c.760$ μg g CO_2) generally show an $c.34$–44% increase in growth (Amthor 1995; Kirschbaum 1994; Long et al. 2004; Poorter 1993). However, the increased photosynthetic rates may be short-lived as some plants grown under elevated CO_2 in open-top chambers acclimate or down-regulate producing less Rubisco, lowering carbon allocation to leaves, decreasing leaf stomatal density, or decreasing expression of specific photosynthetic genes or gene products in response to increased sucrose cycling within mesophyll cells (Long et al. 2004). A meta-analysis of published studies suggests that this effect represents more of an acclimation of the plant to the changed conditions (such as constrained rooting volume or nutrient limitation) rather than a down-regulation per se of gas exchange to former levels (Long et al. 2004). Plants grown under elevated CO_2 in field settings (so-called free-air CO_2 enrichment or FACE experiments) show declines in Rubisco accompanying increased photosynthetic rates, but no change in capacity for ribulose-1,5-bisphosphate regeneration. High growth rates under elevated CO_2 without increased availability of soil nitrogen can lead to lowered tissue nitrogen levels ('nitrogen dilution') and altered protein composition (Newman et al. 2003).

By contrast, C_4 plants, including C_4 grasses, are not limited by CO_2 levels and so exhibit smaller growth responses than C_3 plants to elevated atmospheric CO_2 levels in the range of 10–25% (Long et al. 2004; Wand et al. 1999). Positive growth responses of C_4 grasses to elevated CO_2 often are due to CO_2-induced stomatal closure and enhanced water use efficiency rather than to a direct enhancement of photosynthetic rates. For example, in the C_4 grass *Andropogon gerardii* in a North American tallgrass prairie, reductions in photosynthesis related to water stress were less likely to occur under elevated than ambient CO_2 levels (Knapp et al. 1993; Nie et al. 1992a).

The differential response of C_3 and C_4 plants to elevated CO_2, coupled with a predicted increase in global temperature, has the potential to affect competitive interactions among plants, alter community composition, and elevate grassland ecosystem production by about 17% (Campbell and Stafford Smith 2000; Navas 1998; Polley et al. 1996; Soussana and Luscher 2007). When grown in competition, C_4 species show lower responses to elevated CO_2 at high nutrient conditions than do C_3 grasses or dicots (Poorter and Navas 2003). Given that C_4 plants may be less negatively affected by higher temperatures than C_3 plants, the overall effect of elevated CO_2 coupled with rising global temperature is difficult to predict, especially since elevated CO_2 can vary with abiotic factors (Soussana and Luscher 2007) and affects rates of decomposition and soil nutrient mineralization in complex ways. There clearly are limits on how reliably we can predict vegetation patterns from leaf physiology. Several experimental studies have observed effects of elevated CO_2 on grassland communities in which the proportion of grasses to forbs has declined (e.g. Potvin and Vasseur 1997; Teyssonneyre et al. 2002; Zavaleta et al. 2003). Winkler and Herbst (2004) observed a shift in botanical composition reflecting an increase in legumes in a nutrient-poor, calcareous semi-natural grassland (*Bromus erectus* dominated) subjected to 4 years of elevated CO_2.

Global circulation models (GCMs) project sometimes significant changes in abundances of C_3 and C_4 plants under various climate change scenarios that include elevated temperature and higher atmospheric CO_2. For example, Epstein et al. (2002b) projected an increase in the relative abundance of C_4 grasses by >10% throughout most of the temperate grasslands and shrublands of North and South America at the expense of C_3 grasses (Fig. 4.3). These projections are consistent with predictions under higher temperatures but do not match with the suggestion that C_3 species will benefit at the expense of C_4 species under elevated CO_2. Indeed, some

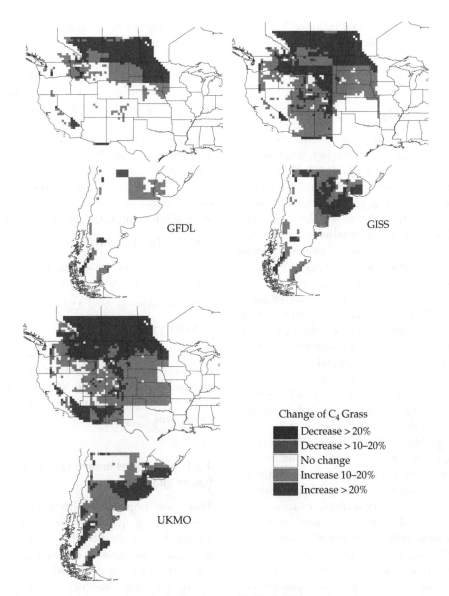

Figure 4.3 Projected absolute percentage change in C_4 grass abundance for double CO_2 scenarios in three general circulation models (GFDL, GISS, and UKMO). Reproduced with permission from Epstein et al. (2002b).

field studies have shown no competitive advantage of C_3 over C_4 grasses under high CO_2 (Owensby et al. 1993). The benefit of elevated CO_2 for C_3 plants appears to be contingent on temperature and moisture; if it is too hot and too dry then C_4 plants maintain their competitive advantage even with elevated CO_2 (Nie et al. 1992b). The importance of multifactor experiments is clear given the interactions between atmospheric CO_2 and other aspects of the climate, such as temperature and precipitation, that are predicted to be affected under global environmental change (Norby and Lou 2004). For example, photosynthetic pathway and precipitation were identified as the most important variables affecting foliage production in a modelling study of semi-humid temperate grassland subjected to various climate change scenarios that included increased growing season precipitation and temperature while maintaining current atmospheric CO_2 levels (Seastedt *et al.* 1994).

4.2 Forage quality

Forage is defined as any plant material, including herbage but excluding concentrates, used as feed for domestic herbivores (see Table 1.2). This technical definition is somewhat narrow, and for an ecological perspective forage should include the plant material in grasslands that is feed for native herbivores (Chapter 9). **Forage quality** is the potential of a forage to produce the desired animal response, with **forage nutritive value** as the nutrient concentrations, digestibility (energy value), and nature of the end products of digestion (by a grazer) (Collins and Fritz 2003). The important components of forage include minerals (nutrient elements), cell-wall carbohydrates (structural carbohydrates), lignin, crude protein, and non-structural carbohydrates (including organic acids, starch, and sugars), proteins, and lipids. There is an important feedback between forage quality and grazer behaviour (see Chapters 9 and 10).

Forage quality is determined ultimately by measuring animal response, such as milk production or weight gain, but such feeding trials are expensive and require extensive labour, animals, and facilities. Laboratory analyses of forage provide information about potential animal response. A typical forage analysis report will include values for neutral detergent fibre (NDF), acid detergent fibre (ADF), crude protein (CP), moisture, and mineral concentration (especially calcium, phosphorus, magnesium, potassium, and sometimes sulfur). NDF is the total fibre or cell wall fraction of the forage following extraction with a neutral detergent solution. ADF is an extract made using 1 N sulfuric acid and is made up of cellulose, lignin, and silica. Hemicellulose is not extracted with ADF but is estimated by subtracting ADF from NDF.

Digestibility is the proportion of dry matter or constituent digested within the digestive tract of the animal (Barnes *et al.* 2003) and can be determined either *in vivo* or *in vitro*. *In vivo* digestibility is measured by feeding animals and weighing fecal dry matter output. *In vitro* digestibility (*in vitro* dry matter disappearance: IVDMD) involves the inoculation of forage in rumen fluid to simulate the ruminant digestive tract. Digestibility is not usually measured by commercial forage-testing laboratories.

Relative feed value (RFV) (Collins and Fritz 2003) provides a single value for comparing forages and is based on the negative correlation between ADF and digestibility and between NDF and voluntary intake (quality of forage that animals consume given an unrestricted supply). Thus,

$$RFV = (DDM \times DMI)/1.29$$

where DDM is dry matter digestibility (%) and DMI is voluntary dry matter intake calculated as:

$$DDM = 88.9 - (0.779 \times ADF)$$

$$DMI = 120/NDF$$

Near-infrared reflectance spectroscopy (NIRS) is a non-destructive technique that allows 90–99% accurate estimations of CP, NDF, ADF, and IVDMD with lower costs than the standard wet lab procedures.

4.2.1 Differences among species in forage quality

There are important differences among species in forage quality, with legumes generally having higher quality than grasses. Legumes and cool-season grasses often have similar ADFV concentrations and digestibility, but cool-season grasses generally have higher NDF. In a comparison between *Medicago sativa* (alfalfa) and *Phleum pratense* (timothy), NDF values were 49% and 66%, respectively and CP values were 16% and 9.5%, respectively (Collins and Fritz 2003). Forage of warm-season C_4 grasses is of lower quality and c.13% lower in digestibility than that of cool-season C_3 grasses because a high percentage of leaf area is occupied by highly lignified and less digestible tissues (vascular bundle, epidermis, sclerenchyma) (Fig. 4.4). Crude protein levels in leaves of warm-season grasses are also lower than in leaves of

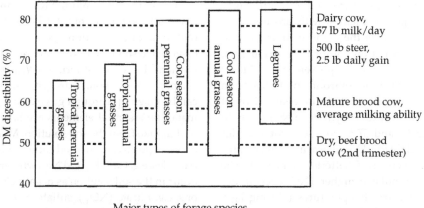

Figure 4.4 Digestibility ranges of major forage types. Dashed lines show forage digestibility levels required to meet energy requirements of different classes of cattle. Reproduced with permission from Collins and Fritz (2003).

cool-season grasses. It has been suggested that C_3 and C_4 grasses may become nutritionally equivalent under elevated CO_2 because of the generally larger response of C_3 species to elevated atmospheric CO_2. However, this hypothesis was not supported in a comparison of five C_3 and C_4 grasses grown under elevated CO_2 (Barbehenn et al. 2004); protein levels decreased in the C_3 grasses but not to the level of C_4 grasses.

Forage quality decreases with plant age/maturity, principally because of a decrease in the leaf:stem ratio (Nelson and Moser 1994). For example, the leaf percentage of total DM declined from 61% to 23% in *Dactylis glomerata* from early vegetative stage to late anthesis (Buxton et al. 1987). Leaves represent the highest-quality part of the forage, and are preferentially chosen by grazers when allowed the choice. The proportion of stems increases as the plant matures through the season and stem and leaf quality declines (Fig. 4.5). In *Dactylis glomerata* the leaf fraction fell more slowly than that of the leaf sheath and stem fractions.

Although plant maturity impacts forage quality more than anything else, maturity is modified by plant environment. Solar radiation is the driving force setting an upper limit for production,

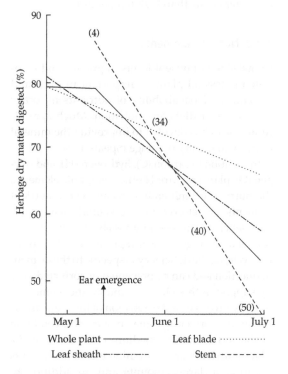

Figure 4.5 Decline in forage quality of S. 37 *Dactylis glomerata* in various plant parts during first growth in the spring. Figures in parentheses are the percentage of stem in the whole plant. Reproduced with permission from Bransby (1981).

modulated by temperature and rainfall. Light intensity and quality affect growth processes, i.e. morphogenesis, giving the plant form and affecting tillering, branching, internode expansion, leaf expansion, and flowering (Nelson and Moser 1994). The influence of environmental factors on forage quality was illustrated in a comparison of total non-structural carbohydrate (TNC) levels in leaves of 128 C_3 and 57 C_4 grasses grown under cool (10/5 °C light/dark) and warm temperatures (25/15 °C), respectively (Chatterton et al. 1989). The C_3 grasses accumulated higher TNC levels under both cool and warm temperatures (312 mg kg^{-1}, 107 mg kg^{-1}, respectively) than did the C_4 grasses (166 mg kg^{-1}, 92 mg kg^{-1}, respectively), mainly due to accumulation of fructan in the leaves, especially in cool conditions. Generally, forage is more digestible at low temperatures because (1) lignification of cell wall materials increases at high temperatures and (2) accumulation of digestible storage products is greater at low than high temperatures.

4.2.2 Nutrient elements

Knowledge of nutrient levels in grasses and associated grassland plants is important in grassland agriculture. Bioavailability of nutrients in forage affects grazer diets and nutrition. Much agricultural research is devoted to improving the mineral nutritional quality of forage (Spears 1994).

In addition to carbon (C), hydrogen (H), and oxygen (O), plants require 14 other inorganic elements, the nutrient (or mineral) elements. These nutrient elements are referred to as **essential elements** if: (1) a deficiency makes it impossible for the plant to complete the vegetative or reproductive stage of its life cycle; (2) the deficiency is specific to the element in question and can be prevented or corrected only by supplying this element; and (3) the element is directly involved in the nutrition of the plant and is not simply correcting some unfavourable condition of the soil or culture medium (Whitehead 2000). **Macronutrients** are those nine essential elements required in large amounts and, in addition to carbon, hydrogen, and oxygen, include nitrogen (N), potassium (K), calcium (Ca), phosphorus (P), magnesium (Mg), and sulfur (S). A further eight **micronutrients** are required in very small, or trace, amounts: iron (Fe), chlorine (Cl), manganese (Mn), zinc (Zn), molybdenum (Mo), boron (B), and nickel (Ni). In addition, sodium (Na) and cobalt (Co) are essential to some plants and are regarded as micronutrients by some authorities (Whitehead 2000).

Apart from carbon, hydrogen, and oxygen, which can be taken up through gas exchange via the stomata, the essential elements are taken in through root uptake from the soil solution. Metals generally occur as positively charged cations adsorbed to soil particles, e.g. Ca^{2+}, K^+, and Na^+, whereas non-metals occur in the soil solution as negatively charged anions, e.g. nitrate (NO_3^-), sulfate (SO_4^{2-}), bicarbonate (HCO^{3-}), and can be readily leached from there. In some soils phosphate forms insoluble precipitates and is adsorbed and held on the surface of compounds containing iron, aluminium, and calcium so that it is retained against leaching and somewhat difficult for plant roots to obtain. Because of this, mycorrhizae are especially important to plants in the uptake of phosphorus as their mycelia increase the surface area available for uptake in the soil compared with non-mycorrhizal roots (Chapter 5). C_4 grasses are obligate mycotrophs especially dependent on soil mycorrhizae for soil phosphorus acquisition in low-phosphorus soils (Anderson et al. 1994; Hetrick et al. 1986).

The concentration of nutrients in the soil of natural and semi-natural grasslands normally reflects the underlying soil parent material, modified by the long-term effects of the environment and climate (Chapter 7). In these soils, recycling, including the death and decay of plants, drives nutrient recycling and plant nutrient uptake. By contrast, the soils of managed grasslands are usually supplemented with nitrogen, phosphorus, and/or potassium fertilizer, and sometimes micronutrients, to supplement deficiencies and replace loss through the removal of plant and animal products. Animal excreta are an important part of the recycling process in grazed grasslands (Chapter 9). Legumes are important and significant components of temperate semi-natural grasslands, and can fix up to 500 kg N ha^{-1} yr^{-1} (Whitehead 2000). White clover *Trifolium repens* is the most important legume in European and New Zealand grasslands, lucerne (alfalfa *Medicago sativa*), red clover *Trifolium pretense*, and birdsfoot trefoil *Lotus corniculatus* in North America, and annual clovers and medics *Medicago* spp. in temperate areas with a pronounced dry season, including parts of Australia.

Acquisition of nutrients occurs principally 5–50 mm back from the root tip. In this region, root hairs (protrusions of individual epidermal cells c.10 μm in diameter and 0.2–1 mm in length) grow into micropores within soil aggregates. Root hairs increase the effective surface absorption area of the root. An abundance of root hairs and fine roots provides the most effective nutrient uptake in limited soils. Nutrient ions close to root hairs move by mass flow or diffusion. The root hairs of grasses are longer and more numerous than those of legumes.

Plant:soil concentration ratios
Differences between elements with respect to accumulation in grassland plants are based on the nature of the soil and interacting environmental factors. There is wide variation between nutrient elements (Table 4.2). Nitrogen exhibits the greatest amount of accumulation (10:1), followed by chlorine (7:1); both are readily taken up from soil solution as NO_3^{-1} or NH_4^+, and Cl^-, respectively. Iron shows the greatest degree of exclusion (0.4:1). With the exception of chlorine, which is involved in water relations, photosynthesis, and ion homeostasis, the micronutrients have lower levels of accumulation than macronutrients.

Tissue nutrient levels
The levels of nutrients in plant tissues range from 1.0–5.3% for nitrogen to close to zero for elements such as sodium and molybdenum (Table 4.2). Nitrogen is the soil-derived essential element with the highest plant tissue level because of its ubiquitous importance in amino acids, proteins, nucleotides, nucleic acids, chlorophylls, and coenzymes. In particular, Rubisco is the most expensive and abundant nitrogen compound in plant tissues.

Table 4.2 Typical nutrient levels in grass tissues, adequate tissue levels (concentrations necessary to maintain vital functions) and critical concentrations (concentration in leaves below which reductions in maximum yield of 10% or more may be expected).

	Typical concentration in soil dry matter	Typical concentration in herbage dry matter: mean and range	Adequate tissue levels (Taiz and Zeiger 2002)	Critical concentrations (Whitehead 2000)	Plant:soil concentration ratio (x:1)
Macronutrients (%)					
N	0.3	2.8 (1.0–5.3)	1.5	2.5–3.2	10.0
P	0.2	0.4 (0.05–0.98)	0.2	0.12–0.46	2.0
S	0.1	0.35 (0.04–0.43)	0.1	0.16–0.25	3.5
K	1.5	2.5 (0.21–4.93)	1.0	1.2–2.5	1.7
Na	0.3	0.25 (0.00–0.39)	10	–	1.0
Ca	1.8	0.6 (0.03–2.73)	0.5	0.1–0.3	0.33
Mg	0.8	0.2 (0.03–0.79)	0.2	0.06–0.13	0.25
Micronutrients ($mg\ kg^{-1}$)					
Fe	35 000	150 (10–2600)	–	<50	0.3
Mn	1 600	165 (6–1200)	50	20	0.1
Zn	150	37 (3–300)	20	10–14	0.25
Cu	30	9 (0.4–214)	6	4	0.3
Cl	500	3500 (500–10 000)[a]	100	150–300	7.0
B	50	5 (1–94)	20	<6	0.1
Mo	2.6	0.9 (<0.02–17)	0.1	<0.1–0.15	0.35

Plant:soil levels and concentration ratios of nutrient elements for temperate grasslands from Whitehead (2000) Tables 3.4 and 3.6, based on >12 400 samples from temperate grasslands in Finland, Pennsylvania and New York (USA), and England and Wales.

[a] Source for Cl range: Barker and Collins (2003).

There is a wide range of tissue concentrations for each element depending upon the stage of maturity of the plant, species differences, seasonal and weather factors, soil type, and the application of fertilizers (Whitehead 2000). Plant maturity is considered to be the single factor with the greatest impact on forage quality, with the biotic and abiotic factors of the environment acting as an important modifier (Baxter and Fales 1994).

Adequate or minimal levels of the essential elements in plant tissues are well known (Table 4.2). These concentrations are necessary to maintain vital functions. The concentrations required for optimum growth, stand longevity, and drought resistance varies among species. The **critical concentration** refers to the concentration in the leaves below which reductions in maximum yield of 10% or more may be expected as a result of inadequate supply of the nutrient element. Much of what we know about critical concentrations in grassland plants is from agricultural work on the principal species in seeded or semi-natural temperate pastures, especially *Lolium perenne*. Critical concentrations for grasses are reasonably well established for the macronutrients, especially nitrogen, phosphorus, and potassium, and are discussed below, except for sodium which is not reported (Table 4.2). Critical concentrations for the micronutrients are less well known, but are discussed in Whitehead (2000). Knowledge of critical concentrations is used for making fertilizer recommendations in managed pastures and for crops. For example, critical concentrations for *Lolium perenne* were established as nitrogen 3.2%, potassium 2.8%, phosphorus 0.21%, sulfur 0.18%, and magnesium 0.07% (Smith *et al.* 1985). Tissue nitrogen levels range from 1.0 to 5.3% (Table 4.2) with actual concentration depending upon soil supply, species, and stage of maturity of the plant. The critical concentration for tissue nitrogen depends principally on stage of maturity. For example, with a 4-week regrowth, critical nitrogen concentration in *Lolium perenne* was 3.2–3.5%, whereas with a 6-week old regrowth it was only 2.5% (Whitehead 2000). Critical concentrations of tissue phosphorus vary from 0.12 to 0.46% in subtropical and tropical grasses, from 0.20 to 0.34% in cool-season grasses and from 0.22 to 0.30% in warm-season grasses (Mathews *et al.* 1998). Critical concentrations of tissue potassium have been reported as 1.6–2.5% for cool-season grasses and 1.2–2.0% for warm-season grasses depending on season and cultivar (Cherney *et al.* 1998). Critical concentrations for sulfur for young grass is 0.16–0.25% (Whitehead 2000). Grasses are rarely deficient in calcium or magnesium, but reported critical concentrations for *Lolium perenne* are 0.1–0.3% for calcium and 0.06–0.13% for magnesium, with the variation reflecting plant age. Management affects critical concentrations of nutrients as grass harvested frequently (e.g. through mowing or haying) or grazed may require higher levels of nutrients to maintain high yield levels.

Species differences in leaf tissue content are illustrated for four temperate grasses and the legume red clover *Trifolium repens* in Table 4.3. Nutrient element content was broadly similar among these temperate C_3 grass species, with no one species having a higher overall level of any one nutrient than the others. The plants in this study were grown as monocultures in field plots treated with NPK fertilizer (Fleming 1963). Tissue nutrient content exceeded minimal levels for most nutrients, but does suggest possible deficiency for some of the micronutrients, most notably boron for all the grasses. Similarity in tissue nutrient levels among co-occurring grasses is not always the case, and Cherney *et al.* (1998) report generally lower levels of tissue potassium in C_4 compared with C_3 grasses (1.2–3.8% vs 1.4–5.2%, respectively) because of the low soil potassium in many tropical regions. Co-occurring forbs have been reported to have higher tissue concentrations of some nutrient elements than grasses, but this varies depending on the particular mixture of the sward. As expected for a nitrogen-fixing legume, *Trifolium repens* in Fleming's (1963) study had almost double the leaf tissue of nitrogen compared with the grasses (Table 4.3).

Varietal or ecotypic differences may significantly affect tissue nutrient concentration, and attempts have been made to breed cultivars with increased concentrations of specific nutrient elements. For example, genetic variation for 8 of 17 essential elements and selectable heritability for tissue magnesium, sodium, and phosphorus, but not for potassium was demonstrated for a *Lolium perenne* breeding pool (Easton *et al.* 1997). In a comparison of

Table 4.3 Nutrient element concentration in leaf tissue (mean of three samples) of four temperate pasture grasses and red clover *Trifolium pretense*

Nutrient	Lolium perenne	Dactylis glomerata	Phleum pretense	Schedonorus pratensis	Trifolium pratense
N (%)	2.1	2.8	2.5	2.6	4.2
P (%)	0.32	0.32	0.13	0.3	0.3
K (%)	2.3	2.6	1.7	2.1	1.7
Ca (%)	0.87	0.57	0.88	0.87	2.1
Mg (%)	0.17	0.15	0.27	0.18	0.3
Fe (mg kg^{-1})	101	67	42	109	134
Mn (mg kg^{-1})	41	105	38	29	136
Zn (mg kg^{-1})	20	23	19	16	42
Cu (mg kg^{-1})	5.0	7.1	4.6	4.9	17
B (mg kg^{-1})	9	10	17	10	26
Mo (mg kg^{-1})	0.47	0.77	0.58	0.60	0.3

From Fleming (1963).

half-sib families of *L. perenne* in southern Australia, significant family variance for yield, magnesium, calcium, and potassium, and family by location interactions were observed for potassium and magnesium indicating genotype × environment effects on tissue mineral levels (Smith *et al.* 1999a).

Tissue mineral concentrations vary considerably among plant parts, both seasonally and depending on stage of maturity. Generally, nutrient concentrations are highest in young compared with older tissues, although the pattern of changes with age can be erratic for the micronutrients. Nitrogen can show a 70% reduction in concentration from early spring to midsummer, and phosphorus and sulfur concentrations can decline by 40–60%. Senescing leaves lose nutrients, especially nitrogen, phosphorus, sulfur, and potassium, through remobilization and/or leaching by rain. Calcium and micronutrient cations show less of a decline, or even a relative increase in concentration in senescing leaves, than the other macronutrients as they are relatively immobile (Whitehead 2000).

4.3 Secondary compounds: anti-herbivore defences and allelochemicals

In common with other plants, grasses produce a large array of secondary compounds that are not directly involved in the primary catabolic or biosynthetic pathways but play an important role in the ecological relationship of grasses with their biotic environment (Fig. 4.6). Many of these compounds are toxicants that affect the digestibility and nutritive value of grasses in forage, or affect herbivore physiology resulting in animal toxicity or stress. Other chemicals, and some of the same ones, are important allelochemicals involved in the stimulation or inhibition of neighbouring plants (see review by Sánchez-Moreiras *et al.* 2004).

The chemical compounds present in non-grass components of forage are not discussed here, except to note that many non-grasses produce toxins that negatively affect forage quality, e.g. plant oestrogens in subterranean clover *Trifolium subterraneum*; tannins in legumes; NO_3^- in the Amaranthaceae (pigweed), Chenopodiaceae (goosefoot), Brassicaceae (mustard), Asteraceae (sunflower), and Solanaceae (nightshade) families; and cyanogenic glycosides in *Prunus virginiana* and *Pteridium aquilinum* (Nelson and Moser 1994).

The secondary compounds of ecological importance in grasses are considered below under four major classes; nitrogen compounds, terpenoids, phenolics, and others, following Harborne (1977). Many of these compounds either play a role in allelochemical interactions or act as feeding deterrents.

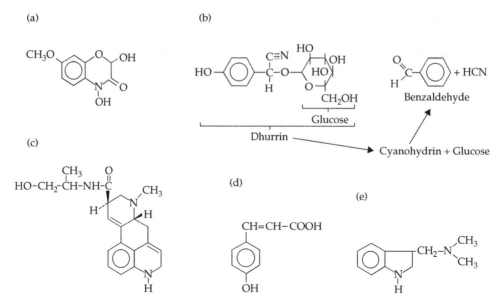

Figure 4.6 Some secondary compounds found in grasses: (a) DIMBOA (2,4-dihydroxy-7-methoxy-1,4-benzoxazin-3-one), a hydroxyamic acid; (b) dhurrin, a cyanogenic glycoside; during cyanogenesis, the molecule is hydrolysed to release hydrogen cyanide; (c) ergonovine, an ergot alkaloid; (d) paracoumaric acid, a phenol; (e) gramine, an indole alkaloid, derived from the aromatic amino acid typtophan. Reproduced with permission from Vicari and Bazely (1993).

4.3.1 Nitrogen compounds

Nitrogen-containing secondary compounds include the alkaloids, amines, non-protein amino acids, cyanogenic glycosoides, and glucosinolates. Glucosinolates are largely restricted to the Brassicaceae and are not considered further. The non-protein amino acid ornithine has been reported from wheat (*Triticum aestivum*), where it appears to act as a feeding deterrent to *Sitobion avenae* (grain aphid) (Ciepiela and Sempruch 1999); otherwise non-protein amino acids are more widespread in other plant groups such as the Fabaceae.

Alkaloids are based on one or more nitrogen-containing rings and include pharmaceutically important compounds such as nicotine, atropine, strychnine, and quinine. These compounds can have dramatic physiological effects on animals. Alkaloids are known from many grasses, especially the genera *Festuca*, *Lolium*, and *Phalaris*, particularly in seedlings and plants under stress. These chemicals act as feeding deterrents to grazers and allelochemicals. For example, indole alkaloids in *Phalaris arundinacea* (reed canary grass) reduce its palatability to sheep, although their concentration decreases with plant age. *Phalaris* staggers is a sometimes fatal condition in sheep due to alkaloids in the tissues of *P. tuberosa* (harding grass) and *P. aquatica* (bulbous canary grass) (Lee et al. 1956). The indole alkaloid gramine (Fig. 4.6) is produced in the youngest leaves of barley (*Hordeum vulgare*) when grown under long photoperiods at high temperatures and acts as a feeding deterrent to aphids and locusts (Ishikawa and Kanke 2000; Salas and Corcuera 1991). Gramine and hordenine have been extracted from the roots of *H. vulgare* act as allelochemicals reducing radicle length and general health and vigour in target species (Sánchez-Moreiras et al. 2004). Fungal endophytes when in mutualistic association with grasses (Chapter 5) produce ergot (Fig. 4.6c) and other alkaloids. Graminivorous insects, especially aphids, are deterred from feeding on endophyte-infected grasses because the alkaloid loline reduces insect growth, survival, and fecundity. Fescue toxicosis and ryegrass staggers are serious conditions affecting large herbivores including cattle and horses when feeding on endophyte-infected *Schedonorus phoenix* and *Lolium perenne* swards (Vicari and Bazely 1993). In these infected grasses, ergopeptine alkaloids accumulate in the leaves, especially the vasoconstrictive ergovaline and the tremorgenic neurotoxin lolitrem B (Gibson and Newman 2001).

Amines are another class of nitrogen-containing molecules synthesized by the decarboxylation of amino acids or by transamination of aldehydes. The plant hormone idoleacetic acid is a tryptamine (tryptophane derivative) and several plants contain aromatic aliphatic amines in their flowers, e.g. in the Araceae. Polyamines have been shown to accumulate in response to developmental and environmental signals (Bouchereau *et al.* 2000), including heat stress in rice (*Oryza sativa*) (Roy and Ghosh 1996) and salt stress in *Bromus* spp. (Gicquiaud *et al.* 2002).

Cyanogenic glycosides, which consist of a nitrile (triple-bonded CN), glucose (or other sugar), and a variable R-group as part of the glucosinolate pathway, occur in *c.*3000 species of higher plants in 110 plant families. Cyanogenic glycosides are hydrolysed through the removal of the glucose molecule by β-glucosidase to release hydrogen cyanide (HCN, prussic acid) when tissue is damaged, e.g. through maceration when grazed. Several forages contain cyanogenic glycosides including the sorghums *Sorghum* spp. (which posses dhurrin, Fig. 4.6), trefoils *Lotus* spp., and vetches *Vicia* spp. (Nelson and Moser 1994). Other grasses reported to contain cyanogenic glycosides include *Cynodon plectostachyus* (Vicari and Bazely 1993), *Dendrocalamus giganteus* (Ferreira *et al.* 1995), and *Pennisetum purpureum* (Njoku *et al.* 2004). Hydrogen cyanide is absorbed through the digestive tract of the grazing animal where it inhibits cytochrome oxidase in the respiratory electron transport chain. Levels of cyanogenic glycosides are highest in young leaf tissues and vary with plant age and growth conditions.

4.3.2 Terpenoids

The terpenoids (isoprenoids) are a large class of volatile secondary compounds (>25 000 compounds) based on five-carbon isoprene (C_5H_8) molecules connected together. Each isoprene unit has a 'head' and a 'tail' end, and isoprene blocks can be joined in many ways. Biosynthetically, terpenoids are derived from the geranyl pyrophosphate pathways, and include mono- (C_{10}), sesqui- (C_{15}), di- (C_{20}), and triterpenoids (C_{30}). The immense variety of terpene compounds that can be built from simple isoprene units include β-carotene (a vitamin), natural and synthetic rubbers, plant essential oils (monoterpenoids such as citral in lemon), the plant growth regulators, abscisic acid (a sesquiterpenoid), the gibberellins (diterpenoids), and steroid hormones (such as oestrogen and testosterone) in animals.

Terpenoids are not widespread in the grasses, although isoprene emissions of 0.02→40 μg g (LDW)$^{-1}$ h^{-1} are reported in *Arundo donax*, *Chusquea* spp., *Phragmites australis* and *P. mauritianum*, and *Triticum aestivum*, and monoterpenoid emissions of <0.08–0.11 μg g (LDW)$^{-1}$ h^{-1} in *Chusquea* spp., *Phragmites mauritianum*, *Secale cereale*, *Sorghum bicolor*, and *Triticum aestivum* (Kesselmeier and Staudt 1999; König *et al.* 1995). Emission of isoprene was reduced under elevated atmospheric CO_2 in *P. australis*; this is thought to be due to inhibition of the expression of isoprenoid synthesis genes and isoprene synthase activity (Scholefield *et al.* 2004). A variety of monoterpenes are reported in emissions from grassland vegetation. Examples include α- and β-pinene, myrcene, and limonene from a *Poa* spp. dominated grassland in the Midwestern USA (Fukui and Doskey 2000), and these plus α-thujene, camphene, sabinene, β-ocimene, and γ-terpinene were reported in emissions from an Austrian grassland (König *et al.* 1995). However, the significance of these emissions for grassland structure and composition is difficult to judge, as expected lifetimes with respect to oxidation by NO_3 are only a few minutes. Terpenoids and other volatile organic compounds are of global importance as their oxidized products are involved in the formation of tropospheric ozone and other photochemically produced oxidants.

The importance of terpenoids to grasses may be as an indirect defensive chemical (Turlings *et al.* 1990). Corn *Zea mays* seedlings were found to release monoterpenes in response to saliva of *Spodoptera exigua* (beet armyworm) larvae. The parasitic wasp *Cotesia marginiventris* was attracted to deposit its eggs into the caterpillars, thus defending the corn against further predation.

Sesquiterpenoid cyclohexane derivatives such as blumenin (e.g. blumenol C, 9-O-(2'-O-β-glucuronosyl)-β-glucoside, structurally similar to abscisic acid) are widespread in mycorrhiza-infected grass roots, most frequently in the tribes Aveneae, Poeae, and Triticeae (Maier et al. 1997). The reason for this fungus-induced accumulation of terpenoids is unclear; the chemicals may play a role in the physiological functions of mycorrhizal fungi (see §5.2.3), or could reflect a stress response to fungal infection by the plant. The highest levels of blumenin were reached 3–4 weeks after inoculation in the cortex of secondary roots of wheat. Blumenin strongly inhibits fungal colonization, suggesting that these compounds might be involved in mycorrhizal regulation by the plant (Fester et al. 1999).

4.3.3 Phenolics

Phenolics are three-ring system compounds constructed from a cinnamic acid derivative and three malonyl CoA molecules as an end product of the shikimic acid pathway. There are >500 types of phenolics in several structural classes including aurones, flavones, anthocyanins, and tannins (polyphenolics). Phenolics occur in all vascular plants. Tannins are important feeding deterrents in many plants, particularly woody plants, because of their astringency (reducing palatability) or indigestibility (protein-binding characteristics), but are not present in the Poaceae and so are not considered here.

The structurally simplest phenolics occurring in grasses are phenylpropanoids, e.g. caffeic, p-coumaric, ferulic, and sinapic acids are reported from the leaves of >70 species of grass (Harborne and Williams 1986). These phenolics occur bound to cell-wall hemicellulose (p-coumaric, ferulic acid), bound in the lipid fraction or linked to hydrocarbon alcohols, fatty acids, or glycerol (caffeic and ferulic acids), or in leaf wax (p-hydroxybenzaldehyde in *Sorghum bicolor*) (Harborne and Williams 1986). When present in leaf wax they are related to resistance to locust feeding. The familiar aroma of freshly mown hay is due to the release of coumarin (1,2-benzopyrone) following the hydrolysis of melilotic acid (dihydro-o-courmaric acid). Artificially synthesized in the laboratory since the late nineteenth century, courmarin is used to make perfumes and flavourings, and in the preparation of anticoagulants and rodenticide. Courmarin has been implicated to be involved in flowering, and inhibiting root development in *Hordeum*, tillering in *Saccharum*, and leaf elongation in *Triticum* (Brown 1981).

Quinones can be produced from phenols by enzymatic oxidation and are compounds with either a 1,4-diketocyclohexa-2,5-dienoid (i.e. p-quinones) or a 1,2-diketocyclohexa-3,5-dienoid (o-quinones) moiety (Leistner 1981). These phenolics do not appear to be widespread in the grasses. Sorgoleone. for example, is found in extracts and exudates of *Sorghum bicolor* where in agricultural fields it acts as an allelochemical reducing the growth of neighbouring weeds (Einhellig and Souza 1992; Sánchez-Moreiras et al. 2004). Mechanistically, sorgoleone acts as an inhibitor of photosynthetic electron transport (Nimbal et al. 1996).

Flavonoids occur in most angiosperm families as intensely coloured flower pigments, giving the orange, red, purple, and blue colours to many fruits, vegetables, flowers, leaves, roots, and storage organs. Chemically, flavonoids are polyphenolic compounds possessing 15 carbon atoms; 2 benzene rings joined by a linear 3-carbon chain.

The chemical structure is based on a C_{15} skeleton with a heterocyclic chromane ring (C) bearing a second aromatic ring (B) (Fig 4.7a). Ring C can occur in an isomeric open form or, more frequently, as a five-membered ring as shown in Fig 4.7a. Ring B is most frequently in position 2 of the heterocylic ring C, although in isoflavonoids it occupies position 3. Over 4000 flavonoids are classified according to the substitution patterns and oxidation state of the heterocyclic ring C, and the position of ring B.

The six major chemical subgroups of flavonoids are: chalcones (mostly intermediates in biosynthesis of other flavonoids), flavones (generally in herbaceous families, e.g. Labiatae, Apiaceae, Asteraceae), flavonol (generally in woody angiosperms), flavanone, anthocyanins, and isoflavanoids (limited mostly to the Fabaceae). Of these, anthocyanins and flavones are considered further below.

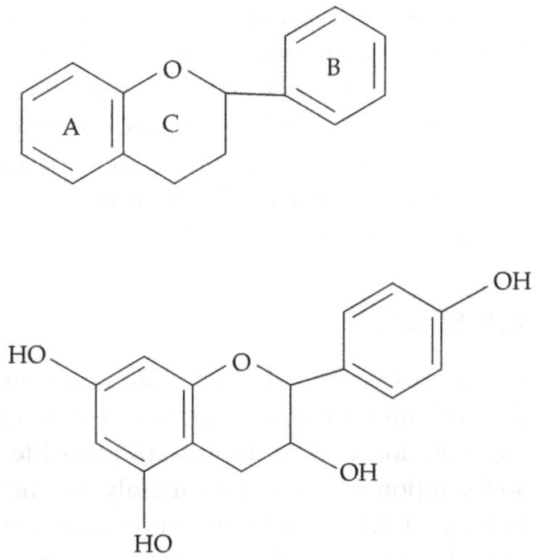

Figure 4.7 (a) The flavonoid skeleton; (b) the flavylium skeleton.

Flavones (e.g. apigenin, luteolin, arthraxin) are heavily oxidized structures based on the flavonoid skeleton (Fig. 4.7a) as both glycosides and aglycones (i.e. whether bound to sugar or not). C-glycoside flavones are characteristic of the grasses, occurring in 93% of species surveyed (Harborne and Williams 1986). Tricin was first isolated from a rust-resistant wheat variety, and has subsequently been found to be common throughout the grass family despite its rarity in other plant groups (Harborne 1967). Tricin extracts from brown rice bran are being investigated as a growth inhibitor of human breast and colon cancer (Cai et al. 2004; Hudson et al. 2000).

Anthocyanins are based on addition or removal of hydroxyl groups or by methylation or glycosylation of a flavylium (2-phenylbenzopyrilium) skeleton (Fig. 4.7b) (Escribano-Bailón et al. 2004). Anthocyanidins (aglycon) are formed when the flavylium skeleton is united with one or various sugars, which in turn can be acylated with an organic acid.

Anthocyanins are intensely coloured and the most important class of pigments responsible for pink, scarlet, red, mauve, violet, and blue colours in petals and leaves of higher plants (Harborne 1967). In grasses, this pigmentation is often obscured by admixture with plastid pigments (chlorophylls), leading to a brown, grey, or black appearance. Nevertheless, the anthocyanins commonly present in the angiosperms as a whole tend to predominate in the grasses too (i.e. cyanidin 3-glucoside—structurally the hydroxyl group in the C_3 ring in the flavylium skeleton (Fig. 4.7b), is replaced by a sugar).

In the Pooideae and Panicoideae, the anthocyanins contain aliphatic acyl groups (e.g. cyanidin 3-(3″,6″-dimalonylglucoside and other malonylyed cyanadin 3-glucosides), whereas in the Ehrhartoideae, Bambusoideae, and Arundinoideae acylated anthocyanins are generally absent and a different pattern of 3-glucosides of cyanidin and peonidin occurs (Fossen et al. 2002).

The anthocyanins in the cereals have been closely investigated (Escribano-Bailón et al. 2004). At least six different anthocyanins have been identified from the cob of purple corn, a pigmented variety of Zea mays. These anthocyanins occur in the epidermal cells and play a role in protection against UV-B radiation and inhibition of alfatoxin production of by the fungal pathogen Aspergillus flavus. Cyanidin-3-glucoside, peonodin-3-glucoside, and small quantities of several other anthocyanins have been identified from the husk of red and black rice grains. Commercially, the anthocyanin pigments are used as food colourants, and, in the field, they inhibit growth of Xanthomonas oryzae, one of the principal pathogens of rice. Similarly, the pigmentation of blue, purple, and red wheat (Triticum aestivum cultivars) is due to the accumulation of anthocyanins, including glucosides and rutinosides, and acylated derivatives of cyanidin and peonidin. Sorghum S. bicolor is one of the few monocotyledons that synthesizes antimicrobial phytoalexins in response to pathogen infection. The phytoalexins of sorghum are pigmented 3-deoxyanthocyanidins, which accumulate in cellular vesicles around the area in the cells under fungal attack.

The functional significance of flavonoids in grasses is also related to grazing deterrence, primarily for insects. For example, the wheat cultivar Amigo is avoided by the aphid Schizaphis graminum because of the presence of tricin and glycosylflavones in the leaves (Harborne and Williams 1986). Other grass flavonoids act as feeding attractants and can even be sequestered and stored in the insect.

For example, 1–2% of the body and wing dry weight of the butterfly *Melanargia galathea* can be sequestered flavones (hydrolysed tricin 7-glucoside). There is some evidence of feeding selection in mammals being due to phenolic levels in grass tissues, but apart from some weakly oestrogenic effect, flavonoids do not appear to affect animal reproduction (Harborne and Williams 1986). An exception might be the mountain vole *Microtus montanus*, which ingests high concentrations of *p*-courmaric and caffeic acids from *Distichlis stricta*, which inhibits winter breeding.

Flavonoids play an important role in the attenuation of solar UV-B radiation mitigating leaf damage (Bassman 2004). Synthesis of flavonoids and other phenolics is due to induced gene expression for the key biosynthetic enzymes. For example, synthesis of hydroxycinnamates (phenolics) was induced by UV-B in *Hordeum vulgare*. By enhancing secondary plant metabolites and thereby affecting plant–herbivore and plant–pathogen interactions, it is postulated that UV-B radiation is an important mediator of multiple trophic interactions in terrestrial plant communities (Bassman 2004).

4.3.4 Other compounds

Hydroxamic acids (4-hydroxy-1,4-benzoxazin-3-ones) are derivatives of the shikimic acid biosynthetic pathway and occur in a wide variety of grasses, including cereal crops, maize, barley, wheat, and some wild grasses (*Deschampsia* spp., *Elymus* spp., *Hordeum* spp., and *Phalaris* spp.), mainly in the Triticeae tribe (Gianoli and Niemeyer 1998; Vicari and Bazely 1993). The main compounds include DIMBOA (2,4-dihydroxy-7-methoxy-1,4-benzoxazin-3-one, Fig. 4.6) and DIBOA (2,4-dihydroxy-1,4-benzoxazin-3-one) inhibitors of mitochondrial metabolism and an inducible defence. Levels increase following leaf damage and deter herbivorous chewing and sap-sucking insects, and bacterial and fungal pathogens. Highest levels occur in seedlings, with concentrations declining with age. In perennials, concentrations vary seasonally, with maximum levels occurring in the summer coinciding with the peak of insect herbivory. Breeding programmes have been established to increase hydroxamic acid levels in cereal crops. The effectiveness of hydroxamic acids against vertebrate herbivores is unknown. DIBOA, DIMBOA, and other hydroxamic acids have been identified from root extracts as allelochemicals inhibiting seed germination and growth from *Elymus repens*, *Secale cereale*, *Triticum aestivum*, *T. speltoides*, and *Zea mays* (Sánchez-Moreiras *et al.* 2004).

4.4 Silicon

Silicon is the second most abundant element in the soil, and, in quartz and other minerals, the substrate for most of the world's plant life. In the soil solution silicon occurs mainly as silicic acid, H_4SiO_4, at 0.1–0.6 mM concentrations similar to that of potassium, calcium, and other major plant nutrients (Epstein 1994). Silicon is readily absorbed by forages with water as $Si(OH)_4$ where it can reach concentrations in the plant up to 10% of dry matter. The normal range in the foliage of forage plants is 400–10 000 mg kg^{-1} (Barker and Collins 2003). Plants which contain >1% silicon and have a silicon/calcium molecular ratio >1 are referred to as **silicon accumulators** (Ma *et al.* 2001), of which the grasses predominate. Soluble silicon concentration reaches a maximum at a soil pH of 8–9 (Prychid *et al.* 2003).

In the plant, silicon is irreversibly precipitated as amorphous silica (SiO_2-nH_2O or 'opal' or silica gel) in **phytoliths**. In grasses, phytoliths occur in the epidermis and subepidermal sclerenchyma of most species. Silica bodies are morphologically diverse in the Poaceae, including dumbbell-shaped, cross-shaped, saddle-shaped, conical-shaped, and horizontally elongated bodies with smooth or sinuous outlines. Silica bodies are abundant in some tissues and can occur as dust particles from plant fragments in the atmosphere where they have been found to be carcinogenic. For example, *Phalaris* species contain high levels of silica bodies in inflorescence bracts. These grasses are known contaminants of cereal crops of north-east Iran where there is also a high incidence of oesophageal cancer (Sangster *et al.* 1983).

Although silicon is not regarded as an essential element (§4.2.1), this view is increasingly being questioned (Epstein 1994, 1999; Richmond and Sussman 2003). The benefits of silicon to plants include improved resistance to pathogens and herbivory, maintenance of stem and leaf rigidity, reduced leaf transpiration through improved water use efficiency, and tolerance of drought and heavy metals (Richmond and Sussman 2003). The 'window' hypothesis suggests that epidermal silica bodies facilitate the transmission of light through the epidermis to the photosynthetic mesophyll or to stem cortical tissue, increasing photosynthesis and plant growth. However, this hypothesis does not appear to have been supported by experimental studies (Agarie et al. 1996; Prychid et al. 2003). Silica bodies in the leaf epidermis do, however, confer resistance to UV-B radiation damage as a result of an increase of phenolic compounds induced by silicon (Li et al. 2004).

The role of silicon in the ecology and co-evolution of grazed grassland ecosystems may be considerable (see Chapter 2). Studies in the African Serengeti showed high concentrations of tissue silicon (averaging 19.6% from a heavily grazed site, 11.9% from a less grazed site) that decreased through the growing season and from the roots to the leaves (McNaughton et al. 1985). In a laboratory experiment, silicon promoted growth of unclipped native grasses and increased leaf chlorophyll content. The authors suggested that natural selection for silicon accumulation was related to grazing exposure. Silicon acts as a defence against herbivory by facilitating abrasion of the mouthparts of herbivores, thus protecting vulnerable plant tissues. Silicon concentrations >2% can also lead to fatal silica urolithiasis in bovine calves. In addition, silicon deposition in tissues may improve the carbon and energy economy of grasses by releasing carbon that would otherwise go into secondary thickening and stiffening of the plant.

4.5 Physiological integration of clonal grasses and mechanisms of ramet regulation

The diverse modular morphology allowing tillering and clonal growth that characterizes many grasses affords the opportunity to forage and explore the local environment (Chapter 3). The extent to which ramets of grasses and other similar clonal plants are able to grow semi-autonomously while remaining physically attached to the mother plant reflects mechanisms of ramet regulation, especially the degree of physiological integration among modules. Morphologically there is a continuum of clonal growth forms ranging from 'phalanx' growth forms with tightly clumped tillers (e.g. bunchgrasses such as *Festuca ovina* and *Schizachyrium scoparium*) to 'guerrilla' forms with widely spreading rhizomes (e.g. *Cynodon dactylon*) (Lovett Doust 1981). No matter what the growth form, all or a subset of the rhizomes in a clone may constitute an **integrated physiological unit** (IPU) in which resources (nutrients, photosynthates, water) may be transferred back and forth. Integration among ramets in this manner allows for (1) equitable distribution of resources among ramets, (2) minimal inter-ramet competition, and (3) optimal efficiency of resource acquisition (Briske and Derner 1998). It has been suggested that high levels of physiological integration would allow compensatory growth following partial defoliation under grazing. However, support for the compensatory growth hypothesis was not found in a study of *Digitaria macroblephara* and *Cynodon plectostachyus* in the Serengeti (Wilsey 2002). Although severed rhizomes produced less biomass than unsevered rhizomes, supporting the idea that physiological integration may be important, this decrease was independent of defoliation. By contrast, physiological integration was found to help ramets of *Psammochloa villosa* survive burial in a sand dune habitat (Yu et al. 2004). Resources transferred from unburied ramets increased the ability of buried ramets to elongate and grow to the surface.

The degree of physiological integration was investigated in the clonal perennial *Panicum virgatum* (Hartnett 1993). Clones of *P. virgatum* growing in North American tallgrass prairie were subjected to combinations of nutrient addition, neighbouring plant removal, and rhizome-severing treatments. There was no effect of rhizome severing or interaction with the other treatments indicating that the integrated physiological unit was smaller than clone size or unrelated to the treatments. Thus, in *P. virgatum*, transfer of resources among ramets may either be short-term or limited to small subsets of ramets within the clone.

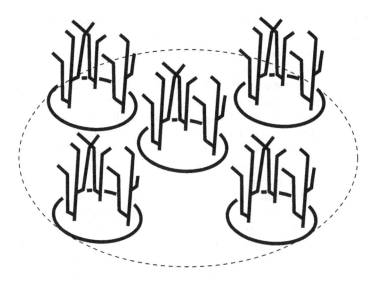

Figure 4.8 Clones of cespitose grasses organized as assemblages of autonomous ramet hierarchies, rather than as a sequence of integrated ramets. The benefits of physiological integration are restricted to individual ramet hierarchies (solid circles), while interhierarchical competition occurs for soil resources beneath the clone as a whole (dashed circle). Reproduced with permission from Briske and Derner (1998).

The study of *P. virgatum* (Hartnett 1993) outlined above, along with other studies (reviewed in Briske and Derner 1998) indicate that complete physiological integration among ramets in a clonal grass is normally limited to young plants. Rather, physiological integration is confined to sets of individual ramet hierarchies (groups of ramets) (Fig. 4.8). Part of the breakdown in physiological integration among ramets is due to disruption of vascular continuity among ramets in older clones. Mycorrhizal connections among roots of neighbouring ramet hierarchies may allow low levels of resource reallocation.

Ramet regulation and hence new tiller recruitment in clonal grasses is under complex physiological and ecological control. Tomlinson and O'Connor (2004) suggest an integrated model in which bud release for new tillers in controlled by the auxin:cytokinin ratio. Red:far-red light ratios (a consequence of the local light environment) affect production and export of auxin from shoots, whereas root nitrogen concentration (itself affected by soil resources) affects cytokinin production. Thus, shifts in local soil nitrogen and the shoot's light environment interact to control the production of new tillers.

CHAPTER 5

Population ecology

> There can be therefore little doubt that this plant generally propagates itself throughout an immense area by cleistogamic seeds, and that it can hardly ever be invigorated by cross-fertilization. It resembles in this respect those plants which are now widely spread, though they increase solely by asexual generation.
>
> Charles Darwin (1877), commenting on rice cutgrass *Leersia oryzoides*

Population ecology seeks to understand the relationship between groups of individuals of a single species in an area (a population) and their environment. The 'father' of plant population ecology was John L. Harper, following the publication of his *Population Biology of Plants* in 1977, and, indeed, much of his work was undertaken in a *Lolium perenne–Trifolium repens* sheep pasture in North Wales (e.g. Sackville Hamilton and Harper 1989; Sarukhán and Harper 1973; Turkington and Harper 1979). Many topics in the population ecology of grasses and grasslands are covered in the book edited by G.P. Cheplick (1998a) (see review in Gibson 1998). In this chapter, some of these important topics are explored. First, reproduction and population growth of grasses is discussed (§5.1), followed by a discussion of how fungi infecting grasses either above or below ground affect their population ecology (§5.2), and finally, ecotypes, polyploidy, hybridization, and genetic structure are covered under the heading of genecology (§5.3).

5.1 Reproduction and population dynamics

5.1.1 Flowering phenology

The timing of flowering (i.e. phenology) in grasses is under genetic and environmental control (particularly daylength). Classic experiments by Calvin McMillan (1956a, 1956b, 1957, 1959b) showed substantial ecotypic variation in flowering time for the dominant grasses across the North American Great Plains (see §5.3 for discussion of ecotypes). Yearly fluctuations in temperature and soil moisture can affect the timing of flowering, even in populations otherwise genetically constrained to a daylength response. A good example is the genetic differentiation and phenotypic plasticity in flowering phenology illustrated by the perennial bunchgrass *Notodanthonia caespitosa* in southern Australia. Flowering in southern populations growing in cool, moist temperate environments was closely tied into daylength responses related to a relatively predictable growing season. By contrast, northern populations growing in hot, semi-arid conditions are frequently exposed to daylengths exceeding the critical value for floral initiation so that once flowering tillers form, temperature and soil moisture determine the rate of reproductive development (Quinn 2000). Similarly, populations of *Danthonia sericea* in the north-eastern USA growing in wet habitats flowered in response to substrate temperature, whereas populations from well-drained upland soils required an additional photoperiod stimulus (Rotsettis et al. 1972).

Prior exposure to low temperatures is necessary to initiate flowering in many grasses, especially festucoid perennials from temperate zones, even under favourable daylengths. This requirement is referred to as **vernalization**, and is necessary every year in perennials. Vernalization temperatures range from −6 °C to *c.*14 °C. The vernalization requirement varies among species and cultivars within species. For example, within *Poa*, *P. alpina* and *P. pratensis* have an obligate vernalization requirement, *P. bulbosa* and *P. palustris* respond to vernalization, but *P. annua* shows no response (Evans 1964).

5.1.2 Anthesis and pollen dispersal

Lodicules in the floret inflate, forcing apart the palea and lemma, allowing the anthers to protrude and pollen to be shed. Generally, species are either morning or afternoon flowering, with some species being nocturnal. For example, Prokudin (in Conner 1986) reported flowering in *Poa* spp. from the same site at 8 a.m.–12 noon for *P. sterilis*, 5–8 p.m. for *P. nemoralis*, and 11 p.m.–12 midnight for five other *Poa* spp.

Grass pollen is predominantly **anemophilous** (wind pollinated). **Entomophily** (insect pollination) is restricted, for the most part, to a few tropical grasses (e.g. *Olyra, Pariana*; Soderstrom and Calderón 1971). For example, honeybees *Apis cerana* were observed to collect pollen and facilitate pollen release in the bamboo *Phyllostachys nidularia* (Huang *et al.* 2002). There are a few observations of entomophily in temperate grasses; for example, pollination by solitary bees (Halictidae) was observed to significantly enhance seed set in *Paspalum dilatatum* (Adams *et al.* 1981). In both cases, the pollen/ovule ratio was low and in *P. dilatatum* the pollen was larger than usual for anemophilous species and dispersed as clusters of grains.

Despite being predominantly wind dispersed, grass pollen often does not travel very far. More than >90% of the pollen was dispersed within 5 m in *Pennisetum glaucum*, and *Zea mays*, although in *Lolium perenne* and *Phleum pretense* >20% travelled >200 m (Richards 1990).

The pollen is captured among trichomes on the stigma, where the grain begins to hydrate immediately. Germination proceeds as the pollen tube emerges through the pore of the pollen grain and grows into the stigma. A vegetative nucleus is located at the tip of the pollen tube, followed by two sperm nuclei. Passage of the pollen tube nuclei to the ovary can be complete within an hour in wheat, or take as long as 22 hours in maize with its long silks (Chapman and Peat 1992). Double fertilization characteristic of the angiosperms involves union of one of the haploid ($1n$) sperm nuclei with the egg to form the diploid ($2n$) zygote. The other sperm nuclei fuse with the two polar nuclei to form the triploid ($3n$) endosperm. Exceptional instances of single fertilization in which the caryopsis has normal endosperm but no embryos are reported for wheat (Evans 1964).

Cross- and self-incompatibility

The control of compatibility in grasses is apparently unique and based on the rejection of self-pollen on the stigmatic surface. Control is enabled through two polyallelic, unlinked loci, $S_{1,2,3...}$, and $Z_{1,2,3...}$. A diploid grass possesses two S and two Z alleles, e.g. $S_1S_2Z_1Z_2$, whereas haploid pollen has one allele from each gene, e.g. S_1Z_2. There are tens of alleles for each gene, allowing hundreds of combinations of alleles of the two genes. Rejection of pollen occurs when alleles of the haploid pollen matches those of the diploid sporophytic stigma; the two must share no alleles to be compatible. Thus, a pollen parent $S_1S_2Z_1Z_2$, would give rise to pollen $S_1Z_1, S_1Z_2, S_2Z_1, S_2Z_2$, all of which would fail if landing on a style of genotype $S_1S_2Z_1Z_2$. By contrast, a pollen parent $S_1S_3Z_1Z_3$ would give rise to S_1Z_1, S_1Z_3, S_3Z_1, and S_3Z_3 pollen of which a quarter fails on the $S_1S_2Z_1Z_2$ style (i.e. S_1Z_1 genotype fails) (Chapman 1996). The incompatibility reaction is retained in polyploids (§5.3). Inhibition of incompatible pollen occurs after the pollen tube tip contacts the stigma, and can take place within as little as 2 minutes of first contact (Heslop-Harrison and Heslop-Harrison 1986).

5.1.3 Breeding systems

There is a wide array of breeding systems in the grasses, from hermaphroditism to dioecism. Perennial grasses are mostly cross-fertile and partially or totally self-fertile. Annual grasses are mainly self-fertile (Gould and Shaw 1983). In some genera there appear to be evolutionary trends from self-incompatible perennials to self-pollinated annuals (e.g. in *Bromus, Hordeum, Elymus, Agropyron, Festuca*, and *Poa*) (Stebbins 1957). Mechanisms to ensure cross-fertilization include late blooming of male spikelets compared with hermaphrodite spikelets in polygamous species in tribes such as Andropogoneae and Paniceae. Entirely hermaphroditic taxa may be proterogynous with the anthers protruding and releasing pollen before the stigmas are visible (e.g. *Alopecurus, Anthoxanthum, Pennisetum, Spartina*). Environmental conditions can modify the mating system, shifting the balance between self-fertilization and cross-fertilization. For example, frost can emasculate wheat and barley ensuring outbreeding (Evans 1964).

Some self-fertile grasses produce cleistogamous flowers in which pollination occurs in closed florets; e.g. *Sporobolus cryptandrus* where terminal spikelets remain partially or totally enclosed in the uppermost leaf sheaths. In *S. vaginiflorus* and some other annuals, lateral cleistogamous spikelets are produced late in the flowering season. Darwin (1877) noted the occurrence of cleistogamy in three grass genera (*Hordeum, Leersia,* and *Sporobolus*) in a survey of this phenomenon. Nowadays, cleistogamy is reported from 321 grass species from 82 genera in all the subfamilies, being most common in the advanced Panicoideae and in temperate zones of the New World (Campbell *et al.* 1983). The genus with the largest number of cleistogamous species is *Stipa* with 46.

A specialized type of cleistogamy is illustrated by the presence of dimorphic axillary cleistogenes (cleistogamous spikelets enclosed by leaf sheaths on the lowermost culm nodes) in at least 22 species from 10 genera: *Calyptochloa, Cleistochloa, Cottea, Danthonia* (7 species), *Enneapogon, Muhlenbergia, Pappophorum, Sieglingia, Stipa*, and *Triplasis* (Campbell *et al.* 1983; Chase 1908, 1918). In these species the axillary seeds are not dispersed, remaining within the leaf sheaths until the flowering tiller senescences. The resulting seed heteromorphism represents an ecological 'bet hedging' strategy for plants. For example, the large, relatively heavy seed produced from the low cleistogamous spikelets in *Triplasis purpurea* can become buried by sand close to the parent plant in the coastal habitat where this species occurs. By contrast, the smaller, 64% lighter, seed produced by chasmogamous spikelets in the upper culms is wind dispersed across the sand surface (Cheplick and Grandstaff 1997).

As an evolutionarily derived feature, cleistogamy has a genetic basis, but its occurrence is also influenced by environmental factors including photoperiod, soil moisture, and temperature. Cleistogamy as a breeding system offers an advantage of efficient and successful pollination, which can enhance fecundity (seed production). For example, plants of *Dichanthelium clandestinum* were noted to produce c.10 times the number of cleistogamous as **chasmogamous** (open-flowered) seeds with c.4 times greater recruitment of cleistogamous-derived seed in the field (Bell and Quinn 1985). An extreme form of cleistogamy is **amphicarpy** in which modified cleistogamous spikelets are borne on underground tillers, e.g. *Amphicarpum purshii, A. muhlenbergianum, Enteropogon chlorideus*. The resulting subterranean seed production in amphicarpic species can provide a safe haven from above-ground disturbances for the plants. For example, placement of seeds below ground in *A. purshii* provides protection from fire in frequently burned Pine Barrens habitats of New Jersey, USA (Cheplick and Quinn 1987). In this species, survival and fitness of seedlings derived from cleistogamous seed was much greater than seedlings from the aerial chasmogamous seed (Cheplick and Quinn 1983, 1987). The **mixed mating system**, in which seeds are produced from spikelets borne either above or below ground, helps maintain genotypic and phenotypic plasticity (Cheplick and Quinn 1986).

Asexual reproduction

Apomixis is reproduction in sexual structures without actual fusion of male and female gametes. Apomixis in grasses is either facultative or obligate. Facultative apomicts can produce offspring both sexually and asexually. Obligate apomicts have lost the capacity for sexual reproduction. An advantage of apomixis is that individuals can live to a great age and cover a large area (Richards 2003). Apomixis is well represented in the grasses (in 62 genera, Czapik 2000), as well as other angiosperms, especially the Asteraceae and Rosaceae. There are two classes of apomixis: **vivipary** in which progeny are produced other than by seed, and **agamospermy** in which reproduction is by seed. In apomictic vivipary, the reproductive structures (e.g. flowers, lemmas, paleas) are transformed or replaced by bulbils or bulblets; it is found in species of *Poa, Festuca, Deschampsia,* and *Agrostis*. This form of vivipary (i.e. **pseudovivipary**) is frequent in grasses of alpine and arctic habitats (e.g. *Poa* *alpina* var. *vivipara*, *Festuca viviparoidea*) some arid habitats (e.g.. *Poa bulbosa*, *P. sinaica*) (see annotated list in Beetle 1980), and environments of large spatial and temporal heterogeneity (Elmqvist and Cox 1996). This is in contrast to **true vivipary**, found most frequently in species such as seagrasses and mangroves inhabiting shallow marine environments, which involves sexually produced offspring with the embryo penetrating the fruit pericarp and growing to considerable size before dispersal (Elmqvist and Cox 1996). Proliferation is sometimes referred to as a form of vivipary but involves the conversion of spikelets above the glumes into a leaf shoot (see discussion in §3.3).

The ecological advantage of pseudovivipary in grasses as a reproductive strategy may be in allowing plants to reproduce quickly in the short growing season of arctic, alpine, and arid environments (Lee and Harmer 1980). These environments are coarse-grained, with parental patches offering the best opportunity for establishment of offspring in both time and space. Potential benefits of pseudovivipary include: the size advantage of bulbils allowing carryover of nutrients and organic matter to the next generation; continuous growth without changing form or energy loss from mobilization and redistribution of reserves for seed production; leaves being ready for assimilation immediately even before dispersal; avoidance of nutrient loss from pollen production; and a likely high establishment probability for propagules. Limitations are those for all forms of apomixis and include restricted genetic exchange and the potentially high energy costs in producing the sometimes large vegetative plantlets. Some species are semi-viviparous, producing both viable seed and viviparous propagules (e.g. *Festuca viviparoidea*).

There are two sorts of agamospermy:

- **Adventitious embryony** has been reported in *Poa* and occurs when the embryo arises in the ovule but outside the embryo sac in the nucellus or integument and the new sporophyte arises directly from sporophyte tissue without a gametophyte; alternation of generations is thus avoided.
- **Gametophytic apomixis** occurs when a $2n$ gametophyte is produced without reduction division and fertilization; e.g. parthenogenesis, the formation of an embryo by mitotic division of the egg cell without fertilization; and may (pseudogamous) or may not (autonomous) require pollination. If pseudogamy, then the male gamete is required for development of the endosperm and embryo, but does not fertilize the ovule.

Agamospermy is most frequent in the Paniceae and Andropogoneae compared with other grass tribes (Brown and Emery 1958); examples include *Poa artica*, *P. alpigena*, *P. alpina*, and *P. pratensis* which are parthenogenetic and pseudogamous (Evans 1946). Agamospermous apomicts of *Poa* species were found to be more successful than amphimicts (i.e. seed produced through fusion of male and female gametes) in more exposed terrain, and in cool environments with short frost-free seasons (Soreng 2000). It is postulated that apomicts are better able than amphimicts to migrate and colonize new areas because apomicts are not limited reproductively with the need to find a mate (pollen). Thus, apomicts appear to be more opportunistic and r-selected than their sexual counterparts.

Dioecy

The presence of two sex forms in populations, one pistillate and seed bearing (i.e. female), and the other staminate and pollen producing (i.e. male) is known as **dioecy** (dioecious from Greek *di* (two) + *oikos* (housed) + ous). Dioecious grasses occur with highest frequency in the continental New World, especially Mesoamerica. Dioecious genera occur in 5 tribes: Arundineae: *Gynerium* (1 dioecious species); Chlorideae: *Buchloe* (1), *Buchlomimus* (1), *Cyclostachya* (1), *Opizia* (2 or 3), *Pringleochloa* (1), *Soderstromia* (1); Eragrostideae: *Allolepis* (1), *Distichlis* (13), *Jouvea* (2), *Monanthochloe* (3), *Neeragrostis* (2), *Reederochloa* (1), *Scleropogon* (1), *Sohnsia* (1); Paniceae: *Pseudochaetochloa* (1), *Spinifex* (4), *Zygochloa* (1), Poeae: *Festuca* (400), *Poa* (500) (Connor et al. 2000). In most populations that have been observed the expected 1:1 male:female sex ratio occurs (Connor et al. 2000). The evolutionary advantage of dioecy appears to be one of an **outcrossing advantage** (avoidance of inbreeding) as niche specialization of males vs females or comparative differences in fitness appear to be lacking (Quinn 1991).

Nevertheless, spatial segregation of the sexes has been observed in some grasses, e.g. *Distichlis spicata* (more females in saline areas) and *Hesperochloa kingii* (more females at wetter sites) (Bierzychudek and Eckhart 1988). As noted above, this does not necessarily demonstrate niche partitioning among the sexes; alternative explanations, such as differential mortality, parental determination of offspring sex ratios, or environmentally determined sex change, may be more likely (Fox and Harrison 1981). For *Distichlis spicata*, sex is genetically determined and gender-specific bias in seedling survivorship appears to be a causal factor in determing the sex ratio in specific microsites (Eppley 2001).

5.1.4 Seed banks, seed dormancy and germination, seedling establishment

The morphology of the grass seed as it relates to dispersal and the resulting seedling were described in §3.7. It is important now to discuss how seeds accumulate in the soil, go through dormancy, germinate, and grow as seedlings.

The seed bank represents the accumulation of seeds in (on) the soil. Seeds (defined very generally to include both seeds and fruits in this context) of an individual species can form a transient seed bank if they live only until the first germination season following maturation, or form a persistent seed bank if they live until the second (or subsequent) germination season (Thompson and Grime 1979). For example, California annual grasslands were observed to possess only a transient seed bank. Few germinable seeds were carried over from one year to the next and the dominant grasses had virtually no carryover (Young *et al.* 1980). Persistent seeds banks, however, allow the seeds produced by the vegetation to build up in the soil. A survey of the literature revealed 61 genera and 99 species of grasses forming persistent seed banks at a density ranging from 1 (*Digitaria ischaemum, Hesperostipa comata, Hordeum vulgare, Sitanion hystrix*) to 9340 (*Zoysia japonica*) seeds m^{-2} (Baskin and Baskin 1998). Another survey revealed grasses as a group ranging in density from 7 (a mixed-grass prairie) to 18 050 (annual grassland) seeds m^{-2} (total seed bank densities 287–27 400 seeds m^{-2} respectively in these two grasslands) (Rice 1989).

A survey of 1725 seedbank studies (45% from grasslands) from north-west Europe recorded 130 of 241 possible grass species known for the region, with 2033 of 4237 (49%) records being for transient species (Thompson *et al.* 1997). Twenty-two grasses were included in the top 100 species ranked by number of records in the database of which *Holcus lanatus, Poa trivialis, Poa pratensis*, and *Festuca rubra* were ranked in the top 10 (3rd, 5th, 8th, and 9th, respectively). Seed density varied by species and depth of sampling. The 5 grasses with the highest mean density in the top 10 cm of soil were *Bromus hordeaceus* (18 110 seeds m^{-2}), *Glyceria fluitans* (17 957 seeds m^{-2}), *Lolium multiflorum* (11 200 seeds m^{-2}), *Poa trivialis* (6479 seeds m^{-2}), and *Alopecurus geniculatus* (6023 seeds m^{-2}).

Despite the apparent longevity of persistent seed banks, few grass seeds are still alive after 5 years of burial in the soil, although experimental studies have reported a small percentage of seed living for longer periods (e.g. 0.3–2.0% of *Poa pratensis, Setaria glauca, S. verticillata, S. viridis* and *Sporobolus cryptandrus* alive after 39 years; Toole and Brown 1946). Longevity of seeds of *Agrostis capillaris* has been reported as >40 years (Thompson *et al.* 1997). In nature, seeds are lost from the persistent seed bank through *in situ* germination, predation, pathogens, or ageing (Fig. 5.1). Mortality follows a Deevey type II or negative exponential curve (indicating little age-related mortality), whereas seed in transient seed banks exhibit a much more rapid initial mortality closer to a Deevey type III curve (Baskin and Baskin 1998). An experimental study in native bunchgrass prairie of Oregon, USA, showed 44–80% of seeds from four species dying within 1 year, with mortality due to three controlled treatments—senescence, disease or vertebrate herbivory—being low (Clark and Wilson 2003). It was postulated that the probable causative agents of most mortality were non-fungal disease (bacteria and viruses), invertebrate predation, competition, and abiotic constraints.

The seed bank represents the living 'memory' of the plant community and the species present include many that grew under previous environmental conditions, especially disturbances. As such, the seed bank provides a buffer against species fluctuations and the risk of local extinction,

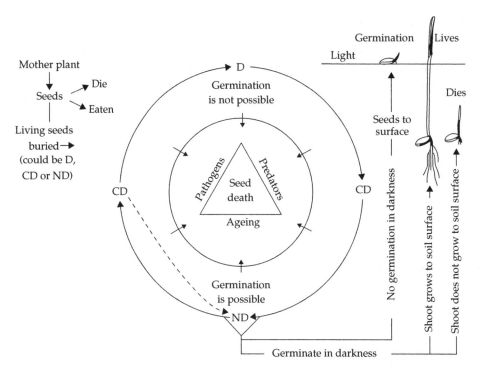

Figure 5.1 General scheme showing factors affecting persistence and depletion of buried seeds of grasses and possible annual changes in dormancy state. D, dormant; CD, conditional dormancy; ND, non-dormant. Reproduced with permission from Baskin and Baskin (1998).

and influences population recovery after disturbance. In undisturbed North American prairie, few of the dominant perennial grasses or forbs occur in the persistent seed bank (Abrams 1988). With a density of >6000 seeds m^{-2} in the top 12 cm of soil, the prairie seed bank is comprised mostly of short-lived opportunistic species exhibiting a 'sit-and-wait' strategy, awaiting suitable conditions for germination (Hartnett and Keeler 1995). In other grasslands, large differences between the composition of the seed bank and the vegetation are frequently observed, principally due to the large numbers of ruderal, 'weedy' taxa in the seed bank. Indeed, the dominant species in the vegetation can be absent from the seed bank. *Lolium perenne* and *Dactylis glomerata*, often dominant in pastures, are poorly represented in the seed bank, whereas *Agrostis* spp. and *Poa* spp. may have extensive seed bank reserves (Rice 1989). Generally, annuals are better represented in grassland seed banks than perennials, and forbs are better represented than grasses. The presence of ruderal species means that the composition of the seed bank can greatly influence the species composition in early succession following a disturbance. Regrowth from vegetative parts of the dominant species (the bud bank) can, however, be more important than recruitment from the seed bank (Benson *et al.* 2004; Virágh and Gerencsér 1988). Nevertheless, the composition of grassland seed banks is generally reflective of disturbances and management regimes (e.g. Kalamees and Zobel 1998; López-Mariño *et al.* 2000), such as rate of nitrogen addition (Kitajima and Tilman 1996). The similarity between the seed bank and vegetation can be complex. For example, a study of Mediterranean grassland showed that the similarity between the vegetation and seed bank declined with increasing altitude but was unrelated to topography and grazing (Peco *et al.* 1998). In species-rich grassland, the number of species represented in the seed bank can be few relative to the vegetation, indicating that most of the species are transient or that seed production is low (Kalamees and Zobel 1998; O'Connor and Everson

1998). In wet semi-natural grassland in Sweden, a species-rich seed bank appeared to contribute little to regeneration of the grassland (Milberg 1993). Seedling establishment was rare in undisturbed vegetation, occurring most frequently in gaps from the recent seed rain. Either way, the number of species in the vegetation is generally positively related to the number of species in the seed bank.

The use of the seed bank in grassland restoration has been suggested. However, although some species may be recruited into restoration from the seed bank, others may have to be naturally or artificially introduced for a successful restoration (Bekker et al. 1997; Milberg 1995). For example, Bossuyt and Hermy (2003) reviewed data from 16 separate studies and noted a decline in total seed density with increasing age since abandonment of traditional grazing or hay-cutting management of European grassland and heathland communities, especially calcareous or alluvial grasslands. Seeds of target species declined with time, and non-target ruderal or competitive agricultural species increased. Similar observations were made by Hutchings and Booth (1996) in a study of calcareous grassland. By contrast, the accumulation of seed in a persistent seed bank can make the eradication or control of invasive species difficult. For example, *Microstegium vimineum* is an annual grass of Asian origin invasive in the forest understory of the eastern USA. It has been estimated to possess a persistent seed bank of 64 seeds m^{-2} in the top 5 cm of soil under patches of flowering individuals (Gibson et al. 2002).

Seed dormancy and germination
Except where attribution to other authors is indicated, the examples of species exhibiting different dormancy and germination characteristics in the following discussion are selected from the longer lists provided by Baskin and Baskin (1998). Seed dormancy is the inability of fresh mature seed to germinate and is very common in the Poaceae. Grass seeds pass through a transition of stages as they go from **dormancy** (D) through **conditional dormancy** (CD: germinating over only a portion of the range of conditions possible for particular species) to **non-dormancy** (ND) (Baskin and Baskin 1998). Of the various types of dormancy exhibited by plants, grass seeds have non-deep **physiological dormancy** (PD), i.e. dormancy can be broken or stimulated by dry storage at room temperature, mechanical injury and/or removal of palea and lemma, or gibberellic acid (GA$_3$). Grasses thus lack morphological or physical dormancy. Some grasses are stimulated to germinate by any one of these conditions (e.g. *Schizachyrium scoparium*), whereas others respond to only one of these conditions (e.g. *Cynodon dactylon* has PD broken only by dry storage). For a grass seed to become part of the persistent seed bank it must be in a state of innate dormancy or enter secondary dormancy. In some grasslands there is little evidence of either, e.g. perennial grasses of the sub-Saharan African savannahs (O'Connor and Everson 1998). Seed of other grasses that are non-dormant when fresh include some winter annuals and summer annuals, and perennials, e.g. *Avena fatua, Milium effusum* (Baskin and Baskin 1998).

Dormancy-breaking requirements in grasses fall into two categories: high summer temperatures or low winter temperatures (Baskin and Baskin 1998). High summer temperatures are required to break dormancy in winter annuals in temperate or Mediterranean climates, and perennials in Mediterranean or tropical-temperate climates, e.g. *Poa annua, Bromus tectorum, Aira praecox, Vulpia bromoides*. Low temperatures are required for summer annual grasses and perennial grasses in temperate regions with dormancy breaking during cold stratification. When first under conditional dormancy, germination is possible only at high temperatures (autumn temperatures are too low). Later, when non-dormant, winter germination occurs under a wide range of temperatures; example perennials include *Dactylis glomerata, Festuca ovina, Panicum virgatum, Poa pratensis*, and *Sorghastrum nutans*, and annuals include *Setaria glauca* and *S. viridis*.

When in a non-dormant state, germination requirements depend upon an often complex interaction of environmental factors. The principal environmental factors are temperature (constant or alternating), light:dark regime, and soil moisture. Mean (± SE) optimum temperature for grass seeds requiring high summer temperature to break dormancy is 16.2 ± 1.1 °C for winter annuals, 27.2 ± 2.0 °C for perennials, and 29.9 ± 0.1 °C for summer annuals. For grass seeds requiring low winter temperatures to come out of dormancy the optimum germination temperatures are 24.8 ± 1.1 °C for summer annuals and 23.5 ± 1.1 °C for perennials (Baskin and Baskin 1998). Some grass seeds require alternating temperatures for germination (e.g. *Agrostis capillaris, Dactylis glomerata, Lolium perenne*). For some, alternating temperatures allow germination in darkness (e.g. *Deschampsia caespitosa, Poa pratensis*). The germination response to alternating temperatures is viewed as a soil depth-detecting mechanism and a canopy gap-detecting mechanism (Thompson et al. 1977). This response occurs because fluctuations in temperature decrease with increasing soil depth or increasing canopy cover. A seed that germinates under alternating rather than constant temperature is thus more likely to be at a shallow depth in the soil and/or under a canopy gap. In either case, germinating under these conditions will enhance the probability of survival of the seedling.

The light:dark requirements for germination are controlled by the phytochrome system. Red (R) light usually promotes germination, whereas far-red (FR) light inhibits it. A low R:FR ratio is experienced under light filtered through green canopies, and reduces the germination of some seeds (e.g. *Apera spica-venti, Elymus repens, Lolium pratense, Poa pratensis, Setaria verticillata*) (Baskin and Baskin 1998), it may act to prevent germination of these species under closed, shady grassland canopies. Many grasses germinate better in light than in the dark (e.g. *Leptochloa filiformis, Muhlenbergia schreberi*), although some require darkness (e.g. *Bromus sterilis*). As with alternating temperatures, the inhibition of germination under leaf canopies acts as a gap-detecting mechanism that has been proposed to allow coexistence of species in grasslands. Experimental work in calcareous grasslands has supported this view (Silvertown 1980a).

Soil moisture level is important for germination as the imbibition of water by the seed is the first step towards germination. Water stress resulting from dry conditions reduces and inhibits germination of grasses. The mean ± SE soil moisture stress reducing germination from 80–100% to 50% in grasses is −0.78 ± 0.09 MPa, with arid land grasses such as *Bouteloua eriopoda* able to maintain germination >50% down to −2.03 MPa. By contrast, some grass seed (e.g. *Zizania palustris, Spartina anglica*—both wetland species) is recalcitrant, dying if its moisture content drops below 20–45%, and in others high moisture content experienced during flooding prevents germination (e.g. *Oryzopsis hymenoides*, Blank and Young 1992). Water stress interacts with temperature and light, controlling germination. For example, germination of *Bromus sterilis* is inhibited by far-red (PFR) light, but only at temperatures <15 °C. Moreover, under low moisture conditions, the inhibitory effect of PFR is increased (Hilton 1984). Thus, light inhibition and moisture stress may delay germination of freshly shed seeds in the autumn, with low temperatures and moisture stress preventing germination until the following spring.

Other factors that affect germination include aspects of the soil chemical environment including soil pH, minerals (especially NO_3^-), salinity, heavy metals, and organic compounds (Baskin and Baskin 1998; Fenner and Thompson 2005). Of particular relevance in fire-prone grasslands is the promotion of seed germination by smoke. As a germination cue, the effects of smoke vary among species. In one study, percentage germination was significantly increased in 8 of 20 species tested (*Austrostipa scabra, Chloris ventricosa, Dichanthium sericeum, Panicum decompositum, P. effusum, Paspalidium distans, Poa labillardieri,* and *Themeda triandra*) (Read and Bellairs 1999). In a similar study, 16 of 22 species tested showed a significant response to either smoke or heat (species showing a positive response to smoke were: *Austrodanthonia tenuior, Eragrostis benthamii, Entolasia leptostachya, Panicum effusum,* and *P. simile* (Clarke and French 2005). These studies suggest that one way in which altered fire regimes change community structure is through stimulation of different species through the effects of smoke and heat on germination.

Seedling establishment

Grass seedlings (see §3.7) face multiple challenges affecting survival and growth. As a result, mortality is usually high, reflecting the small size, soft palatable tissues, and limited reserves. Mortality may result from one of many factors, both abiotic (burial, low light, flooding, drought, fire) and biotic (predation, fungal, bacterial and viral disease, mycorrhizae, intra- and interspecific competition) (Kitajima and Fenner 2000) (Table 5.1). For example, mortality of 24 seedling cohorts of *Bromus tectorum* observed over a 3 year period was principally due to desiccation in the first few months after emergence (Fig. 5.2) (Mack and Pyke 1984). However, at other times in the season, mortality due to disease (smut *Ustillago bullata*), frost-heaving, and herbivory by voles became important, varying in magnitude among sites.

The importance of seedling establishment as a regenerative process in grasslands is variable, depending upon the system. Seedling regeneration can be important following large-scale disturbance such as drought or fire (Glenn-Lewin *et al.* 1990) and locally on small-scale disturbances such as animal burrow systems (Rabinowitz and Rapp 1985; Rapp and Rabinowitz 1985; Rogers and Hartnett 2001a) (see Chapter 9). Catastrophic disturbances such as the Great Drought of 1934 in the US Midwest killed many prairie dominants, allowing extensive seedling recruitment of drought-resistant species such as *Pascopyrum smithii* and ruderals including the grass *Bromus secalinus* and forbs *Conyza canadensis, Lepidium virginicum,* and *Tragopogon lamottei* (Weaver and Albertson 1936). The importance of seedling regeneration in closed, mature, undisturbed grasslands is less certain. Much of the older, 'classic' work on North American prairies provides detailed descriptions of seedlings and the phenology of seedling establishment while also noting the low density of seedlings (Weaver and Fitzpatrick 1932). Even when seeds were artificially sown, seedling density was low and mortality very high due to summer drought, winter frost, or insect herbivory (Blake 1935). More recent work has confirmed seedling regeneration of the dominant species to be a rare and localized event in undisturbed prairie (Benson *et al.* 2004). Significant numbers of seedlings emerge and become established only when openings are created or maintained, and then in moist years (Blake 1935; Glenn-Lewin *et al.* 1990).

Seedlings require a 'safe site' (Harper 1977) that is often different in microhabitat characteristics from that of the adult plant. Appropriate safe sites for seedlings are often rare in grasslands because of the competitiveness of the existing sward (Defossé *et al.* 1997b). Many studies have characterized the safe sites or microsites suitable for seedling establishment in grasslands and these generally involve a temporal or spatial release from competition for resources such as light, moisture, and nutrients (Bisigato and Bertiller 2004; Defossé *et al.* 1997b; Dickinson and Dodd 1976; Romo 2005) or herbivory (Edwards and Crawley 1999). As noted already for North American prairies, gaps

Table 5.1 Fine-scale, primary factors and associated larger-scale environmental determinants affecting seedling establishment of grasses

Primary factors	Primary determinants
Abiotic factors	
Local edaphic conditions (moisture, nutrients, aeration, etc.)	Rainfall patterns and distribution; topography; soil type
Light	Daylength; vegetation structure/type; disturbance levels
Temperature	Regional climate; vegetation structure/type
Biotic factors	
Competition	Population density; vegetation structure/type; disturbance levels
Neighbours (e.g., allelopathy, nurse plant effects)	Population density; vegetation structure/type
Herbivory	Insect/mammal abundance; structural and chemical defences
Litter	Vegetation structure/type; decomposition rates; fire frequency
Pathogens/mutualist/other symbionts	Structural and chemical defences; environmental conditions (moisture, temperature, etc.); availability of inocula
Maternal effects (e.g. seed size heteromorphism)	Environmental conditions present during seed maturation; position of seed maturation on maternal plant

With permission from Cheplick (1998b).

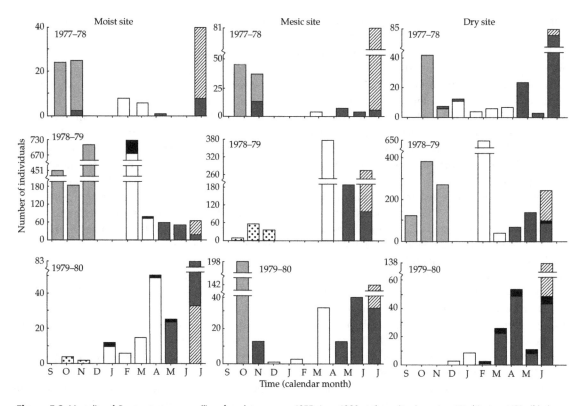

Figure 5.2 Mortality of *Bromus tectorum* seedlings from late summer 1977–June 1980 at three sites in eastern Washington, USA. ([dark grey], desiccation; ([hatched]), smut; ([black]), grazing; ([white]) winter death; ([light grey]), unknown. Reproduced with permission from Mack and Pyke (1984).

can provide the opportunity for seedling regeneration. The size, duration, timing of gap formation, and interactions with other environmental factors such as grazers (Defossé *et al.* 1997a), all interact to affect seedling regeneration success in gaps (Bullock *et al.* 1995). In addition, the propagule rain may limit recruitment into safe sites (e.g. Edwards and Crawley 1999; Foster *et al.* 2004). As noted by Rapp and Rabinowitz (1985) studying seedling regeneration on to small-scale disturbances in prairies, seedling regeneration is individualistic with establishment patterns differing among species. Badger disturbances in tallgrass prairie, for example, have been shown to provide sites for the establishment of distinct suites of 'fugitive' species unable to establish in otherwise undisturbed prairie (Platt 1975).

Grime and Hillier (2000) recognize five regenerative strategies for plants, of which the first four describe successful seedling establishment in specific circumstances:

- **seasonal regeneration** (S): independent offspring (seeds or vegetative propagules) produced in a single cohort
- **persistent seed or spore bank** (B_s): viable but dormant seeds or spores present throughout the year, some persisting >12 months
- **numerous widely dispersed seeds or spores** (W): offspring numerous and exceedingly buoyant in air; widely dispersed and often of limited persistence
- **persistent juveniles** (B_{sd}): offspring derived from an independent propagule but seedling or sporeling capable of long-term persistence in a juvenile state
- **vegetative expansion** (V): new vegetative shoots attached to parent plant at least until established.

It is argued by Grime and Hillier (2000) that the importance of the five regeneration strategies will vary among different types of habitat, for example with W (production of numerous widely dispersed seeds or spores) being the predominant regenerative strategy in early successional habitats whereas persistent seed banks (B_s) were predicted to be more important following disturbance in coppice and heathland habitats. These ideas were tested for the three seed regenerative strategies (S, B_s, and W) in habitats around Derbyshire, UK (Table 5.2). Of interest is the important role of both seasonal regeneration and the persistent seed bank (B_s) in the grasslands compared with early succession habitats where regenerative strategy (W) is more important and wooded habitats where seasonal regeneration (S) is more important. Other investigators have taken a different approach to classifying regeneration strategies in grasslands. For example, seed size appears to be a relevant life-history trait important in understanding the regeneration strategy of species in British calcareous grassland (Silvertown 1981) with large-seeded species (1.0–3.0 mg) germinating in the spring when competition from established species for water and nutrients is high. By contrast, small-seeded species (0.01–1.0 mg) germinate in the autumn when competition is less.

Table 5.2 Relative importance (%) of the regenerative strategies, S (Seasonal regeneration), B_s (Persistent seed or spore bank), and W (Numerous dispersed seeds or spores) in established vegetation of neighbouring habitats in Lathkill Dale, Derbyshire, UK

Habitat	S	B_s	W
Cliffs	38	30	18
Quarry heaps	44	40	18
Calcareous grassland	50	45	5
Acid grassland	49	51	4
Unmanaged calcareous grassland	63	31	0
Scrub	52	21	1
Deciduous woodland	73	30	1

Note: percentages do not sum to 100 because the three categories are not mutually exclusive, and some species could not be classified.

From Grime and Hillier (2000).

5.1.5 Population dynamics

Births, deaths, immigration, and emigration regulate density in a population (Harper 1977). While the total number of individuals may remain fairly constant in a population, there is considerable flux in the number of individuals (Fig. 5.3). Here the implications for studying growth in grass populations are discussed.

Understanding population change can be accomplished using recruitment curves and calculating population growth rate. A **recruitment curve** represents a plot of population size (N) at the next census in the future (i.e. N_{t+1}) against current population size (N_t) for populations governed by density-dependent mortality (Silvertown and Charlesworth 2001). Any population falling on the diagonal represented by $N_{t+1} = N_t$ is at equilibrium. The slope of the recruitment curve (N_{t+1}/N_t) gives the value for **lambda** (λ), the annual (finite) rate of population increase. For example, Watkinson (1990) conducted a 9 year study of the annual grass *Vulpia fasciculata* in two dune systems in North Wales. Over that time, the density of *V. fasciculata* declined. Recruitment curves indicated a negative density-dependent relationship between fecundity and density consistent through time, despite declining overall population density. Finite population growth rates calculated from population estimates in consecutive years indicated a strong positive relationship with percentage cover of bare sand (Fig. 5.4a), which was used to calculate equilibrium population size expected at various levels of sand cover (Fig. 5.4b). Populations of *V. fasiculata* could only be expected to persist where bare sand exceeded 50%. A similar study on the annual grass *Sorghum intrans* again showed a negative density-dependent relationship with fecundity mediated by environmental factors (Watkinson *et al.* 1989). For example, in local areas of poor growth potential, fecundity was too low to sustain the populations in the absence of immigration of seed from outside areas; a rescue effect (*sensu* Hanski and Gyllenberg 1993).

The age- or stage-structure of a population and the probability of individuals in a population changing over time to a different age- or stage state/class (e.g. from a 1 year old plant to a 2 year old plant,

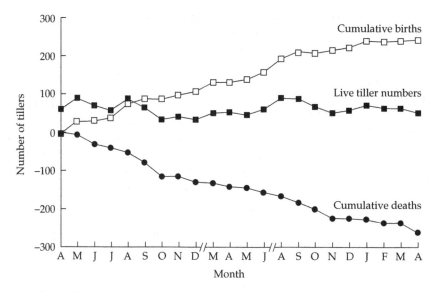

Figure 5.3 Population flux in *Lolium perenne* (number of tillers). Reproduced with permission from Silvertown and Charlesworth (2001).

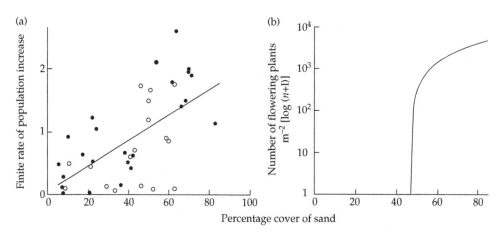

Figure 5.4 The relationship between (a) the finite rate of population increase of *Vulpia fasciculata* and sand cover at two dune systems in North Wales, UK (open and closed symbols) (fitted regression, $r^2 = 0.40$, $p < 0.001$), and (b) estimated equilibrium population density of *V. fasciculata* and sand cover. Reproduced with permission from Watkinson (1990).

or from a seedling to a vegetative tussock, respectively: see Fig. 3.4) can be incorporated into a projection (transition) matrix (Gibson 2002). From this matrix, λ can be estimated by comparing the ratio of the numbers in any one age/size class with the numbers in that class at the previous time interval. An exact estimate for λ can be calculated at equilibrium when a stable state structure is reached as the dominant eigenvalue of the matrix. These types of models have proved useful in understanding the population dynamics of important grassland species in response to environmental factors (e.g. O'Connor 1994) and in allowing the development of best management practices to control the spread of invasive grassland weeds (Magda *et al.* 2004).

The contribution of different parts of the plants life cycle to λ can be calculated as elasticity (e_{ij}) the proportion of λ due to an individual age/stage transition (de Kroon et al. 1986). Thus, elasticity provides a relative measure of a matrix element's contribution to fitness. Matrix model studies of this type show that population growth in grasses can be sensitive to particular age or stage states depending on the species and environmental setting (O'Connor 1993). Year-to-year variation in transitions can markedly affect long-term population growth rates. For example, in *Danthonia sericea* temporal variability affected recruitment of individuals from small into large size classes, with the demographic success of the latter significantly affecting λ (Moloney 1986). Elasticity matrices can be decomposed to compare matrix coefficients that are dominated by different demographic processes, i.e. S (stasis, staying in the same age or stage class from one transition period to the next), G (growth, transitions to larger size-classes), and F (fecundity, reproduction via seed) (Silvertown et al. 1993). For example, in a comparison of six African savannah grasses, high population growth rates were frequently associated with stasis or growth (O'Connor 1994). The most important contributions to population growth rates, however, generally involved the smallest size classes. Environmental variability greatly affects the drivers of population growth, for example in dry years λ in *Hilaria mutica*, a semi-arid tussock grass, was most affected by stasis and retrogression, whereas in moister years fecundity became more important (Vega and Montaña 2004). Fire regime in grasslands affects λ, which in *Andropogon semiberbis* in tropical savannah was in turn predominately affected by small size-class transitions, more so in the absence of fire (61% elasticity) than under burned conditions (50%) (Silva et al. 1991). However, because in reality all these coefficients combine at least two different demographic processes or vital rates (e.g. stasis and growth), one should calculate the elasticity of the vital rates implicit in each matrix coefficient (Franco and Silvertown 2004). A comparison of perennial grasses on this basis shows the importance of these elasticities affecting λ (Table 5.3). Stasis (S) dominates these comparisons because of the importance of the longevity of the clumps of these grasses compared with growth (G). The contribution of the vital rates reflecting fecundity (F) to λ was minor.

The contribution of life-cycle stages to population growth rates in annual grasses is poorly known. However, studies of other annuals suggest an important role for longevity of seed in the seed bank, the transition of individuals from seed to adult plants (Kalisz and McPeek 1992), and fecundity (Fone 1989). Seed survival through the winter, fecundity, and the proportion of seeds escaping predation were important drivers of λ in agricultural populations of the annual grass

Table 5.3 Life-history characteristics of some perennial grasses: finite rate of population increase (λ), lifespan (L: expected age at death), age at sexual maturity (α: average age at which an individual enters a stage class with positive fecundity), net reproductive rate (R_0: average number of offspring produced by an individual over its life span), generation time (μ: mean age at which members of a cohort produce offspring), and elasticities (G: growth, F: fecundity, and S: stasis)

Species	λ	L	α	R_0	μ	G	F	S
Andropogon semiberberis	1.25	8	2.0	2.59	5.0	0.19	0.14	0.68
Aristida bipartita	1.34	6	1.0	6.31	14.7	0.07	0.21	0.72
Bothriochloa insculpta	1.05	11	1.0	1.71	12.7	0.06	0.10	0.84
Danthonia sericea	1.20	41	2.2	15.01	32.7	0.18	0.08	0.74
Digitaria eriantha	1.27	7	3.6	5.81	11.2	0.26	0	0.74
Elymus repens	2.96	6	1.0	25.84	15.5	0.44	0	0.56
Heteropogon contortus	0.91	14	1.0	0.49	6.6	0.19	0.07	0.74
Setaria incrassata	0.94	18	3.9	0.66	5.8	0.27	0	0.72
Swallenia alexandrae	1.00	29	4.0	0.90	28.0	0.03	0.03	0.98
Themeda triandra	1.14	52	2.0	0.94	16.5	0.22	0.06	0.72

Extracted from Franco and Silvertown (2004).

Setaria faberi (Davis *et al.* 2004). In the annual grass *Poa annua*, mid-season reproduction was found to exert a dominant effect on λ; however, the importance of delayed germination and delayed reproduction were shown to increase substantially under increasing density stress (van Groenendael *et al.* 1994). By contrast, plant survival was found to be more important than seed survival affecting λ in populations of the annual savannah grass *Andropogon brevifolius* (Canales *et al.* 1994).

5.2 Fungal relationships

The population biology of grasses and their contribution to plant communities is significantly affected by pathogens. The effect of fungal diseases is usually quite apparent, reducing the survival, growth, and reproduction of host grasses (§5.2.1), but the effects of fungal symbionts are less obvious. In particular, two types of fungi form symbiotic relationships with grasses that can be mutualistic, namely above-ground fungal **endophytes** (§5.2.2) and below-ground fungal **mycorrhizae** (§5.2.3).

Other mutualistic microbes associated with the rhizosphere or phyllosphere of grasses that significantly affect fitness and competitive interactions include non-mycorrhizal fungi, soil bacteria, and viruses (Bever 1994; Malmstrom *et al.* 2005; Westover and Bever 2001). Species of *Azospirillum* bacteria and at least 11 other nitrogen-fixing bacterial genera have been found to be associated with the rhizosphere of several tropical and subtropical grasses, including wheat, sorghum, sugar cane, and maize (Reinhold *et al.* 1986), and some temperate grasses (e.g. *Agrostis stolonifera*, *Calamagrostis lanceolata*, *Elytrigia repens*, and *Phalaris arundinacea* Haahtela *et al.* 1981). These interactions are referred to as **associations** or **diazotrophic rhizocoenoses** and can significantly impact the nitrogen economy of the grass (Baldini and Baldini 2005).

5.2.1 Fungal diseases

Grasses host many important diseases such as rusts, smuts, and ergots that are well known in the major turfgrasses and forage grasses (Couch 1973; Tani and Beard 1997) (Table 5.4). These fungi can have a major impact on the physiology, chemical composition, and population ecology of the grasses, as well as affecting grassland communities and ecosystems, especially when they impact the dominant species (Burdon *et al.* 2006). For example, the flower smut *Ustilago cynodontis* (Ustilaginaceae) sterilized the clonal grass *Cynodon dactylon* by replacing floral structures with fungal stroma, and reduced growth rate, root–shoot ratio, and survival of the grass when in mixture with uninfected plants (Garcia-Guzman and Burdon 1997). Similarly, the systemic floral-smut fungus *Sporisorium amphiliphis* reduced seed production and growth of the perennial grass *Bothriochloa macra*. The interplay between biotic and abiotic environmental aspects was illustrated as the incidence of the fungus in populations of *B. macra* was density dependent towards the edge of the host range and negatively correlated with the frequency of days in the winter with temperature <0 °C (Garcia-Guzman *et al.* 1996). In general, non-systemic rusts and smuts (i.e. not spreading beyond the site of infection) are annual; e.g. *Puccinia graminis* on grasses. By contrast, systemic rusts and smuts can spread through the entire plant host and sporulate on a ramet other than the one originally infected. The fungus can lie latent in the host for some time before sporulating. Infected host plants may show reduced survival and altered growth, particularly distortion and elongation of infected shoots (Anders 1999). Clonal plants may be able to escape the effects of systemic fungi through the production of long rhizomes and stolons or both, and vigorous growth; e.g. clones of *Glyceria striata* infected by the systemic fungus *Epichloë glyceriae* produced fewer tillers but longer stolons than uninfected clones (Jean and Keith 2003).

The evolutionary interaction between grasses and fungi was illustrated in the classic Park Grass Experiment (§6.2.2); plots most heavily fertilized with nitrogen favoured the selection of populations of *Anthoxanthum odoratum* resistant to mildew *Erysphye graminis* (Snaydon and Davies 1972). Over time, selection of populations to host disease resistance in this manner could lead to a negative correlation between disease severity and soil nitrogen levels (Burdon *et al.* 2006).

Fungal pathogens can have complex effects on grassland communities in natural settings. The interaction between susceptibility to soil-borne pathogenic fungi and nematodes affected community mosaic structure of a *Festuca rubra–Carex arenaria* grassland in the Netherlands (Olff et al. 2000). In this grassland, pathogens limited the growth of *F. rubra*, whereas small-scale disturbance by rabbits and ants allowed nematode populations to limit growth of *C. arenaria*. Experimental augmentation with rust *Puccinia coronata* f. sp. *holci* and a leaf-spot fungus *Ascochyta leptospora* to a regenerating *Holcus lanatus* grassland led to variable results comparing the first and second years of succession; in general, however, the fungi (1) reduced abundance of perennial herbs in the short term allowing increased growth of the dominant perennial grasses, thereby leading to a decrease in diversity; (2) decreased abundance of annual species in the succession sequence; and (3) increased diversity when fungal augmentation was targeted against the perennial grasses (Peters and Shaw 1996). A clearer role of soil fungi affecting succession was shown in coastal foredune grasslands where it was shown that levels of soil-borne diseases and disease-tolerant herbaceous plant species both increased following dune stabilization by the dominant, but disease-intolerant, dune-building grass *Ammophila arenaria* (Putten et al. 1993). From an ecosystem perspective, pathogens can reduce production. Experimental exclusion of foliar fungal pathogens (leaf-spot disease and rust fungus) in *Andropogon gerardii*-dominated grassland led to increased leaf longevity and photosynthetic capacity, which in turn led to increased below-ground carbon allocation and root production (Mitchell 2003).

These plant–fungal relationships are complex enough to understand under present-day environmental conditions, but they may be affected by global climate change, altering the disease resistance of some plant species in communities more than others. Novel plant communities may result, with new plant–pathogen relationships (Garrett et al. 2006a). The extent to which these scenarios may be important in grasslands is unclear; however, changes in gene expression of tallgrass prairie dominants *Andropogon gerardii* and *Sorghastrum nutans* in response to simulated climate change has been reported (Travers et al. in Garrett et al. 2006b). The implications of Mitchell's (2003) study of *A. gerardii* and foliar fungal pathogens, described above, are that pathogens can limit ecosystem carbon storage. This limitation may have important consequences for grassland ecosystems to sequester carbon below ground under global climate change.

5.2.2 Endophytes

Clavitaceous fungi (family Clavicipitaceae, [Ascomycota]) grow intercellularly and systemically in the above-ground plant parts of grasses in all subfamilies and most tribes worldwide (Clay 1990b). White (1988) recognizes three categories of grass/endophyte infections (Table 5.5):

- In **type I associations** the host is sterilized by infection and the fungus spreads through spores produced in fruiting bodies (stromata) on leaves or on developing inflorescences. The host plant is rendered sterile.
- **Type II associations** infect warm-season C_4 grasses in the Panicoideae and Chloridoideae with species in the fungal genera *Balansia*, *Balansiopsis*, and *Myriogenospora*. For example, the systemic fungus *Balansia henningsiana* suppresses flowering when infecting the grass *Panicum rigidulum*, with only occasional asymptomatic healthy inflorescences on infected plants (Clay et al. 1989). Type II associations are intermediate between Type I and Type III associations; the host plants produce fertile inflorescences as well as sterile inflorescences with fungal stromata.
- In **type III associations** the host plant is asymptomatic, flowers and sets seed normally, spreading the endophyte vertically through the seed. The sexual stage of the life cycle of the fungi in type III associations has been lost, with the fungi being classified in the asexual genus *Neotyphodium* (syn. *Acremonium*), derived from sexual species of *Epichloë*. Type III associations are limited to cool-season grasses predominantly in the Pooideae.

Most endophytes produce one or more alkaloid classes (e.g. the ergot alkaloid ergovaline, lolitrems, and lolines, Chapter 4) which play a role in defending the host plant against herbivores.

Unlike mycorrhizal symbioses based on the acquisition of mineral resources (see §5.2.3), grass/endophyte associations are based primarily on protection of the host from biotic and abiotic stresses.

Table 5.4 Main diseases of some common forage grasses. X = disease present on the host plant, ▲ = disease particularly detrimental to the host plant

	ALPR	AREL	BRCA	DAGL	ELRE	HOLA	LOMU	LOPE	PHAR	PHPR	POPR	SCAR	SCPR	TRFL
Rusts (*Puccinia* spp.)														
Black rust (*P. graminis*)	X	X	X	X	X	X		X		X	▲	X	X	X
Yellow rust (*P. striiformis*)			X	▲			X	X		X	▲			
Brown rust (*P. recondite*)	X				X		X	X						X
Crown rust (*P. coronata*)	X	X		X	X	▲	▲	▲	X		▲	▲		
Others	X	X		X	X	▲	X	X	X		▲			
Leaf blotches (*Rhynchosporium* spp.)														
R. secalis	X		X	X	X					X		X		
R. orthsporum				▲			▲					X		
Helminthosporium diseases (*Drechslera* spp.)														
D. dactyloides	X		▲	X			▲	▲		X		X	X	
D. siccans				X			▲	▲		X		X	X	
Others	X		X	X	X	X	X	X		X	▲	▲	X	
Mastigosporium leaf flecks														
M. album		▲								X				
M. rubricosum				▲										
Others										X				
Timothy eyespot (*Cladosporium phlei*)										▲				

Disease														
Brown stripe (*Cercosporidium graminis*)	x													
Snow mould (*Microdochium nivale*, *Typhula incarnata*,…)		◄	x						x			x	x	x
Ustilago spp.				x			x		x		x			x
Urocystis spp.							x		x					
Others														x
Powdery mildew (*Blumeria graminis*)	x	x	x	x			x		x		x		x	x
Choke (*Epichloë typhina*)		◄	x	x			x		x					x
Ergot (*Claviceps purpurea*)	x	x		x			x		x		x	x	x	x
Spermospora leaf spot (*S. lolli*)				x			x						x	x
Rasmularia leaf spot (*R. holci-lanati*)														
Ascochyta leaf blight (*A.* spp.)				x			x		x			x	x	x
Neotyphodium spp.							x		x					

Key: ALPR, *Alopecurus pratensis*; AREL, *Arrhenatherum elatius*; BRCA, *Bromus catharticus*; DAGL, *Dactylis glomerata*; ELRE, *Elymus repens*; HOLA, *Holcus lanatus*; LOMU, *Lolium multiflorum*; LOPE, *Lolium perenne*; PHAR, *Phalaris arundinacea*; PHPR, *Phleum pratense*; POPR, *Poa pratensis*; SCAR, *Schedonorus phoenix*; SCPR, *Schedonorus pratensis*; TRFL, *Trisetum flavescens*.

With permission from Peeters (2004).

Table 5.5 Contrasting characteristics of types I, II, and III grass–endophyte associations

	Symptomatic (type I)	Mixed (type II)	Asymptomatic (type III)
Fungus			
Reproduction	Sexual	Both	Clonal
Transmission	Horizontal	Both	Vertical
Propagule	Ascospores	Both	Hyphae in seeds
Host			
Reproduction	Sterile/clonal	Partial sterility	Sexual
Interaction	Pathogenic	Intermediate	± Mutualistic
Infection frequency	Low-moderate	Intermediate	High
Taxonomy	Entire grass family	C_3 pooid grasses	C_3 pooid grasses

From Clay and Schardl (2002), with permission of the University of Chicago Press.

Endophyte infection can confer several benefits to the host plants including enhanced drought tolerance, photosynthetic rate, and growth (growth rate, size, fecundity) (West 1994) which in turn affect the phenotypic plasticity of the grass plant (Cheplick 1997) and its competitive ability against non-infected neighbours (Clay et al. 1993). Studies investigating the demographic consequences of type III endophytes suggest that infection can lead to higher survival, flowering frequency, vegetative tiller production, and biomass (Clay 1990a). Seed production of endophyte infected *Schedonorus phoenix* and *L. perenne* plants was shown to be higher than uninfected plants, and seed of infected plans had a higher germination rate (Clay 1987). Consequently, relative fitness of infected ramets can be higher than in uninfected ramets of the same species. By contrast, by suppressing flowering, type I and type II endophytes can promote vegetative reproduction, and in some cases induce vivipary (Clay 1990b). The mutualistic benefits to the grass host of type III endophyte infection are not always realized. For example, studies with Arizona fescue *Festuca arizonica* have found naturally infected plants to have lower vegetative growth, reduced competitive properties, and lower fitness than uninfected plants (Faeth et al. 2004; Faeth and Sullivan 2003).

Longitudinal studies of endophyte infection suggest that the prevalence of seed-transmitted endophytes can increase over time. In field experiments, infected *Schedonorus phoenix* can suppress other grasses and forbs relative to uninfected fescue and support lower consumer populations (e.g. aphids, rodents). The outcome is that endophyte-infected species in grasslands can substantially alter community and ecosystem structure (Clay 1994, 1997). For example, Clay and Holah (1999) showed a decline in species richness over a 4-year period as *S. phoenix* dominance increased in infected compared to uninfected plots even though total productivity did not differ. Similarly, Spyreas et al. (2001b), while noting a similar relationship in a successional old field dominated by *S. phoenix*, observed a positive relationship between diversity and endophyte infection in mowed plots underlying the complexity of the endophtye–plant interaction.

Although endophytes are widespread in cool-season grasses (e.g. Spyreas et al. 2001a) and infection levels appear to be related to grassland community structure (Gibson and Taylor 2003), detailed population studies are limited predominantly to two agricultural pasture grasses (*Lolium perenne, Schedonorus phoenix*) and the arid grassland native, *Festuca arizonica*. It is becoming clear that the relationship between endophyte and host is more complex than initially thought and depends upon host genotype × environment interactions (Cheplick and Cho 2003; Cheplick et al. 2000; Hunt and Newman 2005). The benefit of the endophyte to the grass host may be most apparent under conditions of environmental stress, such as drought. Consequently, the symbiosis between asexual endophytes and grasses may be best viewed as a mutualism–parasitism continuum similar to the association between plants and mycorrhizal fungi (§5.2.2) (Muller and Krauss 2005).

5.2.3 Mycorrhizae

Mycorrhizae are symbiotic fungal associates occurring in the roots of >90% of plants. Mycorrhizal colonization is widespread in grasses across all habitats, especially in grasslands (Table 5.6). Infection appears to be limited only in wet and fertile habitats. The fungus penetrates the secondary roots during active growth. It grows as an aseptate hypha from a fungal chlamydospore in the soil, or from another infected root, entering the root hairs into the epidermis and cortex, but not the vascular cylinder or meristem. The anatomy of the plant root infected with a mycorrhizal fungus is unaffected but morphologically it is usually less branched. There are several types of mycorrhizae, and, in common with most herbaceous plants, grasses possess endomycorrhizae ('inside mycorrhizae'). Some trees possess endomycorrhizea too (e.g. apples, oranges), but most woody plants possess ectomycorrhizae in which the fungus sheaths the roots.

The fungal spores are the reproductive structures of members of the 'lower fungi', in a monophyletic clade, the phylum Glomeromycota (Schüßler *et al.* 2001). Traditionally, these fungi are classified on the basis of spore characteristics visible under the light microscope. This approach allowed only a relatively limited number of apparently widespread fungi to be identified (Table 5.7), in just 6 genera (*Acaulospora, Entrophospora, Gigaspora, Glomus, Sclerocystis,* and *Scutellospora*) and about 150 species. Molecular analyses have, however, revealed a tremendous diversity of mycorrhizal taxa that is only now starting to be characterized (Fitter 2005).

The fungi produce two characteristic structures in the plants, **arbuscules** which are highly branched haustoria involved in nutrient transfer, and **vesicles** which are terminal hyphal swellings associated with storage. Hence the term frequently used to refer to these mycorrhizae, **vesicular arbuscular mycorrhizae** (VAM). The fungus benefits from the symbiosis through having somewhere to live and assimilation of simple sugars from the plant (up to 20% of carbon fixed by the plant).

The fungal hyphae extend from the plant root, thereby increasing the surface area in contact with the soil. Benefits of mycorrhizal infection to plants include (see reviews in Brundrett 1991; Fitter 2005; Koide 1991):

- increased nutrient and water absorption, especially slow-moving soil phosphorus; absorption of

Table 5.6 Distribution of mycorrhizal colonization in British grass species across different life forms, habitats and soil fertilities

	Percentage of grass species Normally/occasionally colonized	Rarely/never colonized
Life form		
Annual	100	0
Perennial	90	10
Habitats		
Agricultural/managed land	100	0
Sand dunes/beaches	96	4
Inland cliffs/sands	96	4
Scrub/grassland	92	8
Forests (omitting wet forests)	92	8
Forests (with wet forests)	84	16
Bogs/marshes	83	17
Mud flats/salt marshes	80	20
Soil fertility		
Very infertile	93	7
Infertile	86	14
Fertile	78	22
Very fertile	80	20

From Newsham and Watkinson (1998), with permission.

Table 5.7 Mycorrhizal spore densities in a Kansas tallgrass prairie

	Spores kg^{-1} soil (± 1 SE)	Frequency (% of 44 samples)
Glomus constrictum Trappe	800 ± 143	96
Sclerocystis sinuosa Gerd. & Bakshi	600 ± 237	71
Glomus mosseae (Nicol. & Gerd.) Gerd. & Trappe	496 ± 124	100
Glomus etunicatum Becker & Gerd.	340 ± 216	28
Glomus aggregatum Schenck & Smith	222 ± 99	39
Entrophospora infrequens (Hall) Ames & Schneider	216 ± 39	89
Glomus fasciculatum (Thaxer *sensu* Gerd.) Gerd & Trappe	64 ± 28	21
Acaulospora longula Schenck & Smith	60 ± 40	11
Glomus geosporum (Nicol. & Gerd.) Walker	38 ± 17	21
Sclerocystis rubiformis Gerd. & Trappe	36 ± 19	7
Gigaspora gigantea (Nicol. & Gerd.) Gerd. & Trappe	6 ± 3	25
Scutellospora pellucida (Nicol. & Schenck) Walker & Sanders	2 ± 2	14

Reprinted with permission from *Mycologia* (Gibson and Hetrick 1988) © The Mycological Society of America.

soil phosphorus is by far the most important benefit to the plant
• increased root health and longevity associated with altered root architecture and inhibited branching
• increased tolerance to drought, high soil temperatures, toxic heavy metals, extremes in soil pH, transplant shock
• transplant of metabolites from one plant to another via a 'hyphal net'
• production of plant growth regulators and antibiotics (pathogen protection) by the fungus.

In most cases the presence of the mycorrhizal infection in grasses has been shown to have a positive effect on growth and biomass, including fecundity and/or seedling emergence (Newsham and Watkinson 1998). These effects, however, are context dependent. For example, reducing mycorrhizal levels in the soil at two sites reduced fecundity of *Vulpia ciliata* ssp. *ambigua* at one site and increased fecundity at the other (Carey *et al.* 1992). The increased fecundity at one site reflected a reduction in infection of other pathogenic fungi (due to use of the fungicide) which had been negatively affecting plant growth indicating a protective role for mycorrhizal infection (Newsham *et al.* 1995). Cultivated *Avena sativa* was found to benefit from mycorrhizal infection (increased plant lifespan, number of panicles per plant, shoot phosphorus concentration, shoot phosphorus content, duration of flowering, and the mean weight of individual seeds), but infection in wild *A. fatua* had a significantly reduced effect, no effect, or a negative effect on these characters (Koide *et al.* 1988). The difference in response to mycorrhizal infection was attributed to inherent adaptations to nutrient deficiencies in the wild plants defraying a potential benefit from mycorrhizal symbiosis.

A series of studies in tallgrass prairie at Konza Prairie, Kansas, USA, illustrates the importance that mycorrhizae can have in affecting plant growth, mediating competitive interactions, and determining plant community composition. These studies are summarized below.

Greenhouse experiments Native prairie plants grown in sterilized low-phosphorus (<10 mg kg^{-1}) prairie soil either with or without inocula of *Glomus etunicatum* showed high dependency of growth of C_4 grasses and forbs (Table 5.8), with total biomass of some of the C_4 grasses being increased 60–220 fold (Hetrick *et al.* 1988). By contrast, C_3 grasses showed either no dependency on mycorrhizae, or in the case of *Koeleria macrantha*, negative growth in the presence of mycorrhizae. The difference in response of the C_3 and C_4 grasses may be related to

Table 5.8 Mycorrhizal dependency of warm- and cool-season grasses and forbs. Species dependence = $(E^+ - E^-)/E^+ \times 100\%$

	Inoculated (E+) (g dw)	Non-inoculated (E−) (g dw)	Dependence (%)
Warm-season grasses (C$_4$)			
Andropogon gerardii	3.38	0.02	99.5
Bouteloua curtipendula	3.79	0.06	99.5
Panicum virgatum	3.96	0.02	98.4
Sorghastrum nutans	4.31	0.02	99.4
Mean			99.2
Cool-season grasses (C$_3$)			
Bromus inermis	2.27	1.29	43.2
Koeleria macrantha	0.80	0.92	−15.0
Leymus cinereus	3.73	1.99	46.6
Lolium perenne	2.06	1.72	16.5
Pascopyrum smithii	2.05	0.50	75.6
Schedonorus phoenix	3.77	1.67	55.7
Mean			37.1
Forbs (C$_3$)			
Baptisia leucantha	2.55	0.15	94.1
Dalea purpurea	0.88	0.02	97.7
Liatris aspera	0.34	0.03	91.2
Mean			94.3

From Hetrick et al. (1988).

a better ability of C$_4$ grasses to respond to increases in tissue phosphorus, or their coarser roots and lesser ability to aquire soil phosphorus (Newsham and Watkinson 1998). A broader survey confirmed these results and also indicated a low response to mycorrhizal infection in annual grasses and annual and biennial forbs (Wilson and Hartnett 1998). The annuals surveyed were facultative mycotrophs (i.e. colonized by mycorrhizae, but did not show a biomass response to infection) reflective of the generally low infection rate of annual compared with perennial grasses in this tallgrass prairie (c.15% vs c.85%; a contrast to the 100% infection of annuals in British grasses: Table 5.1). Annuals are typical of disturbed habitats where mycorrhizal infection may not be as beneficial as in the more stable, highly competitive conditions of later succession.

Follow-up experiments These investigated the effect of mycorrhizae on competitive interactions between the C$_4$ grasses *Andropogon gerardii* and *Sorghastrum nutans*, and the C$_3$ grasses *Elymus canadensis* and *Koeleria macrantha* (Hartnett et al. 1993; Hetrick et al. 1994). Using two different experimental designs, it was shown in both cases that the presence of the mycorrhizae was required to maintain competitive superiority of the C$_4$ grasses in mixture. In the absence of soil mycorrhizae the C$_4$ grasses were less competitive when soil phosphorus was low.

Field investigations These supported the greenhouse experiments. The production of *A. gerardii* was reduced when soil fungi (including mycorrhizae) were reduced using a fungicide unless supplemental phosphorus was provided (Bentivenga and Hetrick 1991). Labelling experiments showed ^{32}P transport among neighbouring species, consistent with transfer via mycorrhizal hyphal interconnections (Fischer Walter et al. 1996). Suppression of mycorrhizae (to less than 25% of controls) led to decreased abundance of the obligately mycotrophic C$_4$ grasses and compensatory increases in the abundance and diversity of subordinate facultatively mycotrophic C_3 grasses and forbs, while total system biomass remained the same (Hartnett and Wilson 1999; Smith *et al.* 1999b; Wilson and Hartnett 1997).

These coupled greenhouse and field experiments suggest that soil mycorrhizae were playing an important role in mediating the population and community dynamics of tallgrass prairie. Mycorrhizal dependency and demographic and growth responsiveness varied among species and life forms. In particular, the dominant C_4 grasses benefit from their association with the mycorrhizae which allows them to maintain their competitive superiority in the system with low soil phosphorus. However, two related experiments help place these findings into a broader context. First, the effect of soil mycorrhizae on seedling emergence, flowering, and density was more complex than expected given the clear effects on biomass (Hartnett *et al.* 1994). Seedling emergence was unrelated to mycorrhizal status for C_4 grasses (*Andropogon gerardii, Panicum virgatum*), and reduced for C_3 grasses (*Elymus canadensis, Koeleria macrantha*) when the mycorrhizae were suppressed, and variable for forbs. Once established, however, flowering in C_4 grasses was enhanced by mycorrhizae, but only in burned prairie, whereas flowering of a C_3 grass *Dichanthelium oligosanthes*, sedges *Carex* spp., and a forb *Symphyotrichum ericoides* were higher in non-mycorrhizal plots, with the effect being enhanced with burning. Second, parallel experiments to assess mycorrhizal dependency (Hetrick *et al.* 1988), but undertaken in soils with high phosphorus availability (c.40 mg kg^{-1}), did not show high dependency of these C_4 grasses (Anderson *et al.* 1994). The lack of a response to mycorrhizae in high-phosphorus soil underlies the selective advantage of the mycorrhizal symbiosis to phosphorus-limited conditions. Under high phosphorus availability, the plant does not accrue sufficient benefit from the mycorrhizae to justify giving up so much assimilate to the fungus.

Overall, these and other studies elsewhere make it clear that soil mycorrhizae significantly interact with plants in grasslands affecting their population biology, including demographic parameters and fitness, thereby mediating community and ecosystem structure and function. Although mycorrhizal benefits are primarily related to phosphorus acquisition, individual plant response may be dependent upon local conditions (e.g. West *et al.* 1993).

5.3 Genecology

There are three aspects of grass genetics that are of particular relevance for understanding the genetic structure of grasslands in an ecological context (§5.3.4); these are the common development of ecotypes (§5.3.1), the frequent occurrence of polyploidy in the grasses (§5.3.2), and hybridization (§5.3.3).

5.3.1 Ecotypes and metapopulations

An ecotype is a distinct set of genotypes (or populations) within a species, resulting from adaptation to local environmental conditions, capable of interbreeding with other ecotypes of the same species (Hufford and Mazer 2003). The ecotype concept was first developed by Göte Turesson during the 1920s when he studied variation in European species, finding that species grown in a test garden at Åkarp, Sweden retained characteristics of the original location and habitat in which they had been collected (Briggs and Walters 1984). Stapledon (1928) demonstrated the occurrence of ecotypes in the grass *Dactylis glomerata* in response to biotic factors. The later classic work by Clausen, Keck, and Hiesey in the 1940s on species collected from along a 200-mile transect across central California and grown in one of a series of test gardens along the transect confirmed Turesson's observations for a suite of species including the grass *Deschampsia caespitosa* (Lawrence 1945). Other early studies demonstrating ecotypes in grasses included investigations of *Poa pratensis* (Smith *et al.* 1946), *Bouteloua curtipendula* (Olmated 1944, 1945), and *Schizachyrium scoparium* (Larson 1947). Since these early studies, numerous grasses have been shown to exhibit ecotypic differentiation in response to a variety of biotic and abiotic factors. One of the most extensive group of studies of grassland ecotypes is by McMillan (1959b) in which clonal material of 12 prairie grasses was collected from 39 sites from across the North American Great Plains and transplanted to a common garden in Lincoln, Nebraska. Nine of the 12 grasses maintained the flowering phenologies of their original habitat when grown in the common garden. Three patterns of flowering were observed in relation to climate:

- Species in which the source populations were from the north and west flowered earliest and were shorter in stature compared with source populations from the south and east. This pattern was represented by eight species: *Andropogon gerardii, Bouteloua curtipendula, B. gracilis, Koeleria macrantha, Panicum virgatum, Schizachyrium scoparium,* and *Sorghastrum nutans*.
- Species in which source populations from the south flowered earliest and source populations from the east flowered late were represented by one species, *Elymus canadensis*.
- Opportunistic species that flowered at the same time wherever they came from, i.e. no ecotypes, were represented by *Oryzopsis hymenoides, Hesperostipa comata* ssp. *comata*, and *Hesperostipa spartea*.

Subsequent studies by other researchers have confirmed the widespread nature of ecotypic variation in the grasses of the Great Plains and in grasslands worldwide, both within populations and across broad habitat gradients (e.g. Casler 2005; e.g. Kapadia and Gould 1964; Quinn and Ward 1969; Robertson and Ward 1970). Ecotypic differentiation is not restricted to the grasses in grasslands; it is widespread in other taxa in these communities (e.g. Gustafson *et al.* 2002). The importance of small-scale ecotypic variation within species was recognized by McMillan (1959a) when he noted, 'Through natural selection, each stand of true prairie may be fundamentally different from any other stand.' A practical value of recognizing ecotypic variation in grasses is in allowing recognition of the most suitable ecotypes for conservation, restoration, renovation, landscaping, and bioremediation (Chapter 10).

One of the more remarkable examples of ecotypic differentiation in the grasses is the occurrence of ecotypes adapted to soils contaminated with heavy metals around mines, smelters, and refineries (Antonovics *et al.* 1971; Baker 1987). The occurrence of these edaphic ecotypes was first noted in *Agrostis capillaris* colonizing Roman-era lead mines in Wales, UK (Bradshaw 1952). Vegetation on contaminated mine soils is usually sparse, colonized by a low diversity of adapted ecotypes of grasses. Nearby or adjacent uncontaminated soils naturally contain a low percentage (usually <10%) of individuals tolerant of heavy metals. Seed from these few individuals is able to colonize the mine soil, leading to the development of a heavy-metal-tolerant population. Population studies of *Agrostis capillaris* growing on several disused lead mines indicated that plants growing on the mines were morphologically smaller, flowered earlier, and had lower fecundity than plants in populations from nearby uncontaminated habitats (Jowett 1964). Genetic studies of zinc-tolerant *Anthoxanthum odoratum* indicated that tolerance was dominant under a polygenic control system involving a small number of loci (Gartside and McNeilly 1974). Heavy-metal tolerance appears to have high heritability. Provided sufficient genetic variability exists in the source populations, heavy-metal-tolerant ecotypes develop rapidly, within a single generation in some cases (Baker 1987). In some cases metal tolerance of ecotypes appears to be specific to the metals in the contaminated soil (Karataglis 1978), whereas in other cases co-tolerance to one or more metals not part of the selection regime has been reported (Cox and Hutchinson 1979). Metal-tolerant ecotypes have also been reported from naturally contaminated soils; e.g. lead-tolerant *Deschampsia flexuosa* on soils enriched with lead from groundwater leaching of the bedrock (Høiland and Oftedal 1980), and there are also reports of the absence of ecotypic development, i.e. constitutional tolerance (Baker 1987) in grasses colonizing metalliferous mine wastes (e.g. *Andropogon virginicus* on lead/zinc/cadmium mine soil, Gibson and Risser 1982). Heavy-metal-tolerant ecotypes of grasses have been developed for use in revegetating metal contaminated soils, e.g. *Festuca rubra* cv. Merlin for calcareous lead/zinc mine wastes, *Agrostis capillaris* cv Goginan for acid lead/zinc mine wastes, and *A. capillaris* cv. Parys for copper wastes (Smith and Bradshaw 1979).

Ecotypes in grasses can develop rapidly (Bone and Farres 2001). Heavy-metal-tolerant ecotypes have been observed to develop in as few as 30 years (Al-Hiyaly *et al.* 1988). Population differentiation of *Anthoxanthum odoratum* in response to soil liming was observed within 6 years in the Park Grass

Experiment in the UK (Snaydon and Davies 1982). A herbicide-resistant biotype (i.e. genetically distinct individuals, but not in an ecological setting) appeared in *Bromus tectorum* in 2 years following selection under applications of the herbicide primisulfuron (Mallory-Smith *et al.* 1999).

Although the term 'ecotype' has utility in allowing recognition of genetically based groups of ecologically related populations, it should be remembered that the evolutionary unit is the population (Quinn 1978). The Turesson **ecotype concept** tends to imply environmental and habitat uniformity within the recognized ecotypic units. By contrast, the **metapopulation concept** (Hanski 1999), where populations of a species are connected through local dispersal and extinction (see §6.4.3), provides a dynamic, scale-related approach that recognizes the genecological importance of populations.

Metapopulations are defined as a series of populations where regional persistence is governed by patch colonization, extinction, and recolonization (Freckleton and Watkinson 2002). Under this definition of a metapopulation, regional processes dominate over local population processes. The value of the metapopulation concept is that it allows recognition of scale-related population processes. Local population extinction is an expected occurrence in plants existing as part of a metapopulation. From a conservation perspective, land managers may be less concerned with declines or local extirpation of the population of a threatened plant than they might otherwise be, with the knowledge that the population is part of a metapopulation. For example, coastal populations of the dune-building grass *Leymus arenarius* are frequently extirpated as a result of recurrent disturbance, but high levels of seed dispersal by the sea allows rapid recolonization maintaining gene flow consistent with regional metapopulation processes (Greipsson *et al.* 2004). Widespread habitat fragmentation of grasslands (Chapter 1) can destroy the operation of the regional processes necessary to maintain metapopulations. Fragmented grassland patches may take a long time to exhibit the resulting local extinction debt (e.g. some 70 years in Estonian calcareous grasslands, Helm *et al.* 2006).

Not all patchy populations can be defined strictly as metapopulations (Freckleton and Watkinson 2003). When the colonization–extinction processes are not operating, regional ensembles (unconnected local populations across a mosaic of suitable and unsuitable habitat) or spatially extended populations (a single extended population across large areas of habitat) occur. For example, regional populations of the winter annual grass *Vulpia ciliata* occur as a series of discrete local populations but recolonization of extinct populations is rare or absent (Watkinson *et al.* 2000). Hence, the distribution of *V. ciliata* is characterized as a regional ensemble (Freckleton and Watkinson 2002). By contrast, populations of *Avena sterilis* exhibit only local colonization and extinction dynamics characteristic of spatially extended populations (Freckleton and Watkinson 2002).

5.3.2 Polyploidy

Polyploidy occurs when the chromosome number of a plant is a multiple (normally an even multiple) greater than 2 of the base chromosome number (x) of its taxonomic group (Jones 2005). Polyploidy is believed to have played a major role in the evolution of grasses (Stebbins 1956) and affects their population biology (reviewed by Keeler 1998). Most genera have some species with a chromosome number being some multiple of the basic number. About 80% of grass species are of polyploid origin, compared with 30–50% typical of flowering plants as a whole (De Wet 1986).

Chromosome numbers in the Poaceae range from a low of $2n = 6$ in *Iseilema*, a genus of 20 species in Indo-Malaya to Australia, to a high of $2n = 263–265$ in the speciose and widespread *Poa* (De Wet 1986). Common base chromosome numbers in the Poaceae are Pooideae $x = 7$, Chloridoideae and Panicoideae $x = 9$ and 10, Arundinoideae, Ehrhartoideae and Bambusoideae $x = 12$, derived through polyploidy and chromosome reduction from the ancestral complex of $x = 5$, 6, or 7. The most common diploid chromosome numbers in the family are $2n = 20/40$, $2n = 14/28$, and $2n = 18/36$. Diploids based on the ancestral chromosome number constitute <10% of modern grasses. Polyploids can be multiples of the basic ancestral numbers (e.g. *Festuca*, $2n = 14, 28, 42,$

56, 70) or a derivation (e.g. *Bouteloua*, $2n = 20, 40, 60$ derived from a secondary base number $x = 10$).

There are two forms of polyploidy depending on the origin of the duplicated genome: **autopolyploidy**, in which multiple copies of the same genome occur, and **allopolyploidy**, in which several genomes from different species occur in the same individual. At its simplest, autopolyploidy can result from the sexual functioning of non-reduced gametes, e.g. a non-reduced female egg arising from non-disjunction of chromosomes, and allopolyploidy can result from doubling of chromosomes following hybridization. In practice it is more complicated, with the polyploid genome more often than not being a product of multiple occurrences of both allo- and autopolyploidy. For example, the allooctoploid *Pascopyrum smithii* ($2n = 56$) is considered to have arisen after hybridization between two tetraploids (*Elymus lanceolatus* and *Leymus triticoides*, both $2n = 28$) followed by chromosome doubling (Dewey 1975). *Pseudoroegneria spicata* is an example of autopolyploidy where most populations are diploid ($2n = 14$) but some are autotetraploid ($2n = 28$) following spontaneous doubling of their chromosome number (Jones 2005). The general sequence followed in the evolution of polyploids suggested by Stebbins (1947) is that first tetraploids evolved and spread, replacing the progenitor diploids. Subsequently, hexaploids and octoploids form from the tetraploids replacing them. Polyploids are widespread in the Poaceae, implying an ability of newly evolved polyploids to compete successfully for available habitat. A disadvantage of polyploidy is the extensive gene duplication, making it difficult to generate new adaptive systems (Stebbins 1975). Advantages include the conservation of heterosis, or hybrid vigour, and the acquisition of large gene pools through hybridization, allowing polyploids to exploit adaptive properties of one system. Polyploids have greater adaptive amplitude than diploids, allowing them to enter new climatic zones or withstand climatic change. Under such change, diploid progenitors either go extinct or evolve new adaptive systems.

Species of different polyploid levels within a genus in one geographic area form a polyploid complex, believed to play an important role in the evolution of grasslands, especially the North American grasslands (Stebbins 1975). Eight polyploid complexes are recognized in the North American Great Plains belonging to the genera *Bouteloua*, *Buchloe*, genera of the Panicoideae (*Andropogon*, *Bothriochloa*, *Schizachyrium*, and *Sorghastrum*), *Piptochaetium* (represented as only fossil fruits in the Great Plains today), *Stipa*, *Agropyron*, *Elymus*, and *Hordeum*, representing groups of species that evolved and migrated either into or away from the Great Plains over the past $5–10 \times 10^6$ years. For example, the *Bouteloua* polyploid complex of 11 diploids ($2n = 20$), 2 tetraploids, 1 hexaploid, and 6 of mixed diploid and polyploid or aneuploid races, has its centre of distribution in the south-western USA and northern and central Mexico. This distribution implies that the *Bouteloua* complex entered the Great Plains from the south-west. Today, most of the dominant genera of grasses in the Great Plains are polyploid, with diploids found only in a few minor genera including *Sphenopholis*, *Aristida*, *Schedonnardus*, and some species of *Panicum* and *Paspalum*.

Within-species variability in polyploidy levels, known as **polyploid polymorphism**, is reported for 36 grass species (Table 5.9). Keeler and Kwankin (1989) found polyploid polymorphism to occur in 67% (16 of 21 species investigated) of the dominant prairie grasses of North America, significantly more than in common Asteraceae (47%, 10 of 21 species tested) or Fabaceae (22%, 2 of 9 species) species. Furthermore, there was variation in ploidy level among some of the released cultivars of *Panicum virgatum* (i.e. Summer and Kanlow $2n = 36$, Pathfinder, Blackwell, and Nebraska 28 $2n = 54$, Riley and Vogel 1982).

The relevance of polyploid polymorphism to ecological distributions is unclear but appears to occur more frequently in variable, disturbed conditions. Polyploid polymorphism is probably driven by selection and may (1) reflect an adaptive advantage (but not, it appears, for composites or legumes), (2) be a transient feature of a rapidly evolving group, (3) be a neutral feature irrelevant to ecology and evolution (if so, why so prevalent in the grasses?), or (4) reflect cryptic species (Keeler and Kwankin 1989). Habitat differences between diploid and polyploid members of the same species are commonly reported. For example, in

Table 5.9 Grasses reported to exhibit more than three levels of polyploid polymorphism

Genus and species	Ploidy	Chromosome numbers
Agrostis stolonifera	–	28, 35, 42
Andropogon halli	6n, 7n, 10n	60, 70, 100
Anthoxanthum odoratum	2n, 3n, 4n	10, 15, 20
Aristida purpurea	2n, 4n, 6n, 8n	22, 44, 66, 88
Arrhenatherum elatius	2n, 4n, 6n	14, 28, 42
Bouteloua curtipendula	3n, 4n, 5n, 6n, 8n, 10n, 14n	21, 28, 35, 40, 42, 45, 50, 52, 56, 70, 98
B. dactyloides	2n, 4n, 6n	20, 40, 60
B. gracilis	2n, 4n, 6n	20, 29, 35, 40, 42, 60, 61, 77, 84
B. hirsuta	–	12, 20, 21, 28, 37, 42, 46
Dactylis glomerata	2n, 4n, 6n	14, 28, 42
Holcus mollis	4n, 5n, 6n, 7n	28, 35, 42, 49
Koeleria macrantha	2n, 4n, 8n, 10n, 12n	14, 28, 56, 70, 84
Panicum virgatum	2n, 4n, 6n, 8n, 10n, 12n	18, 21, 25, 30, 32, 36, 54, 56–65, 70, 72, 90, 108
Paspalum hexastachyum	2n, 4n, 6n	20, 40, 60
Paspalum quadrifarium	2n, 3n, 4n	20, 30, 40
Phalaris arundinacea	2n, 4n, 5n, 6n	14, 27, 28, 29, 30, 31, 35, 42, 48
Phleum pretense	2n, 3n, 4n	14, 21, 28
Schedonorus phoenix	2n, 4n, 6n	14, 28, 42
Spartina pectinata	4n, 6n, 12n	28, 40, 42, 80, 84
Sporobolus cryptandrus	2n, 4n, 6n, 8n	18, 36, 38, 72
Trisetum spicatum	2n, 4n, 6n	14, 28, 42

Extracted from Keeler (1998).

South Africa there are four varieties or chromosome races of *Themeda triandra*: var *imberbis* is diploid, var. *trachysperma* is mostly diploid, var. *hispidum* is primarily diploid, and var. *burchellii* is primarily polyploid (De Wet 1986). The distribution of these varieties is related to vegetation type, e.g. in transitional forest 81% of the populations are diploid var. *imberbis* and 19% polyploid var. *trachysperma*, in semi-arid bushveld, 50% of populations are diploid var. *imberbis* and *trachysperma* and 50% of populations are polyploid var. *hispidum*, and in savannah 81% of populations are polyploid vars. *hispidum* and *burchellii* and 19% of populations are diploid var. *imberbis*. By contrast, *T. triandra* is polyploid ($2n = 40, 80, 90, 110$) in India, but there is no obvious distribution of ploidy levels.

Andropogon gerardii occurs predominantly as two cytotypes, hexaploids ($2n = 6x = 60$ chromosomes) and enneaploids ($2n = 9x = 90$ chromosomes) (Keeler and Davis 1999; Keeler *et al.* 1987). At the eastern edge of the North American tallgrass prairie, *A. gerardii* populations were found to be exclusively hexaploid; however, the percentage of higher ploidy-level individuals increased to 80% westwards under arid, more variable conditions (Keeler 1990). Morphologically it is not possible to distinguish the cytotypes in the field because of extreme phenotypic plasticity; however, under controlled conditions, the enneaploid can be larger and taller than the hexaploid (Keeler and Davis 1999). Within one site that was examined (Konza Prairie, Kansas), fine-scale mixing of the cytotypes occurred unrelated to fire treatment or soil moisture (Keeler 1992). Similarly, extensive cytotypic variation occurs in *Dactylis glomerata* and *Panicum virgatum* (Table 5.9), that while showing regional and local geographic patterning do so in a manner that is, as yet, not clearly explained at the smallest scales where intermingling of cytotypes can occur (Keeler 1998). *Panicum virgatum* populations, for example, are predominantely early-flowering tetraploids in the north and western areas of the central North American plains, whereas in other areas a diversity of types occur (McMillan and Weiler 1959). In Oklahoma, lowland populations were observed to be uniformly tetraploid ($2n = 36$), with upland populations being a mix of octoploids and ($2n = 72$) aneuploid variants of the octoploids ($2n = 66–77$) (Brunken and Estes 1975).

5.3.3 Hybridization in grasses

As in many other plant groups, hybridization, the combining of genetic material from two or more individuals of different species in their offspring, is common and extensive among some species within genera, and, in some cases, between species of different genera. Over 50 intrageneric grass hybrids were reported by Watson (1990). In some groups, barriers to hybridization appear to be particularly low leading to problems in correct circumscription of species boundaries and the ready breeding of synthetic hybrids (often sterile) in breeding trials, e.g. among species of *Agropyron, Elymus, Sitanion*, and other members of the Triticeae tribe (e.g. Connor 1956; Dewey 1969, 1972). Hybridization has been important in the origin of some domesticated grass cereals, notably wheat and maize. The hexaploid wheat *Triticum aestivum* is the result of a chromosome doubling following hybridization of tetraploid rivet wheat *T. turgidum* with the diploid wild species *Aegilops tauschii*. *Triticum turgidum* itself arose from hybridation between *Aegilops speltoides* and an ancestor of the hexaploid wheat *T. zhukovski* (Dvorák et al. 1998). Maize (corn, Indian corn, *Zea mays* ssp. *mays*, $2n = 20$) offers an even more complex and convoluted story than wheat. *Zea mays* exists only as a domesticated plant, with *Balsas teosinte* (*Zea mays* ssp. *parviglumis*) the wild Mexican and Central American grass generally regarded as its ancestor (Iltis 2000; Matsuoka 2005). Teosinte is the common name for the wild forms of *Zea* and includes three species (*Zea diploperennis, Z. perennis, Z. luxuricans*) and two subspecies of *Z. mays*, i.e. ssp. *huehuetenangensis*, ssp. *mexicana*, and ssp. *parviglumis*, each with a distinct geographic distribution. Over the last 100 years various hypotheses have been put forward to explain the origin of maize (Chapman 1996). Part of the problem in resolving the origin of maize is understanding the transformation of teosinte with its clustered, slender ears into maize with its characteristic single, large ear (the corn cob). Hypotheses have included hybridization of an early cultivar of *Z. mays* in Central and South America 5000–10 000 years ago with teosinte or *Tripsacum*. The most recent view is that maize arose from a single domestication event of teosinte accompanied by a series of rare mutations (the homeotic sexual transformation hypothesis, Iltis 2000) 9000 years ago in southern Mexico. The many landraces that evolved subsequently were the result of repeated introgression between maize and teosinte following the likely exchange of seed among local farmers (Matsuoka 2005).

The ecological and evolutionary relevance of hybridization is that interspecific hybrids coupled with polyploids can persist, even if of reduced fertility, allowing grasses to pass through bottlenecks of partial sterility (Stebbins 1956). The hybrids thus formed may have an ecological advantage over their progenitors in the original or other habitats. The classic example is the origin of *Spartina* × *townsendii* ($2n = 62$) as a new species that arose c.1890 following hybridization of native *S. maritima* ($2n = 60$) with the North American introduced species *S. alterniflora* ($2n = 62$) in Southampton Water in southern Britain in the late 1800s. Although *S.* × *townsendii* is sterile, a fertile amphidiploid arose from it that has been named *S. anglica* ($2n = 120$, 122, 124) and has subsequently spread extensively (Stace 1991). Within habitats, hybrids can show differential spatial patterning occurring in **hybrid zones** separate or distinct from the progenitors. Two models have been proposed to explain hybrid survival in hybrid zones: the **tension zone model** suggests that there is environmentally independent selection against intrinsically unfit hybrids, and the **ecological selection-gradient model** suggests that there are environmentally dependent fitness differences among parental species and hybrids. One study in a calcareous mountain grassland showed that *Prunella grandiflora* × *P. vulgaris* (Lamiaceae) hybrids showed intermediate performance of the hybrids in hybrid zones in support of the ecological selection-gradient model (Fritsche and Kaltz 2000).

5.3.4 Genetic structure of grasses and grasslands

Numerous molecular studies support the genetic basis for ecotypic differentiation discussed earlier (§5.3.2) (see review in Godt and Hamrick 1998). Populations of the tallgrass prairie dominant *Andropogon gerardii*, for example, separated by hundreds of kliometres are genetically distinct (Gustafson *et al*. 1999; McMillan 1959b), as are populations separated sometimes by a matter of metres (Keeler 1992). Indeed, there was greater genetic diversity within than among populations of *A. gerardii* in Illinois, USA (Gustafson *et al*. 2004).

The general level of genetic diversity in grasses, based on allozyme studies, appears higher for grass species, within grass species, and among grass populations compared with other plants (e.g. among population genetic diversity = 27% for grasses, 22% for non-grasses) (Godt and Hamrick 1998). Furthermore, although annual and perennial grasses do not differ in overall genetic diversity, annuals have a higher proportion of polymorphic loci (65% vs 55%) and more alleles per locus (2.65 vs 2.12) than perennials. Grasses with a widespread distribution and those with a mixed-mating or outcrossing breeding system have a higher within-in-population genetic diversity than those with restricted distributions or selfing breeding systems. However, the proportion of genetic diversity accounted for by life-history traits is low in the grasses (1–13%), and so, as in other plant families, much of the genetic diversity and structure is based on the phylogenetic and evolutionary history of each species (Godt and Hamrick 1998).

As noted from Godt and Hamrick's (1998) review of studies outlined above, grasses have high within-in-population genetic diversity. For example, on the basis of morphological markers and observations of self-incompatibility, 170 genets of *Festuca rubra* were identified from $c.84$ m^2 of a Scottish hillside. Although 51% were represented by a single genotype, 90% of genets were represented by a single sample (Harberd 1961). These data indicate that the population of *F. rubra*, which is strongly rhizomatous, was dominated by one dominant genotype which presumably spread vegetatively throughout the area. Nevertheless, sexual reproduction through seed was allowing a high diversity of rarer genotypes to persist. A molecular study using RAPD analysis of *F. rubra* from a comparable mountain grassland in the Czech Republic estimated a substantially higher genet diversity (231–968 genets m^{-2}) of which 68% of 145 different ramets identified were found only once (Suzuki *et al*. 2006). Estimated rates of genet turnover were low (0.1–1% annually) as was seedling recruitment (7–17 genets m^{-2} yr^{-1}), allowing the maintenance of a high level of genet diversity.

Within habitats, the different genotypes of a species are not randomly distributed. Rather, genotypes can show a spatial distribution related to habitat variation. For example, allele–habitat associations related to pH, moisture, and soil depth were demonstrated for *Festuca ovina* in a Swedish alvar grassland (Prentice *et al*. 1995). Subsequent experimental work using nutrient and water additions, indicated that the allele–habitat associations were under a measure of selective control that was maintaining the fine-scale genetic mosaic in these grasslands (Prentice *et al*. 2000).

Landscape-level studies indicate that genetic diversity of grass populations decreases through time as grasslands mature, but is also affected significantly by land-use history and grassland management regime. For example, gene diversity was highest in *Briza media* from grassland fragments with a high proportion of adjacent grassland (Prentice *et al*. 2006). Initially high genetic diversity in younger habitats results from spatial patchiness of genetic variability following initial colonization. Decreasing genetic diversity follows 'environmental sorting' or directional selection of genotypes leading to a loss of patch structure and is consistent with the **Wahlund effect**—the loss of heterogeneity as populations diverge through time.

The change in diversity of genotypes within a population through time affects the coexistence of all species in the community. Aarssen and Turkington (1985b) suggested that a decline through time in an initally high diversity of genotypes in four Canadian pastures occurred as less-fit genotypes were eliminated through competition and grazing. Neighbouring genotypes of *Lolium perenne* and *Trifolium repens* in these pastures were shown to exhibit precisely defined biotic specialization suggesting an ecological sorting of genotypes through natural selection to result in local neighbourhoods of balanced competitive abilities (Aarssen and Turkington 1985a).

There is growing concern over genetic erosion in natural and semi-natural grasslands; i.e. the loss of unique plant genes or genotypes. Genetic erosion can occur as a result of the introduction (by sowing or gene flow) of new cultivars and varieties into grasslands either deliberately or accidentally through management and land-use practices. Sackville Hamilton (1999) demonstrated that genetic erosion was a major problem in UK grasslands with the biggest problems being introgression of genes from improved varieties into native populations and the loss of traditional landraces. However, the few empirical studies available show that although gene flow from cultivars into native populations can occur, the extent to which this can lead to genetic erosion is equivocal (Anttila *et al.* 1998; Van Treuren *et al.* 2005; Warren *et al.* 1988). Indeed, an isozyme study of adjacent 20-year-old plots of *Andropogon gerardii* sown with either local native seed or non-local cultivar seed showed that while gene flow occurred among populations producing introgressed seed, no evidence of gene flow was present in the maternal plants; i.e. the genetically mixed seed was not establishing in the plots (Gustafson *et al.* 2001). By contrast, gene flow from domesticated crops into non-crop wild relatives in the grass family has been implicated for wheat *Triticum aestivum*, rice *Oryza sativa*, pearl millet *Pennisetum glaucum*, and sorghum *Sorghum bicolor* (Ellstrand *et al.* 1999). Indeed the enhanced weediness of *Sorghum halepense* may be the result of introgression of genes from *S. bicolor*. Gene flow from crop to wild plant is currently a big concern in regard to the development and planting of transgenic crops (Ellstrand 2001). Introgressed transgenic DNA constructs have been reported in native landraces of maize *Zea mays* in remote mountain areas of Mexico (Quist and Chapela 2001), although the validity of these findings have been hotly disputed (Metz and Fütterer 2002; Ortiz-Garcia *et al.* 2005).

CHAPTER 6

Community ecology

...a plant community, the central object of study of plant sociology; round this unit knowledge is accumulated for use and classification. The functional relations of the community may be systematized in laws and hypotheses.

What I want to say is what T.S. Eliot said of Shakespeare's work: we must know all of it in order to know any of it.

Watt (1947)

Competition in grassland is necessarily keener and follows a somewhat different course than in forest, owing to the absence of a canopy....Moreover, the dominance of grasses is chiefly a matter of competition for water, and light habitually plays a rôle of quite secondary importance.

Clements *et al.* (1929)

Community ecology considers an assessment of all the species in an area and their interactions and is thus a hierarchical step up from population ecology (Chapter 5). To understand the community ecology of grasslands we need to consider the temporal and spatial patterns exhibited by the species and the processes involved in their interactions with each other and their environment. The history of seeking to understand pattern and process in grassland can be traced back to the pioneering field work of ecologists including Cowles (1899), Clements (1936), and Weaver (see Chapter 1), and the theoretical ideas of Gleason (1917; 1926). The ideas from these early studies were brought together by Alex S. Watt in his Presidential Address to the British Ecological Society (Watt 1947). In his address, Watt described how plant communities, including the acidic Breckland grassland in the UK that he studied for over 30 years (Watt 1940; 1981a), are dynamic in time and space reflecting the growth of the different species. He described plant communities as being patchy and a mosaic of different species at various growth stages; a series of small-scale cyclic replacements. Modern work in grasslands and other habitats has moved these ideas forward (Newman 1982; van der Maarel 1996) allowing us to better understand some of the important patterns and vegetation–environment relationships (§6.1), such as succession (§6.2), in terms of important mechanisms (e.g. competition, facilitation, allelopathy, parasitism, mutualism; §6.3) and current models of community structure (e.g. Tilman's vs Grime's models, metacommunity models, and niche vs neutral models; §6.4). The disturbance regime is an important component of grassland structure and is considered in detail in Chapter 9. The issue of spatial and temporal scaling underlies all attempts to understand grassland communities and is described in §6.5.

6.1 Vegetation–environment relationships

At the largest scales, the distribution of grasslands and their composition is related to climate (Chapter 8). Within grassland communities, clear vegetation–environment relationships are observed at multiple scales. The composition of a particular grassland can be related to one or more factors including local climate, topography, slope, aspect, bedrock and soil, soil moisture and nutrient status, disturbance (Chapter 9), age (time since establishment), and management (Chapter 10). These types of relationship can be summarized and explored using multivariate methods of cluster analysis and ordination (Greig-Smith 1983; Legendre and Legendre 1998; Ludwig and Reynolds 1988). Cluster analysis is used as an objective tool to develop grassland classifications (Chapter 8).

The use of ordination to explore grassland vegetation–environment relationships is illustrated here by reference to two examples, from the African Serengeti grassland and the North American true and upper coastal prairie grasslands.

At the smallest scales, grassland exhibits fine-scale heterogeneity or patchiness that can be related to the response of individual species to their bioitic neighbourhood and their abiotic environment. This was recognized by Watt (1947), and has been the focus of much subsequent research (van der Maarel 1996). For example, small-scale pattern of the vegetation in a dune grassland in Wales was strongly related to the occurrence of tussocks of the dominant grass, *Arrhenatherum elatius*, and the pattern of soil nutrients (Gibson 1988c). Experimental work showed that the soil nutrient patterns were due to nutrient accumulation under the tussocks of *A. elatius* (Gibson 1988a). When this grassland was grazed by Soay sheep, the plant spatial patterns changed as the dominant grasses were grazed close to the soil surface allowing smaller plant species to spread independent of the soil nutrient patterns.

6.1.1 Serengeti grassland communities

The Serengeti grassland in Africa is a tropical/subtropical bunchgrass savannah (§8.2.1 and Plate 4) that occurs over a vast 25 000 km² area of northern Tanzania and southern Kenya. To assess vegetation–environment relationships, the vegetation in 105 stands in the Serengeti National Park and Masai Mara Game Reserve were surveyed in 1975 (McNaughton 1983). These data were analysed by first identifying 17 plant community types from a nearest-neighbour cluster analysis based on proportional similarities calculated between all pairs of stands. Polar ordination was then used to ordinate the 17 plant community types with the relationship of the ordination axis scores to environmental variables assessed through correlation analysis. *Themeda triandra* was the most important species and dominant in 6 of the 17 plant community types. A continuum of shortgrass, mid-to-tall grass, and floodplain grasslands was identified (Fig. 6.1), and compositional variation was most strongly related to grazing intensity and soil

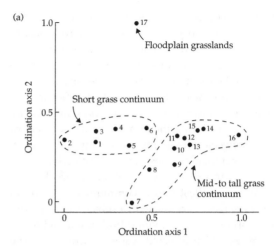

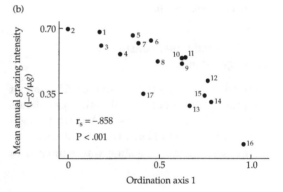

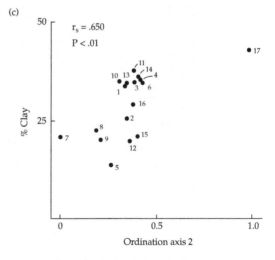

Figure 6.1 (a) Polar ordination of 17 grassland communities from the Serengeti ecosystem, Africa; (b) relation of the first ordination axis to mean annual grazing intensity, where μg denotes standing crop inside exclosures and g the standing crop outside them; (c) relation of the second ordination axis to soil clay content. Reproduced with permission of the Ecological Society of America from McNaughton (1983).

texture. The Serengeti is characterized by large herds of migratory ungulates, and the relationship between vegetation composition and grazing intensity was reflective of the interaction between grazers and the vegetation. Grazing intensity was itself weakly correlated with precipitation, reflecting the habit of the migratory grazers to concentrate in the driest areas during the rainy season. The relationship between soil texture, specifically clay content, and vegetation composition was reflective of the variation in soils overlying volcanic bedrock.

At the smallest scales, community patterns in the Serengeti were related to subsurface Na^+ concentrations and the mound-building activities of termites which affected water infiltration of the soil (Belsky 1983, 1988).

6.1.2 True and upper coastal prairie grasslands

Diamond and Smeins (1988) surveyed 63 remnant true prairie and upper coastal prairie grasslands from the Gulf Coast of Texas, USA north through the Great Plains into North Dakota. At each site, canopy cover of the vegetation was estimated in 25 0.125 m^2 quadrats. The objective was to relate composition, structure, and diversity of the vegetation to environmental variables. A principle components analysis (PCA) ordination of these data revealed two major components of variation in community composition (Fig. 6.2). The first PCA axis illustrated a north–south gradient in composition that was positively correlated with temperature and precipitation, and negatively correlated with soil organic matter. Along this compositional gradient, species richness and the ratio of C_3 to C_4 graminoids increased north to south as some of the dominant species changed in abundance: *Paspalum plicatulum*, *P. floridanum*, *Schizachyrium scoparium*, *Sorghastrum nutans*, *Sporobolus compositus*, *Mimosa microphylla*, *Ratibida columnifera*, and *Ruellia* spp. decreased towards the north, while *Amorpha canescens*, *Andropogon gerardii*, *Symphyotrichum ericoides*, *Carex pensylvanica*, *C. tetanica*, *Helianthus rigida*, *Sporobolus heterolepis*, and *Hesperostipa spartea* increased towards the north. The second PCA axis reflected a soils gradient in the composition of these grasslands, with a negative correlation with percentage clay and pH, but only among the 35 Texas sites. Overall, these grasslands illustrated a continuum in

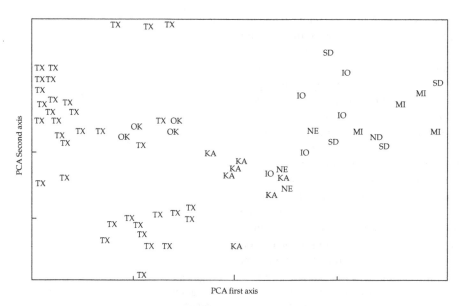

Figure 6.2 Principal components analysis ordination of 63 upland grasslands with the true and upper coastal prairies based on frequency of 42 perennial graminoids. Site locations are labelled by state: IO, Iowa; KA, Kansas; MI, Minnesota; NE, Nebraska; ND, North Dakota; OK, Oklahoma; SD, South Dakota; TX, Texas. Reproduced with permission from Diamond and Smeins (1988).

compositional variation related to the environment across a 192 000 km north–south gradient.

From these very large-scale studies of regional grasslands we can conclude that climatic factors are the principal drivers of the composition of grassland communities at the largest scale. At smaller scales, local factors including soil, topography, and disturbance are important. The specific factors that relate composition to the environment depend upon the scale of analysis (see §6.5).

6.2 Succession

Succession is the change in plant communities through time. **Primary succession** describes the sequence of plant communities on a new site, an area without any prior colonization of plants or animals, e.g. on volcanic lava, a new island arising from the sea, a mud slide, surfaces or sediment exposed by glaciers, or on anthropogenically derived materials such as mine spoil or dredge material. **Secondary succession** refers to the change in communities that occurs in an area after a disturbance, i.e. on a site that already has vegetation cover and soil which is altered by the disturbance, e.g. following fire, flooding, storm damage, clear cutting, or agricultural field abandonment. Except in a very general sense (e.g. the expectation that forests will develop in the eastern USA), the plant communities that develop during succession are not predictable. The sequence of change is not deterministic; succession is a probabilistic process that does not lead to a predetermined end point. This individualistic viewpoint is in contrast to the older climax view that dominated plant ecology and range management (see Chapter 10) during most of the twentieth century. Very generally, grasslands are commonly an intermediate stage in succession, and without frequent disturbance they often convert to forests.

6.2.1 Primary succession

Grasses are common components of the early to intermediate stages of primary succession, especially in areas succeeding to forest (e.g. on unreclaimed coal strip-mines in Oklahoma, USA, grass cover declined with age as the abundance of trees increased; Johnson *et al.* 1982). Open grasslands dominated by 3 m tall *Saccharum sponteneum* occurred as one of the early communities on the Krakatau islands, Indonesia, following the eruption, destruction, and sterilization of this volcanic island in 1883 (Whittaker *et al.* 1989). In some habitats a grass or grasses comprise the dominant vegetation of early succession. For example, estuarine mudflats can be colonized by *Spartina* spp. and often very little else. In north-west Spain, *S. maritima* colonizes mudflats through vegetative reproduction from rhizome fragments transported by tidal currents (Sanchez *et al.* 2001). As individual plants grow they form a mat of individuals that eventually coalesce to form a more continuous sward. Similarly, a *Poa alpina–Cerastium* spp. vegetation type was identified as a pioneer vegetation type on recently deglaciated terrain from the Storbeen glacier foreland, Norway (Matthews 1992).

As noted earlier (§5.3), industrial wastes, especially those contaminated with heavy metals, are colonized by grasses. The flora of these sites is often unique because of the physiological properties of the grasses and the small number of other species that are able to colonize the contaminated soils (Shu *et al.* 2005). For example, in Great Britain, the *Minuartio–Thlaspietum alpestris* community is a local community recognized in the National Vegetation Classification that is restricted to lead mines and outcrops of heavy metals in northern and western Britain (Rodwell 2000). The community comprises an open turf of scattered tussocks of *Festuca ovina*, with patches of *Agrostis capillaris*, and individuals of the herbaceous plants *Minuartia verna* and *Thymus praecox*. The island-like nature of these habitats across the landscape has provided the opportunity to test hypotheses related to the importance of immigration and dispersal in affecting primary succession (Ash *et al.* 1994).

Grasslands often dominate sand dune communities, dominated in the northern hemisphere by the grasses *Ammophila arenaria* (Europe), *A. breviligulata* (North America) or *Uniola paniculata* (US Atlantic coast from Virginia southwards, and US Gulf coast down into Mexico). Some of the earliest ecological studies describing the patterns of primary succession were undertaken in dune habitats (Cowles 1899; Olson 1958); Cowles in particular emphasized the dynamic nature of succession. Primary succession in these habitats follows a pattern of nucleated succession (Yarranton and Morrison 1974) where the first established plants form a focus for the facilitated establishment of later arrivals. This mode of establishment leads to a spatially aggregated distribution of plants (Franks 2003). The rate of primary succession in these coastal dune habitats is extremely variable, depending to a large extent on the frequency of storms which can set back colonization rates of the dune-binding grasses (Gibson et al. 1995). Other factors influencing the rate of primary succession include the presence of soil microorganisms (Putten et al. 1993), local topography, and soil moisture (Ehrenfeld 1990).

6.2.2 Secondary succession

Secondary succession in grasslands has been widely investigated with compositional changes following disturbance described from grasslands the world over. Some of the most valuable studies are those with data collected over a long period of time. Here we consider the insights that can be gained from studies conducted over 30 years (Table 6.1).

It is clear that successional changes following disturbance (or removal of a disturbance, e.g. grazing) can be rapid, but heterogeneous and unpredictable at the species level (e.g. Collins and Adams 1983). A four-stage sequence of secondary succession for grasslands in Oklahoma, USA has been predicted; i.e. (1) 2–3 year weed stage, (2) 9–13 year bunchgrass stage, (3) variable-length perennial grass stages, and (4) mature prairie (Booth 1941). However, examination of a 32 year sequence identified only the pioneer and mature prairie stages (Collins and Adams 1983). Species changes can be rapid in response to local climate (e.g. Albertson and Tomanek 1965). Stable (alternate) states may be reached in response to new conditions, especially if woody taxa invade and establish (e.g. Buffington and Herbel 1965), that will not necessarily be diverted by reinstatement of the natural disturbance regime without intervention (management, e.g. physical removal of woody taxa). Grazing can cause 'arrested successional development' whereby the process of succession is effectively halted (Kemp and King 2001). Late in succession, fluctuations in composition (see below) may occur without a clear directional trend in the vegetation (Collins et al. 1987; Wahren et al. 1994; Watt 1981b). Temporal fluctuations mean that simple monotonic models of succession may not apply. In addition, transient dynamics may be evident as one species and then another assumes temporary dominance in response to new conditions (Fynn and O'Connor 2005). The influence of controlling variables varies considerably among different successional sequences; climatic variation can be of overriding importance (Albertson 1937), but where climate varies little, local factors can be important, such as soil and topography (Bragg and Hulbert 1976). Experimental studies have implicated a hierarchical limitation of controlling variables in support of Liebig's law of the minimum (Fynn and O'Connor 2005); i.e. the resource in shortest supply will be of primary importance followed by the next limiting resource, and so on. To compound matters, the spatial scale of investigation and analysis is important for resolving different features of the successional patterns (Collins 1990). At the smallest, quadrat, scale, variation in species abundance may be high, unpredictable, and chaotic, whereas at larger scales, high potential predictability and movement towards a fluctuating, stable or equilibrial stage may be evident (Fuhlendorf and Smeins 1996). Overall, grassland succession is difficult to predict except perhaps at the physiognomic level with life forms related to climate, and to the time, frequency, and intensity of the last disturbance (Collins and Adams 1983; Collins et al. 1995; Silvertown et al. 2006).

It is worth dwelling for a moment on the results obtained from the Park Grass Experiment (Silvertown et al. 2006). Established in 1856 to test the effects of common farmyard fertilizers, this, the longest-running ecological experiment, has yielded important ecological results. As far as the grassland community is concerned, it was noted that the composition changed rapidly in response to fertilizer treatment over the first few years of the experiment, but that although the species composition continued to change, the proportion of different life forms (grasses, legumes, others) stabilized with respect to soil pH and production early in the twentieth century. The continued changing abundance of individual species suggests independent regulation of species and guild structure. The clear relationship between species composition and fertilizer treatment led to the use of these data in an initial field test of the **resource ratio hypothesis** (§6.4.2).

More recent shorter studies of secondary succession in grasslands reinforce the observations made above from the long-term studies. In addition, these shorter studies are revealing shifts in composition or alterations in the rate of succession in response to projected global climate change (Vasseur and Potvin 1998), and the propensity for non-native species to invade and establish following disturbance (D'Angela et al. 1988; Ghermandi et al. 2004).

In the absence of major disturbance, vegetation changes are non-successional and said to exhibit **fluctuations** (i.e. non-directional irregular changes). Fluctuations are distinguished from successional changes in that they are reversible and there is no net compositional change or invasion of new species (Rabotnov 1974). These compositional shifts can be seasonal or annual cycles (Kemp and King 2001), or occur over a longer term. Seasonal cycles reflect the progression of growth of different species over the course of a year as different suites of species assume dominance at different times. For example, in Australian pastures, perennial grasses (*Phalaris aquatica, Dactylis glomerata, Lolium perenne*) dominate early in the growing season when annual growth starts, whereas annual species (*Vulpia bromoides, Bromus hordeaceus*) including legumes (especially *Trifloium subterraneum*) become more evident in the spring (Kemp and King 2001). As soil moisture declines in the summer the legumes may die, creating gaps in which weedy species (e.g. *Echium plantagineum*) can grow following intermittent summer rain events. A 4 year cycling of the vegetation at Konza tallgrass Prairie, Kansas, USA, was noted, for example, coupled to a 4 year burning cycle (Gibson 1988b). Over the course of these regular cycles, two annual species (*Erigeron strigosus* var. *strigosus* and *Viola rafinesquii*) were noted to change in frequency as they were extirpated during the year of the fire, and increased in abundance over the next 3 years before the next fire. Other grasslands reported to exhibit fluctuations include Russian meadows (Rabotnov 1955), dune grassland in the Netherlands (van der Maarel 1981), sand sagebrush grassland in Oklahoma, USA (Table 6.1; Collins et al. 1987), and the mesotrophic grassland of the Park Grass Experiment (Table 6.1; Silvertown et al. 2006). The changes in abundance of individual species in such systems may oscillate in a chaotic manner (Tilman and Wedin 1991a).

Together, these observations of grassland succession indicate that although species behaviour is individualistic, communities can vary (fluctuate) in a constrained, sometimes chaotic manner, allowing a reconciliation between traditional Clementsian (deterministic) and Gleason (individualistic) viewpoints (Anand and Orloci 1997). In particular, the **integrated community concept** postulates that communities represent the synergism among (1) stochastic processes, (2) the abiotic tolerances of species, (3) positive and negative interactions among plants (§6.3), and (4) indirect interactions within and between trophic levels (Lortie et al. 2004). There is accumulating evidence that these complex processes structure grasslands (Holdaway and Sparrow 2006)

6.3 Species interactions

Interactions between species can be defined as those mechanisms that affect community structure allowing the community to be viewed as possessing emergent properties greater than the sum of the individual plants (van Andel 2005). The various types of interactions among plants growing together in a community depend upon the outcome of the interaction which can be positive (advantageous: +), negative (disadvantageous: −), or indifferent to either or both species (0); i.e.

	Species A	Species B
Competition	−	−
Allelopathy	0	−
Parasitism	+	−
Facilitation	0	+
Mutualism	+	+

6.3.1 Competition

The interaction between plants referred to as competition has been defined in many different ways; Begon et al. (2006) provides a clear

Table 6.1 Examples of long-term (>30 years) studies of secondary succession in grassland arranged in decreasing order of length of the record

Type of grassland (dominant species)	Location	Length of records	Treatments/environmental factors	Reference
MG5 Mesotrophic grassland	Park Grass Experiment, UK	>150 years, ongoing since 1856	Soil pH, nutrients, hay cutting	Silvertown et al. (2006)
Tallgrass prairie (*Andropogon gerardii, Panicum virgatum, Sorghastrum nutans*)	Geary County, Kansas, USA	114 years, 1856–1969	Tree cover assessed on burned and unburned sites on different soils	Bragg and Hulbert (1976)
Semidesert grassland (*Artistida* spp., *Bouteloua eriopoda, Hilaria mutica, Scleropogon brevifolius, Sporobolus flexuosus*)	Jonarda Experimental Range, New Mexico, USA	105 years, 1858–1963	Abundance of dominant shrubs (*Flourensia cernua, Larrea tridentata, Prosopis glandulosa*) assessed by soil type	Buffington and Herbel (1965)
Tallgrass prairie (*Schizachyrium scoparium*)	Kansas Flint Hills, Kansas, USA	54 years, 1928–1982	Annual burning in different seasons of ungrazed prairie	Towne and Owensby (1984)
Southern tall grassveld (*Heteropogon contortus, Themeda triandra, Tristachya leucothrix*)	Ukulinga research Farm, KwaZulu Natal, South Africa	> 50 years, 1950–ongoing	Influence of nutrients, productivity, and soil pH	Fynn and O'Connor (2005)
Desert grassland range	Santa Rita Experimental Range, Arizona, USA	50 years, 1904–1954	Abundance of four woody plants (*Prosopis velutina, Opuntia fulgida, O. spinosior, Larrea tridentata*) surveyed in response to climate, grazing and rodents.	Humphrey and Mehrhoff (1958)
Subalpine grassland (*Poa hiemata*)	Bogong High Plains, Australia	48 years, 1946–1994	Cattle grazed and ungrazed plots compared.	Wahren et al. (1994)
Semi-arid grassland (*Hilaria belangeri*)	West Texas, USA	45 years	Response following release from overgrazing and drought.	Fuhlendorf and Smeins (1996)

Sand sagebrush grassland (*Artemisia filifolia*, *Schizachyrium scoparium*, *Bouteloua gracilis*, *Andropogon hallii*, *Sporobolus cryptandrus*)	USDA-ARS Southern Plains Experimental Range, Oklahoma, USA	39 years, 1940–1978	Cattle grazed and ungrazed pasture	Collins et al. (1987)
MG1 *Arrhenatherum elatius* grassland	Bibury Verges, UK	> 38 years, 1958 and ongoing.	Mowing	Dunnett et al. (1998)
Festuca ovina, *Hieraceum pilosella*, *Thymus drucei* grassland	Grassland 'A', East Anglian Breckland, UK.	38 years, 1936–1973	Grassland with rabbits excluded	Watt (1981b)
Tallgrass (*Andropogon gerardii*, *Panicum virgatum*, *Schizachyrium scoparium*, *Sorghastrum nutans*) prairie	University of Oklahoma Grassland Research Area, Oklahoma, USA	32 years, 1949–1981	Three ungrazed plots, one protected, two initially plowed.	Collins and Adams (1983); Collins (1990)
Sagebrush (*Artemisia tridentata*)–grass range	Upper Snake River Plains, Idaho, USA	30 years, 1936–1966	259 ha burned in 1936 and sheep grazed from 1938 onwards.	Harniss and Murray (1973)
Shortgrass (*Bouteloua gracilis*–*Bouteloua dactyloides*), mixed grass (*Schizachyrium scoparium*), and tallgrass (*Andropogon gerardii*) prairie	Hays, Kansas, USA	30 years, 1932–1961	Three ungrazed prairie areas subject to two severe droughts.	Albertson and Tomanek (1965)

definition: 'Competition is an interaction between individuals, brought about by a shared requirement for a resource and leading to a reduction in the survivorship, growth and/or reproduction of at least some of the competing individuals concerned.' This interaction is quite complex as it can involve members of the same species (**intraspecific competition**) or be between different species (**interspecific competition**), and it involves identifying the cause (the limited resources) and the net effect (yield or fitness reduction). Mechanistically, competition requires some intermediary, a resource that both competitors need; both respond to changes in the abundance of the resource (e.g. light, soil moisture). It is important to identify and measure the suppressive effect that one individual has on another (**net competitive effect**) and the response of an individual to competition from others (**net competitive response**) (Goldberg 1990). Competition can be **symmetric**, in which case the effects are proportional to the size of individuals concerned, as it usually is for competition for belowground resources. Alternatively, competition can be **asymmetric,** in which case the 'winner' wins out of proportion to its size, as in the case of light competition where a plant overtopping another shades neighbours no matter what size they are or how large it is (it simply has to place its leaves between the sun and their leaves).

Competition as a phenomenon and a process has been foremost in the understanding of community structure in grasslands since the work of Clements et al. (1929). Their work was undertaken in North American prairie and involved growing plants together in the field and greenhouse starting either from seedlings or from transplanted sod for several seasons. The general conclusion was that competition was of paramount importance in determining composition and structure of the mature prairie. Of the factors tested, competition for soil water was determined to be of greatest relative importance, with competition for light second, and nutrients last. The position of species in the canopy was of particular importance in determining the factors competed for. Dominant species occupying the uppermost layers of the canopy were competing primarily for light, whereas soil moisture was more important for competition between the subdominant species. The outcome of competition between seedlings of different species was determined primarily by emergence time and growth rate—the species that started growth first and grew the fastest had an advantage, except that tolerance to low soil moisture status (particularly drought) could overcome a late start and slow growth, allowing drought-tolerant species to become dominant. Their work foreshadowed later work in suggesting the mechanistic importance of the ratio between supply and demand of resources. They also pointed out that the competitive balance among species can be significantly altered by grazing, which tends to disadvantage the taller dominants and favour the shorter subdominants.

In the 1960s the role of competition in agricultural systems, especially pastures, gained prominence with the development of the de Wit replacement series design accompanied by a mathematical analysis that allowed niche relationships between competing individuals to be quantified on the basis of yield reduction (de Wit 1960). This work, along with Harper's recognition of its ecological importance in pastures (Harper 1978), spawned an enormous worldwide research effort that continues to this day. The pre-eminent models of community structure are based on competitive interactions revealed from studying grasslands (§6.4). Discussing the processes involved in the performance of natural and sown pastures in the introductory chapter in their book on competition and succession in pastures, Tow and Lazenby (2001) noted, 'Plant competition is an important factor controlling these processes' but went on to comment that 'the nature of such competition is not yet fully understood.'

Competition in grasslands involves biotic factors (e.g. pollinators, Anderson and Schelfhout 1980) and abiotic factors including light above ground (e.g. Dyer and Rice 1997) and soil nutrients and soil moisture below ground (e.g. Sharifi 1983), or, in multispecies mixtures, an interaction among factors (Schippfers and Kropff 2001, see §6.4). The competitive ability of grasses is related to their particular growth form, with laminate leaves positioned to limit self-shading and a diffuse, adventitious root system providing a large surface area close to the soil surface (Chapter 3). The competitive effects of grasses were summarized by

Lauenroth and Aguilera (1998) as including (1) an alteration of light quantity and quality under shaded leaves with photosynthetic photon flux density, infrared radiation, and shortwave radiation decreasing with increasing with leaf area, and (2) ability to deplete soil nutrients and soil water from the surrounding rhizosphere with redeposition of nutrients under plants (e.g. Gibson 1988a). These effects enable grasses to reduce soil nutrients (Tilman 1982; Tilman and Wedin 1991b) and water (Eissenstat and Caldwell 1988), and shade out their neighbours (e.g. Skeel and Gibson 1998). The competitive responses of grasses include (1) increased tillering in response to high red:far red radiation, (2) differential proliferation of root systems and elevated uptake capacity in response to spatial heterogeneity in nutrient distribution, and (3) reduced ability to respond to increased water availability after drought.

The competitive balance among species in grasslands is altered by fire (e.g. Curtis and Partch 1948), grazing (e.g. Briske and Anderson 1992), and other forms of disturbance (Chapter 9), climate (temperature, rainfall, sunshine, and now how their effects on competition are changing with global climate change, Polley 1997), the presence of soil mycorrhizae (Chapter 5) and parasites (§6.3.3), the production of allelochemicals (§6.3.2), and the genetic identity of the competitors (Gustafson *et al.* 2002; Helgadóttir and Snaydon 1985). Numerous field studies in grasslands have shown an increase in growth of supposedly suppressed species when competition is relaxed, e.g. by removing potential competitors, indicating the effect of competition on affecting niche relationships (e.g. Mueggler 1972).

A general review of competitive relationships among grassland plants is provided by Risser (1969), who concluded optimistically that

If for any given species the seed size and number, time of germination, rate of vegetative reproduction, rate of growth, maximum number and size of individuals attained under optimum conditions, soil level at which roots operate, time of and conditions for initiation of root and shoot growth, and any allelopathic considerations are known, a reasonable prediction can be made concerning the success of that species relative to any other for which the same information is known.

6.3.2 Allelopathy

Allelopathy is defined as 'any direct or indirect harmful or beneficial effect of one plant (including microorganisms) on another through production of chemical compounds that escape into the environment' (Rice 1984). The chemical nature of some of the allelochemicals produced in grasses is described in §4.3 (and see Gibson 2002). Some of the first experimental demonstrations of allelopathy involved chemicals produced from the roots of grasses, i.e. *Triticum aestivum* (wheat) and *Avena sativa* (oats) (Hierro and Callaway 2003). The role of allelopathy as a mechanism structuring plant communities, including grasslands, remains controversial. The exclusion of annual grasses (*Avena fatua, Bromus rigidus,* and *B. mollis*) in zones around stands of *Brassica nigra* was famously attributed to the release of toxic allelochemicals into the soil by *B. nigra* (Bell and Muller 1973). The early stages of old-field succession to prairie in Oklahoma, USA were reported to be hastened by the production of allelochemicals by pioneer weed-stage species including *Sorghum halepense* which produced self-inhibitory phytoxins. Some of these species also produce phenolics which inhibit nitrifying and nitrogen-fixing bacteria, allowing colonization of species with low nitrogen requirements such as *Aristida oligantha* (Abdul-Wahab and Rice 1967). More recently, allelopathy has been clearly implicated in the invasion of exotic species into grasslands. For example, the Eurasian thistle *Centaurea diffusa* inhibited the growth of North American bunchgrasses (*Festuca idahoensis, Koeleria macrantha,* and *Pseudoroegneria spicata*) from Montana grasslands significantly more than it did congeneric species (*Festuca ovina, Koeleria laerssenii,* and *Agropyron cristatum*) from its native Eurasian habitat (Callaway and Aschehoug 2000; Hierro and Callaway 2003). The North American species had no effect on *C. diffusa* but was negatively affected by the Eurasian species.

6.3.3 Parasitism

Parasites depend on their host to ensure their own fitness, whereas the host does not need the parasite. **Holoparasites** are heterotrophic, lacking chlorophyll and dependent on their host for both root and shoot products. **Hemiparasites** can photosynthesize and obtain products only from roots of their host. Grassland parasites include the holoparasite vine *Cuscuta* and the hemiparasite annuals *Agalinis, Rhinanthus,* and *Tomanthera,* and the perennial *Striga.* The genus *Striga* includes 11 species regarded as serious agricultural pests, especially of the C_4 grass cereals maize, millet, sorghum and sugar cane, and the C_3 rice (Cochrane and Press 1997). Most parasites are generalists infecting many different hosts; *Rhinanthus minor* is reported parasitizing >150 hosts from 18 families (30% in Poaceae) in European grasslands (Gibson and Watkinson 1989). These parasites can greatly reduce the fitness of individual plants, and can affect community structure through shifting the competitive balance among species (Press and Phoenix 2005). For example, sowing the generalist hemiparasite *Rhinanthus alectorolophus* into experimental grassland plots reduced host plant biomass, especially that of the grasses, and especially in plots with low functional type diversity (Joshi *et al.* 2000). *Rhinanthus* spp. in grasslands in Britain and Italy reduced productivity by 8–73% and shifted the compositional balance among species by causing a decline in grass abundance, allowing the proportion of dicotyledons to increase (Davies *et al.* 1997). *Rhinanthus minor* preferentially infests fast-growing grasses, reducing their competitive dominance, and in doing so increases plant diversity and reduces community productivity (Bardgett *et al.* 2006).

6.3.4 Facilitation

Facilitation is the modification of the abiotic environment by a plant so that it becomes more suitable for the establishment, growth, and/or survival of other physiologically independent plants either in space or in time (Brooker *et al.* 2008; van Andel 2005). Thus, plants may, for example, improve the soil conditions for future plants or act as nurse plants sheltering seedlings of other plants by improving the light, humidity, temperature, or soil moisture and nutrient conditions. In these conditions, the positive benefits that must occur to allow recognition of facilitation must outweigh the negative competitive effects of close proximity to another plant such as low light intensity under the canopy of the nurse plant (Holmgren *et al.* 1997). Facilitation forms one end of a continuum of interactions between plants, with competition at the other end (Callaway 1997). To fully understand the nature of these interactions it is important not only to recognize the phenomena, but also to recognize the underlying mechanistic basis for the interaction. Much of the ecological theory that has been developed to understand community structure gives primacy to competitive interactions and lacks adequate inclusion of facilitation as an important process (Bruno *et al.* 2003; Cheng *et al.* 2006; Lortie *et al.* 2004).

In grasslands, tall grasses can provide shelter and hence facilitate the establishment of seedlings. For example, recruitment of seedlings of *Schizachyrium scoparium* seedlings in Kansas grassland was facilitated in the presence of neighbours, but only in low-productivity sites where the neighbours were less competitive (Foster 2002). Similarly, neighbouring plants facilitated the establishment of *Arabis hirsuta* and *Primula veris* in nutrient-poor limestone grassland (*Mesobrometum*) in Switzerland (Ryser 1993). By contrast, other species such as *Plantago lanceolata* and *Sanguisorba minor* established slightly better in gaps rather than in the presence of neighbours, indicating a beneficial effect of avoiding competitive stress. The interplay between competition and facilitation is illustrated in Patagonian steppe where shrubs (mostly *Mulinum spinosum*) offer aerial protection for seedlings of *Bromus pictus*, but only when root competition for water by established grasses was experimentally reduced (Aguiar *et al.* 1992). Moreover, when the shrubs are young, they are not surrounded by zones of competitive grasses and so at this stage offer protection and facilitate grass establishment (Aguiar and Sala 1994).

Facilitation through improved soil conditions was illustrated in experimental grasslands in Minnesota, USA, where nitrogen-fixing legumes (*Lespedeza capitata* and *Lupinus perennis*) stimulated overyielding from other herbaceous perennials (*Achillea millefolium* and *Monarda fistulosa*) when the dominant grasses were absent. This facilitative effect of the legumes contributed to a positive diversity–productivity relationship in the grassland (Lambers *et al.* 2004). Grasses may also facilitate each other through a combined positive effect on microclimate (Kikvidze 1996).

6.3.5 Mutualism

This type of interaction benefits both partners. Important mutualistic interactions in grasslands include the interaction with fungal mycorrhizae and endophytes (see §5.2 for details), bacterial rhizobia, and pollinators (§5.1 for details on grass pollination). *Rhizobium* is a soil bacterium that lives in nodules induced to grow on the root hairs of leguminous (Fabaceae) plants. *Trifolium repens*, the most important non-grass of seeded ryegrass–clover (*Lolium perenne–T. repens*) pastures worldwide (Chapter 8), is a legume. The *Rhizobium* bacteria fix atmospheric nitrogen, making it available to the legume as NH_4^+ (§7.2). Some of this nitrogen becomes transferred to non-legumes in the community through one or more of several mechanisms: (1) decomposition of donor plant debris, (2) sloughing off of cortex cells and subsequent uptake of mineralized nitrogen, (3) mycorrhizal hyphal connections (§5.2), or (4) uptake of nitrogen compounds from legume root exudates (Payne *et al.* 2001). The additional nitrogen available to legumes alters the competitive balance among grasses and legumes in pastures (Davies 2001), providing a competitive advantage to the legume in nitrogen-limited grasslands, especially under conditions of frequent defoliation (grazing) (Davidson and Robson 1986). Indeed, the *Rhizobium–T. repens* mutualism is critical and determines to a large extent the coexistence of *L. perenne* and *T. repens* in low-input grazed pastures and provides a useful, relatively simple, model system for understanding grassland communities (Schwinning and Parsons 1996). In more complex natural grasslands, nitrogen-fixing legumes play a similar role in the nitrogen budget of the system. In North American tallgrass prairie, for example, legumes vary widely in nodulation and nitrogen-fixing ability, with greater nitrogen fixation observed in early compared to late successional species (Becker and Crockett 1976), and are more abundant in nitrogen-limited annually burned grassland than in less frequently burned grassland (Towne and Knapp 1996). The mutualistic relationship between bacteria and legumes contributes to strong facilitative relationships between legumes and other grasses (Lee *et al.* 2003; Rumbaugh *et al.* 1982; Spehn *et al.* 2002).

Plant–pollinator interactions are important in grasslands for ensuring fertilization of the non-grass members of these communities. Many plant–pollinator interactions are non-specific, but, when they are, the plant species concerned can face reproductive failure in the absence of the pollinator. Even non-specific plant–pollinator interactions can be compromised in fragmented habitats where pollinator abundance can be low (Kwak *et al.* 1998). For example, plants of the butterfly-pollinated perennial *Dianthus deltoides* (Caryophyllaceae) in meadow fragments received fewer insect visitors and had lower seed set than in unfragmented habitat (Jennersten 1988). Studies in European calcareous grasslands support the view that fragmentation disrupts plant–pollinator services, decreasing both plant fitness and insect diversity (Steffan-Dewenter and Tscharntke 2002). Competition among co-occurring plants for pollinators may be important in structuring grassland communities, and there is some evidence from studies of flowering phenologies in support of this view (Lack 1982).

6.4 Models of grassland community structure

The relevance of two alternative models of community structure has been debated in the ecological literature for decades; i.e. the community-unit hypothesis (Clements 1936) and the continuum hypothesis of Whittaker (1951) derived from Gleason's (1917; 1926) individualistic hypothesis. The **community-unit hypothesis** suggests that communities are discrete, repeatable units, readily recognized across the landscape, and was based in part on Clements's familiarity with the North American grasslands. Interestingly, Gleason, based in Illinois, was familiar with the same grasslands as Clements. The continuum view suggests that plant communities change gradually in composition and structure along environmental gradients, and disallows the recognition of distinct species associations. Modern views of the plant community subscribe to the **continuum hypothesis** (see §6.1), although vegetation classification is still clearly based on the ability to recognize and name fairly discrete plant communities (Chapter 8). Descriptive broad conceptual models such as these are valuable for generating hypotheses regarding the processes that determine community structure in grasslands. Some of the modern developments of these models are described later in this chapter; i.e. niche-based metapopulation and metacommunity models (§6.4.3), and neutral models (§6.4.4). First, however, two alternative mechanistic models are described.

Grime's **CSR model** and Tilman's **R* model** are two mechanistic models of community structure that have been proposed to allow us to understand how species can coexist and form communities in grasslands. Both were developed principally in grasslands, but have application in other vegetation types too. In addition, both are based on niche theory because they assume that species traits represent evolutionary adaptations to the environment. The trade-off between traits, e.g. allocation to root or shoot growth, is regarded as an essential mechanism employed by species in their interactions with other species. This is in contrast to models based on neutral theory that asserts functional equivalence (§6.4.4).

6.4.1 Grime's CSR model

The CSR model is an expansion of the earlier r- and K-selection theory where species were seen to posses adaptations to deal with environments limited by a species intrinsic population growth rate (r) or environmental carrying capacity (K) (MacArthur and Wilson 1967). In the CSR model, three different plant life histories represent a continuum of adaptations and trade-offs to environmental features of competition (C), 'the tendency of neighbouring plants to utilize the same quantum of light, ion of a mineral nutrient, molecule of water, or volume of space'; stress (S), 'phenomena which restrict photosynthetic production'; or disturbance (R)—for ruderals, 'partial or total destruction of biomass' (Grime 1979). Competitors are species with life-history adaptations enabling them to maximize relative growth rate (RGR$_{max}$) and rapidly make use of abundant above- and below-ground resources. Stress tolerators are species with low RGR$_{max}$ that can withstand low resource levels and low disturbance. Ruderals are species that can exploit newly created, low-stress, highly or frequently disturbed habitats. These three strategies are arranged in a triangular ordination (Fig. 6.3) with individual species placed at the location where they fit with respect to the CSR strategies. Thus, in early successional habitats species occur that can best deal with the conditions of high disturbance are located in the lower right-hand corner.

The CSR model was developed to provide a mechanistic explanation of plant communities from a range of habitats, although much of the experimental data was from investigations of British grasses. The model helps us understand the role of plant strategies in structuring grassland communities (Joern 1995). For example, Wolfe and Dear (2001) found the CSR model to be a useful conceptual representation for understanding the population dynamics of species in ley pastures in Australia. Pasture species were noted to be located mainly near the centre of the CSR ordination (Fig. 6.3) where competition, stress, and disturbance are all equally important. Annual legumes in these leys (e.g. *Trifolium subterraneum*) were considered ruderal species because of their ability to withstand and colonize disturbances. The dominant grass in

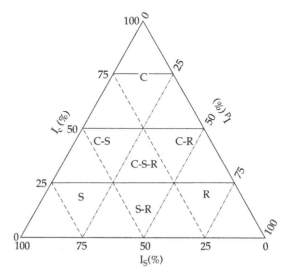

Figure 6.3 Grime's CSR triangular model describing the various equilibria between competition (I_c), stress (I_s), and disturbance (I_d). Reproduced with permission from Grime (1979).

these pastures, *Lolium perenne*, was considered to possess the characteristics of a competitor due to its ability to compete well with *T. subterraneum* but shallow roots lending susceptibility to droughts, and also the characteristics of a ruderal because of its free-tillering growth habit, and ability to re-establish well from seed and recover from grazing (Wolfe and Dear 2001).

As a trait-based scheme, the CSR approach allows an integrative evaluation of the life-history characteristics of the species in grassland to be summarized and interpreted. The relevant traits of many European species are available (Grime *et al.* 1988). Such a 'CSR signature' (Hunt *et al.* 2004) provides a useful framework for summarizing vegetation patterns, attribute responses, and environmental gradients. Data from the long-term Bibury verges (Table 6.1) looked at in this way showed that over 44 years the grassland had remained relatively stable between CSR and C/CSR types, but nevertheless showed significant declines in the abundance of C-species corresponding to increases in S- and R-species (Fig. 6.4). A long-term decline in spring and summer precipitation had allowed the increase of a number of species typically associated with relatively dry conditions, including *Achillea millefolium*, *Galium verum*, *Knautia arvensis*, *Schedonorus phoenix*, *Phleum bertolonii*, *Taraxacum officinale* agg., and *Vicia sativa*.

Categorizing species according to the CSR scheme can provide a subtly different and more informative viewpoint of grassland vegetation than use of traditional Raunkiaer life forms (Raunkiaer 1934) or simple morphological or regenerative traits. An assessment of semi-natural upland grassland in Norway (dominant grasses included *Agrostis capillaris*, *Deschampsia flexuosa*, *Nardus stricta*, *Phleum alpinum*, *Poa annua*) showed a significant correlation between CSR plant traits and a primary soil-fertility gradient of compositional variation in the vegetation; the latter is represented by a detrended correspondence analysis (DCA) ordination (Vandvik and Birks 2002). The CSR strategies explained a significant and unique proportion of the floristic variance (10.8%), with only soil factors explaining a greater percentage (16.8%). The soil fertility gradient in the vegetation was paralled by an increase in stress-tolerant species.

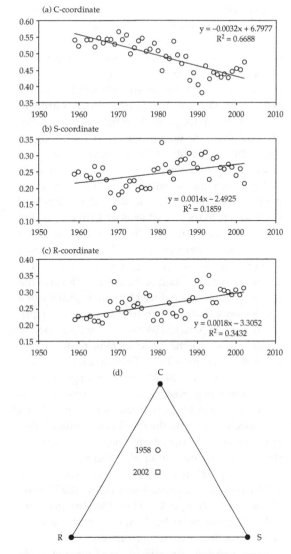

Figure 6.4 Vegetation at the Bibury verges, UK, 1958–2002. (a–c) Changes with time in the individual C, S, and R coordinates (I_c, I_s, I_d in Fig. 6.3, respectively), with linear trendlines, equations and R-squared values; (d) movement in the whole CSR signature of vegetation sampled in 1958 and in 2002 (data taken from the fitted trendlines shown in (a–c); this triangle shows the whole of CSR space, bounded by the limits C = 0 to 1, S = 0 to 1, and R = 0 to 1; the Cartesian distance between the two points is 0.14 units. Reproduced with permission from Hunt *et al.* (2004).

6.4.2 Tilman's R* model

David Tilman's resource ratio (R*) model of community structure (Tilman 1982, 1988) was developed initially from observations of aquatic diatom communities, with the full fleshing out of the model coming from his work in tallgrass oak savannah at Cedar Creek, Minnesota, USA, and data from the mesotrophic grassland of the Park Grass Experiment, UK. This model gives primacy to the outcome of competitive interactions among species. Individual species are assumed to compete below ground for nutrients (principally nitrogen) or water and above ground for light. Species differ in their ability to use resources and in doing so lower the availability of the resource in the environment. Species growth rates depend on their ability to acquire the resources, and the best competitor is the species able to draw down levels of the resource to the lowest level and survive at that level. The competitive ability of a species based on the total integrative effect of all plant traits is known as R^*, defined more precisely as 'the level to which the concentration of the available form of the limiting resource is reduced by a monoculture of a species once that monoculture has reached equilibrium', i.e. in practical terms, 'the concentration of available resource that a species requires to survive in a habitat' (Tilman 1990). Thus, when the level of a resource exceeds the R^* value for a species then its growth rate is positive, and when resource levels are less than R^* growth rate is negative. When two species compete, they will both reduce levels of the resource in the environment (above ground this is achieved by overtopping and shading one's competitor) with the stronger competitor prevailing by reducing resource levels below the R^* level of the weaker competitor. Thus, the stronger competitor is predicted to be the species with the lower R^* for a particular limiting resource.

In savannah grassland at Cedar Creek, Minnesota, Tilman noted that species appeared to compete principally for two resources, nitrogen and light. His model suggests that species differ in their ability to exploit particular ratios of the two resources. The model is illustrated graphically by comparing zero net growth isoclines (ZNGIs—resource levels of no net population growth) of competing species with respect to nitrogen and light (Fig. 6.5). Experimental confirmation of the model came from R^* estimates from measurements of perennial grass species along a fertility gradient (Tilman and Wedin 1991b) and observing the outcome of competition among these species (Wedin and Tilman 1993). R^* was found to be predicted fairly well by a number of plant traits, notably root biomass which explained 73% of the variance in below-ground R^*. In the mesotrophic grassland of the Park Grass Experiment (Table 6.1), Tilman (1982) noted that relative dominance of major taxonomic groups and individual species in response to differential resource availability brought about by the fertilization of the various experimental plots was entirely consistent with predictions of the R^* theory (for example, Fig. 6.5).

Tests of predictions of the R^* model are few, with only 26 of 1333 citations to the model reporting 42 tests (Miller et al. 2005). Of these, only 5 involved terrestrial systems and only 1 involved grasses or grassland vegetation outside the original Cedar Creek grassland that Tilman used as the basis for his work; i.e. a fertilizer experiment conducted in *Cynosurus cristatus* hay meadows in Somerset, UK, which supported the prediction that nutrient ratios rather than absolute amounts are important in determining vegetation composition and diversity (Kirkham et al. 1996). One additional experiment to Miller et al.'s (2005) review tested predictions of the R^* model with respect to invasion of *Bromus tectorum* into a western North American grassland (Harpole 2006; Newingham and Belnap 2006). Based on prior experimental work, it was predicted that *B. tectorum* would be a poor competitor for soil phosphorus or potassium, or when salt stress was high. Resource ratio theory predicts that decreasing the supply of phosphorus or potassium or increasing the supply of sodium would favour the native species, *Hilaria jamesii* (Fig. 6.6). Field and greenhouse experiments in which levels of these nutrients were amended supported this prediction, consistent with the R^* model. By contrast, estimated R^*s did not correctly predict the outcome of competition among two native (*Pseudoroegneria spicata, Helianthus annuus*) and a non-native species (*Centaurea maculosa*) in semi-arid rangeland (Krueger-Mangold et al. 2006).

Although the CSR and R^* models appear to work well in the grasslands for which they were first proposed, their extension to explain community structure in other grasslands is less clear. The competitive interactions at the heart of these models and the equilibrial stage of community

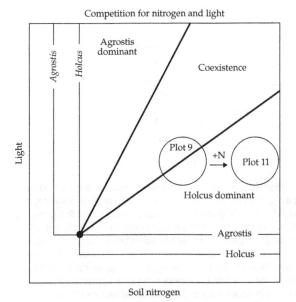

Figure 6.5 Support for the R* model from the Park Grass Experiment, UK. The figure shows zero net growth isolines (ZNGIs) of *Agrostis capillaris* and *Holcus lanatus* for limiting nitrogen and light. *Holcus* is dominant in sites with high biomass and is the superior competitor for light and *Agrostis* is the superior competitor for nitrogen. The circles show the variation in resource availability for two plots (numbers 9 and 11) and the shift in resource availability brought about through fertilization leading to *Holcus* dominance. Reproduced with permission from D. Tilman: *Resource Competition and Community Structure.* © (1982) Princeton University Press.

development required to produce the predicted outcomes are infrequently obtained because of the common occurrence of perturbations, spatial and temporal heterogeneity, and interactions with herbivores (Joern 1995). The two models have generated heated discussion in the literature between Grime and Tilman (Grime 2007; Tilman 2007) and others (Craine 2005; Grace 1990). Much of the discussion has involved semantics, although there are real differences in the traits and evolutionary trade-offs necessary to confer competitive superiority (Grace 1990), with the CSR model emphasizing the importance of maximum growth rates (RGR_{max}) allowing resource capture, whereas the R* model emphasizes the ability to drawdown resources (i.e. R*). Competition is given primacy in the R* model with competition assumed to be intense under all conditions, but for different resources. By contrast, in the CSR model competition decreases in intensity and importance with increasing stress and decreasing production. Recent ideas incorporate facilitation (see §6.3.4) as an important mechanism structuring communities in arid and semi-arid environments, including grasslands such as Mediterranean semi-arid steppe (Maestre *et al.* 2003). The **stress-gradient hypothesis** suggests that facilitation is important in harsh environments with competitive interactions becoming more important in benign environments (Bertness and Callaway 1994). Acceptance of the predictions of this hypothesis is equivocal (Lortie and Callaway 2006; Maestre *et al.* 2005, 2006). The empirical experimental data overwhelmingly support a rather simple low stress–high competition and high stress–high facilitation sort of pattern. An alternative suggestion is that competitive interactions predominate at the extremes of environmental stress gradients, with positive interactions (facilitation) being important at medium levels of environmental stress (Cheng *et al.* 2006; although see Michalet *et al.* 2006).

Overall, it is apparent that, taken together, the two theories of Grime and Tilman allow the outcome of competition to be predicted depending upon the ability of species to pre-empt resources, with the ability to acquire leaf area dominance and root length dominance conferring competitive superiority for light and soil resources, respectively (Craine 2005). However, the theories do not take spatial and temporal resource heterogeneity adequately into account.

6.4.3 Metapopulation and metacommunity models

The idea that populations of a species are connected to each other through processes of dispersal and local extinction led to development by S. Levins, later modified by Hanski (1999), of the metapopulation concept where a metapopulation can be defined as a regional assemblage or 'population of populations' (see §5.3.1). At the regional scale, incorporation of scaling processes and colonization/extinction factors implicit in the metapopulation concept provided the impetus for development

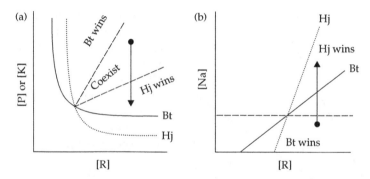

Fig. 6.6 Predicted responses to soil amendments according to the R* model: (a) Competition for two limiting resources. *Bromus tectorum* (Bt) is a poorer competitor than *Hilaria jamesii* (Hj) for P and/or K, but is a superior competitor for some other resource (R). Bt dominates at high [P] or [K], but addition of soil amendments that decrease the availability of P or K should decrease relative abundance of Bt (represented by vector). Dotted line is the zero net growth isocline for Hj: the set of resource ratios for which its growth exactly balances its loss. It will have a positive growth rate for points above and to the right of its isocline, and negative growth rate for points below and to the left of its isocline. Solid line is the isocline for Bt. Dashed lines separate regions of coexistence and competitive exclusion. (b) Competition for one limiting resource and an environmental stress factor. Bt is more negatively affected by salt stress than Hj, but is a superior competitor for some other resource (R). Increasing concentration of Na favours Hj (represented by vector). Coexistence in this case is possible, if [Na] is spatially heterogeneous and includes levels that allow both species to persist in isolated patches. Solid and dotted lines are the isoclines for Bt and Hj. Dashed line separates regions of competitive exclusion. Reproduced from Harpole (2006) with kind permission from Springer Science and Business Media.

of the **core–satellite species (CSS) hypothesis** (Hanski 1982). The CSS hypothesis predicts a bimodal distribution of species occupancy frequencies allowing the recognition of core (widely distributed and abundant) and satellite (rare and patchy) species. The CSS hypothesis assumes homogeneity of patch structure, semi-automatous population fates, similarity of migration and emigration patterns of all species in an assemblage, and a rescue effect (colonization from large populations can maintain small populations from local extirpation) (Gibson *et al.* 2005). Species populations are assumed to fluctuate stochastically within a region in response to stochastic environmental fluctuation. The shape of species occupancy distributions can reflect biological mechanisms important in structuring the communities concerned including the presence of a large, regional species pool, the importance of local filters, and patterns of niche preemption by dominant species. On this basis, the CSS hypothesis shares similarities with Grime's classification of species in ecosystems as dominant, subordinate, or transient (Gibson *et al.* 1999; Grime 1998).

The CSS hypothesis has been found to adequately describe the spatial and temporal dynamics of species regional abundance in Kansas tallgrass prairie. Bimodal species occupancy distributions were found to predominate when data from 19 sites were plotted over an 8 year period (Collins and Glenn 1991). Bimodality was also observed among soil types and at several spatial scales (Collins and Glenn 1990; Gotelli and Simberloff 1987). However, at the largest scales, unimodal patterns of species occupancy were observed consistent with Brown's (1984) niche-based **resource-use model**. The resource-use model postulates that as habitat variability increases across large regions only a very few generalist species will be common, allowing relatively large numbers of rarer, more specialized species to occur. Surveys of the flooding Pampa grassland of Argentina (Chapter 8) did not find a distinct group of core species; rather, c.70% of the vascular flora was composed of satellite species (Perelman *et al.* 2001), consistent with the resource-use model.

An extension of metapopulation theory is the recognition of **metacommunities**; sets of local communities that are linked by dispersal of multiple potentially interacting species (Leibold *et al.* 2004). The value of the metacommunity concept is that it provides a framework for thinking about how the relevant processes that occur within local communities (e.g. niche-based processes) are mediated by processes, such as dispersal, operating at larger, regional scales. This approach emphasizes the importance of spatial dynamics in affecting species success, and has particular relevance for fragmented grassland communities embedded in a landscape of different habitats. Although a number of studies acknowledge the importance of the metacommunity concept in understanding grasslands (Bruun 2000; Davis *et al.* 2005), empirical tests of the concept are few. The importance of dispersal limitation, a tenant of metacommunity theory, was illustrated in a field experiment in which resident and non-resident native species were sown into a low-productivity grassland; resident species were able to recruit into the grassland, but non-residents could only establish following a cover-reduction disturbance (Gross *et al.* 2005). The interplay between local biotic processes and regional dispersal limitation consistent with metacommunity theory has been found in seed addition experiments in oak savannah in Minnesota, USA (Foster and Tilman 2003), and seminatural subalpine grassland in Norway (Vandik and Goldberg 2006).

6.4.4 Neutral models

The unified neutral theory of biodiversity and biogeography has been proposed to describe the expected distribution of species' relative abundance patterns based on the premise of functional equivalency among all individuals in a community or metacommunity (Hubbell 2001). Any differences in species traits do not, according to the neutral model, lead to any differences in per capita demographic rates (i.e. vital rates of birth, death, and migration: see §5.1.5) among species or individuals. Species differentiate in relative abundance through ecological drift (demographic stochasticity). Therefore, community assembly would be considered a random process driven by differences in propagule abundance. Support for the neutral model has come from tropical forest communities, but not from grasslands. The distribution of relative species abundances should fit a zero-sum multinormal distribution to be consistent with neutral theory, and this was found to be the case with data from old field grasslands at Cedar Creek, Minnesota, USA (Harpole and Tilman 2006). However, these data also fit a log-normal distribution consistent with niche theory. Moreover, species with low estimated R^* values (see §6.4.2, above) generally had a greater abundance than species with higher R^* values in these old field grasslands, in younger old fields also at Cedar Creek, and in mature tallgrass prairie at Konza Prairie, Kansas, USA; consistent with community structure following species niche differences and not the neutral theory. In addition, species abundance patterns changed in response to nitrogen addition in a way reflective of their R^* values, again inconsistent with neutral theory. Another experimental study at Cedar Creek showed strong, non-random intraguild competition as the dominant C_4 grasses were shown to most effectively resist the invasion of other C_4 grasses; a result consistent with niche theory and inconsistent with community assembly resulting from a random drawing from the regional species pool as predicted by neutral theory (Fargione *et al.* 2003).

Neutral theory predicts that random fluctuations in birth, death, immigration, and dispersal among species can lead to patterns of dominance among species, such as those characterizing species–area and species–time relationships. However, a test of the predictive capability of neutral models to simultaneously simulate species–area and species–time relationships observed in Kansas grasslands was unsuccessful (Adler 2004). Rather, a functional-type model that included assumptions of niche differences among species was more successful. Failure of the neutral model was considered as being due to the variable environment of grasslands favouring shorter-lived plants than the tree-dominated communities that inspired the neutral theory. The **stochastic niche theory** attempts to reconcile aspects of neutral theory with classical trade-off-based niche theories by assuming that invading species can establish and survive to maturity only if they can survive stochastic mortality while surviving on resources unused by the resident species (Tilman 2004). Support for this idea comes from experimental studies in which grassland species were experimentally sown into existing vegetation plots. The inhibitory effect of resident species was strongest on invaders in the same functional group, consistent with a prediction of the stochastic niche theory that resident species should resist invasion by species that they are most similar too.

6.5 Summary: an issue of scale

As discussed in this chapter, several models have been proposed to assist in explaining community structure and composition of grasslands. They all have virtue to some degree. Can they all be right? Yes; it is an issue of scale. Three aspects of spatial scale have relevance: the **grain** (size of sample unit), **extent** (geographic area over which comparisons are made), and **focus** (area of inference) (Rahbek 2005; Scheiner *et al.* 2000). The spatial scale at which communities are examined affects what we observe about the system. For example, fire can be viewed as a disturbance or a stabilizing factor in tallgrass prairie depending on whether the data are viewed qualitatively or quantitatively (Allen and Wyleto 1983). In this example, the observational scale affects the pattern that we see, but the results inform us about the hierarchical levels of organization implicit in the system. (For more on the way in which we view disturbances in grassland, see Chapter 9.) Similarly, the effects of soil type, burning, and mowing were shown to be scale related in Kansas tallgrass prairie (Gibson *et al.* 1993). Within a soil type, mowing effects were greater than burning, whereas between soil types, burning effects were greater than mowing. In the Serengeti grasslands (see §6.1.1) species richness is strongly related to climate and topography, but the specific patterns depend on spatial scale, particularly grain and extent (Anderson *et al.* 2007). Between a spatial grain of 1 m and 10^3 m, the determinants of species richness changed from niche relationships to heterogeneity, but only at spatial extents ≥150 km. The implications are that the importance and relevance of underlying processes and mechanisms vary with scale. Indeed, the **species-pool hypothesis** (Zobel 1997) makes this point explicitly. This phenomenological model posits that local variation in species richness reflects variation at large scales, i.e. the species found at the patch level, are constrained by availability from the larger, regional species pool. Support for the species-pool hypothesis was found from a survey of semi-natural grasslands in Sweden (Franzén and Erikkson 2001).

At the smallest scales, competition, facilitation, and other neighbourhood interactions are predominant. At these scales, the CSR and R* models have relevance, and grassland communities can be observed to follow a number of simple assembly rules that can restrict community development and limit species coexistence. In other words, certain groups, or guilds, of species, often related to functional types (e.g. graminoids, forbs, legumes) can be identified (Wilson and Roxburgh 1994; Wilson and Watkins 1994). At larger scales, metapopulation models such as the CSS hypothesis can operate, and at the largest, regional scales, models such as the resource-use model appear to be more important.

CHAPTER 7

Ecosystem ecology

The fundamental concept appropriate to the biome considered together with all the effective inorganic factors of its environment is the ecosystem. And, in regard to grassland ecosystems...The dynamic equilibrium maintained is primarily an equilibrium between the grazing animals and the grasses and other hemicryptophytes which can exist and flourish although they are continually eaten back.

(Tansley 1935, using the term 'ecosystem' for the first time)

Ecosystem ecology is the study of the biotic and abiotic components of the environment; in other words, the study of living organisms and their interaction with their environment. It was introduced as a concept to ecology by Tansley (1935). At this level of study, less emphasis is placed on individual species or communities; rather, the focus is on the integrated components of the environment and their involvement in fundamental processes of energy flow and material cycling. Ecosystems can be viewed a comprising three subsystems connected by the transfer of energy and matter: the **plant subsystem,** the **herbivore subsystem,** and the **decomposer subsystem.** The initial fixation of carbon through photosynthesis via green plants characterizes the plant subsystem (see §7.1.2). The recycling of dead organic matter (detritus) represents the decomposer subsystem (see §7.3). The role of herbivores is discussed in Chapter 9.

In this chapter, three important ecosystem processes as they relate to grasslands are described, i.e. productivity (carbon and energy flux) (§7.1), nutrient cycling (§7.2), and decomposition (i.e. the ecosystem recycling process) (§7.3), followed by a description of grassland soils (§7.4).

7.1 Energy and productivity

7.1.1 Energy flow

Energy is the ability to do work and is referred to as the 'currency' of ecosystems as it expresses the capability of the organisms at a particular trophic level to store and use the materials necessary for life. The SI unit of energy is the joule (J), where one joule is the power required to produce or transfer one watt (W) of power continuously for one second.

The sun is the ultimate source of all energy for grasslands and is harnessed by **primary producers** (autotrophs, the green plants) and transferred to higher trophic levels of consumers including herbivores, carnivores, insectivores, and omnivores, or to the decomposer pool through above- and belowground detritus. **Secondary production** refers to the biomass attained by heterotrophic organisms. In grasslands, the **primary consumers** include large herbivores such as cattle, bison, and antelope, as well as smaller herbivores represented by a wide variety of small mammals (e.g. voles, mice, shrews) and invertebrates (e.g. grasshoppers, caterpillars) (Chapter 9). **Secondary consumers** in grasslands include a variety of carnivores such as the coyotes in American prairies and lions in African savannah, as well as small organisms such as grasshoppers and other arthropods (Chapter 9). At each transfer between trophic levels, most of the energy is lost from the system as heat. Organic molecules that contain much of the chemically stored energy for organismal 'work' and 'maintenance' are oxidized in many cellular respiration pathways, and carbon is lost to the atmosphere as CO_2 or transferred to decomposers and incorporated into the soil organic matter (§7.3).

Grasslands exhibit low transfer efficiency of energy from one trophic level to the next. North American grasslands capture <1% of the solar radiation in both ungrazed and grazed grasslands (Fig. 7.1, Sims and Singh 1978c). According to that study, about three times as much of the captured energy flows below ground as above ground. In the absence of grazing, 13 times as much energy is contained below ground as in above-ground live shoots. Energy input and output is approximately balanced in the various compartments. Grazing alters the energy balance. When North American grassland was grazed, the rate of energy loss exceeded energy capture by $c.34$ kJ m^{-2} d^{-1}, especially below ground (Sims and Singh 1978c). A comparison among temperate grasslands of North America showed a somewhat greater efficiency of energy capture for mountain and mixed-grass grasslands ($c.1\%$) compared with shortgrass and tallgrass prairies ($c.0.7\%$). The lowest energetic efficiency has been reported in desert grassland (0.17 and 0.14% in ungrazed and grazed desert grassland, respectively). Cool-season dominated grasslands tend to be more efficient in energy capture than warm-season dominated grasslands (Sims and Singh 1978b) although tropical grasslands show a higher efficiency than temperate grasslands. Energy efficiency in Indian tropical grasslands ranged from 0.23% for a semi-arid grassland to 1.66% for a dry subhumid grassland (Lamotte and Bourliére 1983). High-input pastures are even more energetically efficient as high nitrogen levels from fertilizer increase primary and secondary production. A comparison of annual energy content in a mixed-grass prairie and a tallgrass prairie (Table 7.1) shows higher conversion efficiency and

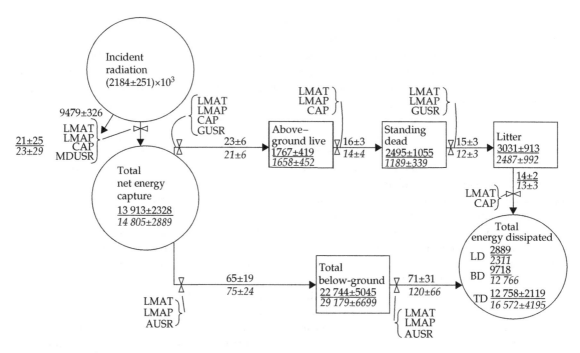

Figure 7.1 Energy flow through the producers of an 'average' North American grassland ecosystem during the growing season. Numbers in square boxes are the mean standing crop (kJ m^{-2}) of the primary producer compartments; numbers in circles are energy inputs and outputs (kJ m^{-2}) to the system; numbers on arrows are flow rates in kJ m^{-2} day^{-1}. Values in roman type are for ungrazed grasslands; values in italics are for grazed grasslands. 'Bow-ties' on the arrows represent abiotic variables related to energy and material flow; LMAT, long-term mean annual temperature; LMAP, long-term mean annual precipitation; CAP, current annual precipitation; GUSR, growing season usable incident solar radiation; MDUSR, mean daily usable incident solar radiation during the current growing season; AUSR, annual usable incident solar radiation. Reproduced with permission from Sims et al. (1978).

production above ground in the tallgrass prairie, but the opposite below ground. Energy transfer from forage to secondary production (animal biomass elaborated over time) was also higher in the tallgrass prairie, although overall efficiency was only 0.006%. As noted below in the discussion of production, water availability is the main factor limiting energy transfer.

7.1.2 Productivity and carbon

Carbon storage in grasslands is high compared with other ecosystems (c.30% of global soil carbon stocks; Table 1.9) and is driven by energy flow (§7.1.1) and productivity. Carbon from primary productivity is actively sequestered into relatively stable soil carbon pools. **Primary productivity** is the 'rate at which energy is stored by green plants in the form of organic substances' (Redmann 1992) and is generally expressed as **net primary productivity (NPP)** in terms of the amount, or biomass, of carbon accrual, per unit area per unit time (e.g. $g\,m^{-2}\,d^{-1}$), after subtracting respiration (R) losses during the measurement period from the total energy fixed (i.e. **gross primary productivity, GPP**). Thus,

$$NPP = GPP - R$$

Net primary productivity is equivalent to net photosynthesis. **Production** is 'the amount of dry matter obtained by integrating NPP over a specified period of time' (e.g. annual production in $g\,m^{-2}$). Above-ground production is taken to be equivalent to NPP in warm-season grasslands which die back to the soil surface at the end of each season. **Standing crop** (biomass) is the weight of organic components in an ecosystem at any moment, usually expressed as dry weight or ash-free dry weight per unit area (e.g. $g\,m^{-2}$), and probably includes both live and dead biomass from both the current and previous seasons.

In grasslands, **above-ground net primary productivity (ANPP)** is usually estimated from biomass measurements obtained using the harvest method of: (1) peak live biomass (which will generally include standing dead biomass produced in the same year); (2) peak standing crop; (3) maximum minus minimum live biomass; (4) sum of positive increments in live biomass; (5) sum of positive increments in live and dead biomass plus litter; and (6) sum of changes in live and dead biomass adjusting for decomposition (Hui and Jackson 2005). Method (1) is the most commonly used in the absence of grazing. Detailed methods regarding the number and size of sample units

Table 7.1 Annual energy content (MJ) and conversion efficiencies for primary (grazable forage) and secondary (animal gain) production in relation to total and photosynthetically active radiation in a mixed-grass prairie (Texas) and in a tallgrass prairie (Kansas). Arrows indicate the conversion efficiency between components, e.g. 0.1% from total solar radiation and above-ground forage

System component	Mixed prairie		Tallgrass prairie	
	Energy content	Conversion efficiency	Energy content	Conversion efficiency
	(MJ/ha)	(%)	(MJ/ha)	(%)
Solar energy				
Total radiation	10 319 708		9 227 033	
Photosynthetic radiation	4 573 047	0.1	4 087 414	0.17
Primary production (plant growth)		0.22 0.50 0.002		0.39 0.38 0.006
Above-ground (forage)	10 320		16 188	
Below-ground (roots)	41 076		19 364	
Total plant biomass	51 396	2.0	35 504	3.6
Secondary production (animal gain)	206		579	

From West and Nelson (2003).

are described by Milner and Hughes (1968). Non-destructive methods include ^{14}C tracer studies and CO_2 exchange methods (Redmann 1992).

Problems in interpreting published ANPP values from harvest methods lie in whether or not the measurements include standing dead material (litter), a certain proportion of which comes from plant tissue that was produced and has since died in the current year. Losses, or even stimulation, due to herbivory may be significant and should also be accounted for in measuring NPP. Estimates of ANPP based on maximum peak live biomass may significantly underestimate ANPP by 2–5-fold, except in certain grasslands, e.g. temperate steppe grasslands (Scurlock et al. 2002).

The time of year when measurements are taken is important as there are seasonal trends in tiller and leaf production, growth, and death (Figs 3.5 and 7.2) that need to be taken into account (Radcliffe and Baars 1987). Seasonal biomass accumulation may show a unimodal or bimodal pattern. In temperate cool-season grasslands there is often a midsummer drop in production related to low soil moisture, followed by a resumption of growth in the early autumn as temperatures drop and precipitation picks up. By contrast, in North American warm-season grasslands (and probably other similar grasslands elsewhere, such as the South African grasslands), the seasonal course of above-ground biomass shows that >80% of annual ANPP occurs during the initial 2 months of the growing season with peak biomass occurring in mid-summer (Knapp et al. 1998). North American grasslands dominated by both cool- and warm-season grasses show a bimodal pattern of production reflecting early season growth of the cool-season grasses and late-season growth of the latter (Sims and Singh 1978a).

Below-ground productivity (BNPP) is difficult to measure (see §3.5), and many estimates based on harvest data (literally digging up roots, separating them from the soil, drying, and weighing) underestimate fine root production, and nearly all studies of root production ignore losses from organic exudates and sloughed root materials (Redmann 1992). Most estimates of BNPP are derived from either: (1) sequential biomass sampling or (2) turnover coefficients coupled with biomass estimates.

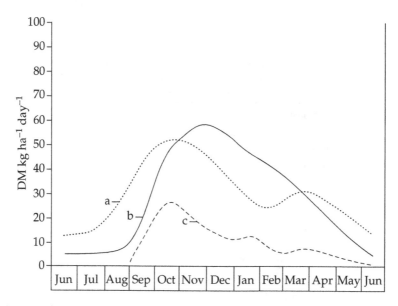

Figure 7.2 Seasonal patterns of production in New Zealand pastures: a, North Island with warm winters and dry summers; b, South Island with cold winters and autumns and drier than the North Island; c, a very dry site in Central Otago, South Island with severe season-long moisture stress. Reproduced with permission from Radcliffe and Baars (1987), Copyright Elsevier.

A general estimate of BNPP using simple environmental variables is (Gill et al. 2002):

$$BNPP = BGP \times \frac{LiveBGP}{BGP} \times turnover$$

where

$$BGP = 0.79(AGBIO) - 33.3(MAT + 10) + 1289$$

is the below-ground biomass ($g\,m^{-2}$); AGBIO is the peak above-ground live biomass ($g\,m^{-2}$); MAT is the mean annual temperature (°C) of observed below-ground biomass; and LiveBGP/BGP = 0.6. Turnover is the proportion of roots that are produced or die each year and is best estimated (Gill et al. 2002) as

$$turnover = 0.0009\,g\,m^{-2}(ANPP) + 0.25\,year^{-1}$$

The strength of this algorithm ($r^2 = 0.54$, $p = 0.01$) when tested against a large global database of global grassland NPP allows estimates of BNPP to made for large areas in the absence of actual below-ground measurements.

Estimates of maximum standing crop from grasslands across the world range from <$500\,g\,m^{-2}$ in xeric-dry grasslands to >$5000\,g\,m^{-2}$ in cool-temperate sub-Antarctic island grasslands (Table 7.2, average = $1571 \pm 256\,g\,m^{-2}$). Total NPP from these sites ranges from 278 to $5581\,g\,m^{-2}$ (average = $1853 \pm 365\,g\,m^{-2}$) and is on average $282\,g\,m^{-2}$ higher than standing crop, reflecting the within-season die-off of living tissue. Above-ground standing crop is significantly correlated with NPP (Table 7.2 data, $p < 0.0001$, Spearman's R = 0.86), although below-ground standing crop is not correlated with below-ground NPP ($p = 0.35$, R = 0.24). Litter is an important component of grassland ecosystems (see §7.3) and averages $251 \pm 40\,g\,m^{-2}$, although the amount is not correlated with any measure of standing crop or annual production. The main pattern that emerges from the data in Table 7.2 is that production is positively related to precipitation, but not necessarily to temperature. Although some of the sites with the lowest production are the most arid, hot, and dry, the sites with highest production are humid, but either hot (e.g. India) or cool (e.g. Antarctic islands). Productivity of Mediterranean and semi-arid grasslands is determined primarily by the length of time during the growing season when adequate soil moisture is available (Biddiscome 1987). Indeed, the relationship between ANPP and precipitation is generally linear for grasslands worldwide (Paruelo et al. 1998), including savannahs and other tropical tree–grass systems (Scholes and Hall 1996). In rangelands of the African Sahelo-Sudanian zone and the Mediterranean basin, for example, each millimetre of rainfall was estimated to produce $1\,kg\,ha^{-1}$ and $2\,kg\,ha^{-1}$ of consumable dry matter, respectively (Le Houerou and Hoste 1977). A linear relationship between ANPP and precipitation was observed in North American grasslands up to 500mm, after which increases in biomass levelled out with increasing growing season precipitation (Sims and Singh 1978b), probably reflecting the increasing importance of other limitations such as light or nutrients.

Interannual seasonal variation in ANPP can be substantial, and ranged from 279 to $785\,g\,m^{-2}$ (mean 527.5 ± 26.9) in annually burned lowland tallgrass prairie in Kansas, USA, largely in response to variation in precipitation (Fig. 7.3). Within temperate grasslands, interannual variation in ANPP was found to be greater than differences between sites and related to growing season precipitation and early season soil temperature. Indeed, more than 60–70% of the variation in ANPP has been attributed to growing season precipitation in both British semi-natural grasslands and North American grasslands (Radcliffe and Baars 1987; Sims and Singh 1978c).

Temporal variation in ANPP at the Kansas tallgrass prairie site (Fig. 7.3) was greatest in annual and 4 yearly burned sites (coefficient of variation 25% and 29%, respectively) compared with infrequently burned sites (CV 10–12%). ANPP in tallgrass prairie is limited by the interaction between three resources: soil moisture, nitrogen levels, and light. The occurrence of fire, topographic position, and herbivory indirectly affect ANPP as they affect these resources, the abundance of plant tissue available for photosynthesis, or both factors. For example, ANPP increases following fire in spring as light limitation is reduced, and ANPP is higher in lowlands where there is more soil moisture. Nitrogen limits ANPP, especially in

Table 7.2 Standing crop biomass (gm^{-2} dry matter) and annual net primary production (gm^{-2} of biomass) of natural grasslands from across the world

Grassland type	Location	Canopy Green	Canopy Dead	Canopy Total	Litter	Below ground	Total biomass	Root/shoot ratio	Net production Above ground	Net production Below ground	Net production Total
Mixed grass community	Pilani, India	76	27	103	31	86	230	0.8	217	61	278
Shallow loam	South Africa			208		71	279	0.3			
Desert grassland	New Mexico, USA			105	44	225	374	2.1			
Sand	Nylsvley, South Africa			171		226	397	1.3			
Ungrazed grassland	Khirasara, India	201	178	379	40	205	624	0.5			
Sehima nervosum community	Ratlam, India	363	316	679	275	873	1827	1.3	433	399	832
Sehima nervosum community	Jhansi, India			1408	226	333	1967	0.2	1019	497	1516
Mixed grass community	Kurukshetra, India	424	306	730	231	1040	2001	1.4	617	785	1402
Dichanthelium annulatum grassland	Ujjain, India	457	422	879	423	925	2227	1.1	520	464	984
Mixed prairie	Cottonwood, South Dakota, USA	184	210	301	452	1520	2273	5.0	433	269	702
Agrostis magellanica mire	Marion Island			639		2024	2663	3.2			
Heteropogon contortus community	Sagar, India	572	518	1090	433	1381	2904	1.3	914	937	1851
Flooding Pampa	Argentina	222	640	862	140	1956	2958	2.3	532	496	1028
Mixed prairie	Matador, Canada	131	504	560	268	2383	3211	4.8	447	677	1124
Desmostachya bipinnata upland	Veranasi, India			2360	145	788	3293	0.3			
Mixed prairie	Hays, Kansas, USA	131	234	560	268	2383	3211	4.8	422	288	710
Mixed prairie	Dickinson, North Dakota, USA	192	504	672	797	2168	3637	3.2	580	391	971
Deschampsia klossii tussock grassland	New Guinea			3431		421	3852	0.1			
Festuca contracta community	South Georgia			2535	140	1642	4317	1.7	492	350	842
Sesbania bispinosa community	Kurukshetra, India	1921	1440	3361	331	900	4592	0.3	2143	998	3141
Desmostachya bipinnata community	Kurukshetra, India	838	740	1578	227	2868	4673	1.8	862	1592	2452
Mixed grass community	Kurukshetra, India	1974	1268	3242	300	1167	4709	0.4	2407	1131	3538
Eragrostis nutans lowland	Varanasi, India			3296	152	1282	4730	0.4	3396	1161	4557
Chionochloa antarctica tussock grassland	Campbell Island			2717		2322	5039	0.9			
Poa cookii tussock grassland	Marion Island			2574		3236	5810	1.3			
Poa cookii community, crest	Marion Island			4458		2001	6459	0.4			
Poa foliosa tussock grassland	Macquarie Island			3510	101	4800	8411	1.4	1911	3670	5581
Mean ± SE		549±168	522±177	1853±365	251±40	1571±256	3210±388	1.6±0.3	1021±217	833±201	1853±365

From Coupland (1993b), where original sources are provided. Copyright Elsevier

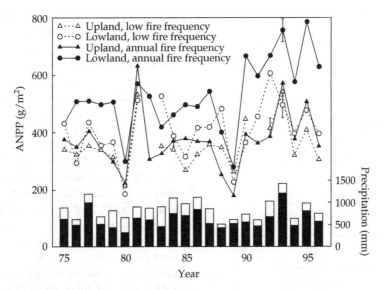

Figure 7.3 Seasonal variation in annual ANPP (g m^{-2} ± SE) in upland and lowland areas of tallgrass prairie subject to either annual or low-frequency fire on Konza Prairie, Kansas, USA. Vertical bars show annual precipitation and growing season precipitation (April–September, lower shaded portion of bars). Reproduced with permission of Oxford University Press, Inc. from Knapp et al. (1998).

frequently burned areas as burning volatilizes soil nitrogen. In the absence of frequent fire, detritus accumulates (see §7.3) leading to light limitation reducing ANPP, except in low-rainfall years when soils can remain moist longer under a 'thatch' of litter, allowing for reduced surface evaporation and enhanced ANPP.

Many grasslands are non-equilibrium systems in which ANPP is controlled by multiple interacting resources (light, nitrogen, and water) (Fig. 7.4). ANPP exhibits a 'pulse', a large short-term increase (c.30% compared with annually burned sites), on sites that are burned after long-term fire exclusion, due to the sudden input of additional light to the system. In the absence of fire, inorganic and mineralizable nitrogen builds up, which can be utilized under conditions of adequate light availability following fire (Blair 1997). This response in ANPP cannot be sustained and ANPP returns to lower levels within a few years. The **transient maxima hypothesis** has been postulated to explain this type of productivity response to the change in relative importance of limiting resources in non-equilibrial systems (Seastedt and Knapp 1993). In tallgrass prairie, light limits ANPP in long-term unburned sites, whereas after fire, nitrogen becomes the most important limiting resource; hence the switch.

Herbivores remove material from the plant canopy, allowing more light to penetrate to the soil surface, warming the soil, enhancing ANPP, or at least allowing compensatory regrowth if grazing is not too intense. Perhaps more importantly, grazing by large ungulates is selective and patchy, producing spatially heterogeneous levels of ANPP. Higher grazing intensity, or greater stocking density, generally decreases productivity, especially in the most productive grasslands. Grazing can, however, increase ANPP (Milchunas and Lauenroth 1993). For example, in studies of a *Phalaris tuberosa*—*Trifolium repens* New Zealand pasture grazed by Merino sheep it was observed that NPP was highest at 20 sheep ha^{-1} compared with 10 and 30 sheep ha^{-1} (Vickery 1972). Grasslands with a long evolutionary history of grazing display the smallest effect of grazing on ANPP; in these grasslands species composition is more likely to be affected by grazing than by productivity (Milchunas and Lauenroth 1993). Herbivore biomass itself and net secondary productivity are positively related to ANPP, which is regarded as an integrative ecosystem variable

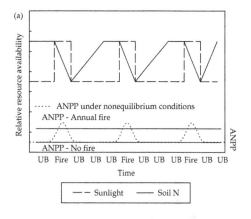

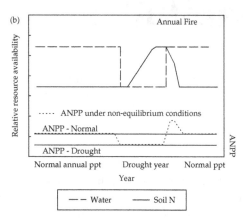

Figure 7.4 Idealized view of how the interaction between multiple limiting resources (light, nitrogen, or water) produce transient maxima responses in net primary production in tallgrass prairie: (A) Under an annual fire regime, soil nitrogen limits production in all but drought years, whereas light availability limits production with fire excluded. Under non-equilibrium conditions of intermittent fire, the availability of light and nitrogen vary at different temporal scales. Nitrogen accrues slowly in the soil when fire is excluded, whereas light limitations are alleviated rapidly by fire but become important in years when fire is absent. In years with a fire following long-term fire exclusion, both light and nitrogen become relatively non-limiting enhancing production. (B) A similar scenario occurs with water, nitrogen availability, and drought years. When water limits production in drought years, soil nitrogen increases as microbial mineralization of nitrogen continues in relatively dry soils. When normal precipitation returns, both water and nitrogen availability are high and production is maximal. In both cases, the maximum ecosystem response occurs not at equilibrium conditions of annual fire, no fire, or average precipitation, but under non-equilibrium conditions when the availability of resources is altered by variability in fire or climate. Reproduced with permission of Oxford University Press, Inc. from Knapp et al. (1998).

that influences processes at other trophic levels (McNaughton et al. 1989). Secondary production varies spatially and temporally in response to variations in primary production with production of the small organisms (e.g. grasshoppers) being influenced by grazing by the large ungulate consumers (Meyer et al. 2002).

In grasslands, up to six times more energy is captured and trapped below ground than above ground (Sims and Singh 1978c). As a result, 22–80% of ANPP was transferred underground in the sites reviewed in Table 7.2, contributing to the levels of below-ground NPP (BNPP). At these sites, BNPP ranged from $61\,g\,m^{-2}$ in a mixed grass community to $3670\,g\,m^{-2}$ in sub-Antarctic tussock grassland (mean $833 \pm 201\,g\,m^{-2}$). Average root/shoot ratio across these sites is 1.6, reflecting the observation that 2–4 times more plant biomass occurs below ground than above ground (Rice et al. 1998; Risser et al. 1981). Of this, most of the biomass is in the upper soil layers (e.g. 85% in the top 40 cm in Kansas tallgrass prairie, Rice et al. 1998). In contrast to ANPP, which is closely tied to moisture, BNPP is more closely related to temperature. In North American grasslands, solar radiation and temperature accounted for 80–90% of the variability in root biomass with BNPP ranging from 148 to $641\,g\,m^{-2}\,yr^{-1}$ increasing with decreasing long-term mean annual temperature (Sims and Singh 1978b; Sims et al. 1978). Grazing increased BNPP in these grasslands and increased root/shoot ratios, especially on cooler grasslands. Root turnover rates (calculated as annual increment divided by peak below-ground biomass: 0.18 for mixed-grass, 0.30 for tallgrass, 0.49 for shortgrass prairies) were positively related to useable incident solar radiation and reflect the dynamic nature of the below-ground ecosystem, especially in the shortgrass prairies. In addition to the temperature dependency of BNPP, variability was found to be more sensitive than other carbon fluxes to between-year differences in precipitation, and shortgrass steppe in North America was more responsive to alterations in precipitation than tallgrass prairie (McCulley et al. 2005).

In North American tallgrass prairie, fire increases BNPP but decreases tissue quality (Rice et al. 1998; Risser et al. 1981). Grazing effects on

BNPP are highly variable, although a global survey showed the grazing increasing root mass + 20%, leading to a positive effect in 61% of sites where grazing had a negative effect on ANPP (Milchunas and Lauenroth 1993). The fraction of total NPP transferred below ground (f_{BNPP}) varies from 0.40 to 0.86 and was found to be the least in savannah and humid savannah grasslands and the greatest in cold desert steppes, decreasing linearly with increasing mean annual temperature and precipitation (Hui and Jackson 2005). Each 1 °C increase in temperature corresponded to a 0.013 decrease in f_{BNPP}. Within grasslands, f_{BNPP} was most variable in savannah and humid savannahs and least variable in cold desert steppes, although there was no within-site relationship to annual temperature or precipitation. Overall, geographic variability in f_{BNPP} was generally greater than within-site variability. These patterns of below-ground biomass allocation suggest that there are general constraints on species responses and adaptations to environmental conditions across grassland ecosystems.

Our understanding of energy and material cycling in grasslands has been facilitated greatly through the use of ecosystem simulation models. These include the ELM Grassland Model of the IBP programme of the 1970s (Risser et al. 1981; Risser and Parton 1982), the PHOENIX model (O'Connor 1983), the Hurley Pasture Model (Thornley 1998), and the CENTURY model (Parton et al. 1996; Parton et al. 1988). These models have enabled predictive quantitative assessment of the relative importance of different components (e.g. species, productivity, precipitation, temperature) on ecosystem processes. Comparing standard runs of a model against runs in which the value of certain parameters are varied allows the sensitivity of these parameters to be assessed. Model simulations provide the only way for the effects of decadal time scales to be assessed. An application of many of these models has been to predict the effects of global environment change, particularly temperature, precipitation and elevated CO_2, on ecosystem parameters.

Seastedt et al. (1994) used the CENTURY model to test how various combinations of C_3 and C_4 grasses would respond, in terms of productivity, to changes in temperature and precipitation. A combination of C_3 and C_4 grasses was shown to be most sensitive to these changes. Communities consisting only of C_3 grasses were the least sensitive. Thus, the composition of the vegetation influences the outcome of climatic change.

The effects of elevated CO_2 on nitrogen mineralization were evaluated using the Hurley Pasture Model (Thornley & Cannell, 2000). It was found that the extra carbon fixed under elevated CO_2 is used to capture and retain extra soil nitrogen, but there is a decadal time lag in the latter, meaning that any new equilibrium under elevated CO_2 will take a very long time (Thornley and Cannell 2000). These models and their derivatives or spin-offs have allowed an enhanced understanding of the dynamics of nutrient cycling in grassland, and their findings are incorporated as appropriate in the discussion below.

7.1.3 Productivity relationships with diversity, invasibility, and stability

The relevance to grasslands of the following ideas related to the relationship between productivity as an ecosystem function and community issues of diversity, invasibility, and stability (Chapter 6) have been proposed and are discussed in this section.

- There is a unimodal, humped-back, relationship between diversity and productivity.
- The **shifting limitations hypothesis** (SLH) proposes that limitations imposed by the regional species pool on local colonization and hence diversity will be greatest in sites of moderate productivity, but will decline as production increases due to the increasing importance of competitive exclusion.
- The **dynamic equilibrium model** (DEM) predicts that non-equilibrium conditions in low- to-moderately productive habitats allows high diversity to contrast with equilibrial conditions under high production.
- The **biodiversity-ecosystem function** (BDEF) hypothesis predicts that loss of biodiversity will lead to loss of ecosystem function.
- Decreased diversity leads to reduced ecosystem stability.
- Invasibility is negatively related to diversity (see discussion of scale dependency in Chapter 6).

Generally, there is a unimodal or humped-back relationship between diversity and productivity in terrestrial ecosystems (Waide *et al.* 1999), including grasslands. The most diverse grasslands are those with intermediate levels of productivity. Mechanistically, this relationship is considered by Grime (1979) to reflect a competition–stress/disturbance gradient (see §6.4.1). At low levels of production, only a few species can withstand the stresses of unproductive habitats (or high disturbance), whereas only a few species dominate at high levels of production as a result of competitive exclusion. Production increase due to fertilization leads to a decline in species richness in mature and newly restored grassland (Baer *et al.* 2003; Gibson *et al.* 1993; Rajaniemi 2002; Suding *et al.* 2005). The humped-back relationship was shown clearly for 14 sites in northern England, with the grassland sites occupying the most diverse sites of intermediate production (Al-Mufti *et al.* 1977). A survey of 281 old, permanent grasslands in western and central Europe revealed a similar pattern and showed that the humped-back relationship could also be obtained by plotting diversity against major soil nutrients (nitrogen, phosphorus, and potassium) instead of biomass (Janssens *et al.* 1998). However, within grasslands, diversity–productivity relationships are probably more complex than was first thought. For example, the extent to which the importance of competition varies along the productivity gradient is unclear, and the balance between regional and local processes is debated. Indeed, Tilman (see chapter 6) argues that competitive intensity is independent of resource availability. It is also argued that only a decrease in richness is observed in most fertilization addition experiments, and that many experiments survey only an incomplete portion of the production gradient (Pausas and Austin 2001). Moreover, the decline in species diversity with increasing production may be a sampling artifact reflecting the physical exclusion of small species from sample plots as large species increase in biomass with increased resources at the high end of production gradients (Oksanen 1996).

Several additional hypotheses have been subsequently proposed to explain the diversity–production relationships (Grace 1999). Among these, the **shifting limitations hypothesis** (SLH) proposes that limitations imposed by the regional species pool (itself the product of evolutionary and historical processes) on local colonization and hence richness will be greatest in sites of moderate productivity, but will decline as production increases due to the increasing importance of competitive exclusion. In other words, there will be a shift from regional to local control of diversity along a gradient of increasing productivity. Similarly, the **dynamic equilibrium model** (DEM, Huston 1979) predicts non-equilibrium conditions (slow population dynamics and slow rates of competitive exclusion) in low-to-moderately productive habitats, allowing high diversity to contrast with equilibrium conditions under high production with high population growth and exclusion rates. Support for the SLH was found in a multispecies sowing experiment along a natural productivity gradient in a Kansas, USA, grassland dominated by early successional grasses (*Bromus inermis, Schedonorus phoenix, Poa pratensis, Andropogon virginicus*) (Foster *et al.* 2004). In this study, the humped-back diversity–productivity relationship was not observed, rather, diversity declined with increased production, with higher richness in the presence of disturbance especially at low levels of production.

A corollary of production-diversity relationships is the **biodiversity-ecosystem function (BDEF)** hypothesis which posits that loss of biodiversity (species, genotypes, etc.) will lead to loss of ecosystem function (e.g. amounts, fluxes, and stability of materials, biomass, and energy). The BDEF hypothesis has obvious conservation concerns for grasslands as biodiversity declines worldwide due to habitat loss, fragmentation, and other factors (see Chapter 1); however, the extent to which it holds true is equivocal. In 100 studies, 74% of grassland experiments found a positive BDEF effect on productivity and 44% on decomposition (Srivastava and Vellend 2005). Indeed, 51% of BDEF research through 2004 was conducted in grassland, mainly north-temperate grassland. Whereas positive BDEF effects may be found within grasslands, comparisons among grasslands more frequently do not observe this effect because of confounding environmental factors.

The mechanisms through which biodiversity can maintain ecosystem function relate to stocks and flux types of ecosystem functions, and stability of ecosystem functions, respectively (Table 7.3). Disentangling these different mechanisms is difficult, in part because of experimental difficulties in manipulating and contrasting realistic and random extinctions (Allison 1999). Removal experiments, for example, can illustrate the differential capacity of species for functional compensation. In experimental grassland assemblages in Switzerland, statistical effects of diversity (portfolio and sampling effects: Table 7.3) played a role in production (Spehn et al. 2000). During establishment, insurance effects increased above-ground resource use, with legumes and non-legumes interacting to enhance biomass. This experiment was one of seven sites across Europe, all of which showed a decline in production with a loss of diversity (Hector et al. 1999). The role of different functional groups appears critical for understanding production–diversity relationships. For example, high-diversity grassland plots at Cedar Creek, Minnesota, USA, were the most resistant to drought because they contained key drought-tolerant species (two grasses, *Andropogon gerardii*, *Sorghastrum nutans* and two milkweeds, *Asclepias syriaca* and *A. tuberosa*) that increased in biomass as other species declined (Tilman 1996).

Species loss may be non-random, as certain species and indeed entire functional groups may be more susceptible than others to the stresses associated with habitat alteration and management. Non-random species loss can have a different effect on declines in production compared with random species loss. For example, many European mesic grasslands have shifted from intensive to low-input management with considerable local species loss. When experimental grassland mesocosms were exposed to 2 years of high-intensity management (fertilizer and cutting) followed by 2 years of

Table 7.3 Potential mechanisms behind positive effects of diversity on function in single-trophic level systems

Proposed mechanism	Description
Stocks and flux types of ecosystem functions	
Niche complementarity	Niche differentiation between species or genotypes allows diverse communities or populations to be more efficient at exploiting resources than depauperate ones, leading to greater productivity and retention of nutrients within the ecosystem
Functional facilitation	A positive effect of one species on the functional capability of another will lead to an increase in function in more diverse communities
Sampling effect (positive selection effect)	This effect combines probability theory with species-sorting mechanisms. When there is positive covariance between the competitive ability of a species and its per capita effect on ecosystem function, the probability of including a dominant, functionally important species will increase with diversity
Dilution effect	Lower densities of each species or genotype in high-diversity communities may reduce the per capita effects of specialized enemies such as pathogens (e.g., via reduced transmission efficiency) or predators (e.g., via reduced searching efficiency). In essence, specialized enemies create frequency-dependent selection among species or genotypes
Stability of ecosystem functions	
Insurance effects	Species that are redundant in functional roles or capacity respond differently to stressors, allowing maintenance of net community function after perturbation
Portfolio effect	Independent fluctuations of many individual species may show lower variability in aggregate than fluctuations of any one species, much as a diversified stock portfolio represents a more conservative investment strategy than would any single stock. This effect does not require any interactions between species
Compensatory dynamic effects	Negative temporal covariance between species abundances create lower variance in their aggregate properties, such as total biomass

From Srivastava and Vellend (2005). Reprinted with permission, from *Annual Review of Ecology, Evolution, and Systematics*, Vol 36 © 2005 by Annual Reviews www.annualreviews.org.

low-intensity management (reduced fertilizer and cutting frequency), loss of extinction-prone species reduced biomass production 42–49%, and loss of all species except those with a low extinction risk (which also had the greatest biomass production) reduced production 2–35% (Schläpfer et al. 2005). If species had been lost randomly from these plots the production loss would have been greater, at 52% and 26–54%, respectively. At Konza Prairie Kansas, USA, ANPP was unaffected by non-random loss of rare species as it was compensated for by production of the dominant species, but ANPP declined when dominant species were removed (Smith and Knapp 2003). The dominant species were the main controllers of ecosystem function in this grassland and were able to provide a short-term buffer against non-random species loss, although loss of complementary interactions among rare and uncommon species may contribute to additional species loss.

Another important aspect of the relationship between ecosystem function (production) and diversity is Elton's (1958) classic idea that decreased diversity leads to decreased stability. Numerous studies in grasslands have attempted to test this idea, with mixed results (reviewed in Tilman 1996). Studies from temperate pastures suggest that complex sown mixtures are more stable, and hence more profitable in terms of animal performance, than pastures sown with simple one- or two-species mixtures (Clark 2001). One of the most comprehensive tests of the diversity–stability relationship is a 13-year study of plant species diversity along a production gradient in Minnesota grassland dominated by *Elymus repens, Andropogon gerardii, Poa pratensis*, and *Schizachyrium scoparium* (Tilman 1996; Tilman and Downing 1994). Variability in community biomass (as a measure of stability) was significantly dependent on diversity, even through a major drought, supporting Elton's diversity–stability hypothesis. Indeed, species-rich plots returned to pre-drought production levels quicker than species-poor plots. The most diverse plots were also those with the lowest stability of abundances of individual species, suggesting an important role in compensatory competitive release among disturbance-resistant species. Thus, in diverse communities, disturbance-resistant species are able to increase in abundance in response to reductions in abundance of disturbance-susceptible species. In low-diversity communities, there are few species available to compensate for the loss of susceptible species and so the system exhibits low stability. These findings were corroborated by experiments in which 168 plots in the same grassland were seeded to one of five different levels of perennial grassland species richness (1, 2, 4, 8, or 16 species). Despite considerable year-to-year climatic variability altering the abundance of species and productivity, the most diverse plots were the most stable (Tilman et al. 2006b). A 24 year study of two Inner Mongolia steppe sites dominated by the perennial grasses *Leymus chinensis* and *Stipa grandis* (Bai et al. 2004) provides further support. In this study, compensatory responses arise among both species and functional groups (annuals and biennials being the least abundant and most variable, perennial grasses the opposite), and ecosystem stability increased with an increase in the organizational hierarchy (species, to functional groups, to community). There is evidence that stability in production varies among grasslands; for example, ANPP of an alpine meadow on the Quighai-Tibetan plateau was less variable than annual precipitation or temperature, and more stable than grasslands in southern Africa and Israel (Zhou et al. 2006).

Another idea of Elton (1958) is that invasibility is negatively related to diversity as more species use more resources, lessening the opportunities for an invasive species to colonize. Tests of this hypothesis in grasslands have been mixed, although the results from manipulative experiments generally show that the more diverse communities are the most resistant to invasion (reviewed in Hector et al. 2001). In a California annual grassland, Elton's hypothesis was supported as entire functional groups were lost faster than expected by chance as diversity declined in experimental plots reducing resilience (resistance to invasion) (Zavaleta and Hulvey 2004). By contrast, experiments with the invasive annual *Centaurea solstitianis* in grassland mesocosms, also a California annual grassland species, and of *Centaurea maculosa*, a perennial invader of the western USA, suggested that it was functional group diversity rather than species diversity that can confer resistance to invasion (Dukes 2001).

7.2 Nutrient cycling

7.2.1 A general nutrient cycling model for grasslands

Nutrient cycles describe and summarize the movement of nutrients (Chapter 4) between and within biotic and abiotic components of an ecosystem (Clark and Woodmansee 1992). The important components, in addition to the plants, include the litter, large and small animals, decomposer microorganisms, soil, and atmosphere. Nutrient cycles in grasslands can be described as being open or closed depending on the nutrient and the level of exchange with outside reservoirs. Most nutrient cycles are open to some extent, as there are inputs from or losses of nutrients to outside reservoirs. The nutrient cycles of managed pastures with fertilizer input, a high stocking density, and animal exports are relatively open. Inputs and outputs may occur simultaneously, so the level of a particular nutrient may remain relatively constant over time. The nitrogen cycle (§7.2.2) is relatively open, as nitrogen input comes from biological fixation and atmospheric deposition and NO_3^--nitrogen compounds in the soil are highly soluble, allowing ready leaching and loss. By contrast, the phosphorus cycle is relatively closed because of the relative insolubility of most forms of phosphorus in the soil. Nutrient cycles with high levels of internal recycling are relatively closed. Levels of internal cycling (translocation) can be high, and can account for 95–98% of total nitrogen flow in natural grasslands (Clark and Woodmansee 1992). In annual grasslands there is no internal recycling and storage in herbage from one season to next as herbaceous annual vegetation dies (except seeds) at the end of the growing season. Nutrients in the current year's vegetation in annual grasslands are derived from mineralization of plant and soil organic matter (Jones and Woodmansee 1979). The above-ground herbage in C_4 grasslands also dies back at the end of the growing season, but the below-ground plant organs remain alive and function in retaining and recycling nutrients.

The transfers and fluxes among the soil, plant, and animal components are central to understanding nutrient cycling in grassland ecosystems (Fig. 7.5). These transfers include the conversion of insoluble to soluble forms and their interaction with microbes in the soil; inputs from the atmosphere (wet and dry deposition), fertilizers, manures, animal feed, and biological nitrogen fixation; and losses such as leaching and run-off, volatilization, denitrification (reduction of NO_3^- to gaseous nitrogen by certain soil fungi and bacteria), NO and N_2O lost during mineralization (the 'leaky pipe' or 'hole-in-the-pipe' model; see §7.2.2), and movement off-site via grazers. Internal cycling of nutrients includes surficial transfers (e.g. overland flow, wind redistribution); plant uptake, recycling, and release; animal and invertebrate transfers; and litter and microorganism transfers. The role of herbivores is very important in grassland nutrient cycles. Ruminants, for example, rely on nutrients transferred from the plants, and they return nutrients via excreta and urine. There is an important interaction of the biological components with geochemical cycles that mediates localized cycling and turnover rate. Human activities increase cycling rates through fossil fuel burning and the production of fertilizers. As management of grasslands becomes more intense, so nutrient input from fertilizer, silage, and animal feeds, and removal in animal products becomes more important. The focus in the following discussion is on nitrogen and phosphorus dynamics as these two nutrients are those most frequently limiting in grassland soils.

7.2.2 Nitrogen

Nitrogen is generally the most important soil nutrient in grassland ecosystems because of its importance in amino acids and their derivatives, especially the nitrogen-rich enzyme Rubisco (Chapter 4). For example, nitrogen is the principal nutrient limiting productivity and affecting species composition in the soils of the savannah grasslands of Cedar Creek, Minnesota, USA (Tilman 1987) and in North American tallgrass prairie (Blair *et al.* 1998; Risser *et al.* 1981).

Nitrogen pools
Globally, gaseous nitrogen (N_2) is the largest nitrogen pool, followed by the soil organic pool, of which the largest fraction (usually >95% of the

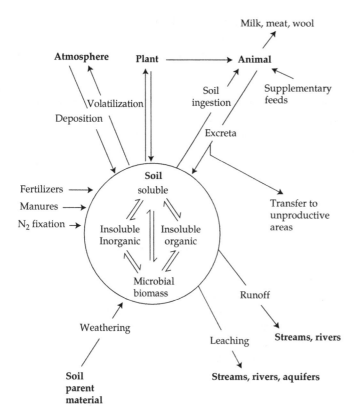

Figure 7.5 Major transformations involved in the cycling of nutrients in grassland ecosystems. Reproduced with permission from Whitehead (2000).

nitrogen pool) is organic nitrogen in soil organic matter that has remained in the soil for long periods of time (Fig. 7.6). The active nitrogen pools in soil are less stable than atmospheric nitrogen and closely tied to soil microbial biomass and activity. In US tallgrass prairie, >98% of nitrogen in the ecosystem occurred in the soil pool (1148 g m^{-2} ungrazed, 1213 g m^{-2} grazed, respectively), even at peak standing crop (Risser and Parton 1982). Soil inorganic nitrogen (NO_3^- and NH_4^+) was a very small fraction (<0.5%) of total soil nitrogen, with the level of NH_4^+ c.10 × that of NO_3^-. In plants, nitrogen levels were greatest in litter, especially when grazed (7.3 g m^{-2} compared with 4.5 g m^{-2} ungrazed). Nitrogen content of live shoots and roots was similar (c.2–4 g m^{-2}), although 13–52% higher in the roots and 35–83% higher under grazed conditions. By comparison, soil organic nitrogen ranged from 200 to 500 g m^{-2} in annual grasslands, with nitrogen levels in above- and below-ground plant parts being 3.5–8.0 g m^{-2} and 2.0–8.0 g m^{-2}, respectively (Jones and Woodmansee 1979). In neotropical savannahs of Venezuela, nitrogen was similarly 20–80 times greater in the soil than in the plant biomass, with only 1.5–2% of the total nitrogen in the live above-ground biomass (Sarmiento 1984).

Biological nitrogen fixation

Free-living bacteria (*Azospirillum*, *Azotobacter*, *Beijerinckia*, *Clostridium*), Cyanobacteria (*Nostoc*, *Tolypothrix*, *Calothrix*), and symbiotic *Rhizobium* bacteria associated with grassland legumes (especially the clovers *Trifolium repens*, *T. pretense*, and *T. subterraneum* in managed pastures) can fix substantial amounts of atmospheric nitrogen (see §6.3.5). Rates of nitrogen fixation generally range from 0.1 to 10 g N m^{-2} yr^{-1} in grasslands and temperate savannahs, and 16.3 to 44.0 g N m^{-2} yr^{-1} in tropical savannahs (Cleveland *et al.* 1999). *Rhizobium* in clover from New Zealand pastures have been

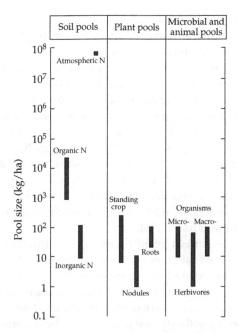

Figure 7.6 Relative size of nitrogen pools in grassland ecosystems. Reproduced with permission from Floate (1987), Copyright Elsevier.

reported to fix $c.18$ g N m^{-2} yr^{-1} (Floate 1987). The usual range of symbiotic nitrogen fixation with clover in grasslands is 20–40 g N m^{-2} yr^{-1}, although in natural and semi-natural grasslands the amount is usually much less, <2.5 g N m^{-2} yr^{-1} (Whitehead 2000). Nitrogen fixed by the clover–*Rhizobium* symbiosis becomes available to the plants through decomposition of plant tissue or following animal consumption and subsequent excretion of nitrogen in faeces or urine. Pasture production systems rely heavily on *Rhizobium*-fixed nitrogen for maintaining soil fertility, whereas legumes are rarely a dominant component of natural grasslands and some may have low levels of nodulation. Nitrogen-fixing legumes are more abundant in early seral communities than in late successional grasslands, and those that do occur in mature soils may not fix nitrogen (O'Connor 1983). Nitrogen fertilization of pastures reduces the level of nitrogen fixed by the clover, enhances plant tissue nitrogen concentrations, and promotes mineralization of soil nitrogen. Nitrogen fixation by free-living soil bacteria can be up to 1.5 g N m^{-2} yr^{-1} but is often 0.1–0.2 g N m^{-2} (Whitehead 2000). Soil crusts of *Nostoc muscorum* in tallgrass prairie can fix up to 1 g N m^{-2} yr^{-1} (Blair et al. 1998). In tropical savannahs, the soil bacterium *Azospirillum* has been reported to fix 0.9–9.0 g N m^{-2} yr^{-1} (Lamotte and Bourliére 1983). The roots of most tropical grasses are associated with high densities of bacteria allowing appreciable nitrogen fixation. For example, *Azospirillum lipoferum* associated with *Panicum maximum* can fix $c.4$ g N m^{-2} (Kaiser 1983).

Inorganic nitrogen inputs and affects of chronic deposition

Additional nitrogen is available to plants through **direct input** of available nitrogen, e.g. fertilizers or through bulk precipitation (1^{-2} g N m^{-2}). Over a 12 year period, inorganic nitrogen inputs in wetfall in tallgrass prairie averaged 0.46 g N m^{-2} with approximately equal amounts of NH$_4$-nitrogen and NO$_3$-nitrogen, and bulk precipitation (which includes dryfall) averaging 0.6 g N m^{-2} yr^{-1} (Blair et al. 1998). The anthropogenic production of fertilizers through the Haber–Bosch process (an industrial method for synthesizing ammonia directly from nitrogen and hydrogen), fossil fuel combustion, and biological nitrogen fixation through cultivation causes the fixation of $c.156$ Tg N yr^{-1} globally (Galloway et al. 2004; Vitousek et al. 1997), exceeding the estimates of natural biological nitrogen fixation on land (107–128 Tg N yr^{-1}). This is a large and significant increase in nitrogen input to terrestrial ecosystems. In the Netherlands, which has the highest rate of nitrogen deposition in the world (4–9 g N m^{-2} yr^{-1}), heathlands are converting to species-poor grasslands and forest (Aerts and Berendse 1988). Species-rich calcareous grasslands (*Festuco–Brometea*) in Great Britain and Denmark have seen a decrease in diversity associated with an increase in *Brachypodium pinnatum* as a result of nitrogen deposition since the 1970s. Neutral–acidic and acid grasslands are similarly affected as nitrogen deposition affects competitive relationships among species and can cause soil acidification (Bobbink et al. 1998). A species richness decline of one species per 4 m^2 quadrat for every 2.5 kg N ha^{-1} of chronic nitrogen deposition were observed in acid *Agrostis–Festuca* grasslands in Great Britain (Stevens et al. 2004). At the mean level of chronic nitrogen deposition for Europe (17 kg N ha^{-1}), a 23% decline in species richness was observed. Low-nutrient habitats such as infertile, acidic grasslands are particularly susceptible to these nitrogen-driven changes, leading to a shift towards species associated with high nutrient availability (Emmett 2007). In some of these European grasslands, particularly those with a long history of elevated nitrogen deposition, the effects of the additional nitrogen provided through deposition are unclear, as the nitrogen is stored in the system with only minor increases in the natural levels of output fluxes through volatilization and leaching (Phoenix et al. 2003).

Nitrogen flow and cycling
Nitrogen is available to plants through uptake from the inorganic nitrogen pool in the soil solution as NH_4^+ or NO_3^-, as monomers containing bioavailable **dissolved organic nitrogen** (DON) (e.g. amino acids, amino sugars, nucleic acids), or as NH_4^+ via symbiotic nitrogen fixation (Schimel and Bennett 2004). There are large pools of nitrogen in **soil organic matter** (SOM), especially in grasslands (see above and Chapter 1). The availability of nitrogen to plants depends upon **mineralization** by soil microorganisms of otherwise immobilized organic nitrogen in SOM to form NH_4^+ that is nitrified to form NO_3^-. Mineralization occurs through **depolymerization** of SOM that releases nitrogen-containing monomers into the soil solution which are either taken up directly by plants or microbes. Nitrification by soil microbes oxidizes NH_4^+ into nitrite (NO_2^-) and then NO_3^-. Nitrogen mineralization rates have been observed to be higher in temperate grasslands (1.76 mg N kg^{-1} soil d^{-1}) compared with temperate woodlands or agricultural systems (7.34 and 3.97 mg N kg^{-1} soil d^{-1}, respectively) (Booth et al. 2005). Generally, nitrogen mineralization rates are positively correlated with soil microbial biomass, carbon, and nitrogen concentrations. Grassland SOM is inherently more mineralizable, with a lower carbon:nitrogen ratio than woodland SOM.

The pattern of nitrogen cycling through an ungrazed tallgrass prairie is illustrated in Fig. 7.7. The major rate-limiting step is the input of nitrogen to the system and its availability to plants in the inorganic soil solution. Plant uptake and translocation is greatly affected in this system by seasonality (which affects temperature), fire regime, and water availability (Blair et al. 1998). Temperature and water supply are particularly important in affecting both plant growth rates and microbial activity. Nitrogen is translocated from rhizomes (46% of tissue nitrogen) and roots (55%) above ground in the spring as new tissues grow, and a correspondingly large translocation of nitrogen from leaves back below ground occurs as plant senesce at the end of the growing season (e.g. 58% of *Andropogon*

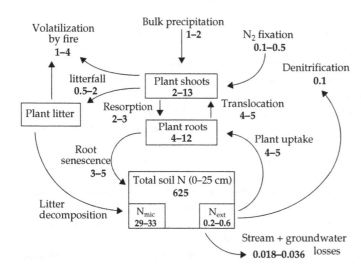

Figure 7.7 Conceptual model of nitrogen cycling in ungrazed tallgrass prairie. Units for standing stocks (boxes) are g m^{-2} and fluxes (arrows) are g m^{-2} yr^{-1}. N_{mic} is microbial biomass nitrogen: N_{ext} is KCl extractable NH_4^+- and NO_3^--nitrogen. Reproduced with permission of Oxford University Press, Inc. from Blair et al. (1998).

gerardii leaf nitrogen). This internal recycling is maximized in drought years when productivity is low (Fig. 7.3). Frequent fire causes nitrogen limitation of tallgrass prairie as accumulated detritus is consumed and nutrients are volatilized, leading to low rates of net nitrogen mineralization even though total nitrogen levels in the soil may not be affected (see §7.1.2 and Fig. 7.4). Rapid plant growth occurs following fire under conditions of high soil surface solar radiation and warm soil temperature. Consequently there is an increase in microbial and plant demand for nitrogen, and although root mass may increase below ground, tissue nitrogen concentrations are lower than in infrequently burned grassland. Grazing by large native herbivores (i.e. *Bison bison*) in this system increases rates of nitrogen cycling and availability that can lead to higher tissue nitrogen concentrations and delayed end-of-season below-ground translocation. Domestic grazing of managed pastures returns up to $2.0\,g\,m^{-2}\,yr^{-1}$ of nitrogen to the soil in excreta, depending on stocking rate, time spent in the grazed area, composition of the sward, levels of fertilizer application, and the availability of supplemental feed (Whitehead 2000). Animal dung and urine in locally nitrogen-rich patches can input nitrogen in excess of $50\,g\,m^{-2}\,yr^{-1}$. It was estimated that 75% of NO_3^- in the groundwater of New Zealand pastures was derived from the urine of grazing animals (Quin 1982).

Plants can take up DOM in the form of amino acids or other organic nitrogen compounds directly from the soil or via soil microbes including mycorrhizae (Henry and Jefferies 2003; Thornton 2001). Amino acids can be the dominant soluble form of nitrogen in acidic, low-productivity grassland, such as Arctic tussock grass systems. Free amino acids occurred in higher amounts in an unimproved *Agrostis capallaris–Festuca ovina* grassland than in an improved *Lolium perenne* grassland in Europe (11.41 µg/g soil vs 8.33 µg/g soil, respectively, Bardgett *et al.* 2003). Unimproved grasslands had a greater percentage of the more complex amino acids (e.g. arginine, histidine, and phenylalanine) compared with simple organic forms (e.g. glycine and alanine) that predominate in improved grassland. There are species-specific preferences in the uptake of inorganic vs organic forms of nitrogen and preferential use of chemical forms of DOM of varying complexity that relate to the competitive success of different plant species in grasslands of varying fertility (Weigelt *et al.* 2005). Perhaps the one factor preventing amino acids from being a more important source is the fact that microbes surrounding roots appear to be equally well adapted, or better, at consuming these molecules. Further, root herbivory by small animals such as nematodes may cause roots to 'bleed' amino acids, and this uptake by plants may reflect a strategy to neutralize or at least minimize losses.

Nitrogen losses
Nitrogen is lost from grasslands in solution as soluble NO_3^-, by volatilization by fire, volatilization of NH_4^+ from excreta, denitrification in basic soils, and removal of plant tissues by grazers or as hay. Loss of nitrogen through hydrologic fluxes and volatilization are generally minor in the absence of fire. Hydrologic loss of dissolved nitrogen from Konza tallgrass prairie (Fig. 7.7) was only 0.02–$0.04\,g\,N\,m^{-2}\,year^{-1}$, equivalent to only 0.01–6% of annual precipitation inputs (Blair *et al.* 1998). By contrast, loss of nitrogen from hill pastures in New Zealand through surface and subsurface was measured as 1.0–$1.4\,g\,N\,m^{-2}\,year^{-1}$ (Lambert *et al.* 1982). Urine patches are a significant source of leached NO_3^-. Grazers remove the greatest amount of nitrogen in intensively managed pastures as meat, milk, wool, or forage. These losses are controlled by stocking rates. For example, 17% of nitrogen ingested by steers is retained in tissues. Nitrogen can be transferred from pasture to bedding grounds, e.g. in one study 22% of faeces were deposited in 3% of the area. In annual California grassland, $1.1\,g\,m^{-2}$ of nitrogen losses per year accrued from grazing. Volatilization losses are exacerbated by grazers as ammonia is released from animal excreta as well as directly from soils, decomposing litter, and plants. Leaching losses are also exacerbated by grazers as, although 90% of urine is urea- and amino-nitrogen immediately available to plants, >50% of urea-nitrogen can be volatilized in warm–dry conditions (Jones and Woodmansee 1979). Some 20–80% of total nitrogen can be lost through volatilization of ammonia from slurry (animal faeces) following application to the surface of grassland (Whitehead

2000). Volatilization losses increase with increasing temperature and in soils with high pH and low cation exchange capacity.

Denitrification is the microbial reduction of NO_3^- by soil bacteria to volatile forms (NO_x and N_2) and is negligible in natural grasslands because of the usually low levels of soil NO_3^-, except under flooded conditions (Clark and Woodmansee 1992). In managed grassland, denitrification can lead to nitrogen loss in undrained conditions especially when concentrations of NO_3^- are high following the application of fertilizer or locally in urine patches.

Nitrogen is lost from the soil through emissions of NO and N_2O. The microbial and ecological factors that regulate these emissions have been conceptualized in the '**leaky pipe**' or '**hole-in-the-pipe (HIP)**' model (Davidson et al. 2000; Firestone and Davidson 1989). In this analogy, loss of NO and N_2O is considered as fluid in a leaky pipe with regulation due to (1) the rate of NH_4^+ oxidation (i.e. nitrification) and/or NO_3^- reduction (i.e. denitrification) considered as the 'flow' of nitrogen through the pipe, and (2) the relative proportions of NO and N_2O produced by the nitrifying and denitrifying bacteria (i.e. the size of the 'holes' in the pipe through which NO and N_2O leak). These rates themselves are regulated primarily by soil moisture content and diffusivity, with NO emissions predominating in dry soils and N_2O in wetter soils, and dentrification proceeding to N_2 in very wet soils (Davidson and Verchot 2000). For example, in Danish grass–clover pastures (*Lolium perenne–Trifolium repens*) of different ages, nitrification was the main source of N_2O emissions which varied under changing environmental conditions but were lowest during wet conditions in May when oxygen availability was low (Ambus 2005). In western Brazil, forest clearing of moist terra firma tropical forest to cattle pasture (planted to the forage grass *Brachiaria brizantha*) markedly altered the source and amounts of NO and N_2O emissions from soils consistent with the leaky pipe model (Neill et al. 2005). Emissions of N_2O increased in the first 3 years after forest clearance (1.7–4.3 to 3.1–5.1 kg N ha^{-1} yr^{-1}, respectively), but declined to lower levels (0.1–0.4 kg N ha^{-1} yr^{-1}) in older forests. Mean annual NO emissions from forests were 1.41 kg N ha^{-1} yr^{-1} whereas from pastures NO emissions dropped to 0.23 kg N ha^{-1} yr^{-1}.

Nitrification was the source of NO emissions from forest soils, whereas it was not a major source in pasture soils. Denitrification was not a major source of N_2O in forest soils whereas it was in pasture soils, but only when NO_3^- was available. Soil moisture levels and nitrogen availability were important in regulating N_2O and NO emissions but other factors including soil carbon levels appeared important too. By contrast, nitrification was the major pathway for NO emissions (0.17–1.7 kg N ha^{-1} yr^{-1}) from northern Texas, USA, *Prosopis glandulosa* savannah, c.2% of annual nitrification rate reflective of a significant, albeit modest, leak from the pipe (Martin et al. 2003).

7.2.3 Phosphorus

Second only to nitrogen, phosphorus is often limiting in grasslands; e.g. calcareous *Festuca–Avenula* grasslands (Morecroft et al. 1994) and the long-running Park Grass Experiment (Wilson et al. 1996) in England, pastures of the Natal region of South Africa (Tainton 1981b), and some annual California grasslands (Woodmansee and Duncan 1980). In some grasslands, however, phosphorus does not appear to limit productivity so long as the dominant plants co-occur with their mycorrhizal symbionts (e.g. Kansas, USA, tallgrass prairie; see §5.2.1 and Hartnett and Fay 1998). In managed grasslands, fertilizer (principally superphosphate) and animal feed increase the amount of phosphorus in the ecosystem, increases pasture growth, and increases the rate of phosphorus cycling. The phosphorus cycle is relatively closed compared to others, with annual input and losses being relatively minor, as the main pool of phosphorus is in the SOM (Fig. 7.8). Particular characteristics of the phosphorus cycle are (1) the lack of a gaseous phase, and (2) the readiness with which available (soluble) phosphorus is rendered unavailable and the slowness with which it is again released.

Phosphorus inputs and cycling
The primary inputs are atmospheric deposition (wetfall and dryfall), mineral inputs from parent rock materials (especially the apatite group of minerals), and fertilizer. However, except for fertilizer, these inputs are usually minor over the

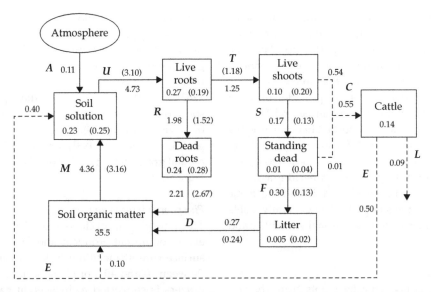

Figure 7.8 Phosphorus budget under chronic grazing and 7 years after grazing ceased (numbers in parentheses) of flooding Pampa grassland in Argentina. Numbers denote average nutrient contents (boxes: g m^{-2}) and mean daily nutrient flows (arrows: mg m^{-2} day^{-1}) over winter/spring season of peak plant productivity (150 days). Transfer processes are: A, atmospheric wet deposition; U, root uptake; T, net translocation; S, shoot senescence; R, root death; F, litter fall; C, consumption by herbivores; D, decomposition (roots/litter); E, urine/faeces excretion; M, net mineralization; V, volatilization; L, livestock removal. Note that the decomposer subsystem was not assessed. Reproduced with permission from Chaneton et al. (1996).

short term. Atmospheric deposition, mainly rainfall, provides 0.02–0.15 g m^{-2} yr^{-1} originating as dust and burning of plant materials and fossil fuels (Whitehead 2000). Fertilizer is applied to grasslands usually as superphosphate (Ca(H$_2$PO$_4$)$_2$ + CaSO$_4$.2H$_2$O) or triple superphosphate (Ca(H$_2$PO$_4$)$_2$) as an individual fertilizer or in combination with nitrogen or potassium as mono- or diammonium phosphate (NH$_4$H$_2$PO$_4$ or (NH$_4$)$_2$PO$_4$, respectively) or rock phosphate (Ca$_3$(PO$_4$)$_2$/apatite). In the soil, phosphorus occurs in inorganic and organic forms. The major source of available phosphorus for plants comes from mineralization of SOM; plant, animal, and microbial materials; and a labile inorganic pool.

Inorganic phosphorus includes phosphates bound to calcium, iron, or aluminium, which are important as regulators of phosphorus concentrations in the soil solution. Phosphorus is poorly mobile in soil solution. Much of the phosphorus in the soil is precipitated and insoluble or occluded and chemisorbed. This phosphorus is in equilibrium with the soil solution, and removal of phosphorus from the soil solution by plants stimulates its release from inorganic and organic sources. The fraction of the soil phosphorus which can enter the soil solution by isoionic exchange is referred to as the **labile inorganic phosphorus**. The soil solution equilibrates with the labile fraction of adsorbed phosphorus. Phosphorus also becomes immobilized with amorphous oxides and hydrous oxides of iron and aluminium as coatings on clay minerals through the process of phosphate fixation.

The organic phosphorus pool in SOM is larger than the inorganic phosphorus pool and consists of active, slow, and passive fractions. Organic phosphorus builds up as phosphorus becomes immobilized in organic matter following uptake in plants and microbial activity and depends directly on biological consumption (Parton et al. 1988). Reported levels of organic phosphorus in grasslands include 11.4 g m^{-2} in ungrazed tropical savannah in India (compared with 119.8 g m^{-2} in nearby natural forest; Lamotte and Bourliére 1983), 35.5 g m^{-2} in subhumid temperate grassland of Argentina

(Fig. 7.8), and 150–350 g m^{-2} (0.1–2.0 g m^{-2} available) in annual grasslands (Jones and Woodmansee 1979). Microbial biomass is important for the redistribution and accumulation of organic phosphorus. Mineralization by microbes of organic phosphorus from litter and dead roots can provide sufficient phosphorus to plants (Woodmansee and Duncan 1980). Decomposers are important in regulating the organic phosphorus pool; for example, uptake to the organic phosphorus pool was shown to be 4–5 times that of plant uptake in semi-arid grasslands (Cole et al. 1977). Levels of organic and inorganic phosphorus in the soil profile are generally stable over the long term, despite seasonal variability and dynamics.

Phosphorus uptake
Phosphorus is taken up by plants from the soil solution in the surface soil as soluble phosphate ions ($H_2PO_4^-$, HPO_4^{2-}) and is dependent on root distribution. Phosphorus uptake is aided by endomycorrhizal fungi which increase the surface area available for uptake, particularly in some mycorrhiza-dependent warm-season grasses from phosphorus limited grasslands (see §5.2.3). Some root exudates may promote desorption or dissolution of phosphorus, increasing local availability. Phosphorus concentration in plants is dependent on species, age, and physiological condition of the plants (see §4.2.2). The soil solution is replenished daily with phosphorus from the pool of labile inorganic phosphorus, which is itself replenished from mineralization of labile organic phosphorus and some leaching of standing dead and litter. The labile inorganic phosphorus pool was reported as the sole source of phosphorus for live plants (i.e. 18%, into roots) and microbes (82%) in tallgrass prairie (Risser et al. 1981). Indeed, soil microbes have a relatively high phosphorus requirement (average concentration of 1–2%), so phosphorus in the soil can be effectively immobilized by the microorganisms. Mineralization of organic phosphorus occurs when the organic carbon:phosphorus ratio is <200:1 in phosphorus-deficient soils or <100:1 for soils richer in phosphorus, and rates increase with temperature. Mineralization rates of 1.3–1.5 g m^{-2} are reported from annual California grasslands (Woodmansee and Duncan 1980) and 3.0–6.4 g m^{-2} for semi-arid grasslands in Colorado, USA and Saskatchewan, Canada, respectively (Cole et al. 1977). Uptake rates of 0.49 g m^{-2} yr^{-1} are reported from semi-arid grasslands and 4.96 g m^{-2} yr^{-1} from dry subhumid grasslands in India (Lamotte and Bourliére 1983), 0.45–0.75 g m^{-2} yr^{-1} (150-day growing season basis) from a subhumid temperate grassland of Argentina (Fig. 7.8), 0.54 g m^{-2} yr^{-1} in tallgrass prairie (Risser et al. 1981), and 1.0–1.5 g m^{-2} yr^{-1} in annual grasslands in California (Woodmansee and Duncan 1980).

Phosphorus returns and losses
The phosphorus cycle is closed except for the export of animal products, and to a lesser extent, surface run-off including loss of excess fertilizer. Between 10 and 40% of phosphorus consumed by grazers is converted to liveweight gain or milk; the rest is returned in excreta (Whitehead 2000). Phosphorus levels in animal excreta are determined mostly by the amount consumed and the phosphorus concentration in consumed plant tissues. The organic phosphorus content of faeces is c.0.06 g per 100 g of feed eaten, except at low concentrations of phosphorus in feed where this relationship breaks down and more phosphorus is retained. Only trace amounts of phosphorus occurs in animal urine. Phosphorus is lost from the grassland ecosystem when animal products are removed, or even by moving animals from daytime pastures to sleeping paddocks at night. The export of phosphorus in animal products can amount to as much as 36% of ingested phosphorus in dairy farming and 10% for wool and lamb production, and can transfer 0.7–1.1 kg per stock unit in excreta (Gillingham 1987). Intensively managed grassland grazed by dairy cattle lost 1.2 g P m^{-2} yr^{-1} in the UK (Whitehead 2000). Phosphorus outputs through cattle removal exceeded inputs through atmospheric deposition by 86% leading to a net loss of 0.09 mg P m^{-2} day^{-1} over the winter-spring season in grazed flooded Pampas (a subhumid temperate grassland of Argentina; Chaneton et al. 1996). Grazing can thus exacerbate phosphorus deficiency in phosphorus-limited grasslands and hasten nutrient cycling rates. Surface run-off can occur, reported as 0–0.05 g P m^{-2} in the San Joaquin annual grasslands of California. Organic

phosphorus is released from plant tissues and litter through decomposition releases. Organic phosphorus from dung and plant tissue contributes mostly to soil organic phosphorus reserves, with some immediately available to plants, although most decomposes slowly unless buried by dung beetles or earthworms. Phosphorus returns from plants to soil of $0.32\,\mathrm{g\,m^{-2}\,yr^{-1}}$ are reported from humid grassland and $3.03\,\mathrm{g\,m^{-2}\,yr^{-1}}$ from dry sub-humid grassland in India (Lamotte and Bourliére 1983). In tallgrass prairie 70% of phosphorus was leached or lost from standing dead plants in the first year (Risser et al. 1981). Phosphorus losses, especially losses of fertilizer, can lead to significant downstream effects in aquatic systems (Jones and Woodmansee 1979).

Combustion of plant tissues during fire leads to a return of phosphorus in the ash and hence negligible losses of soil phosphorus, except when ash is redistributed by wind. Fire, though, leads to increases in mineralization rates of SOM and can lead to increased labile phosphorus both over the short and long term and to increased plant uptake (Ojima et al. 1990; Risser et al. 1981). This pulse of available phosphorus can also stimulate nitrogen fixation. In those systems with very low phosphorus, nitrogen limitation appears to be constrained, so low phosphorus constrains the availability of nitrogen (Eisele et al. 1989). This limitation effect of phosphorus on nitrogen may only occur in organic wetlands in temperate zones and in systems with young (high phosphorus) soils, but could be important in some of the ancient soil grassland areas of Australia and elsewhere (T.R. Seastedt, personal communication).

7.3 Decomposition

Compared with some other ecosystems, grasslands have a low rate of decomposition and high stores of organic matter. Coupled together, these two features help characterize the grassland ecosystem and the soils that develop there (see §7.4).

7.3.1 Decomposers and decomposition rate

Decomposition is the breakdown of organic detritus into mineral nutrients that then become available for uptake by plants. Up to 90% of primary production in grasslands enters the decomposer subsystem (Úlehlová 1992).

The decomposer community includes a diverse group of microorganisms (the microflora: fungi and bacteria) and invertebrate animals classified on the basis of size (microfauna: nematode and protozoa; mesofauna: Collembola, Acari, Enchytraeidea, Isoptera, Diplura, Symphlua, Protura, Diptera; and macrofauna: earthworms, Diplopoda, Isopoda, Coleoptera, Mollusca, Orthoptera, Dermaptera) (Úlehlová 1992). These animals are heterotrophs, deriving their energy from SOM and utilizing organic compounds. High densities of microorganisms occur close to roots feeding off of the organic compounds released through root exudates into the soil. The chemical nature of root exudates varies among plant species. Both bacteria and fungi play a primary role in decomposition in grasslands with populations in temperate grasslands of the order of 10×10^6 bacteria $\mathrm{g^{-1}}$ soil and 3000 m fungal hyphae $\mathrm{g^{-1}}$ soil (Swift et al. 1979). Mean fungal biomass in the soils of natural grassland in North America ranged from 18 to $88\,\mathrm{g\,m^{-2}}$ in the top 10 cm of soil and up to $334\,\mathrm{g\,m^{-2}}$ at depths down to 30 cm (Úlehlová 1992). In the North American shortgrass prairie, 95% of the energy flow through the detrital food chain is mediated by these microbes (Hunt 1977). Bacteria were estimated to mineralize the largest amount of nitrogen in a shortgrass prairie ($4.5\,\mathrm{g\,N\,m^{-2}\,yr^{-1}}$) followed by the soil fauna ($2.9\,\mathrm{g\,N\,m^{-2}\,yr^{-1}}$) and fungi ($0.3\,\mathrm{g\,N\,m^{-2}\,yr^{-1}}$) (Hunt et al. 1987). Bacteria-feeding amoebae and nematodes were responsible for 83% of the nitrogen mineralized by the fauna, and altogether the fauna were responsible for 37% of nitrogen mineralization. Faunal population counts in temperate grasslands are of the order of 500×10^6 microfauna $\mathrm{m^{-2}}$, 37000 mesofauna $\mathrm{m^{-2}}$, and 1400 macrofauna $\mathrm{m^{-2}}$ (Swift et al. 1979).

Earthworms are important in Eurasian grasslands, reaching a high biomass that can exceed that of the herbivores. Their activity promotes deep, non-stratified brown soils (see §7.4.6, Chernozems), whereas the North American prairies have a similar structure despite the absence of native burrowing earthworms. Tropical savannahs have a higher density of bacteria ($55 \times 10^6\,\mathrm{g^{-1}}$ soil), and a lower faunal density than temperate grasslands (microfauna 30 000 m^{-2}, mesofauna 6900 m^{-2}, and macrofauna 200 m^{-2}) (Swift et al. 1979). Termites and dung beetles dominate the macrofauna of tropical savannahs, leading to soil turnover and structuring equivalent to the effects of earthworms in temperate grasslands (Swift et al. 1979). The total biomass of soil microorganisms in the top 10 cm of soil among grasslands worldwide ranges from 27 to 940 g m^{-2}, with that of grasslands exceeding croplands (Úlehlová 1992). These soil biota feed on detritus, including the faeces and carcasses of herbivores and predators from both herbivore subsystem, and the decomposer subsystem itself.

The two important stages in decomposition are (1) physical breakdown, i.e. fragmentation and size reduction by macroarthropods, and (2) microbial oxidation of organic matter for energy and nutrients. The latter proceeds in three stages:

- rapid loss of readily decomposable organic substrates such as root exudates, lysates, proteins, starches, and celluloses by moulds and spore-forming bacteria
- decomposition of organic intermediates by a variety of microorganisms
- decomposition of resistant plant constituents such as lignin by actinomycetes and other fungi.

The decomposer subsystem is explicitly recognized as a submodel in grassland models that simulate carbon flow(see §7.1.2). For example, in the ELM Grassland Ecosystem model, inputs to the decomposer submodel come from primary and secondary resources represented in the primary producer, and insect and mammalian consumer submodels (Hunt 1977; Risser et al. 1981). In the CENTURY model (Parton et al. 1996), the decomposer and SOM submodel simulates inorganic and organic carbon and nitrogen soil dynamics. The submodel is controlled by soil surface temperature and soil water content. The Hurley Pasture Model (Thornley 1998) has a soil and litter submodel with inputs from plant litter and grazing animal (sheep) faeces divided explicitly into a rapidly degrading metabolizable fraction, a slowly degrading cellulose fraction, and a resistant lignin fraction. The BOTSWA savannah model (Furniss et al. 1982; Morris et al. 1982) is unique in including consumption by termites, which were partitioned into litter feeders and humivores (consuming humus and small fragments). A sensitivity analysis showed termite activity to be one of the most important factors regulating the litter decay in *Burkea africana* South African savannah (Furniss et al. 1982).

In grasslands, most organic input to the decomposer subsystem is from root biomass; the second largest input is from microbial litter. This is in contrast to forests, where the large amount of litterfall is the principal input. Movement through the decomposition cycle can represent the major flux of a mineral, e.g. 60–70% of annual plant nitrogen flux (10–24 g m^{-2}) is through the decomposer subsystem rather than the herbivore subsystem (4–16 g m^{-2}) in New Zealand hill pastures (Lambert et al. 1982).

Decomposition rate is expressed as the decay rate constant (k), the annual fractional weight loss of particular detrital components as a negative exponential decay function:

$$\ln\left(\frac{X_t}{X_0}\right) = -kt$$

where X_t and X_0 are the amounts of each constituent at time t and time 0 respectively (Olsen 1963). Mean residence time can be calculated as $3/k$ to provide an estimate of the time taken for 95% of the constituent to decompose, or $5/k$ corresponding to the time taken for 99% of the constituent to decompose (Swift et al. 1979). Decomposition estimates among different ecosystems for turnover of the litter component (k_L yr^{-1}) were 1.5 and 3.2 for temperate grassland and savannah, respectively, in between the rates for temperate deciduous forest and tropical forest (0.77, 6.0, respectively). Comparable $3/k_L$ values for these ecosystems were 4, 2, 1, and 0.5, respectively (Swift et al. 1979). Decomposition in boreal forest and tundra ecosystems are lower than in temperate deciduous forest. A comparison of decomposition rates among grasslands in North America shows k estimates being highest in tallgrass prairie and lowest in mixed-grass prairie (Table 7.4). Regional comparisons indicate that mean annual precipitation (positive relationship), mean annual temperature (positive), and percentage soil clay content (negative) together explain 51% of the variance of k for total SOM decomposition across the US Great Plains, with mean annual

Table 7.4 Half-life ($t_{1/2}$) and decomposition rate (k) for cellulose decomposition during the growing season for 6 North American grasslands

Site	$t_{1/2}$ (days)	k (% per day)
Tallgrass prairie (OK, USA)	40	1.73
Shortgrass prairie (CO, USA)	55	1.26
Tallgrass prairie (MO, USA)	78	0.88
Desert grassland (MN, USA)	80	0.87
Mixed prairie (SD, USA)	85	0.82
Mixed prairie (Canada)	250	0.28

From Risser et al. (1981).

precipitation alone explaining 31% of the variance (Epstein et al. 2002a). The lack of a strong effect of temperature on decomposition rates across the Great Plains is because of the water-limited nature of these grasslands. Soil texture was important because of its effect on soil moisture and the availability of organic substrates in the soil profile.

Efflux of CO_2 from the soil reflects microbial respiration and is used as a measure of soil microbial activity. In temperate grasslands, CO_2 efflux occurs throughout the year following the season pattern of soil temperature reaching a maximum in June and July (Swift et al. 1979). Midsummer microbial respiration can be limited by soil moisture, whereas earlier and later in the year CO_2 efflux is temperature limited. By contrast, in tropical savannahs, soil microbial activity is limited by soil moisture during the dry season but high efflux during rains leads to rapid decomposition of non-woody NPP.

As noted above for the US Great Plains, the most important rate-limiting factors for decomposition are soil temperature and moisture. For example, decomposition in semi-arid regions occurs as bursts associated with microbial activity during favourable periods of soil temperature and moisture (Dormaar 1992). Decomposition in temperate grasslands is limited by soil moisture in tussocks through most of the year, but proceeds rapidly under high temperatures when moisture levels are sufficiently high, leading to the characteristically low SOM content of some grasslands. Semi-arid grassland soils have low SOM content because of high decomposition rates, not just because of low production (Dormaar 1992). Decomposition rates peak after flushes of microbial activity when otherwise dry soils are rewetted in moist conditions. Decomposition in tropical savannahs is predominantly moisture limited, with litter accumulating during the dry season. Decomposition is then rapid when rain falls.

Within a climatic region, decomposition rates are positively related to soil nitrogen and plant litter quality, and negatively related to the amount of plant cover. The decomposition rate of *Schizachyrium scoparium* litter in old fields in Minnesota, USA, was significantly correlated with percentage of soil nitrogen, but not fertilizer addition (Pastor et al. 1987). Different species contribute to a 'rhizosphere effect' through root exudates, which stimulate decay, and a 'drying' of the rhizosphere, which lowers microbial activity in the presence of living roots. Decomposition rates also differ among species because of differences in production (i.e. carbon input), anatomy, and chemical composition of tissue, especially the carbon:nitrogen and carbon:phosphorus ratios, lignin, and carbohydrate content (Dijkstra et al. 2006). Invasive species can alter the composition of the soil flora community by offering a new detrital chemical spectrum, thereby altering system decomposition (Vinton and Goergen 2006). For example, the invasive exotic *Schedonorus phoenix* caused shifts in detritivore communities that tended to increase litter decomposition rates in old fields in Oklahoma, USA, (Mayer et al. 2005). By contrast, invasion by *Aegilops triuncialis* was shown to slow above-ground litter decomposition in a serpentine grassland (Drenovsky and Batten 2006). Alteration of the decomposer community through invasion of exotic fauna can also alter decomposition rates. The exotic earthworm *Pontoscolex corethrurus*, for example, was shown to increase leaf litter decomposition rates in tropical pastures in Puerto Rico by elevating rates of litter consumption/digestion or microbial activity (k with earthworms = 2.94, without = 1.97; Liu and Zou 2002).

Decomposition rate declines with depth in the soil. The surface 15 cm of soil has >60% of the roots in tallgrass prairie soil (Weaver et al. 1935). The decline in decomposition rate with depth corresponds to a comparable decline in root biomass, temperature and soil moisture. Soil organic matter also decreases with depth but the ratio of SOM to root organic matter increases with depth, as does the ratio of soil nitrogen to root nitrogen. The discrepancy in the pattern of decline in SOM and root biomass with depth reflects the variability in decomposition rates as well as transport of fine particles of recalcitrant organic matter deeper in the soil profile. It has been proposed that the decline in soil moisture with increasing depth in the soil profile promotes bacteria and fungi that synthesize lignin-like decomposition-resistant compounds (Gill and Burke 2002). The resulting accumulation of decomposition-resistant compounds at depth exacerbates the decline in decomposition rate that would be due to a simple decrease in soil moisture and temperature. In support of this notion, decomposition rate constants (k_{mass}) of fresh root material of *Bouteloua gracilis* in Colorado, USA, shortgrass steppe were $0.26\,yr^{-1}$ at 10 cm and $0.14\,yr^{-1}$ at 100 cm with $5/k$ residence times of 19 and 36 years respectively (Gill and Burke 2002). Lignin content increased by 62% at 100 cm compared with no change compared with the initial amount after 33 months.

Other local effects that can affect decomposition rates in grasslands include fire regime and topography. Frequently burned tallgrass prairie tends to be nitrogen-limited (see §7.2.2) and has been shown to exhibit higher decomposition rates than unburned grassland, leading to a rapid release of nitrogen as decay proceeds (O'Lear et al. 1996). Moreover, in this system, decomposition was recorded to be faster in shallow-soil, upland sites compared with deep-soil, lowland sites and may have reflected the development of seasonally anaerobic soils in the lowland areas accompanied by rapid dry-down during the growing season when plants reduce soil moisture to very low levels.

Decomposition rate varies through the season in grasslands, reflecting seasonal patterns of production and root growth. Below-ground carbon allocation was greatest in the spring in New Zealand dairy pastures, as was root decomposition rate, time of optimal soil moisture, and temperature conditions (Saggar and Hedley 2001). Lower decomposition rates in the summer may have been related to drier conditions and limited supply of nitrogen and phosphorus to the roots. Conversely, high rates of litter decomposition during the dry season in Nigerian savannah were attributed to consumption ($790\,kg\,ha^{-1}$ in December–March) by fungus-growing termites (Macrotermitinae) (Ohiagu and Wood 1979).

There is concern that atmospheric nitrogen deposition (see §7.2.2) may increase decomposition rates through stimulation of plant production. Conversely, there could be positive feedback if increased nitrogen inputs increase litter quality (i.e. lower carbon:nitrogen ratio) or alter species composition. Altered decomposition depends both on the level of nitrogen addition and on litter quality (Knorr et al. 2005). Long-term studies do not support changes in decomposition rate in response to nitrogen addition in grasslands (Aerts et al. 2003). Variations in decomposition rates among species with nitrogen addition appear to be primarily related to the amount of carbon input (positive relationship) and root nitrogen concentration (negative relationship) (Dijkstra et al. 2006).

7.3.2 Litter

Plant litter (mulch, dead plant debris on the soil surface) is an important input to the decomposer subsystem in the absence of frequent fire. The amount of litter in grasslands varies widely (Table 7.2), with amounts up to $1000\,g\,m^{-2}$ in tallgrass prairie (Weaver and Rowland 1952). Litter adds to SOM, decreases soil bulk density, protects the soil surface from wind and water erosion, reduces light levels at the soil surface, increases microscale humidity, lowers air movement and convective heat exchange with the atmosphere, and causes the soil to warm more slowly in the spring. An equilibrium in the accumulation of litter can be reached in semi-arid–dry-subhumid climates where litter biomass does not exceed annual above-ground NPP. In neotropical *Trachypogon* savannahs in Venezuela, above-ground biomass increases each year for 5 years after fire, after which point an equilibrium is reached at $c.1200\,g\,m^{-2}$ as decomposition equals production (Sarmiento 1984). There are slight oscillations during each annual fire cycle as production exceeds decomposition at the start of each growth cycle. Decomposition rates were observed to oscillate during the year around 1.8–30 mg g^{-1} d^{-1}, comparable to values calculated for two Michigan secondary grasslands of 1.3–13.6 mg g^{-1} d^{-1}. Generally, 40–70% of the litter disappears over a 2 year period (Deshmukh 1985 and references therein).

In subhumid regions litter accumulation in the absence of fire can retard growth through shading of new shoots (Knapp and Seastedt 1986) which reduces energy by up to 14% over the growing season, reduces production by 26–57% (Weaver and Rowland 1952), and can delay growth by up to 3 weeks in the spring. By contrast, in California prairie, the litter layer provides a favourable environment for seed germination and seedling establishment (Heady et al. 1992). Litter accumulation provides fuel, enhancing the probability of fire. Some species such as the invasive annual bromes (*Bromus* spp.) in North American prairies, increase fire frequency as their senesced biomass increases fuel loads. High production levels of tallgrass prairie dominants also lead to large amounts of litter, enhancing the probability of fires, and benefiting these species as competitors are killed. Thus, litter can both inhibit the growth of prairie dominants and also enhance the optimal conditions (frequent fire) to ensure their continued dominance. Together, modelling these observations suggests that litter generates non-linear dynamics and chaos in tallgrass prairie, inducing self-generating disturbances leading to high spatial heterogeneity and diversity levels (Bascompte and Rodríguez 2000).

The litter/fire interaction regulates several ecosystem processes. The transient maxima hypothesis (see §7.1.2) explains how an absence of fire and the consequent increase in litter results in continued nitrogen mineralization in the presence of lower plant uptake that becomes higher than in the absence of a litter layer, because the enhanced moisture and temperature are more conducive to microbial activity. Thus, inorganic nitrogen accumulates in the soil until the next fire, then ANPP is highest when light and nitrogen are not limiting.

7.4 Grassland soils

Soil is the thin mantle of weathered material and organic matter that supports terrestrial life. Its properties reflect the climate, vegetation, fauna, topography, and parent material over the time span through which it has formed, as well as recent human activity. The 'typical' grassland soil has been described as having the following features (Acton 1992):

Surface layer: dark in colour (dark brown, black or dark grey), 15–30 cm thick, friable with cloddy structure, high in bases and nutritive elements.

Underlying layer: paler in colour, more yellow, grey or reddish-brown, cloddy in structure, but could be blocky, prismatic or columnar, sometimes platy.

By comparison, in desert soils the dark, humus-rich surface layer is lacking; in forest soils the surface layer is a mull or modor humus with a thin surface layer that has a more grey or reddish subsurface layer; mountain soils have a 'fluffy' surface layer due to the accumulation of organic matter; and wetland soils also have thick layers of decomposing organic material due to high plant productivity and reduced decomposition (Acton 1992). The high organic matter content of grassland soils relative to other terrestrial ecosystems is the result of high productivity of the characteristic dominant species under dry conditions that limit decomposition. At least for North America and many areas in the northern hemisphere, it is important to emphasize the time since glaciation, resulting in less weathering (oxidation) that also contributes to high organic matter content.

The International Union of Soil Scientists officially adopted the World Reference Base for Soil Resources (WRB) as the system for soil classification in 1998 (Driessen and Deckers 2001). The WRB supersedes the former FAO–UNESCO Soil Classification System. Under the WRB scheme, there are 30 Reference Soil Groups aggregated into 10 sets. The sets bring together mineral soils in the Reference Soil Groups on the basis of shared soil-forming factors. Grasslands occur in many of these soil groups but are predominantly found in seven of these sets described below and outlined in Table 7.5. The 'typical' grassland soil is a Chernozem, a member of set 8, the steppic mineral soils (see §7.4.6). Soils that do not occur primarily under grassland are not described, e.g. Andosols in set 3 which are predominantly volcanic. The soils in sets 6–9 are referred to as **zonal soils,** as their properties are

Table 7.5 Occurrence of grasslands on Reference Soil Groups of the World Reference Base for Soil Resources (WRB)

Set #	Environment	Reference Soil Group	Characteristics	Occurrence of grasslands
SET #2 Mineral soils whose formation was conditioned by human influences	(not confined to any particular region)	Anthrosols	Plaggic surface layer of incorporated animal bedding.	Extensive areas in Europe under crops or grass
SET #3 Mineral soils whose formation was conditioned by their parent material	Developed in residual and shifting sands	Arenosols	Well drained sandy soils, low fertility.	Hot subtropics of Australia and India, tropical savannas of Central Africa and Brazil.
	Developed in expanding clays	Vertisols	High clay content, clays shrink causing surface cracks when dry.	Worldwide especially tropical areas with a dry season, e.g., Blue Nile region of Sudan.
SET #4 Mineral soils whose formation was conditioned by the topography/physiography of the terrain	Elevated regions with non-level topography	Regosols	Deep, well-drained, skeletal, non-differentiated, immature soils. Ochric surface horizon.	Worldwide, mainly eroding arid areas, dry tropics.
SET #6 Mineral soils whose formation was conditioned by Climate.	(Sub-)humid tropics	Ferralsols Nitisols Acrisols Lixisols	Deep, red, intensely leached and weathered soils.	Tropical savannas
SET #7 Mineral soils whose formation was conditioned by Climate.	Arid and semi-arid regions	Solonetz	Natric horizon with accumulated salts toxic to many plants and limited water percolation and root penetration.	Scattered on loess throughout semiarid grasslands and subhumid grasslands otherwise dominated by Set # 8 soils.
SET #8 Mineral soils whose formation was conditioned by Climate.	Steppes and steppic regions	Chernozems Kastanozems Phaeozems	Dark, brown-black, deep fertile soils with accumulations of calcium carbonate at depth	Large areas of tallgrass, fescue prairie, and shortgrass prairie in N. America, steppe in Eurasia
SET #9 Mineral soils whose formation was conditioned by climate.	(Sub-)humid temperate regions	Planosols	Bleached, light coloured upper horizons over clayey lower horizons	Seasonally waterlogged flatlands with grassland or scattered trees

From Driessen and Deckers (2001). Sets and Reference Soil Groups of negligible or minor importance for the development of grasslands are omitted; these include Set # 1 Organic soils, Histosols, Set # 3 Andosols, Set # 4 Fluvisols, Gleysols, and Letosols, Set # 5 Cambisols, Set # 6 Plinthosols and Alisosl, Set # 7 Solonchaks, Gypsisols, Durisols, and Calcisols, Set # 8 Podzols, Albeluvisols, Luvisols, and Umbrisols, and Set # 10 Cryosols.

determined by the climate, whereas **intrazonal soils** result from the predominance of local factors, such as the Anthrosols in set 2 and Arenosols in set 3. Soils which are too young to reflect site-specific characteristics in their profile, such as very early successional soils, are referred to as **azonal soils**.

7.4.1 Set 2. Mineral soils conditioned by humans: Anthrosols

These soils occur in any environment and vary widely as a result of characteristics arising from human activities (anthropedogenic processes) including addition of organic materials, household wastes, irrigation, or cultivation. These materials become incorporated into a surface horizon that overlies the original buried soil. Anthrosols with a 'plaggic' surface horizon (i.e. Plaggic Anthrosol, or under the older FAO scheme, Femic Anthrosols) are typically planted to rye, oats, barley, or potato, but increasingly used in Europe for silage, maize production, or grass, yielding 9–12 kg ha^{-1}. The plaggic horizon can originate from the traditional medieval practice of spreading used animal bedding (straw) as 'organic earth manure' on arable fields. This practice raises the surface of fields by 0.1 cm yr^{-1} and has been carried out in some areas of Europe for >1000 years. The plaggic surface horizon is >1 m thick in places, black or brown in colour, and in some places can include sods incorporated from nearby heathland which raises the clay content. The resulting soil is well drained, pH 4–5, organic carbon 1–5%, carbon:nitrogen ratio 10–20 (higher when the plaggic horizon is black rather than brown), high phosphate content, and cation exchange capacity (CEC) 5–15 cmol(+)/kg soil. Plaggic anthrosols cover >0.5 × 10^6 ha in Europe, especially in the Netherlands, Belgium, and Germany.

7.4.2 Set 3. Mineral soils conditioned by parent material: Arenosols and Vertisols

Mineral soils conditioned by parent material and supporting grassland include Arenosols on sand and Vertisols on expanding clays.

Arenosols (from Latin *arena*, sand) occur on (1) residual sands following prolonged weathering of quartz-rich rocks including granite, quarzite, and sandstone, (2) aeolian sands deposited by wind in dunes, beaches, or fluvial environments, and (3) alluvial sands deposited by water. As azonal soils, Arenosols are not characteristic of a particular climate, although they are common in arid environments. Arenosols are extensive, covering c.900 × 10^6 ha worldwide, and are characterized by a pale yellowish or reddish-brown to dark brown A horizon over a B horizon of coherent sand to sandy loam that is red, yellow, or brown, massive, porous and compact, overlying either weathered rock, a massive clay, or a hard indurated layer. Arensols generally have low moisture-holding capacity, low fertility with low organic matter content (1% at the surface decreasing with depth), low CEC, and low base saturation, and are moderately to strongly acidic.

Arenosols are the major soil type under grasslands in the hot subtropics of Australia (e.g. associated with spinifex grasses such as *Triodia* spp. and *Plectrachne* spp.) and India, and occur scattered throughout tropical savannah of central Africa, Brazil (including the Campo Cerrado—see Chapter 8), and the Guiana Shield in South America (Acton 1992). In these areas, the soils are typically very old, over a highly weathered substrate, contain a high content of sesquinoxides of iron and aluminium, are extremely acidic (surface pH <4.5) and are referred to as **ferralic arenosols**.

Vertisols (from the Latin *vertere*, to turn) are heavy, clay-textured soils (≥30% clay content) which develop in tropic semi-arid to (sub)humid and Mediterranean climates with seasonal rainfall and a dry season of several months. Vertisols occur on 335 × 10^6 ha worldwide in low landscape positions on sediments with high clay content. Characteristically, Vertisols have a vertic horizon; i.e. a 'subsurface horizon rich in expanding clays and having polished and grooved ped surfaces ('slickensides'), or wedge-shaped or parallelepiped structural aggregates formed upon repeated swelling and shrinking' (Driessen and Deckers 2001). The clays shrink, forming cracks from the surface downwards in the dry season, allowing materials from the surface to fall into the cracks (i.e. the surface is said to be 'self-mulching'). Water enters the cracks in the wet season, causing the clays to expand, swell, and become plastic and sticky. These soils have a dark surface layer that is grey, brown, or red in colour to 1 m depth. Organic matter content is 1% or more and the soil pH is 6.0–8.0. The vegetation on Vertisols is limited to grasses and slow-growing deep-rooted trees. These soils occur within grasslands on all continents, e.g. Blue Nile region of Sudan; edge of Kalahari desert, southern Africa; Deccan plateau, India; inland plains of eastern Australia; Pampas of Argentina; southern and coastal prairies of Texas, USA; and glacial lacustrine clays in the northern Interior Plains of North America. Vertisols have limited agricultural use as they can only be tilled for a short time in the season between the wet and dry period, and are therefore used mostly as pasture.

7.4.3 Set 4. Mineral soils conditioned by topography: Regosols

This group contains soils that do not fit within the other Reference Soil Groups. They are very weakly developed mineral soils in unconsolidated materials with just one diagnostic horizon, i.e. an ochric surface horizon (thin, lacking fine stratification, either light colour or thin, or low in organic matter, or massive and very hard when dry). These soils are extensive in eroding areas in all climate zones without permafrost, covering 260×10^6 ha worldwide, 50×10^6 ha in the dry tropics, and 36×10^6 ha in mountainous areas. Regosols intergrade with other soils especially in arid grazed areas of the Midwestern USA, northern Africa, the Near East, and Australia.

7.4.4 Set 6. Mineral soils conditioned by a wet (sub)tropical climate: Ferralsols, Nitisols, Acrisols, and Lixisols

These soils underlie tropical savannahs and are generally deep, intensely weathered, and red to yellow in colour, formed over geologically old substrates. They differ depending upon the mineralogy of the substrate and the extent and nature of leaching. Horizons tend to be diffuse with a crumby structure due to intense termite activity. These soils cover extensive areas in the wet (sub)tropics and, although mostly supporting forest, are overlain by grassland in many areas of savannah. Just as there is a climatic progression of savannah types depending upon the influence of soil moisture and reflected in the abundance of trees, so too there is comparable, albeit complex, variation in soils (Montgomery and Askew 1983). For example, the most extensive soils underlying moist savannahs are Ferralsols and Acrisols, whereas soils in drier savannahs can also include a wide variety of soils including Lithosols (shallow soils over rock) and Vertisols (see §7.4.2). There is also disagreement on whether the differences between savannah and forest soils are the result of the influence of the vegetation or whether soils are one of the independent causative factors along with climate that determine savannah/forest boundaries. In some areas (e.g. Colombia and western Nigeria), the same soil series has been mapped on both sides of the forest/savannah boundary, whereas in other areas (e.g. Brazilian Campo Cerrado) significant differences in texture and fertility occur across this boundary (Montgomery and Askew 1983).

Ferralsols are deep, intensely weathered, red or yellow soils on geologically old substrates in the tropics. The name comes from the high content of iron (Latin *ferrum*) and alum left in the profile following downward leaching by water of silica and aluminosilicates. As a result these soils have a low pH and low fertility with deficiencies in trace elements and calcium; the soil horizon boundaries are diffuse, partly as a result of termite activity. Ferralsols occur over 750×10^6 ha in the humid tropics including parts of the African savannahs and the Campo Cerrado, chaparral, and savannahs of South America, especially Brazil and Venezuela, and in the uplands of Madagascar. These soils occur alongside Acrisols and Nitisols, with the former in lower topographic positions or on more acidic rock (e.g. granite) and the latter over more basic rock (e.g. dolerite). Ferralsols are also classified as Oxisols (U.S. Soil Survey Staff 1975) and latisols (UK) in older literature.

Nitisols are deep, well-drained, dusky red or dark red tropical soils, with diffuse horizons (due to intense termite activity, as in Ferralsols). They are characterized by a clayey 'nitic' (from the Latin *nitidus*, shiny) subsurface horizon with strongly angular, shiny ped faces due to alternating microswelling and shrinking, pH 5.0–6.0, high CEC compared with other tropical soils because of the high clay content (>30%), and a deep and porous structure allowing deep rooting and good drainage characteristics. They are the most productive of the tropical soils occurring over weathered intermediate to basic rock. Nitisols occur over 200×10^6 ha under tropical rain forest or savannah on level to hilly terrain, >50% of it in tropical Africa, and elsewhere in tropical Asia, Central America, South America, and Australia. Nitisols are classified as Kandic groups of Ulfisols and Alfisols in the USA (U.S. Soil Survey Staff 1975).

Acrisols (from the Latin *acris*, sharp or acidic) have a dark, reddish-brown or strong yellow surface horizon due to release of iron oxides from the silicates. Formed from old, acid rocks in warm humid climates, they occur predominantly under forest, but support tropical savannahs of South America and Africa, and grasslands mixed with forest in the western USA. Intense leaching leads to accumulation of clay in the pale B horizon (i.e. presence of an argic horizon characterized by the presence of higher amounts of clay than in the overlying horizon). Acrisols have a lower base saturation than Lixisols, but are more intensely weathered, which limits agricultural productivity. Acrisols are classified as Ultisols in the USA (U.S. Soil Survey Staff 1975).

Lixisols (from the Latin *lix*, ashes or lye) cover 13% of the earth's land surface, mostly deciduous–coniferous forest, deciduous forest, and broad-leaved evergreen forest, and are a minor component of soils in tropical and subtropical grasslands and savannah—e.g. domimant in grasslands of subtropical highlands, veld, temperate dry shortgrass steppe, and tussock grasslands of South Africa. In South America, Lixisols occur in the Campo Cerrado in Brazil. Lixisols are formed from base-rich rocks and so have relatively high pH and base saturation but are not as strongly weathered as the Ferralsols. Organic matter is well mixed in a dark A horizon. Ferralsols are classified as Alfisols in the USA (U.S. Soil Survey Staff 1975).

7.4.5 Set 7. Mineral soils conditioned by a (semi) arid climate: Solonetz

Solonetz (from the Russian *sol*, salt, and *etz*, strongly expressed) soils contain free soda (Na_2CO_3), are strongly alkaline (pH >8.5), and characterized by possession of a natric B horizon within 100 cm of the soil surface, i.e. a 'subsurface horizon with more clay than any overlying horizon(s) and high exchangeable sodium percentage; usually dense, with columnar or prismatic structure' (Driessen and Deckers 2001). Solonetz soils have a thin, loose litter layer over 2–3 cm of black humified material. The surface A horizon <15 cm, is black or brown, and granular over the natric horizon. Calcium or gypsum salts can accumulate below the natric horizon and sodium and other salts become more concentrated with depth. The soil is slowly permeable, water can pond on the surface, and the natric horizon hinders downward movement of water and root penetration. Solonetz soils occur in semi-arid, temperate, subtropical areas with a steppe climate (dry summers, annual precipitation <400–500 mm) over 135×10^6 ha worldwide on flat, gently sloping grasslands with loess/loam or clay. These soils are frequently scattered in semi-arid and subhumid grasslands in areas otherwise dominated by Chernozems, Kastonozems, and Phaeozems, or with Arensols and Vertisols in subtropical and tropical areas. For example, in North America Solonetz soils occur on morainal and lacustrine deposits of the Interior Plains, and in Argentina on loessial deposits in the eastern Pampas.

7.4.6 Set 8. Mineral soils conditions by a steppic climate: Chernozems, Kastonozems, and Phaeozems

These are the soils that are often depicted as 'typical' grassland soils, especially the Chernozems, with their deep, dark, rich, and fertile A horizon, which underlie the northern prairies of North America and the steppes of Eurasia.

Chernozems (from the Russian *chernyi*, black and *zemlya*, soil) are characterized by their black or nearly black A 'mollic' horizon 25 cm thick (range 8–50 cm), usually flushed free of carbonates. There is a brown, prismatic B horizon, 15–60 cm, devoid of $CaCO_3$ above a yellowish-brown layer of accumulated calcium carbonate (starting within 200 cm of the soil surface). Organic matter content is moderately high, 3–7% at the soil surface, decreasing with depth, pH is neutral at the surface (range pH 5.5–7.5) increasing with depth, high CEC, high in montmorillonite, and base-saturated with calcium. Soil fauna can be very active, contributing to homogenization of the soil. Chernozems typically form on loess and support northern tallgrass prairie and native fescue prairie in North America and northern steppe in Romania east through the Ukraine into Kazakhstan and western Siberia. They cover 230×10^6 ha worldwide, especially in the middle latitude steppes of Eurasia and North America, and can grade into Kastonozems towards warm areas in the south and Phaeozems towards warm, humid regions. These soils vary considerably, becoming calcic in eastern North America and eastern Europe, and leached and degraded in grassland–forest transition areas in association with Greyzems. Chernozems have high inherent fertility (regarded by some as the best soils in the world), are well drained, with production limited by the short growing season. They have been extensively ploughed for cultivation (especially wheat, barley, maize, and more recently for soybean production) or used for livestock grazing. Chernozems are classified as Mollisols in the USA (U.S. Soil Survey Staff 1975).

Kastanozems (from the Latin *castanea*, chestnut and Russian *zemlya*, soil) have a brown to dark-brown, thin, neutral-slightly, organic matter rich (2–4%), alkaline surface horizon over a brown, prismatic, lime-free 'mollic' layer, itself over a strongly calcareous olive-brown layer with concretions of secondary carbonates. The horizons are sharply differentiated. Kastanozems occur over 465×10^6 ha worldwide, especially in the Great Plains of North America on the eastern side of the Rocky Mountains from southern Canada to the Gulf of Mexico, extensive areas in the intermountain area west of the Rocky Mountains from Washington State south to Arizona, in South America in the Pampas of Argentina, Uruguay and Paraguay, and the Eurasian shortgrass steppe of southern Ukraine, southern Russia, and Mongolia south of the Chernozem belt. Kastanozems border Chernozems and Phaeozems along cooler, moister margins and desert soils along warmer, drier areas. Th high fertility of these soils (less than that of Chernozems) is limited by aridity; they are cultivated for small grains and forage. Kastanozems were earlier called 'chestnut soils', and are classified as ustolls and borolls in the Mollisol Order of the US Soil Taxonomy (U.S. Soil Survey Staff 1975).

Phaeozems (from the Greek *phaios*, dusky, and Russian *zemlya*, soil) are soils of the wet steppe (prairie), similar to Chernozems and Kastonozems in structure, but more intensively leached so the dark upper horizons are less base-rich and lacking secondary carbonates in the upper metre of soil. The mollic surface horizon is thinner (30–50 cm) and less dark (brown-grey) than in the Chernozems, and overlies a yellowish-brown to grayish-brown, often mottled, blocky B horizon extending to 1 m or more, and an underlying yellowish-brown, leached carbon horizon, with gradual boundaries between each. The soil pH is slightly acid, increasing with depth and then sharply so to a pH of 7 or 8 in the carbon horizon. Phaeozems occur over steppe or forest, are fertile, and are cultivated to cereals or pulses or pastured for cattle. Occupying 190×10^6 ha worldwide, Phaeozems extend to 70×10^6 ha in the subhumid lowlands and easternmost Great Plains of the USA, 50×10^6 ha in subtropical Pampas of Argentina and Campos of southern Brazil and Uruguay, and 8×10^6 ha in north-eastern China. Generally, Phaeozems occur on the humid side of the Chernozem belt. Phaeozems are classified as udolls in the USA (U.S. Soil Survey Staff 1975) and as podozolized and leached chernozems in the former Soviet Union.

7.4.7 Set 9. Mineral soils conditioned by a (sub) humid temperate climate: Planosols

Planosols (from the Latin *planus*, flat) are characterized by a bleached, light-coloured sandy (or coarser) eluvial surface horizon reflecting periodic water stagnation, over a dense, slowly permeable clayey subsoil. There is an abrupt textural change

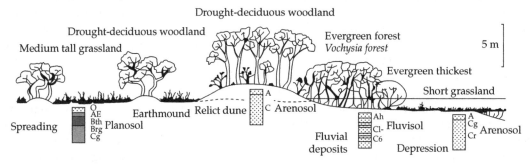

Figure 7.9 Characteristic vegetation formations and soils in the northern Pantanal of Poconé, Brazil. The solid line shows the mean flooding height during the rainy season, the dashed line during the dry season. Reproduced with permission from Zeilhofer and Schessl (2000).

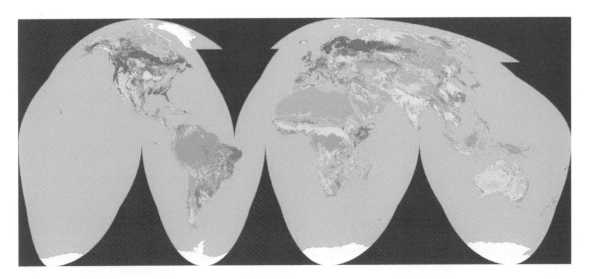

IGBP Legend

Class_Names

Evergreen needleleaf forest
Evergreen broadleaf forest
Deciduous needleleaf forest
Deciduous broadleaf forest
Mixed forest
Closed shrublands
Open shrublands
Woody savannas
Savannas
Grasslands
Permanent wetlands
Croplands
Urban and built up
Cropland/natural vegetation
Snow and ice
Barren or sparsely vegetated
Water

Plate 1 Global land cover map. Seventeen land cover classes represent regional vegetation types and mosaics that occur at the 1 km resolution (Loveland *et al.* 2000). Available from http://edcsns17.cr.usgs.gov/glcc/glcc.html. Data courtesy of the U.S. Geological Survey/EROS, Sioux Fall, SD.

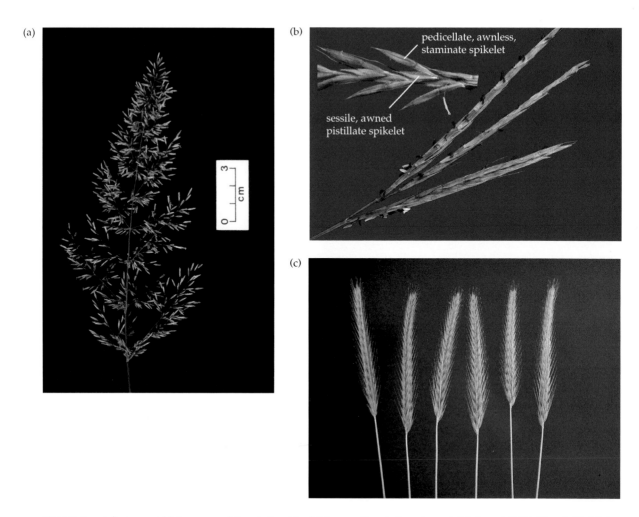

Plate 2 Grass inflorescences: (a) An open, much branched panicle of *Calamagrostis porterii* ssp. *insperata* (photograph © D.J. Gibson), (b) digitate raceme (sometimes referred to in this species as a false panicle) of *Andropogon gerardii* (x1); the inset (x2) shows a partially dissected inflorescence unit with labelled pedicellate and sessile spikelets (photograph © D.L. Nickrent), and (c) spikes of *Hordeum pusillum* (x0.6) with three spikelets per rachis node; one central sessile spikelet and two pedicelled lateral spikelets (photograph © D.L. Nickrent).

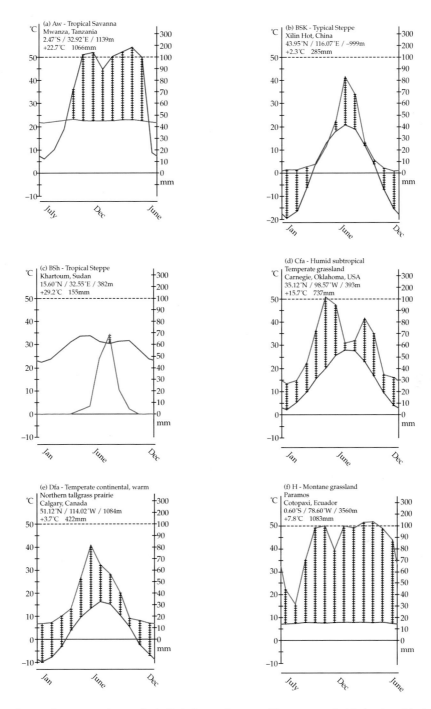

Plate 3 Climate diagrams for representative grasslands. Each diagram shows monthly temperature (red line) and precipitation (blue line). The legend for each panel indicates Köppen climatic region, location, longitude and latitude, elevation, average annual temperature, and precipitation sum. Note that the astronomic summer is in the middle of each diagram, June for northern hemisphere and December for the southern hemisphere. Adapted from Lieth *et al.* (1999).

Plate 4 Tropical savanna, Serengeti National Park, Tanzania. Köppen climatic region Aw (see §8.2.1). Photograph © Sam McNaughton.

Plate 5 Shortgrass prairie steppe, Pawnee Prairie, Colorado, USA. Köppen climatic region BSk (see §8.2.2). Photograph © David Gibson (1980).

Plate 6 Typical steppe from north-eastern Inner Mongolia, China, dominated by *Stipa* spp. (*S. gobica*, *S. krylovii*, *S. bungeana*, or *S. orientalis*). Köppen climatic region BS (see §8.2.2). Photograph © Zicheng Yu (1987).

Plate 7 Meadow steppe from Inner Mongolia, China. Köppen climatic region BS (see §8.2.2). Photograph © Hongyan Liu.

Plate 8 Mitchell (*Astrebla* spp.) grassland, Longreach Downs, Queensland, Australia. Köppen climatic region BW (see §8.2.2). Photograph © Steve Wilson.

Plate 9 Spinifex grassland (*Triodia* spp.), Carter Ranges near Middleton, Queensland, Australia. Köppen climatic region BS (see §8.2.2). Photograph © Steve Wilson (2006).

Plate 10 Grazed campos grassland of the Rio de la Plata grassland in southern Uruguay (31°54′S, 58°15′W) dominated by C_3 and C_4 grasses including *Nassella neesiana, Stipa charruana, Coelorachis selloana,* and *Cynodon dactylon*. For details see Altesor *et al.* (2006). Köppen climatic region Cfa (see §8.2.3). Photograph © Gervasio Piñeiro.

Plate 11 North American tallgrass prairie. Konza Prairie Biological Station, Kansas, USA dominated by C_4 perennial grasses *Andropogon gerardii* and *Sorghastrum nutans*, and a high diversity of herbaceous forbs, including *Asclepias tuberosa*, the butterfly milkweed, shown here. Köppen climatic region Cfa (see §8.2.3 and §8.2.5). Photograph © David Gibson (1987).

Plate 12 Subalpine snow tussock *Chionochloa antarctica* grassland from 360 m asl on the east slope of the Hooker Hills on the sub-Antarctic Auckland Island in the southern Pacific Ocean (Vitt 1979). Köppen climatic region H (see §8.2.6). Photograph by Dale Vitt, reproduced with permission of the National Research Council of Canada.

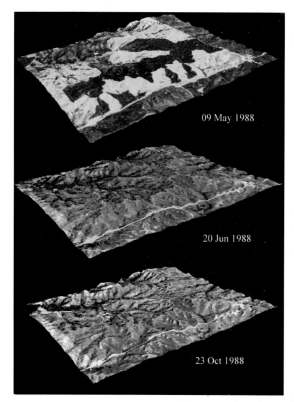

Plate 13 Landscape scale heterogeneity illustrated using false-colour SPOT satellite images draped over a digital elevation model showing the effects of burning tallgrass prairie (Konza Prairie, Kansas, USA) (see §9.1). Red areas represent areas of high vegetative cover. Burned watersheds show up clearly as black areas shortly after burning in early May, whereas this discrimination is less apparent later in the season. Note the western boundary (left) of the Konza Prairie with private lands that are grazed more heavily. Reproduced with permission of Oxford University Press, Inc from Briggs et al. (1998).

Plate 14 Curtis Prairie, Wisconsin, USA. This tallgrass prairie restoration is the oldest prairie restoration in the world and was started in 1935 (see §10.3.4). Photograph © David Gibson (1993).

from the upper to lower horizons within 100 cm of the soil surface. Planasols are found in seasonally waterlogged flatlands in subtropical or temperate, semi-arid and subhumid regions with light forest of scattered trees or sparse grassland (Fig. 7.9). They are poor soils, planted to rice or fodder crops or used for grazing, and cover 130×10^6 ha worldwide in Latin America (southern Brazil, Paraguay, Argentina), southern and eastern Africa (Sahelian zone, east and southern Africa), the eastern USA, south-east Asia (Bangladesh, Thailand), and Australia.

7.4.8 Relationships between grassland vegetation and soil

The continuum of both vegetation and soils and the graduation of one soil type into another occur in much the same way that vegetation types grade into each other, and are represented as a mosaic of differently scaled patches across the landscape. The soil–vegetation continuum is largely related to climate, topography, and underlying geology which alter the soil moisture status and mineralogy. For example, the tropical floodplain vegetation of the northern Pantanal of Brazil is a type of hyperseasonal savannah (i.e. more frequently flooded than the seasonal savannahs of the Campo Cerrado elsewhere in the region (Fig. 7.9 and see §8.2.1). Grasslands of the northern Pantanal dominate areas of poor drainage or prolonged inundation, with medium grasslands dominated by *Panicum stenodes* and *Axonopus purpusii* or *Andropogon hypogynus* and *Axonopus leptostachyus* occurring over soils classified as Cambisols, Solonetzs, Planosols, or Alisols (Zeilhofer and Schessl 2000). In areas flooded >5 months of the year, short grasslands dominated by *Reimarochloa acuta*, *Panicum laxum* and *Hydrolea spinosa* occur over Arenosols, Cambisols, Vertisols, Acrisols, and Gleysols.

CHAPTER 8

World grasslands

Grasses constitute an important part of the vegetation of most temperate countries, forming large masses of verdure on plains and hillsides, and giving to the landscape that hue on which the eye can longest gaze untired... Anne Pratt (1873).

Grasslands cover a vast area, occupying $41-56 \times 10^6$ km^2, or 31–43% of the earth's land surface (Chapter 1). Spread across every continent except Antarctica, grasslands vary in composition and physiognomy depending upon regional and local variations in climate, underlying geology, soil moisture, and disturbance regime (principally grazing and fire regimes). The term encompasses the sparse open communities of desert grasslands as well as the lush 'sea of grasses' of the South American Pampas. Summarizing this variation adequately is a challenge.

Daunting though the task may be, a classification system for the world's vegetation types, including grasslands, is necessary. It provides a framework for communication, allowing researchers, conservationists, and others to understand what each is talking about, and also allows management and preservation decisions to be made on a sound basis. Above all, a vegetation classification scheme provides a framework for 'pigeonholing' all the data and information that we collect about the natural world. A clear classification system allows one to readily conceptualize a unit of vegetation, analogous to the hierarchical classification scheme of taxonomic nomenclature that allows us all to understand which organism someone is talking about. The importance of an adequate classification scheme is readily apparent if we consider the inadequacy of trying to conserve biological diversity on a species-by-species basis. Although it is sometimes necessary to try and save particular species that might be on the verge of extinction, the sheer magnitude of biological diversity (somewhere between 10 and 100×10^6 species) means that protecting most of them requires a broad-based approach that protects habitats and species assemblages rather than individual species.

In this chapter, the approaches to describing and classifying the world's grasslands are described. A general overview of the world's grasslands based on their climatic relationships is presented. In the last part of the chapter examples of regional grassland classifications that provide a fine level of resolution are discussed, to illustrate the level of detail required to encompass local variation.

8.1 Ways of describing vegetation

In developing and using grassland classifications three important contrasts have to be considered: (1) natural vs cultural vegetation, (2) existing natural vs potential natural vegetation, and (3) floristic vs physiognomic classifications.

The **natural vs cultural vegetation** contrast describes recognizing vegetation as either natural (apparently unmodified by human activities) or cultural (planted or actively maintained by humans, such as croplands or pastures). Since most grassland has been impacted by humans to some degree, this can be a problematic and arbitrary distinction to make. Nevertheless, the Ecosystems of the World book series provides separate classifications of natural vs managed grasslands to deal with this distinction (Breymeyer 1987–1990; Coupland 1992a, 1993a).

The **existing natural vs potential natural vegetation** contrast refers to the development of classifications based on grassland as it is found today compared with the grassland that is hypothesized to occur without human interference under present climatic and edaphic conditions, i.e. the potential climax community. The problems with the climax view of vegetation are well understood (Chapters 6, 10) and include the observations that vegetation change is not as predictable as expected under the climax theory, that vegetation dynamics are non-equilibrial, and that successional change that might lead to a climax community is non-deterministic (Pickett and Cadenasso 2005). Nevertheless, classifications based on potential natural vegetation form the basis for traditional range management techniques (Chapter 10). A good example of a widely used classification based on potential natural vegetation is Küchler's scheme which both classified and mapped US vegetation at 1:3 168 000 and 1:750 000 scales (Küchler 1964). For example, Küchler's (1974) map of potential vegetation of the state of Kansas, in the Great Plains region of the USA, describes nine grassland communities (**phytocenoses**) within the prairie.

Vegetation can be classified on the basis of the flora or the physiognomy. The **flora** of an area refers to the species that make up an area, whereas the **physiognomy** describes the structure (height, size, and growth form) of all the species together. The vegetation is the product of the flora and the physiognomy. Floristic classifications use the **association** as the basic floristic unit, which is defined as 'a plant community type of definite floristic composition, uniform habitat conditions and uniform physiognomy' (Mueller-Dombois and Ellenberg 1974). Associations sharing diagnostic species are grouped into higher units such as **alliances, orders and classes.** The European Zürich–Montpellier or Braun-Blanquet association/habitat system is the classic example of a floristic classification (Westhoff and van der Maarel 1973). The United Kingdom National Vegetation Classification is an example of a floristic classification currently in use; it includes 33 grassland communities in three main types of grassland (i.e. mesotrophic grasslands, calcicolous grasslands, and calcifugous grasslands) (Rodwell 1992).

The **formation** is the basic unit for physiognomic classifications and is a community type defined by dominant growth-forms and broad environmental relations (Whittaker 1973). Physiognomic classifications include Walter's characterization of the world's ecosystems in which, at the largest spatial scale, grasslands and savannahs are grouped with deciduous forests into Zonobiome II (humido-arid tropical summer rain region with red clays or red earths [fersialitic soils]) and steppes with cold deserts in Zonobiome VII (arid-temperature climate with Chernozems to Serozems [raw soils]) (Breckle 2002). **Zonobiomes** are ecological climatic zones, a **biome** being a large and climatically uniform environment. An advantage of this scheme is that each zonobiome is clearly defined by climate and corresponds well to soil types and vegetation.

Since Walter's scheme provides a classification of the *potential* natural vegetation of the earth it does not recognize managed grasslands or pastures artificially produced and maintained in areas that would otherwise be under different vegetation. For example, managed ryegrass–clover grasslands occur in large areas of Europe that were originally broadleaved deciduous forest (Zonobiome VI).

Defined by climate, Walter's nine zonobiomes are characterized by soil type and vegetation of a particular growth form and structure (i.e. zonal soil type and vegetation type, respectively). Floristic details become important in this hierarchical scheme at the scale of **biogeocenes**, which correspond to plant communities or associations, e.g. *Stipa–Bothriochloa laguroides* steppe of the eastern Argentinian Pampas, part of Zonobiome VII. The advantage of a physiognomic classification scheme is that it allows for a fast, efficient way to categorize vegetation that can be linked to remote sensing signatures, and easily ground truthed without the need for floristic expertise. Comparisons between regions that have few or no species in common can be readily made. As with Walter's classification described above, many approaches combine both floristic and physiognomic elements at different scales, with an increasing emphasis on floristics at hierarchically smaller scales.

Modern systems of vegetation classification are spatially hierarchical, with the **ecoregion** as one of the largest fundamental units. Hierarchical systems are those which allow for different levels of detail depending upon the level of resolution. Smaller units are embedded within larger units.

At the upper level of the hierarchy, an **ecoregion** is defined as a relatively large unit of land and water delineated by the biotic and abiotic factors that regulate the structure and function of the communities within it (Maybury 1999). The ecoregion provides a unit of geography that is more relevant than political units for organizing and prioritizing conservation planning efforts. **Communities** are the basic management unit and are defined as assemblies of species that co-occur in a defined area at certain times and that have the potential to interact with one another (Maybury 1999). By managing communities, we protect many species including the charismatic ones, preserve unique sets of interactions among species, preserve numerous ecosystem functions, and have available an important tool for systematically characterizing current patterns and future changes in the condition of ecosystems and landscapes.

8.2 General description of world grasslands

Grasslands are described here in relation to climate, generally regarded as being the most important factor related to the development of natural grasslands. The widely used **Köppen climate classification system** (modified by Trewartha 1943) uses letters to describe 6 major climatic regions, with 24 subclassifications based on temperature (6 types) and precipitation (4 types). The Köppen system has limitations, but allows large areas with climatic similarities to be delineated (Williams 1982) and is widely used as a basis for classifying climates (Stern et al. 2007). Grasslands that occur in the categories described in Köppen's system are described below (Fig. 8.1), with representative climate diagrams in Plate 3. Types of climate that do not have grasslands, or have only minor

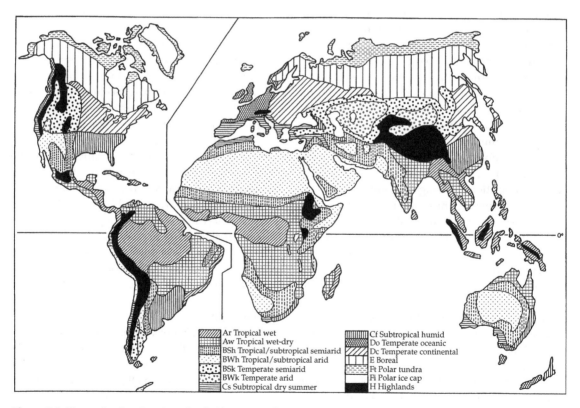

Figure 8.1 Köppen climatic regions. Reproduced with permission from Bailey (1996).

grasslands, are omitted (e.g. Köppen Group E Polar Climates). It is noteworthy that under this scheme, grasslands are generally limited to climatic regions characterized by either a winter dry or summer dry season. Grasslands characterize the tropical wet Aw (Savannah), dry BS (Steppe), and BW (Desert) climate types, but also occur associated with C (Moist subtropical mid-latitude) and D (Moist continental mid-latitude) climates. Grasslands occur as alpine meadows in H (Highland, Montane) climates, but are absent from P (Polar) climates.

The correspondence between grasslands classified according to Köppen climatic types, Bailey and WWF ecoregions, the International Vegetation Classification (IVC), and the Chinese Vegetation habitat classification system (§8.4.3) is shown in Table 8.1. The correspondence among different schemes is not exact, but the crosswalk provided in Table 8.1 allows approximate comparisons to be made.

8.2.1 Type A. Tropical moist climates (savannahs)

Tropical grasslands of warm regions occur in a climate characterized by a dry season in winter (hence the symbol Aw, tropical winter), with 70% or more of annual precipitation (1–1.5 m yr^{-1}) falling in the summer (Plate 3a). Precipitation is usually at least 6 cm per month, but at least one month has <6 cm. In the dry season, precipitation is like that of a desert. Frequent fires occur during the winter dry season. The average temperature in the coldest month is at least 18 °C, with a temperature range of 16 °C. Soils include Acrisols, Ferralsols, Lixisols, and Nitisols (§7.4.4). Characteristic animals include giraffe, antelope, wildebeest, zebra, lion, hyena, general ungulates, and carnivores. Savannah occurs to a lesser extent in tropical monsoon climates (Am) and tropical wet climates (Af), particularly after disturbance. A detailed description of the climate of tropical savannahs is provided by Nix (1983).

There is debate on the definition of savannah (Bourlière and Hadley 1983; Sarmiento 1984). Nevertheless, savannah vegetation is dominated by tall, often tussocky, coarse grasses (at least 80 cm high, up to 3 m in mesic, productive areas), some shrubs and low trees, with the trees in clumps giving a park-like appearance. Dense forests occur along stream courses and on floodplains. The density of trees increases towards margins of tropical rainforests, and herbaceous cover decreases as the amount of bare ground increases towards deserts. Growth patterns are closely associated with the alternating wet and dry seasons. The grasses brown in the dry season, becoming inedible and burning readily. Recurrent fire is a natural factor in savannahs, but may also be started by people. Herbivores migrate following rains or fire or both to feed on palatable young growth.

Tropical savannahs occupy vast areas in the southern continents; for example, 65% of Africa, 60% of Australia, and 45% of South America (Huntley and Walker 1982). Examples of savannah vegetation include the Llanos of the Orinoco Valley, in Colombia and Venezuela, the Campos Cerrados in the tablelands of central Brazil, the Veld of South Africa, the Serengeti of Africa, grasslands in southern and south-east Asia (including parts of India, Burma, and the Malay Peninsula), and Australian savannah.

The Llanos of the Orinoco floodplain cover $c.0.5 \times 10^6$ km^2 in a depression stretching from the Andes to the Orinoco river, with the vegetation composition classified by water availability (Table 8.2) reflecting winter flooding and local elevation which varies by only 1–2 m between the highest and lowest points. About 65% of the Venezuelan savannahs are dominated by arrow-grass (*Trachypogon plumosus*, *T. vestitus*) and typically also include *Axonopus canescens*, *A. anceps*, *Andropogon selloanus*, several species of the genus *Aristida*, *Leptocoryphium lanatum*, *Paspalum carinatum*, *Sporobolus indicus*, *S. cubensis*, sedges of the genera *Rhynchospora* and *Bulbostylis*, legumes including *Cassia*, *Desmodium*, *Eriosema*, *Galactia*, *Indigofera*, *Mimosa*, *Phaseolus*, *Stylosanthes*, *Tephrosia*, and *Zornia* (Pérez and Bulla 2005), and scattered trees, mostly *Byrsonima crassifolia* (*manteco*) and the *Curatella americana* (*chaparro*). Cattle and buffalo, sometimes with wild capybara, graze the Llanos. These grasslands are under severe threat, as 71% of Venezuelan savannahs have been converted to cropland including 5000 km^2 converted to *Pinus caribaea* plantations (Pérez and Bulla 2005)

Table 8.1 Crosswalk among alternative world grassland classifications

Köppen climate groups[a]	Bailey ecoregions (domain–division)[b]	IVC LEVEL 2 (Formation Subclass)[c]	WWF ecoregions[d]	Chinese Vegetation-habitat Classification System[e]
A Tropical and humid climates	Aw—Savannah	Humid tropical–savannah division (410)	2.A. Tropical Grassland, Savannah & Shrubland	Tropical and Subtropical Grasslands, Savannahs, and Shrubland	Tropical Shrubby Tussock
B Dry climates	BSh–(hot) Tropical/subtropical semi-arid: Desert & Steppe	Dry-tropical/subtropical steppe & desert division (310, 320)	3.A. Warm Semi-Desert Scrub & Grassland	Deserts & Xeric Shrublands (Warm)	Typical & Desert Steppe
	BSk–(cold) Temperate semi-arid grasslands/desert	Dry-temperate steppe & desert division (330,340)	3.B. Cool Semi-Desert Scrub & Grassland	Deserts & Xeric Shrublands (Cool)	Typical & Desert Steppe
C Subtropical climates	Cf–Humid subtropical	Humid temperate–Mediterranean division (260)	2.B. Mediterranean Scrub & Grassland	Mediterranean Forests, Woodlands, and Shrub	
	Cs–Subtropical Dry Summer	Prairie Division (250)[f]			
D Temperate climates	Dca, Dcb–Temperate continental grasslands,	Prairie Division (250)[g]	2.C. Temperate & Boreal Grassland & Shrubland	Temperate Grasslands, Savannahs and Shrublands	Meadow Steppe, Temperate Meadow, Marsh
H Highland climates[g]	H–Montane grasslands	Montane divisions of tropical and temperate grassland climates (M410, M310,M320,M330,M340, M250, M260)	4.A. Tropical High Montane Vegetation	Montane Grasslands and Shrublands (Tropical, Temperate & Boreal)	Alpine Meadow, Alpine Steppe, Alpine Desert
		4.B. Temperate & Boreal Alpine Vegetation		

[a] Trewartha (1943). [b] Bailey (1998). [c] Faber-Langendoen et al. (2008) [d] Olson et al. (2001) [e] See Tables 8.3 and 8.6. [f] Köppen did not recognize the prairie division as a distinct climate type. Bailey's ecoregion classification system represents prairie as the arid sides of the Cf, Dca, and Dcb Köppen types. [g] Highland climates are not part of the Köppen scheme, but are recognized by Bailey (1998) to accommodate low-temperature, high-altitude variants of climates in similar latitudes.

Table 8.2 Classification of Llanos savannha grassland in Colombia and Venezuela

Inundable (flooding) savannahs
Mesosetum savannah
Andropogon savannah
Humid savannahs
Leptocoryphium lanatum savannah
Trachypogon ligularis savannah
Dry savannahs
Trachypogon vestitus–Axonopus purpusii savanna
Paspalaum pectinatum savannah
Trachypogon vestitus savannah

From Coupland (1992c).

and 30 000 km² affected by disturbances associated with the oil industry. A number of African grasses, including *Brachiaria mutica*, *Hyparrenia rufa*, *Melinis minutiflora*, and *Panicum maximum* are aggressive invaders.

The Campos Cerrados of south and central Brazil occupy $c.2 \times 10^6$ km² and although they contain some woody elements (*campos sujos*), part of the region is open, grassy savannah (*campos limpos*) dominated by genera including *Echinolaena*, *Elyonurus*, *Paspalum*, *Trachypogon*, and *Tristachya* (Coupland 1992c; Rawitscher 1948). There are also *campos de murunduns*; savannahs with termite mounds, covered with trees. The vegetation shows continuous variation in physiognomy and species composition from arid, almost treeless grassland (tree density <1000 ha⁻¹, basal area $c.3$ m² ha⁻¹) to the wooded *cerradão* (tree density 3000 ha⁻¹, basal area <30 m² ha⁻¹) that is related to seasonality, topography, and soil fertility (Goodland and Pollard 1973; Sarmiento 1983). In this region annual rainfall is 750–2000 mm with a 3–5 month winter dry season (Sarmiento 1983). The Campos in the Pantanal of Mato Grosso is a 140 000 km² floodplain in the south-western part of Brazil bordering Bolivia and Paraguay. This seasonally flooded area contains a wide variety of herbaceous plants as well as grasses including *Paspalum almum* and *P. plicatulum*, which are considered valuable forage for cattle. These areas are maintained as pasture by ranchers using fire, machetes, and axes, but are being invaded by a number of exotic species including *Vochysia divergens* (*cambará*, Vochysiaceae), which vigorously spreads into pastures and can form monospecific stands locally called *cambarazais* (Nunes da Cunha and Junk 2004). In southern Brazil, southern Paraguay and north-eastern Argentina, Campos grassland occurs under a subtropical mesothermal climate (see §8.2.3).

The Serengeti grassland in Africa (see §6.1.1 and Plates 3a and 4) is a widely studied tropical/subtropical bunchgrass savannah, part of the widespread African savannahs (for an overview see Menaut 1983). The Serengeti National Park in Tanzania and the adjoining Masai Mara Game Reserve in Kenya cover 13 000 km² of the larger 25 000–30 000 km² regional Serengeti ecosystem defined by the movement of the large migratory herds of wildebeest *Connochaetes taurinus*, zebra *Equus burchelli*, and eland *Taurotragus oryx*, as well as Thompson's gazelle *Gazella thomsonii*, buffalo *Syncerus caffer*, and topi *Damaliscus korrigum*. The area regularly burns in the 2–7 month dry winter season, which keeps encroachment from tropical forest in check. The vegetation is a grassland/savannah mosaic characterized in most areas as medium-height *Themeda triandra* grassland with compositional variation related to grazing intensity and soil texture (McNaughton 1983). The dominant *T. triandra* (red oat grass) is a tufted perennial widespread in all warm and tropical regions of the Old World; in East Africa it constitutes 16% of the grasslands (Skerman and Riveros 1990). In areas of the Serengeti with >900 mm annual rainfall *Hyparrhenia filipendula* dominates the savannah understorey. Across the African savannahs as a whole, the co-dominance of trees and grasses is related to availability of resources (water, nutrients) and disturbance regimes (fire, herbivory). In regions where mean annual precipitation (MAP) is <$c.650$ mm woody cover increases linearly with MAP whereas, where MAP is >$c.650$ mm, moisture levels are sufficient for tree growth and grass dominance is maintained by disturbance (Sankaran *et al.* 2005).

Veld (originally a Dutch term, meaning field) refers to the undulating grassy plateaus of South Africa and Zimbabwe. Locally, the terms 'high veld' and 'low veld' distinguish between cooler, high-rainfall regions above $c.1500$ m and hot, more arid regions below about 750m, respectively (Tainton and Walker 1993). As an example, mid-elevation high veld savannahs are dominated largely by Andropogoneae (*Cymbopogon plurinodis, Diheteropogon filifolius, Heteropogon contortus,* and *Themeda triandra*) and Paniceae (*Brachiaria serrata, Digitaria eriantha,* and *Setaria flabellata*). Grazing allows an increase in the forb component (e.g. *Helichrysum* spp. *Senecio* spp.), a replacement of the dominant grasses with other grasses (e.g. *Aristida congesta, Chloris virgata, Eragrostis* spp., *Sporobolus capensis, S. pyramidalis, Urochloa panicoides*), and an increase in tree density (*Acacia* spp., *Chrysocoma ciliata, Pentzia globosa, Stoebe vulgaris*) (Tainton and Walker 1993).

Grasslands in southern and south-east Asia, including parts of India, Burma, and the Malay Peninsula
Most tropical and subtropical savannahs in this region are secondary, having invaded areas abandoned from cultivation following clearing of the original rain forest and wet sclerophyll forest. For example, $c.2 \times 10^6$ km² of south-east Asia is grassland dominated by *Imperata cylindrica* between altitudes of 300 and 700m. In more limited areas above 900m *Arundo madagascariensis* dominates (Singh and Gupta 1993; Skerman and Riveros 1990). These secondary savannahs develop on deep basaltic soil and spread with the burning associated with shifting cultivation. The scattered trees associated with these savannahs vary with region, but include *Pinus merkusii* in northern Sumatra; *Borassus flabellifer, Eucalyptus alba,* and *Casuarina junghuhniana* in eastern Java, north-eastern Bali, Sumba, and Timor; and *Albizia procera, Careya* spp., *Clerodendrum serratum, Dillenia ovata, Grewia* spp., *Phyllanthus emblica,* and *Ziziphus rugosa* elsewhere. The most widespread grassland in India is the *Sehima nervosum–Dichanthium annulatum* grassland in the region that includes the extensive peninsular area (Misra 1983). Peak growth occurs in September following the June/July monsoon. After maturation in October, the grasses remain dormant during the ensuing 8 month dry season (Singh and Gupta 1993). In Sri Lanka, savannahs occur which are dominated by *Cymbopogon nardus* var. *confertiflorus, Imperata cylindrica, Themeda arguens,* and *Sorghum nitidum* in a narrow belt between the extensive montane grasslands and dry evergreen forests.

Australian tropical savannahs occur in an arcing belt at $c.17\,°$S in northern Western Australia parallel to the northern edge of the continent, and then extending southward through eastern Queensland to about 29°S. Rainfall of 500–1500 mm yr⁻¹ occurs mainly during the summer (December–March). There is a mosaic of savannah communities ranging from monsoon tallgrass savannah on low-fertility soils dominated by *Themeda triandra, Schizachyrium fragile, Sorghum* spp., and *Chrysopogon fallax,* through tropical and subtropical tallgrass savannah (black spear grass savannahs) with *Heteropogon contortus* and *H. triticeus,* to midgrass savannah dominated by *Aristida* spp. on infertile soils or *Dichanthium sericeum, Bothriochloa decipiens, B. bladhii,* and *Chloris* spp. on moderately fertile cracking black clays (Woodward 2003). The trees in these savannahs are generally dominated by Eucalypts (*Eucalyptus* spp.) or wattles (*Acacia* spp.). The extent of woodland development corresponds for the most part to fire frequency with fire exclusion allowing succession to closed eucalypt-dominated forest (Gillison 1983).

8.2.2 Type B. Dry climates

In climate type B, potential evaporation and transpiration exceed precipitation. The subtype BS is a semi-arid climate or steppe, covering 14% of the earth's land surface; BW (the W is from the German *Wüste,* desert) is an arid climate or desert. Dry B climates extend from latitude 20° to 35° north and south of the equator and in large continental regions of the mid-latitudes, often surrounded by mountains. Areas of BS grassland generally surround real desert, separating it from more humid zones. Temperature range is 24°C, with annual precipitation ranging from <10 cm in the driest regions to 50 cm in moister steppes.

Two climate subtypes within BS that include grasslands are BSh and BSk (h from the German *heiss,* hot; k from *kalt,* cold):

- **BSh:** Tropical/subtropical semi-arid desert and steppe grasslands; low-latitude steppe, Chaparral, Grassland hot. Temperatures exceed at least 18°C

with all months >0 °C. Potential evaporation exceeds precipitation, especially through the summer, and through early autumn in years with low rainfall (<1.5 cm in September, <3 cm in October) (Plate 3c). Soil moisture recharge occurs November–March. This climate type occurs near desert at c.13–15 °N latitude in Africa at the southern border of arid climate, 20 °S latitude in southern Africa, borderlands of the Australian desert, portions of southern South America, portions of India, borders BWh (Dry-tropical, average annual temperature > 18 °C) climates of northwestern Africa, Saudi Arabia, and western India.
- **BSk**: Temperate arid grasslands; mid-latitude steppe, Grassland cold). Cool and dry, temperature <18 °C, at least one month <0 °C, soil water deficiency June–August, and through September and October if precipitation under 1.5 cm or 3 cm, respectively (Plates 3c and 5). Soil water recharged following precipitation January–March, or starting earlier in November–December if precipitation >4 cm or 3 cm, respectively. This climate type occurs near desert in the western great plains of the USA and south central Canada, and borders BWh areas of the Caspian Sea eastward to China/Mongolia. The desert steppes of Russia and China are described below.

Steppe grasslands, named from the Russian *stepj*, consist of low-growing, shallow-rooted bunchgrasses. Cover can be relatively even to widely spaced. Scattered small thorny trees or bushes are interspersed with the grasses. The major grasslands classified as steppe include the eastern Eurasian steppe (of Russia, Mongolia, and China), and the shortgrass prairies of the North American Great Plains (§8.2.5).

The Russian steppes have a cold, dry climate with short, sparse, xerophytic bunchgrasses which support grazing with low carrying capacity. The Himalayas to the south block warm, moist air from the Indian Ocean, so there is very little precipitation. Nothing blocks northern arctic winds, though, so the winters are very cold and windy. The bunchgrass habit is advantageous under these arid conditions, with buried vegetative buds also providing protection against grazing and trampling from wild and domesticated ungulates. Snow and dust accumulate among the densely packed tillers, promoting moisture and nutrient retention. The above-ground organs exhibit xeromorphic adaptations including narrow, more or less folded leaves with stomata located on the inner surface. Lavrenko and Karamysheva (1993) recognize four types of steppe (corresponding to Bews' 1929 and Boonman and Mikhalev's 2005 classifications, with the addition of desert steppe) that successively replace each other from north to south as the climate becomes more arid (decreasing precipitation, increased and warmer growing season, shorter frost-free period).

- **Meadow steppe** (= forest steppe) occurs in the region of semi-humid climate near the Black Sea between forest to the north and true steppe to the south. This transitional vegetation zone is characterized by turf-forming grasses <35 cm, e.g. *Festuca valesiaca* and *Koeleria gracilis*, plus species characteristic of true steppe, e.g. *Stipa* spp.
- **True or typical steppe** occurs across $1.43 \times 10^6 \text{km}^2$ in arid to semi-arid conditions south and south-east of the meadow steppe. These treeless plains are dominated by several species of *Stipa* <60 cm, e.g. *S. pulcherrima*, *S. lessingiana*, *S. capillata*, and *S. zalesskii* and *Festuca valesiaca* in the Black Sea–Kazakhstan subregion. Forbs are common and vary in abundance depending on local conditions.
- **Desertified bunch-grass and dwarf half-shrub–bunch-grass (semi-desert) steppe** occurs as a crescent-shaped area across large parts of Kazakhstan and north of the Caspian Sea in European Russia. Semi-desert is very arid and although still dominated by *Stipa* spp, thorny shrubs, e.g. *Artemisia pauciflora*, become increasingly important.
- **Desert dwarf half-shrub–bunch-grass steppe** is hyperarid (Köppen climate BSh, above) and characterized by a different suite of low-growing *Stipa* species, e.g. *S. caucasia* ssp. *glareosa*, *S. gobica*, and *S. klemenzii*, and forbs like *Allium polyrrhizum* and *Iris bungei*.

Species occurring throughout the range of aridity from meadow steppe to desert steppe are referred to as **euryxerophytes**, and include *Festuca valesiaca*, *Stipa capillata*, *S. krylovii*, and *S. lessingiana*. The steppe thus represents a north–south transition associated with increasing aridity, comparable to the east–west transition in the North American Great Plains. Associated with the steppe transition from meadow steppe to semi-desert there is a decrease in species diversity from 40–50 to 12–15 species m^{-2}, a decrease in the height of the grass canopy from 80–100 cm to 15–20 cm, and a decrease in canopy cover from 70–90% to 10–20% (Lavrenko and Karamysheva 1993). Russian steppe is grazed by livestock and cut for hay. Millions of acres of steppe were ploughed for agricultural cropping in the 1950s and 1960s, but much now lies fallow. Old abandoned fallow fields take 10–15 years to return to grassland dominated by characteristic steppe species including *Festuca valesiaca* and *Stipa* spp. (Boonman and Mikhalev 2005).

Grasslands in China (see §8.3.3) range from the plains of the north-east to the Tibetan plateau in the south-west (i.e. within the latitudinal range 35–50°N) of which *c*.80% of the area is occupied by zonal steppe and 20% by azonal meadows. To the north-west they merge with the grasslands of Mongolia and the former Soviet Union. Perennial, xerophilous, caespitose, or rhizomatous grasses dominate these grasslands, especially *Stipa* spp. As a result, these grasslands have low basal cover and forbs are less abundant than in the less arid Russian steppe to the west. The Chinese Vegetation-habitat Classification System (Zizhi and Degang 2005) (§8.3.3) recognizes three subclasses of temperate steppe and two subclasses of desert. A general summary of the characteristics of steppe according to an older but comparable classification by Ting-Cheng (1993) into five zonal types is shown in Table 8.3. As with the Russian steppe there is a significant decrease in diversity and production with increasing aridity from meadow steppe through typical steppe to desert steppe. The distribution of Chinese steppe is more complex than

Table 8.3 Characteristics of grassland steppe in China

Characteristic	Meadow steppe	Typical steppe	Desert steppe	Shrub steppe	Alpine steppe
Precipitation (mm)	350–500	280–400	250–310	380–460	450–700
Accumulated temperatures[a]	1800–2500	1900–2400	2100–3200	2400–4000	<500
Foliage cover (% of soil surface)	50–80	30–50	15–25	30–60	30–50
Species diversity (no. m²)	17–23	14–18	8–11	6–10	8–10
Forage yield (t ha^{-1})	1.5–2.5	0.8–1.0	0.2	0.5	0.2–0.35
Grazing capacity (sheep ha^{-1})	2.9–2.5	1.0–1.5	0.65	0.6	0.4

[a] Accumulated temperatures = annual sum of daily mean temperatures > 10 °C.

Important species (in order of decreasing importance) include:

Meadow steppe	Steppe species	*Stipa baicalensis, Filifolium sibiricum, Leymus chinensis*
	Meadow species	*Calamagrostis epigeios, Lathyrus quinquervius, Hemerocallis minor*
Typical steppe	Steppe species:	*Stipa grandis, S. krylovii, S. breviflora*
	Legumes	*Caragana microphylla, Astragalus melilotoides, Medicago ruthenica*
Desert steppe	Steppe species:	*Stipa caucasia* ssp. *glareosa, S. gobica, S. klemenzii*
	Desert species	*Reaumuria songorica, Calligonum mongolicum, Salsola passerine*
Shrub steppe	Steppe species:	*Stipa bungeana, Bothriochloa ischaemum, Themeda triandra*
	Shrubs	*Vitex negundo, Ziziphus sponosa*
Alpine steppe	Steppe species	*Stipa purpurea, Festuca ovina, Poa alpina*
	Cushion plants	*Androsace tapete, Arenaria musciformis, Thylacospermum caespitosum*

From Ting-Cheng (1993).

that of the Russian steppe, with **meadow steppe** occurring in the north-east bordering **forest steppe**. **True/typical steppe** is located centrally on the Inner Mongolian plateau, west of the meadow steppe bordering **shrub steppe** to the south and **desert steppe** to the east (Plates 6 and 7). The desert steppe is transitional to the desert zone to the north-east, **alpine steppe** to the west, and true steppe to the south and east. Shrub steppe is only questionably a grassland as it consists of shrub islands among a matrix of perennial xerophilous and broadleaved herbs and secondary thermophilous shrubs (especially members of the Betulaceae, Caprifoliacea, Oleaceae, Rhamnaceae, Rosaceae, and Thymelaceae) (Ting-Cheng 1993). **Alpine steppe** covers 29% (377 000 km^2) of the Tibetan plateau and is dominated by xeric perennial microtherms (*Stipa purpurea, S. subsessiliflora*) and alpine cushion species (*Androsace tapete, Arenaria musciformia, Oxytopis microphylla*) (Miller 2005).

The grasslands in China are an important grazing resource, with most of the grazing occurring in typical steppe, semidesert steppe, or desert areas. Sheep are the principal grazers. Much of these grasslands (tens of millions of hectares) are degraded and unusable for grazing and are being desertified at an increasing rate as a result of a combination of increases in population, overextension of agriculture, and overgrazing by domesticated animals (National Research Council 1992). In Tibetan steppe, for example, pasture degradation increased from 18 to 30% of total land area between 1980 and 1990 (Miller 2005).

Mongolian steppe covers 1 210 000 km^2, 80.7% of the land area of the country, with dry steppe grassland occurring over 410 000 km^2 and Gobi desert steppe and desert 580 000 km^2 (Suttie 2005). These grasslands occupy the transitional inner and central Asian zone between forest and desert areas. As in China and Russia, the steppe is characterized by grasses, predominantly species of *Stipa* and *Festuca*, with few legumes, and shrubs towards the desert. Dominants in the steppe zone include *Agropyron cristatum, Artemisia frigida, Carex duriuscula, Cleistogenes squarrosa, Elymus chinensis, Koeleria macrantha, Potentilla acaulis*, and S*tipa capillata,* (Suttie 2005). Mongolian steppe has been extensively grazed for centuries by pastoral herders (see §10.1) with livestock including cattle, yak in the highest areas, horses, camels, sheep, and goats. Overgrazing occurs close to urban centres, but the pastures are generally in good condition, and even undergrazed in some remote areas.

The southern grasslands of the Argentinian/Chilean Patagonia region (between 39° and 55°S) can be classified as BSk transitioning into more arid desert (BWk) to the north. The grasslands comprise treeless semi-arid grass and shrub steppe characterized by xerophytes (Cibils and Borreli 2005). Grass–shrub steppe communities have about 47% cover, with the most important grasses being *Achnatherum speciosum S. humilis, Poa ligularis, P. lanuginose, Festuca argentina*, and *F. pallescens* combined with several shrubs; i.e. *Adesmia campestris, Berberis heterophylla, Colleguaya integerrima, Mullinum spinosum, Senecio filaginoides, Schinus polygamus*, and *Trevoa patagonica*. Historically, the only large herbivore was the guanaco (*Lama guanicoe*: a camelid similar to a llama), but since the late nineteenth century commercial sheep grazing has significantly modified these grasslands as unpalatable woody plants (e.g. *Senecio filaginoides*) replaced palatable grasses (e.g. *F. pallescens*). Adaptive management plans (§10.2.4) have been proposed for sustainable sheep production and to control and prevent desertification (Cibils and Borreli 2005).

The characteristic vegetation of deserts (BW) is a sparse cover of shrubs or dwarf shrubs. However, semideserts receiving between 150 and 400 mm of annual precipitation often include perennial grasses as an important component of the dominant cover (Woodward 2003). For example, c.500 000 km^2 of arid, xerophytic tussock grasslands dominated by 0.5–1 m high Mitchell grass (*Astrebla* spp.) occur over cracking clay plains from the north of Western Australia to southern Queensland (McIvor 2005; Moore 1970, 1993) (Plate 8). The tussocks are widely spaced (c.0.6 m apart) with bare ground or short annual (*Iseilema membranaceum, I. vaginiflorum, Dactyloctenium radulans, Brachyachne convergens*) and perennial (commonly *Aristida latifolia* and *Eragrostis xerophila*) grasses and forbs (*Sida* spp.) in between. Scattered shrubs or trees occur in some areas and include *Acacia farnesiana, A. cana, Terminalia volucris*, and *T. arostrata*. These grasslands are stocked with sheep and are the most productive arid grazing lands in Australia, but overgrazing in large areas has lead to dominance by *Triodia basedowii*, a species of poor forage quality. Even more extensive than Mitchell grasslands in Australia are the spinifex grasslands (*Triodia* spp.) that are widespread in arid desert areas (Plate 9).

In North America, desert grassland occurs in areas of arid to semi-arid climatic conditions (mean annual precipitation 230–460 mm in the USA, up to 600 mm in Mexico) extending from west-central and south-eastern Arizona, southern New Mexico, and western Texas in the USA into central Mexico down as far as the northern part of the State of Mexico. Four subregions are recognized (Schmutz *et al*. 1992):

- **High desert sod-grass** dominated by sod-grasses, *Bouteloua gracilis* and *Hilaria belangeri* in the USA and *B. gracilis* and *B. scorpioides* in Mexico.
- **High desert bunchgrass** on mesic high elevation slopes, valley and mountainsides at elevations from 1100 to 1700 m. The grasses are taller than in the high desert sod-grass subregion and include *Bouteloua curtipendula*, *B. eriopoda*, *B. hirsuta*, *Digitaria californica*, *Elyonurus barbiculmis*, and *Eragrostis intermedia*.
- **Chihuanuan desert grassland** in south-eastern Arizona, southern New Mexico and far-western Texas south into eastern Mexico. Shrubby (*Acacia constricta*, *Flourensia cernua*, *Larrea tridentata*) with grasses scattered to abundant, especially *Bouteloua eriopoda*.
- **Sonoran desert grassland** in west-central and south-central and south-eastern Arizona in the USA and into eastern Sonora and western Chihuahua in Mexico. Heavily invaded by spineless deciduous shrubs (*Gutierrezia sarothrae*, *Isocoma tenuisecta*); few grasses (*Aristida divaricata*, *A. hamulosa*, *A. longiseta*, *Bothriochloa barbinodis*, *Bouteloua curtipendula*, *B. rothrockii*, *Digitaria californica*) now remain.

The reasons for the increase in cacti and shrubs that occurred in the North American desert grasslands in the early twentieth century are unclear but are probably related to overgrazing by livestock, climatic change, fire suppression, and the influence of rabbits and rodents (Schmutz *et al*. 1992).

8.2.3 Type C. Moist subtropical mid-latitude (humid mesothermal) climates

Temperate deciduous forests predominate, but temperate mid-latitude grasslands such as southern tallgrass prairie, Pampas, and Campos are also in this climate type, with the North American prairies south of Nebraska lying on the arid western side extending into the subtropical climate at lower latitudes. The temperate grasslands in these humid subtropical climates (Cfa, continental forest warm) have hot humid summers with many thunderstorms. Winters are mild, with precipitation during this season coming from mid-latitude cyclones. Average temperature in the coldest month is −3 to 18 °C, and the average in the warmest month is >10 °C. Precipitation is 25–75 cm yr^{-1} with no distinct dry season (Cw and Cs by contrast are winter and summer dry, respectively). This climate type occurs in central areas of continents. If on a mountain, pole-side slope 900–1500 m, equator-side slope 1100–1800 m. A good example of a Cfa climate is the south-eastern USA, where examples with grasslands occur in mixed prairie westward through all of Oklahoma (Plate 3d), except the panhandle, as well as central and east Texas (§8.2.5). The temperature corresponds to those of adjacent humid climates, forming the basis for two types of prairies: temperate and subtropical.

In South America, temperate subhumid grasslands occur over 700 000 km^2 from 28 to 38 °S in an area of central–eastern Argentina, Uruguay, and southern Brazil around the Rio de la Plata (Soriano 1992) bordering the Atlantic coast. Within this region there are two subregions (Burkart 1975):

- the 500 000 km^2 humid Pampas occupying mainly the Argentine province of Buenos Aires, eastern La Pampa, and southern parts of Córdoba, Santa Fé, and Entre Ríos provinces, divided into several regional subdivisions including the rolling Pampas (so called for their gently rolling topography) and the extensively flooded 60 000 km^2 flooding Pampas.
- the Campos of Uruguay and southern Rio Grande do Sul (Brazil) (Plate 10).

The entire region is essentially a vast continuous plain with rocky outcrops, hills, or mesa relief limited to 500 m elevation above the plain. Native C_4 tussock grasses dominate the Pampas in the warm season, for example, the bunchgrasses *Bothriochloa laguroides, Briza subaristata*, and *Paspalum dilatatum* in the flooding Pampas and *B. laguroides* and *Nassella neesiana* (C_3) in the rolling Pampas. *Paspalum quadrifarium* is a tall tussock grass that is a strong determinant of community structure in the flooding Pampas, reducing the chances for exotic species invasion (Chaneton et al. 2005). C_3 grasses dominate the Pampas in the cool season, including members of the Agrosteae, Aveneae, Festuceae, Phalarideae, and Stipeae tribes (Suttie et al. 2005). The flooding Pampas are grazed by almost 3.5×10^6 cattle. The more subtropical Campos to the north of the Pampas are C_4 dominated; for example, in the Mesopotamia region of Argentina, the Campos include tussock prairies dominated by *Andropogon lateralis* and shortgrass grasslands dominated by 30–40 cm high *Paspalum notatum* and *Axonopus compressus* (Pallarés et al. 2005; Soriano 1992). The subtropical to temperate climate of southern Brazil where Campos grassland occurs is theoretically conducive to the development of forest. Grassland is considered relictual from cooler, drier conditions 10 000 years BP, maintained since then by (possibly anthropogenic) fires (Overbeck et al. 2005a) and grazing (Oliveira and Pillar 2004), although little is known about the composition of these grasslands before the introduction of domestic herbivores. Much of Rio de la Plata region is now under intensive cultivation (wheat, maize, sorghum, soybeans) that followed the introduction of cattle and sheep ranching by European settlers beginning in the sixteenth century. Natural and semi-natural grasslands remain a dominant landscape feature only in the flooding Pampas and the Uruguayan Campos (Soriano 1992) (Plate 10). Cattle grazing leads to increased diversity of subordinate species, especially forbs and exotics species, by reducing the dominance of the major species (Altesor et al. 2005; Facelli et al. 1989; Rodríguez et al. 2003) allowing the spread of unpalatable weeds and prostrate species such as *Paspalum notatum* and *Axonopus affinis* (Altesor et al. 1998). In the absence of grazing the tussock grasses grow to 40–50 cm tall, whereas under grazing the short turf may be <5 cm tall. Charles Darwin (1845) recorded in his journal the common occurrence of European thistles in disturbed areas of the Pampas around Buenos Aires '...there is little good pasture, owing to the land being covered by beds either of an acrid clover, or of the great thistle'.

Humid subtropical regions of the USA, especially Alabama, Mississippi, and Texas, have grassland instead of forest in areas underlain by easily decomposed marls, i.e. islands of highly productive prairie soils. Known as 'blackland prairies', these grasslands are dominated by *Schizachyrium scoparium* and occur scattered among hardwood and pine/hardwood communities. Once widespread (>3000 km² in south-western Arkansas), these communities are now almost entirely lost as a result of agriculture (Peacock and Schauwecker 2003).

8.2.4 Type D. Moist continental mid-latitude (temperate, humid microthermal) climates

These mid-latitude temperate grasslands (Dfa, moderate continental warm) include northern tallgrass prairie. The temperature average in the coldest month is −3 °C to 18 °C, the average in the warmest month is >10 °C, and at least 4 months are >10 °C (Plate 3e). Precipitation is 25–75 cm per year with the 'f' in the symbol indicating that this climate is wet all seasons. The cold winters allow only a 140–200 day growing season. These cold–snowy climates are characterized by frozen ground in the winter, and generally support forest. This climate type occurs in eastern and midwestern USA (Atlantic coast to the 100th meridian, north of the type C climate (§8.2.3) supporting tallgrass prairie (see below). Similar grasslands occur in east central Europe (the Danube plains of south-eastern Russia), northern China, and Manchuria. In the eastern USA much of this area is naturally covered with temperate deciduous forest, but is maintained as managed pasture (see §8.2.7) as a result of anthropomorphic clearing and grazing. Similarly, the temperate grasslands of Eurasia occur within otherwise deciduous broadleaved or coniferous boreal forest zones and are regarded by Rychnovská (1993) as semi-natural; i.e. originating after clearing initiated during the Neolithic period, unseeded, autochthonous, and maintained by management (see §8.2.7).

In the southern hemisphere, this climate type is generally limited to high-altitude regions such as the Andes, and comparable areas in Africa, Australia, New Zealand, and New Guinea (Moore 1966). Similarly, in the northern hemisphere, Ellison (1954) described a *Achnatherum lettermanii* dominated grassland in sheep-grazed range of the subalpine vegetation at 3000 m asl of the Wasatch Plateau, Utah, USA.

8.2.5 The North American Great Plains: a special case

The grasslands of the central North American Great Plains cover 3.022×10^6 km^2 (c.12.5% of the land area of North America) east of the Rocky Mountains in southern Canada and the USA, and east of the Sierra Madre Oriental in north-eastern Mexico (Lauenroth *et al.* 1999). The Great Plains are unique in occurring in three contiguous climate zones (BS, C, and D) and are thus classified as principally dry subhumid (34%) and semi-arid (32%) (Lauenroth *et al.* 1999) (Plates 3, 5 and 11). In addition, these grasslands are topographically simple compared with the complex Eurasian grasslands, which cover many international boundaries and cross several major mountain ranges and two large areas of desert. The Great Plains area has been likened to an upturned bowl or craton, rimmed by the relatively young (Miocene) Rocky Mountains in the west and the ancient (Permian), much eroded Appalachian Mountains to the east cutting off free atmospheric circulation. The Mississippi river drains the great valley at the centre of this bowl (Dix 1964).

Precipitation depends upon moist air masses from the Pacific Ocean and Gulf of Mexico interacting with cold air from the Arctic and the mountains. Annual precipitation thus ranges from 125 mm along desert margins in the south-western USA and northern Mexico to 1000 mm along contacts with deciduous forest in the east and south-central USA. Grassland vegetation extends into the forest region south of Lake Michigan as dry air flow patterns wedge east, creating the 'prairie peninsula' (Transeau 1935). Summer is the wet season, with 50% of rainfall occurring in April–July; however, summer and autumn droughts are common. Winter is the dry season, but winter snowfall (<10 days with at least 25 mm south of the Texas–Oklahoma border to >80 days north of the line through central Iowa, northern Nebraska, and along the Colorado–Wyoming border) is important, protecting plants from cold, dry winter air, and providing a soil recharge during spring melt. The frost-free season ranges from <100 to >300 days. Temperatures range from an annual mean <2 °C in the north to >18 °C in the south with mean growing season warmest month temperatures of 16–28 °C. In addition to the geographic climatic variability, there are wide annual fluctuations in precipitation and temperature. The subtropical limit crosses the region east–west through central Oklahoma and the Texas panhandle.

The strong environmental gradients lead to large-scale vegetation gradients across the Great Plains (Diamond and Smeins 1988) (see §6.1.2). Shortgrass regions (BS steppe) run north–south in a westerly band merging into mixed-grass prairie of C (south of Nebraska) and D climates, and tall-grass regions in the east. Canopy height increases from <20 cm to >200 cm moving from west to east. A strong north–south gradient occurs, with C_3 species most important in the cool (mean annual temperature <2 °C), dry (mean annual precipitation ≤500 mm) north and C_4 species most important in the warmer, wetter, south. The C_4 species contribute to above-ground NPP throughout the region, with the C_3 taxa being primarily limited to the north-western two-thirds of the region. Most C_3 species are mid-height or short and so are out-competed by tallgrasses for light in wet areas. The positive correlation between temperature and precipitation means that the C_4 species dominate wet areas because these areas are also warm, and the C_3 species dominate the coolest areas, which are also relatively dry (Lauenroth *et al.* 1999).

The Great Plains can be divided into two regions, the **true prairie** in the east and the **mixed prairie** to the west. A broad ecotone divides the two regions, running north and south of a line crossing through central Nebraska, Kansas, and Oklahoma where annual precipitation varies from c.50 cm in the north to c.75 cm in the south. As noted before, the true prairie becomes increasingly more humid receiving more precipitation towards the east while the mixed prairie becomes increasingly arid towards the Rocky Mountains in the west.

The true prairie dominants are *Hesperostipa spartea* north of Kansas, and *Schizachyrium scoparium* and *Sporobolus heterolepis* in upland communities (Dodd 1983). Important associated species are *Bouteloua curtipendula* and *Koeleria macrantha*. *Andropogon gerardii*, *Sorghastrum nutans*, and *Panicum virgatum* are regarded by many as the typical tallgrasses of the true prairie and dominate extensive, mesic areas throughout reaching up to 2–3 m in height when in flower (Weaver 1954) (Plate 11).

Three main upland communities can be recognized (Dodd 1983):

- *Schizachyrium scoparium* 55–90%, with *Andropogon gerardii*, *Sporobolus heterolepis*, *Hesperostipa spartea*, and the exotic *Poa pratensis* as common associates in central parts of the region.
- From Kansas northwards *H. spartea* dominates, especially in xeric habitats (50–80% cover); associates include *S. scoparium*, *A. gerardii*, *Koeleria macrantha*, and *Bouteloua curtipendula*.
- *Sporobolus heterolepis* dominates dry uplands, with similar associates.

Three major lowland communities can be recognized:

- *Spartina pectinata* dominates in wet lowlands throughout with a very dense canopy allowing few other associates.
- *Panicum virgatum* (Nebraska south and southeast) and *Elymus canadensis* (especially in the west and northwards) in less mesic areas that *S. pectinata* communities.
- In the driest lowlands dense stands of *A. gerardii* dominates with *S. nutans*, *P. virgatum*, *H. spartea*, and lesser amounts of *S. scoparium*.

Extensive local variation occurs in response to interactions among topography (affecting soil nutrients and moisture), disturbance regime (especially fire), and grazing (originally bison, cattle since European settlement) (Collins and Steinauer 1998). Bison have been described as a **keystone species** for the tallgrass prairie (Knapp *et al.* 1999), while a suite of other animals including badgers, antelope, pocket gophers, ground squirrels, and other small mammals affect small-scale vegetation heterogeneity (e.g. Gibson 1989; Platt 1975; Reichman and Smith 1985) (see Chapter 9).

Mixed prairie

The mixed (or mixed-grass) prairie is the largest grassland association in North America, covering up to 900 000 km^2 (Küchler 1964; Risser *et al.* 1981) with a latitudinal range of about 2800 km from 52° to 29°N (south-eastern Alberta, southern Saskatchewan and south-western Montana to Texas) (Coupland 1992b). The name refers to the more-or-less equal mixture of mid- and short-grasses (Weaver and Albertson 1956). The midgrasses *Hesperostipa comata*, *Pascopyrum smithii*, *Sporobolus cryptandrus*, and *Koeleria macrantha* and shortgrasses *Bouteloua curtipendula* and *Bouteloua dactyloides* are present throughout, although the vegetation is not uniform over its range (Weaver and Clements 1938). Coupland (1961) recognized five community types in the Canadian mixed prairie:

- **Mesic eastern areas:** *Hesperostipa curtiseta*, *Elymus lanceolatus* ssp. *lanceolatus* dominant with *Hesperostipa comata* and *Pascopyrum smithii* as associates; together making up 75% of the plant cover.
- **More arid regions:** *Bouteloua gracilis* dominates with *H. curtiseta* and *E. lanceolatus* ssp. *lanceolatus* as associates.
- **Xeric sandy loams:** *H. comata* and *Bouteloua gracilis*.
- **Impermeable soils:** *B. gracilis*, *P. smithii*, and *E. lanceolatus* ssp. *lanceolatus*.
- **High moisture-holding soils:** *E. lanceolatus* ssp. *lanceolatus* and *Koeleria macrantha*.

Shortgrass prairie (shortgrass disclimax, shortgrass plains, shortgrass steppe: Plate 5)

The westernmost 280 000 km^2 of the mixed prairie is regarded by some authors as a grassland type in its own right (Carpenter 1940; Lauenroth *et al.* 1999; Lauenroth and Milchunas 1992; Risser *et al.* 1981), although earlier ecologists regarded it as a disclimax part of the mixed prairie (Weaver and Albertson 1956; Weaver and Clements 1938).

Shortgrass prairie extends from the Colorado–Wyoming border at 41°N south to 32°N in western Texas, and extends east from the foothills of the Rocky Mountains no further than 100°W in the Oklahoma panhandle (Lauenroth and Milchunas 1992). Soil disturbance, overgrazing, and drought have led to a complete change in the vegetation of the mixed prairie in this area, with the tall- and mid-grasses being lost leaving shortgrasses including *Bouteloua gracilis*, and *B. dactyloides* in central and northern regions, and *Hilaria belangeri* or *Hilaria mutica* in the south and southwest. Overgrazing has also allowed tree and shrub invasion by *Prosopis*, *Acacia, Condaloa*, and *Larrea*, accompanied by high cover of *Opuntia*, giving a brushland in overused, arid areas of the southern mixed prairie, especially in the shortgrass region.

In the south-western USA and north central Mexico, the mixed prairie gives way to the 207 565 km² desert grassland, an open, sparse community dominated by *Bouteloua eriopoda* and *H. mutica* with *B. rothrockii, Aristida divaricata*, and *A. purpurea* (Risser *et al*. 1981).

The grasslands of the North American Great Plains have been extensively altered as a result of fragmentation, cultivation, overgrazing, exotic species invasion (e.g. the spread of Russian thistle *Salsola tragus* and Canada thistle *Cirsium arvense*), and urbanization (Bock and Bock 1995). Only about 9.4% of the original prairie remains as grassland (White *et al*. 2000).

8.2.6 Type H. Highland climates (montane grasslands)

Temperature decreases with increasing altitude, $c.0.6\,°C\ 100\,m^{-1}$; as a result, zones of climate in highland regions change with elevation roughly corresponding to a poleward shift in latitude. Seasons only occur in highlands, however, if they also exist in the nearby lowland regions. Thus, in type A climates (tropical regions), zones of increasingly cooler temperatures corresponding to Köppen's C, D, and E climates occur, but without the seasonality. Otherwise, the seasons and wet and dry periods are the same as those of the biome that they are in. There is thus no specific highland climate as there is for the other climate types; rather, the specific highland climate is a modification of the local condition (Trewartha 1943).

At high elevations, grasslands can occur above the treeline to form subalpine or alpine meadows. The elevation of the treeline decreases towards the poles and is lower on north-facing than on south-facing aspects in the northern hemisphere and vice versa in the southern hemisphere.

An example of grassland in a highland climate is the *campos de altitude* (high-altitude grasslands) of the south-eastern coastal highlands (>1800–2000 m asl) of Brazil. The *campos de altitude* is classified as Cwb by Safford (1999), reflecting a dry winter, cool summer (warmest month <22°C), warm temperate climate. This grassland comprises a continuous matrix of bunchgrasses (*Cortaderia, Calamagrostis, Andropogon*) and bamboo (*Chusquea* spp.) with scattered shrubs (especially species of *Baccharis, Vernonia*, various *Eupatorieae, Tibouchina, Leandra*, and Myrtaceae), and short, often stunted, trees (e.g. *Escallonia, Maytenus, Rapanea, Roupala, Symplocos, Weinmannia*). The grasslands of the *campos de altitude* are similar floristically and physiognomically to other montane vegetation of the region; for example, the much higher and more extensive tropical *páramos* of the equatorial Andes in Venezuela, Columbia, and Ecuador (Safford 1999). In the *páramos* of Cotopaxi, Ecuador, for example, where the average annual temperature is a cold 7.8°C, *Calamagrostis, Cortaderia, Festuca*, and *Stipa* are characteristic genera (Ramsay and Oxley 1997) (Plate 3f). A tussock growth form of the grasses is characteristic of the alpine grasslands, as are giant rosette plants, e.g. *Espeletia* (Asteraceae) and *Puya* (Bromeliaceae) in the Andes, *Senecio* (Asteraceae) and *Lobelia* (Campanulaceae) in East Africa, *Cyathea* and other tree ferns in Malaysia, and *Argyroxiphium* (Asteraceae) in Hawaii.

Bews (1929) noted the change in tropical bunchgrass savannah to temperate grassland with altitude. Above the frost line, tropical species decline and are eventually lost. Andropogoneae (*Andropogon, Schizachyrium* spp.) remain dominant, mixed with Paniceae (e.g. *Paspalum, Digitaria*), but grow as small, hard tussocks. With increasing altitude, phylogenetically primitive temperate genera appear and become dominant, e.g. *Bromus, Danthonia, Festuca*, and *Poa*. In the Andes,

Calamagrostis and *Agrostis* are important. Legumes and bulbous monocots (e.g. lilies) are also important in these grasslands.

Temperate montane grasslands are frequently dominated by species of *Festuca, Poa, Calamagrostis,* and *Agrostis*. Grassland extends well below the treeline because of sheep grazing; these areas would otherwise be shrubby or forested. In Great Britain, for example, the commonest form of montane vegetation is a form of low-diversity grassland maintained by sheep grazing and frequent frost-heaving. Common grass taxa include a viviparous form of sheep's fescue *Festuca ovina* f. *vivipara, Agrostis capillaris, Deschampsia flexuosa,* and *Nardus stricta* (Pearsall 1950) mixed in with a few sedges (e.g. *Carex bigelowii*), herbs (e.g. *Galium saxatile, Potentilla erecta*) and bryophytes (*Hypnum cupressiforme, Pseudoscleropodium purum, Rhytidiadelphus squarrosus*) (Rodwell 1992). The communities represented in these montane and submontane habitats depend upon underlying soil variation (particularly base-richness) and soil moisture, snow-lie, altitude, and aspect.

In China, 14.8% (580000 km^2) of total grassland area is classified as alpine steppe (see §8.3.3) and is dominated by cold-resistant grasses and composites of which the most important are *Stipa purpureum, S. subsessiflora, Festuca ovina* subsp. *sphagnicola, Orinus thoroldii, Carex moorcroftii, Artemisia stracheyi* and *A. wellbyi*. Alpine meadows cover 16.2% (630000 km^2) of total grassland area and are dominated by cold-resistant grasses and forbs including *Kobresia pygmaea, K. humilis, K. capillifolia, K. myosuroides, K. littledalei, K. tibetica, Carex atrofusca, C. nivalis, C. stenocarpa, Blysmus sinocompressus, Poa alpina, Polygonum viviparum* and *P. macrophyllum*. Both of these montane grasslands are important for livestock with a carrying capacity for sheep grazing of 3.73 and 0.98 ha sheep^{-1} yr^{-1}, respectively (by comparison, temperate steppe has a carrying capacity of 1.42 ha sheep^{-1} yr^{-1}) (Hu and Zhang 2003).

In the southern hemisphere, the composition of high-altitude alpine grasslands of New Zealand is similarly related to altitude (with the treeline at *c.*1200–1300 m), drainage, and snow cover; four species of 0.1–1.5 m tall snow tussock (*Chionochloa* spp.) distinguish these communities (Mark 1993). Other important grasses in these communities include *Festuca novae-zelandiae* and *Poa cita*. Similar snow tussock grasslands occur at 270–500 m altitude under the harsh climate of the Auckland Islands (Godley 1965) (Plate 12).

8.2.7 Managed, semi-natural grasslands

In many areas of the world where grassland occurs it does not represent the natural vegetation. These **semi-natural** grasslands have developed on land that has been cleared of trees or drained of marshes. When grasslands such as these occur on land not used or not fit for crop cultivation they are referred to as **permanent grasslands** or **secondary grasslands**. These grasslands, subject to various degrees of intensive management, are in contrast to the natural grasslands discussed above that are subject to little or no deliberate direct modification. In addition, grassland that is sown (seeded) and part of a crop rotation is a **ley** (Old English *lēah*, an alternative word for meadow). Whether a semi-natural grassland is sown or not, grassland mown for hay is a **meadow** (Old English *mæd*, from the verb *māwan*, to mow) and one used exclusively or predominantly for grazing is a **pasture** (from the Latin *pasco*, to graze) (Table 1.2).

Compositionally, semi-natural grasslands vary depending upon geographic location, soil type, climate, and management and are often floristically very rich. Nevertheless, the main grasses, in Europe for example, represent only about 20 species (Table 8.4). Since the origin of these grasslands was by clearing existing non-grassland vegetation and not through planting or seeding, most of the species are from within the region. Often the species existed before clearance on the margins of other habitats including the woods, forests, bogs, marshes, high mountains, and coastal areas. Others, such as *Arrhenatherum elatius*, probably immigrated from a Mediterranean–Atlantic origin (Scholz 1975). Others have evolved races (subspecies, ecotypes, varieties) *in situ* adapting to the new grassland habitat. For example, tetraploid ecotypes of *Anthoxanthum odoratum* are widely distributed throughout Europe and are believed to have arisen by hybridization between the diploid *A. odoratum* and the diploid *A. odoratum* ssp. *alpinum*.

Table 8.4 Principal meadow and pasture grasses of Europe

Scientific name	Common name
Agrostis gigantea Roth.	Redtop bent
A. stolonifera L.	Creeping bent
A. capillaris L.	Browntop bent
Alopecurus pratensis L.	Meadow foxtail
Anthoxanthum odoratum L.	Sweet vernal grass
Arrhenatherum elatius (L.) J. & C. Presl.	False oat grass
Cynosurus cristatus L.	Crested dogtail
Dactylis glomerata L.	Cocksfoot
Festuca rubra L.	Red and chewings fescue
Helictotrichon pubescens (Huds.) Bess. ex. Pilger	Hairy oat grass
Holcus lanatus L.	Yorkshire fog
Hordeum secalinum Schreb.	Meadow barley
Lolium multiflorum Lam.	Italian rye grass
L. perenne L.	Perennial rye grass
Phleum pretense L.	Timothy
Poa pratensis L.	Meadow grass
P. trivialis L.	Rough meadow grass
Schedonorus phoenix (Scop.) Holub.	Tall fescue
S. pratensis (Huds.) P. Beauv.	Meadow fescue
Trisetum flavescens (L.) Beauv.	Golden oat grass

From Scholz (1975).

Secondary grasslands in the UK
Secondary grasslands in Europe, and the UK in particular, have a long history going back several thousand years. These early secondary grasslands are semi-natural as the plants are not seeded, but they are artificial and require maintenance to stop reversion to scrub or forest. The primary forests were destroyed by cutting and fire, with pastures developing first as cattle were grazed in the woodlands. In Britain, archaeological evidence supports forest clearance dating from the Neolithic (Sheail *et al*. 1974). Clearing continued through Roman times and by the time of the Norman Conquest records in the Domesday Book (completed in 1086) indicate that 1.2% of the land in England was meadow ($c.1210\,km^2$), and $c.1/3$ of England ($c.360\,000\,km^2$) was in pasture supporting $c.648\,000$ oxen, 1×10^6 cattle, and 2×10^6 sheep (Rackham 1986) (see Fig. 1.3). Pastures were extremely valuable land, and not just for feeding livestock. Animal faeces were collected to be spread as farmyard manure. From this the practice was developed of folding animals that had fed on pasture during the day on arable land at night to avoid the trouble of handling the dung. Meadows were valuable, providing hay that could be stored for animals, especially from January to April when animals were needed for ploughing but the grass in pastures was not growing. Changing agricultural practices have led to a considerable loss in the extent of semi-natural grassland communities during the second half of the twentieth century (Blackstock *et al*. 1999).

The floristic composition of these semi-natural swards in the UK is determined by soil fertility (pH and nutrient status) and drainage modified by use (grazing intensity, haying). With respect to fertility, highly productive, fertile pastures have the highest proportion of *Lolium perenne* and *Trifolium repens*, with increasing amounts of *Agrostis* spp, *Holcus lanatus*, and *Festuca rubra* as fertility decreases (Fig. 8.2). *Festuca rubra* dominates the least fertile pastures along with *Anthoxanthum odoratum* and *Holcus mollis* on acid soils and *Brachypodium pinnatum* and *Zerna erecta* on calcareous soils. Forbs are also abundant on infertile soils and include various composites (Asteraceae), *Plantago* spp., *Ranunculus* spp., and *Rumex acetosella* on acid soils (Green 1990).

The ploughing and re-seeding of permanent grassland as leys is known from the sixteenth century, but the practice gained momentum in the 1930s with the introduction of scientifically balanced seed mixtures, improved fertilizers, and methods of drainage. In 1939 the British Parliament made available grants of £2/acre (£4.9/ha, equivalent to $c.\$400$/ha at current prices) to farmers for ploughing up grasslands >7 years in age and >0.81 acres in extent (Sheail *et al*. 1974). Leys are floristically depauperate compared with natural ancient grassland as they are seeded with a restricted range of species and are infrequently >20 years old (many being on a 4-year rotation with crops).

Leys and seeded pastures
Cultivated or sown pastures are essentially a crop that supplements forage available from natural vegetation. The choice of pasture species depends upon the local environment, physical conditions

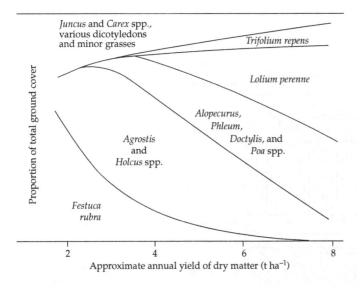

Figure 8.2 Typical composition of semi-natural grasslands at various levels of fertility (indicated by yield) in the UK. Reproduced with permission from Green 1990.

(slope, rockiness, etc.), economic modification costs (e.g. for drainage or irrigation), and the intended use of the pasture. Different species are suitable for different purposes such as silage, grazing, or forage. Generally, the sown species include both annual and perennial grasses and legumes of temperate and/or tropical/subtropical origin. In Europe, the main sown species are those native species common in semi-natural grassland (Table 8.4). Elsewhere, many species are introductions and include *Dactylis glomerata, Lolium perenne, L. multiflorum,* and *Schedonorus phoenix*—the most common planted grasses worldwide. For example, these grasses and other cool-season C_3 grasses have been introduced in temperate areas of North America, Japan, South Africa, and southern Australia and New Zealand (Balasko and Nelson 2003; Edwards and Tainton 1990; Ito 1990) where they are used there in cool–high rainfall regions. In tropical regions a different suite of warm-season C_4 grasses are used, e.g. in tropical regions of South Africa *Pennisetum clandestinum* (from central Africa), *Paspalum* spp. (e.g. *P. dilatatum, P. urvillei* from South America), *Chloris gayana* (probably from India), plus native species (e.g. *Eragrostis curvula, Digitaria eriantha*) are sown. In southern North America, *Cynodon dactylon, Paspalum notatum, P. dilatatum, Pennisetum purpureum* and *Bothriochloa saccharoides* are introduced grasses that are planted along with native prairie grasses (*Andropogon gerardii, Sorghastrum nutans, Panicum virgatum*) (Redfearn and Nelson 2003). The most widely used temperate legumes are varieties of *Trifolium repens* (white clover), *T. pratense* (red clover), *Medicago sativa* (alfalfa, lucerne), and *Lespedeza* spp. Legumes adapted for growth with warm-season grasses in tropical and subtropical regions include *Stylosanthes guianensis* (stylo), *Alysicarpus vaginalis* (alyceclover), and *Arachis glabrata* var. *glabrata* (rhizome peanut) (Sollenberger and Collins 2003).

Notwithstanding the sown species, the botanical composition of leys depends upon sward age, soil type, fertilizer regime, and management. In England and Wales, for example, the amount of *Lolium perenne* is of key importance, with a high content, especially in an old sward, indicative of fertile soil and intensive use (Green 1990). *Poa annua* and *P. trivialis* are common early, short-lived invaders. Unsown species typically comprise 45% ground cover 8 years after sowing, and in leys >20 years old, the cover of preferred species (i.e. those seeded) is <20% (Fig. 8.3).

A large number of cultivars of the principal seeded species are available to allow land owners to sow seed mixtures optimal for local conditions and needs. Cultivars are released after breeding programmes, and research aims at (1) increased yield, (2) good early-season growth, (3) good aftermath

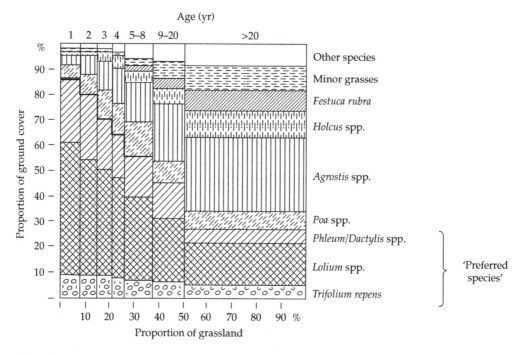

Figure 8.3 Average botanical composition of sown grassland in England and Wales according to age group. Reproduced with permission from Green (1990).

following haying, (4) production at a high level over the full life of the ley, (5) a high proportion of green leaf to stem, and (6) persistency of the strain (Moore 1966). As examples, Aberystwyth S.23 is the best known and widely used *Lolium perenne* cultivar worldwide, and KY 31 is the principal cultivar of *Schedonorus phoenix* planted across 14×10^6 ha in the eastern USA (Ball *et al*. 1993).

Amenity grasslands

Grasslands with recreational, functional, or aesthetic value without the principal aim of agricultural productivity are referred to as amenity grasslands and are considered in greater detail elsewhere (e.g. Dunn and Diesburg 2004; Fry and Huang 2004; Turgeon 1985). When intensively used and maintained the grasslands are referred to as **turfgrass areas** or **turfgrass facilities** (Waddington *et al*. 1992). The relatively few grasses important in these areas are designated **turfgrasses**. In the UK alone, the total turfgrass area was 4278.5 km² in 1977 (Shildrick 1990). Worldwide, it is estimated that turfgrass areas comprise 1–4% of the total land area depending on the country. Uses of turfgrass areas include sports facilities (especially golf courses), domestic lawns, ornamental lawns, road verges, waterway banks, and parks and open spaces. The grasses chosen for these low-diversity areas depend upon climate (temperate vs tropical), soil, and, most importantly, the intensity of use and maintenance. Turfgrass areas subject to the highest intensity of use and wear include sports facilities (especially golf putting greens and tees), tennis courts, cricket pitches, and dog-racing tracks. As an example, North America is divided into a number of turfgrass zones that reflect a north–south division based on the suitability of cool-season festucoid (C_3) vs warm-season (C_4) grasses, and a west–east division reflecting arid or semi-arid vs humid climate with different grasses being the most suitable for the different zones (Shildrick 1990). Worldwide, cultivars of *Agrostis stolonifera*, *Lolium perenne*, *Poa pratensis*, and *Schedonorus phoenix* are useful cool-season

Table 8.5 Turfgrass suitability for the Netherlands. High figures represent good performance in each respect

Species	Resistance to:					Suitability for:		
	Drought	Excessive moisture	Shade	Wear	Winter hardiness	Sports grounds	Lawns	Road verges
Agrostis canina	5	8	7	4	9	4	9	7
A. capillaris	8	6	6	5	9	4	9	8
A. stolonifera	8	8	5	4	9	4	8	7
A. vinealis	9	4	5	3	9	3	5	8
Cynosurus cristatus	6	6	4	6	5	4	4	5
F. filiformis	9	4	6	5	8	3	5	9
F. rubra: fine rhizomes (2n =42)	8	7	8	6	8	6	9	9
F. rubra: coarse rhizomes (2n=56)	7	7	8	5	9	5	7	9
Festuca rubra ssp. *fallax*	8	6	8	5	8	5	9	9
Lolium perenne (turf)	7	6	4	9	6	9	7	6
L. perenne (pasture)	7	6	4	8	6	8	6	5
Phleum pratense ssp. *bertolonii*	4	8	4	6	9	6	6	6
P. pratense ssp. *pratense* (pasture)	6	8	4	7	10	7	5	5
Poa nemoralis	7	4	7	3	9	1	3	4
P. pratensis	8	7	5	8	10	8	8	7
P. trivialis	3	9	7	5	8	4	5	6
Schedonorus phoenix	8	9	6	6	7	4	3	3

From Shildrick (1990).

turfgrasses and *Cynodon dactylon* and *Zoysia* spp. are useful warm-season turfgrasses. The turfgrasses are all from within three grass subfamilies, i.e. the Chloridoideae (major turfgrass genera: *Cynodon, Zoysia*; minor genera: *Buchloë, Bouteloua*), Panicoideae (*Axonopus, Paspalum, Pennisetum,* and *Stenotaphrum*), and Pooideae (major genera: *Agrostis, Festuca, Poa, Lolium, Schedonorus*; minor genera: *Agropyron, Bromus, Cynosurus, Phleum, Puccinellia*) (Turgeon 1985). Table 8.5 shows turfgrasses suitable for the Netherlands and illustrates the merit of different species for various uses. Despite the artificial nature of these amenity grasslands, they are included in phytosociological grassland classifications (e.g. the *Digitario ciliaris–Zoysietum japonicae* lawns described from North Korea: Blažková 1993). Ecological principles help in the understanding and management of amenity grasslands (see papers in Rorison and Hunt 1980), and the structure and dynamics of these systems can help in the understanding of ecological concepts (e.g. assembly rules, Roxburgh and Wilson 2000; Wilson and Watkins 1994).

8.3 Examples of regional grassland classifications

Regional vegetation classifications that identify grasslands have been developed worldwide, many based on the IVC described above (see §8.2). Three contrasting regional grassland classification systems are described here. Others include Australia's National Vegetation Information System (Cofinas and Creighton 2001, and http://audit.ea.gov.au/ANRA) and the Vegetation Types of South Africa (Low and Robelo 1995).

8.3.1 The US National Vegetation Classification System

The US National Vegetation Classification System (USNVC) (Grossman *et al.* 1998) is a regional vegetation classification system that has been adopted by US federal agencies. Within this hierarchical scheme the formation levels and above are explicitly physiognomic and the two lowest levels, alliance and association, are floristic. Alliances are physiognomically uniform groups of plant associations sharing one or more dominant or diagnostic species, roughly equivalent to the Society for American Foresters 'cover types' (Eyre 1980). Associations (defined in §8.1) represent the finest level of the hierarchy (Grossman *et al.* 1998) (e.g. Box 8.1). The USNVC includes over 5000 vegetation associations and 1800 vegetation alliances described for the conterminous USA.

Grasslands are classified in the USNVC within the graminoid growth form subclass of the Herbaceous Vegetation Formation Class. The

---

Box 8.1 An example of an association description from the USNVC

V.A.5.N.a. Tall sod temperate grassland

Andropogon gerardii—(Sorghastrum nutans) **Herbaceous Alliance (A.1192)**

Big Bluestem—(Yellow Indiangrass) Herbaceous Alliance

Andropogon gerardii—Panicum virgatum—Schizachyrium scoparium—(Tradescantia tharpii) Herbaceous Vegetation

Big Bluestem—Switchgrass—Little Bluestem—(Tharp's Spiderwort) Herbaceous Vegetation

Dakota Sandstone Tallgrass Prairie

Unique Identifier: CEGL005231

Element concept

Summary: This sandstone tallgrass prairie community is found in the eastern central Great Plains of the United States. Stands occur on moderate to steep dry-mesic slopes and ridgetops of the Dakota Sandstone region of north-central Kansas and adjacent Nebraska. The loam soils range from shallow and somewhat excessively well-drained to moderately deep and well-drained. The parent material is sandstone and sandy shale. Tall graminoids are the dominant vegetation in this community although trees and shrubs may be widely scattered. The most abundant species include *Andropogon gerardii*, *Schizachyrium scoparium*, and *Sorghastrum nutans*. *Bouteloua curtipendula*, *Bouteloua gracilis*, and *Sporobolus compositus* are common graminoid associates. *Amorpha canescens*, *Symphyotrichum ericoides*, *Echinacea angustifolia*, *Calylophus serrulatus*, *Psoralidium tenuiflorum*, *Mimosa microphylla*, *Oligoneuron rigidum* and *Tradescantia tharpii* are typical forbs of this community. *Clematis fremontii*, *Oenothera macrocarpa* and *Talinum calycinum* are unusual species of this area.

Environment: This community is found on moderate to steep, dry-mesic slopes and ridgetops. The loam soils range from shallow and somewhat excessively well-drained to moderately deep and well-drained, formed in material weathered from sandstone and sandy shale. The parent material is primarily Dakota sandstone.

Vegetation: Tall graminoids are the dominant vegetation in this community, although trees and shrubs may be widely scattered. The most abundant species include *Andropogon gerardii*, *Schizachyrium scoparium*, and *Sorghastrum nutans*. *Bouteloua curtipendula*, *Bouteloua gracilis*, and *Sporobolus compositus* are common graminoid associates. *Amorpha canescens*, *Symphyotrichum ericoides*, *Echinacea angustifolia*, *Calylophus serrulatus*, *Psoralidium tenuiflorum*, *Mimosa microphylla*, *Oligoneuron rigidum*, and *Tradescantia tharpii* are typical forbs of this community. *Clematis fremontii*, *Oenothera macrocarpa* and *Talinum calycinum* are unusual species of this area

Element distribution

Range: This sandstone tallgrass prairie community is found in the eastern central Great Plains of the United States on moderate to steep dry-mesic slopes and ridgetops of the Dakota Sandstone region of north-central Kansas and adjacent Nebraska.

(NatureServe 2007a, 2007b)

presence of woody strata is used to define groups within subclasses. Each group in the USNVC is divided into a Natural/seminatural or Cultural subgroup before being further divided into Formations based on dominant lifeform physiognomy and hydrology. Of 4149 Associations described in 1997, 882 (21%, more than any other) are in the Perennial Graminoid subclass and 11 (<1%) in the Annual Graminoid and Forb subclass. In the Great Plains alone, 105 Alliances in 20 Formations represent grasslands in the Perennial Graminoid subclass (NatureServe 2007a, b). The *Andropogon gerardii–Panicum virgatum–Schizachyrium scoparium–(Tradescantia tharpii)* Alliance is presented as an example from the Perennial Graminoid subclass in Box 8.1. Beyond the Great Plains additional grasslands are recognized, e.g. *Quercus stellata, Quercus marilandica)/Schizachyrium scoparium* Wooded Herbaceous Alliance (V.A.6.N.q), a sandstone glade of southern Illinois that is dominated by the grass *Schizachyrium scoparium* (Faber-Langendoen 2001).

The complementary US Terrestrial Ecological Systems Classification provides 'meso-scaled' units as a basis for analysing vegetation patterns, habitat usage by animals and plants, and systems-level comparisons across multiple jurisdictions (NatureServe 2003). A terrestrial ecological system is wider in scope than a vegetation classification and is defined as a group of plant community types (associations) that tend to co-occur within landscapes with similar ecological processes, substrates, and/or environmental gradients. The US Terrestrial Ecological Systems Classification also provides systematically defined groupings of USNVC Alliances and Associations. Of 599 ecological system types, grasslands are included in 166 types (28%) that are predominantly herbaceous, savannah, or shrub steppe. The parallel ecological systems of Latin America and the Caribbean classification identify 694 types, of which 198 (28%) are predominantly herbaceous, savannah, and/or grassland (Josse *et al.* 2003).

8.3.2 EUNIS: the European Nature Information System

EUNIS (http://eunis.eea.eu.int/index.jsp) is a standardized habitat classification with 1200 hierarchically arranged natural ecological units (referred to as **habitats**). The classification integrates environmental factors with predominant vegetation and is linked to 928 Alliance units from the more traditional phytosociological European Vegetation Survey (Rodwell *et al.* 2002). At each level of the hierarchy, factsheets provide a full description of the characteristic features of the habitat along with lists of synonyms and relationships with habitat types from older classification schemes. EUNIS is used by EU member states as a basis for designating conservation areas (Leone and Lovreglio 2004).

Grasslands are principally recognized in EUNIS in Habitat E (Grasslands and lands dominated by forbs, mosses or lichens), as one of eight habitats in Level 1 of the classification. Within the grassland habitat, seven categories are recognized at Level 2:

E1 dry grasslands
E2 mesic grasslands
E3 seasonally wet and wet grasslands
E4 alpine and subalpine grasslands
E5 woodland fringes and clearings and tall forb stands
E6 inland salt steppes, sparsely wooded grasslands
E7 sparsely wooded grasslands.

The distinction among categories at Level 2 is based on the presence or absence of trees, climate zone, salinity, and soil moisture (Fig. 8.4). Similar criteria allow distinction of finer-scale hierarchical categories within Level 2 grasslands. For example, 7 types of mesic grassland are distinguished based on management regimes, whereas 12 types of dry grassland are distinguished based on soil characteristics and biogeographic region. Some grassland habitats are also recognized within Habitat B (Coastal habitats): B1.4 coastal stable dune grassland, which includes B1.45 Atlantic dune [*Mesobromion*] grassland and B1.49 dune Mediterranean xeric grassland; B1.84 dune-slack grassland and heaths; and B2.41 Euro-Siberian gravel bank grasslands.

As with the distinctions among Level 2 grasslands, the grasslands of coastal habitats are distinguished from other habitats on the basis of physiognomic characteristics.

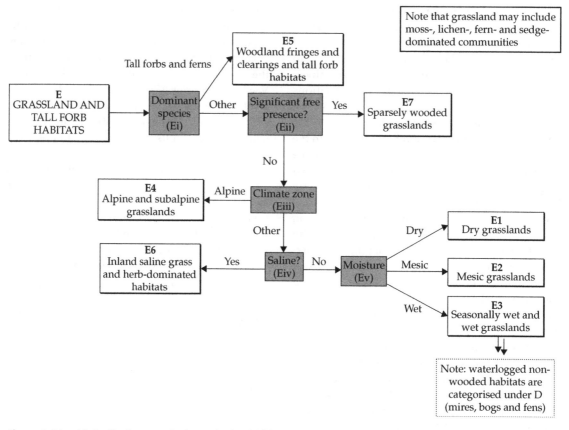

Figure 8.4 Level 2 classification categories for grassland and tall forb habitats according to the EUNIS Habitat Classification (http://eunis.eea.eu.int). See text for details.

8.3.3 Grassland classification in China

China has 7.5% of the world's grassland ($>3.92 \times 10^6 \, km^2$), in third place behind Australia and Russia (Chapter 1). Usable grassland covers over $3.30 \times 10^6 \, km^2$, 34.5% of the national land area (Zizhi and Degang 2005). Because of a long history of grassland utilization, much effort has been directed towards developing systems of grassland classification. The Vegetation-habitat Classification System (VCS) or China Grassland Classification System is one such system.

Vegetation-habitat Classification System
The VCS is a non-numerical system whereby vegetation stands are placed into categories based on the subjective assessment of the surveyor. It was used for the Chinese National Survey of Grassland Resources conducted from 1980 to 1990. Four grades or levels are identified in the VCS (Table 8.6):

• **First grade:** Nine classes based on temperature and vegetation features. The largest class is Temperate Steppe (18% of grassland area) and is divided into three subclasses.

• **Second grade:** 18 subclasses are identified from division of the classes based on climate and vegetation features. The largest subclass is the Temperate Typical Desert subclass of $>450\,000\,km^2$ in the Temperate Desert Class. Areas classified as Temperate Typical Steppe and Alpine Typical Steppe each cover $>410\,000\,km^2$.

• **Third grade:** Subclasses are divided into groups according to morphological groups of grasses,

Table 8.6 Grassland classes in the Chinese Vegetation-habitat Classification System

Class (number of subclasses)	Area (km²)	%	Lifeforms	Example dominants
Temperate Steppe (3)	745 375	18.98	Xerocole and fascicular grasses and subshrubs	*Leymus chinensis*, *Stipa* spp., *Festuca ovina*, *Artemisia* spp.
Temperate Desert (2)	557 342	14.19	Super-xerocole shrubs and subshrubs	*Seriphidium terrae-albae*, *Artemisia soongarica*, *Salsola passerine*, *Caratoides lateens*
Warm Shrubby Tussock (2)	182 730	4.65	Medium height grasses and some forbs	*Bothriochloa ischaemum*, *Themeda triandra*, *Imperata cylindrica* var. *major*
Tropical Shrubby Tussock (3)	326 516	8.31	Hot season grasses	*Miscanthus floridulus*, *Imperata cylindrica* var. *major*, *Heteropogon contortus*
Temperate Meadow (2)	419 004	10.68	Perennial temperate and medium-humid mesophytic grasses	*Achantherum splendens*, *Arundinella hirta*, *Calamagrostis epigeious*, *Bromus inermis*, *Poa pratensis*, *Miscanthus sacchariflorus*
Alpine Meadow (1)	637 205	16.22	Cold resistant forages	*Kobresia* spp. *Carex* spp., *Poa alpina*, *Polygonum viviparum*
Alpine Steppe (3)	580 549	14.77	Cold resistant grasses and sedges	*Stipa purpureum*, *Carex moorcroftii*, *Artemisia stracheyi*
Alpine Desert (1)	75 278	1.92	Cold and drought resistant plants	*Rhodiola algida* var. *tangutica*, *Seriphidium rhodanthum*, *Ceratoides compacta*
Marsh (1)	28 738	0.73	Sedges and grasses	*Carex* spp., *Scirpus yagara*, *Phragmites communis*, *Triglochin palustre*
Other (not-classified)	375 588	9.55		
Total	3 928 326	100.00		

From Zizhi and Degang (2005).

e.g. Tall Herbaceous Group, Medium Herbaceous Group. The groups represent divisions based on the foraging attributes of the plants for livestock utilization.
- **Fourth grade:** The basic unit of this classification is 276 types named according to dominant species.

Yaxing and Quangong (2001) used the VCS in combination with remote sensing technology to classify and evaluate grazing capacity of grasslands in north-east Tibet. Their study identified 4 classes (Bush Meadow, Alpine Meadow, Alpine Steppe, and Desert Steppe) that were subdivided into 10 groups and 32 types. Along with an assessment of grazing capacity they were able to identify areas of overgrazing and grassland deterioration. Zhenggang *et al.* (2003) identified 14 VCS grassland types in Gansu province, north-west China. Each type was subsequently evaluated in terms of ecological, social, and productive value to allow development of a conservation management plan.

CHAPTER 9

Disturbance

In grassy plains unoccupied by the larger ruminating quadrupeds, it seems necessary to remove the superfluous vegetation by fire, so as to render the new year's growth serviceable.

Charles Darwin travelling from Bahia Blanca to
Buenos Aires, Brazil in 1832 (Darwin 1845).

On the 14th day of April of 1935,
There struck the worst of dust storms that ever filled the sky.
You could see that dust storm comin', the cloud looked deathlike black,
And through our mighty nation, it left a dreadful track.

From Oklahoma City to the Arizona line,
Dakota and Nebraska to the lazy Rio Grande,
It fell across our city like a curtain of black rolled down,
We thought it was our judgement, we thought it was our doom.

From 'The Great Dust Storm (Dust Storm Disaster)',
on the album Dust Bowl Ballads by Woody Guthrie,
RCA/Victor, 1940.

Disturbance, defined as 'any relatively discrete event in time that disrupts ecosystem, community, or population structure and changes resources, substrate availability, or the physical environment' (White and Pickett 1985) is an integral and important natural phenomenon in grasslands worldwide. The importance of disturbance is that it changes fundamental properties of grasslands at a variety of spatial and temporal scales. Disturbance resets secondary succession (see §6.2.2). Together, the different types and frequencies of disturbance that characterize grassland define the **disturbance regime**. Each type of disturbance varies in its spatial distribution, frequency and intensity, predictability, area or size affected, and interaction with other disturbances. Consequently, characterizing and understanding disturbances are critical for understanding grasslands.

Disturbances can be **exogenous** or **endogenous** depending upon whether they are initiated within or outside the community, respectively. Examples of exogenous disturbances include the effects of flooding, drought, and fire. Endogenous disturbances include the senescence and death of dominant plants, perhaps through disease. However, endogenous and exogenous factors represent endpoints of a continuum, and distinctions among these two types can be difficult to make (White and Pickett 1985). For example, is a community change initiated by overgrazing considered an endogenous factor if herbivores are a natural part of the system, or exogenous if the grazers are, say, migratory, moving into the community from elsewhere only when the forage levels rise above some threshold value? Several aspects of natural disturbance regimes are best viewed as a continuum including the relative discreteness of disturbance events in time and space (i.e. patch size) (White and Pickett 1985).

Disturbances can also be categorized as **biotic** or **abiotic** depending on their source, with herbivory and disease being examples of the former and fire and drought examples of the latter.

In this chapter, the effects of disturbances on grassland are described. The focus is on the response to fire, herbivory, and drought. Other disturbances that can impact grassland include pollution (e.g. heavy metals, Chapter 5) and the invasion of non-native, exotic species (Chapter 1). The use of disturbance, specifically fire and grazing, as a management tool is discussed in Chapter 10.

9.1 The concept of disturbance

There are three particular problems in understanding the significance of disturbance in grassland. First, in most grassland today there are few areas where the disturbance regime currently approaches that of historical conditions. Most former grassland is either heavily managed, fragmented, does not support the historical biotic agents of disturbance (e.g. native animal herbivores), or is altered following invasion by exotic species (Chapter 1). Secondly, there is a semantic issue of defining exactly what constitutes a disturbance. On the one hand, the occurrence of fire, for example, clearly alters grassland ecosystems in many ways (see §9.2) and so meets the definition of disturbance given earlier. However, since fire is necessary to maintain many types of grassland and prevent them becoming woodland, it can be argued that it is a lack of fire rather than fire itself that should be considered a disturbance (Evans et al. 1989). Few investigators, however, subscribe to this view. Third, identifying a factor as an agent of disturbance depends upon the spatial and temporal scales focused on by the observer as well as the variable under consideration (Allen and Starr 1982). Again, using fire as an example, at the stand level, a single fire that causes little if any change in diversity if most species simply resprout, even though it obviously changes biomass, may not be recognized as a disturbance. Within stands, a fire may be considered a disturbance as fire temperature heterogeneity due to variations in fuel load (Gibson et al. 1990b) can differentially affect the ability of species to resprout, impacting local alpha-diversity (i.e. diversity within a particular area). Similarly, grazing effects can be scale-dependent and vary across spatial scales. In temperate Pampas grassland of eastern Argentina, grazing by domestic herbivores (cattle) promoted invasion of exotic plants enhancing community richness, whereas at the landscape scale, grazing reduced compositional and functional heterogeneity (Chaneton et al. 2002).

To understand disturbance it is necessary to separate cause and effect (van Andel et al. 1987). Disturbance in grasslands should be viewed as a cause, or a mechanism, that leads to an effect, or a change in the system at some level. Thus, as a simplistic example, fire can be viewed as a disturbance that causes plant material (fuel) to burn up. The resulting effects of a fire (see §9.2) can include reduced above-ground plant biomass, local species loss, altered species composition, lower soil albedo, higher soil temperature, and the volatilization and release of nutrients as ash. Furthermore, a disturbance, although it may itself be a fairly discrete event in time, can cause a cascade of effects that may reverberate through the system for a long period of time and in different trophic levels. Thus, the altered nutrient status of the soil following volatilization of biomass during a fire can affect nutrient uptake dynamics and resource availability, which through altered plant growth rates can ultimately affect plant–plant interactions (e.g. the balance between competition and facilitation) thereby altering the fitness and relative abundance of the species in the community.

Issues of pattern and scale are important in understanding disturbance. Communities can be viewed as a hierarchical nesting of patches representing disturbances at different scales. Such a gap-phase dynamic model has its origin in the ideas of Watt (1947) (see Chapter 6) developed more recently by Allen and Starr (1982) and Pickett and White (1985). Although grassland may be dominated by matrix-forming species, e.g. *Androgogon gerardii*, *Sorghastrum nutans*, and *Schizachyrium scoparium* in North American tallgrass prairie, both large- and small-scale disturbance events lead to a patchwork mosaic of areas in different stages of succession following one or more disturbances (Fig. 9.1). Large-scale disturbances include fire, drought, and grazing; small-scale disturbances include prairie-dog towns and bison wallows. The effects and interactions of these disturbances maintain the grassland. The resulting effect of the suite of disturbances that affect grassland produces a temporally and spatially variable mosaic of patch types across the landscape that varies in extent at different scales of resolution. For example, at the landscape scale, satellite imagery can be used to resolve heterogeneity resulting from both seasonal differences in vegetative phenology and the effects of land management such as grazing or burning (Plate 13).

The **disturbance heterogeneity model** (DHM; Kolasa and Rollo 1991) suggests that disturbance

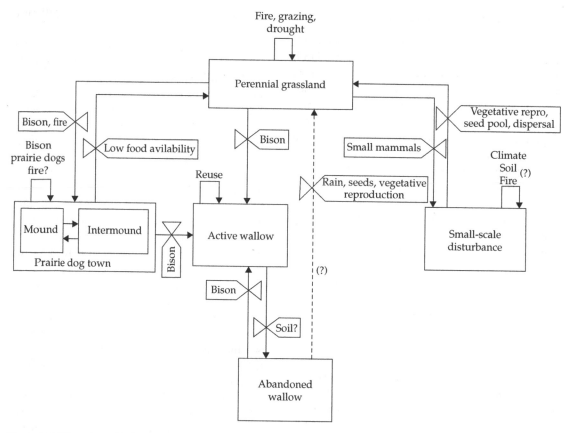

Figure 9.1 Schematic model of patch dynamics and interactions among patches in North American tallgrass prairie emphasizing the role of animals, particularly bison. Boxes represent identifiable patches in the landscape; asymmetrical bow ties contain the factors creating the patches; arrows indicate the direction of interactions within and between patches. Reproduced with permission from Collins and Glenn (1988).

increases community heterogeneity between patches so long as the disturbance is small relative to the size of the community (e.g. an animal burrow). Disturbances are thus a mechanism for generating and maintaining spatial heterogeneity in communities, i.e. the mean degree of dissimilarity in communities from one point to another. This phenomena is evident in grasslands from observing the grazing lawns (McNaughton 1984) and wallows (Polley and Collins 1984) of large herbivores such as North American bison, and burrow systems of small mammals such as pocket gophers or badgers (Gibson 1989). By contrast, large-scale, frequent, exogenous disturbances such as fire were shown to decrease diversity and heterogeneity as they were lowest in most frequently burned sites, increasing with a decreased frequency of disturbance in North American tallgrass prairie (Collins 1992). In a local and regional comparison of the effects of burning and grazing, the most homogeneous sites at the local level (0.01 ha) were the most heterogeneous at the regional scale (22 500 ha) (Glenn et al. 1992). Grazed or grazed and burned sites were the most heterogeneous at the regional scale, whereas undisturbed sites were the most heterogeneous at the local level.

The **intermediate disturbance hypothesis** (IDH) is a mechanism that was originally developed for explaining species coexistence under disturbance in marine systems (Connell 1978). This hypothesis posits that species diversity is maximized under intermediate levels (intensity) of disturbance and intermediate time spans (frequency) since disturbance. A few highly competitive species dominate at low levels of disturbance; at the other end of the scale, few species are able to tolerate high levels of disturbance. For the IDH to operate, competition is assumed be a major factor structuring communities (see §6.3.1) as it does in Grime's and Tilman's models of community structure (see §6.4) and the regional species pool is assumed to be larger than the number of species that can coexist in one patch (Collins and Glenn 1997). Patterns consistent with the IDH have been found in several grasslands (e.g. Leis *et al.* 2005; Martinsen *et al.* 1990; Vujnovic *et al.* 2002) as well as other habitats (Shea *et al.* 2004). However, disturbance frequency and intensity are independent and can be unrelated. For example, in North American tallgrass prairie, species richness was lowest on annually burned sites compared with sites burned less frequently, whereas richness reached a maximum at an intermediate time since the last fire (Collins *et al.* 1995).

The importance of the pattern exhibited by the IDH is that it can be used to infer process. In this context, the effect of disturbance as an extinction-causing event in grasslands should be viewed separately from the response to disturbance where recovery following disturbance is a balance between immigration and extinction (Collins and Glenn 1997). Considering the spatial extent of disturbances, the IDH is a between-patch mechanism in which the area under consideration (e.g. a particular grassland) includes patches of different ages or disturbance frequency (Wilson 1994). At a larger scale, the IDH could be applied to a comparison of grasslands each under a different disturbance regime. The IDH can also be applied to different trophic levels within grassland as disturbance can either affect all trophic levels similarly or affect a single trophic level (Wooton 1998), although studies specifically addressing this issue are lacking.

It is important to recognize the interactive nature of different types of disturbance (Collins and Glenn 1988). Grassland disturbances rarely occur in isolation. The net effect of one type of disturbance is contingent on its interaction with one or more other components of the disturbance regime. Fire, for example, is less likely to ignite and spread in grasslands where the available fuel load is low because of heavy grazing or drought. By contrast, the often lush growth of the vegetation in a burned area of the landscape can attract herbivores, thereby increasing the local intensity of grazing as a disturbance. Fire thus has a 'magnet effect' on grasslands in affecting the distribution of grazers across the landscape (Archbold *et al.* 2005). The importance of the interactions among disturbances was illustrated in North American mixed prairie where it was a combination of natural disturbances (grazing, fire, prairie dogs, bison wallows) that maximized diversity (Collins and Barber 1985). Similarly, in the Serengeti grassland of East Africa, grazing is a major factor influencing the grassland but its effects interact with other environmental factors including soils, fire, and the overlying tree canopy in savannah areas, to influence alpha-diversity (McNaughton 1983) (see §6.1.1).

9.2 Fire

9.2.1 The occurrence of grassland fire

Fire is an important component of the disturbance regime in grasslands worldwide and is widely used as a management tool (see §10.1.3). Indeed, most fires worldwide occur in C_4 grass ecosystems (Bond *et al.* 2005). Grasslands are both fire-derived and fire-maintained, although the balance between the two is often unclear and open to debate. For fire to be important in the spread of grassland, there must be a dry season allowing fuels to become sufficiently flammable to allow fires to spread, and the terrain must allow winds to fan and carry the fire across the landscape (Axelrod 1985). For example, woody plants readily invade the North American tallgrass prairie in the absence of frequent fire, implicating an important and critical role of fire in promoting stability of this system, especially in the moist eastern sections (Kucera 1992). The climate of the Great Plains, which includes frequent droughts, readily allows the spread of landscape-scale lightning-set fires. The large amount of dry standing dead matter that accumulates (Chapter 7), often over many seasons, provides a readily combustible fuel. Nevertheless, despite the clear role of fire in shaping the tallgrass prairie historically, it is regarded as one part of the ecological complex of the region that included the dominant influence of climate and other secondary factors such as grazing (Borchert 1950; Weaver 1954). Worldwide, fire is implicated in the spread of grassland as the grasses evolved (Chapter 2).

Natural grassland fires are predominantly started by lightning, although volcanic eruptions and sparks from boulders rolling down slopes can also occasionally ignite them. Scattered, isolated trees in grassland provide foci for lightning strikes, with the ignition and spread of fires being favoured during periods of drought, hot seasons, and dry descending warm winds (föhn winds). For example, lightning-strike fires occurred from $6\,yr^{-1}\,10\,000\,km^{-2}$ in mixed-grass prairies to $92\,yr^{-1}\,10\,000\,km^{-2}$ in pine–savannah in the Northern Great Plains of North America, burning 0.004–1158 ha each (Higgins 1984). Studies of South African vegetation showed that the vegetation types subject to the highest frequencies of lightning strikes (e.g. 'sour' and 'mixed' grassland, and moist savannah–woodland) were those that were the most tolerant and that required the most frequent fires, compared with other vegetation types, suggesting evolutionary adaptation of the plant populations to an historically high lightning-set fire regime (Manry and Knight 1986). In the Thunder Basin National Grasslands, Wyoming, USA, fires caused by lightning were most frequent in July, a time of year when drought is common, allowing fires to ignite and carry even in overgrazed pastures when fuel loads were low (Komarek 1964). Thus, despite the clear importance of anthropogenically set fire (see below), the importance of natural grassland fires cannot be overemphasized. Grassland ecosystems evolved 6–8 Mya as decreasing atmospheric CO_2 provided the conditions favouring evolution of the C_4 photosynthetic pathway (Chapter 4). The climate of grasslands, including the occurrence of dormant periods, dry seasons, and periodic droughts, is conducive to the ready ignition and spread of fire. Natural fires must have been important in allowing the spread of these grasslands at the expense of forests long before the appearance of modern humans (Keeley and Rundel 2005).

Historically, fires have also been started anthropogenically. Anthropogenic fires caused by primitive peoples would have started from escaped 'kept' fires borrowed from natural fires, but as humans learned to make fire 30 000–40 000 years ago the intentional use of fire started. There is clear evidence that fire was widely used in many ways that altered the landscape and plant communities in which humans lived, including to harvest food (grain, nuts, fruit); to drive, herd, and expose game; to reduce snake populations; to enrich the soil; to clear land for agriculture; to allow travel; to allow predators and enemy raiding parties to be more easily seen; and as a defensive and offensive weapon to drive fire into enemies (Pyne 2001). Gleason (1922, p. 80) noted, 'prairie fires were set annually by the Indians in the autumn months to drive game from the open prairies into the forests, where it was more easily stalked.' To a large extent there was interdependence between humans, fire, and the landscape. For example, in Australia, landscape burning by indigenous people (so-called 'firestick farming'), encouraged the growth of nutritious grasses which attracted kangaroos that were, in turn, hunted as game (Murphy and Bowman 2007). Many anthropogenically set fires escaped and burned across large areas. These fires helped maintain a patchwork of grassland at different stages of post-fire succession across the landscape. Conversely, the cessation of fire following the development of settled agrarian societies and agricultural development after European contact led to decreases in grassland area and replacement by forests. For example, a 60% decrease in area of prairies in Wisconsin, USA occurred in 25 years (1829–1854) after the cessation of fires set by Native Americans (Chavannes 1941). Similarly, replacement of unburned grasslands by scrub has occurred globally (Pyne 2001).

The importance of fire in grassland varies with latitude and climate. Many grasslands are fire-dependent, being maintained in the presence of frequent, recurrent fire. Examples of fire-dependent grassland include the vast grasslands of South Africa, Australia, and the North America Great Plains. Indeed, model simulations suggest that in a world without fire, large areas of humid C_4 grassland in South Africa and Australia would probably revert to closed forest (Bond et al. 2005). In general, grassland with low levels of production due to aridity or low temperature are less fire-dependent than grassland in more humid regions (Kucera 1981). For example, fire may play a minor role in the maintenance of the semi-arid shortgrass plains and steppe of the Serengeti or the Great Plains of the western USA (see Chapter 8). In South Africa, grasslands receiving >650 mm rainfall are fire-dependent and susceptible to woody plant invasion in the absence of fire, whereas those receiving less precipitation show no trend of changing composition when fire is excluded (Bond et al. 2003). In other grasslands periodic fires occur, but other ecological factors may be of greater importance. For example, edaphic factors are considered to be of primary importance by some researchers in savannah grassland in the New World tropics which occur under conditions of low nutrients and fluctuating water regimes. However, even in these grasslands, fire is clearly one of the several ecological factors important at least in the maintenance if not in the establishment of the grassland communities. It is of relevance to note the changing views of researchers on the role of fire. New World tropical savannahs were considered to be of climatic origin by the early nineteenth-century naturalists such as von Humboldt and Shimpfer. By the mid-twentieth century many researchers considered these savannahs to be of anthropogenic origin after their abandonment from swidden agriculture (which itself includes fire as a management tool to clear land). In the 1950–1960s, the edaphic viewpoint outlined above prevailed. More recently, a more pluralistic view taking edaphic, climatic, paleoclimatic, and cultural factors into account has been proposed (Scott 1977), recognizing the importance of fire, including lightning fires even during the wet seasons (Ramos-Neto and Pivello 2000). Similarly, at the largest scales, the grasslands of the North American Great Plains are controlled by climate (Bochert 1950), but fire acts to control woody plant invasion and greatly affects the vegetation at the more local level (Weaver 1954).

Whatever the cause, the natural fire frequency in grasslands is uncertain. Estimates are based on the accounts of early explorers and settlers, fire scars from grassland with sufficient trees, and charcoal incorporated into lake sediments or soils. It is likely that the North American prairies, for example, burned at least every 5–10 years, although in areas of dissected topography and natural fire breaks such as the Rolling Plains and Edwards Plateau of Texas if may have been every 20–30 years (Wright and Bailey 1982).

9.2.2 The nature of grassland fires

Grassland fires range from low, slow-burning back fires, to conflagrations travelling at 3–4 km h^{-1} with flame heights >3 m. Initially they require some wind, which allows them to spread at a rate approximately the square of wind velocity (Daubenmire 1968). However, fires generate their own wind as the rising convection column of hot air draws in air (and thus oxygen) into the blaze. If the prevailing wind is light (<5 km h^{-1}) and the fuel load light, this indraught can contain the fire (Cheney and Sullivan 1997). Firewhirls or whirlwinds can develop as fires build up near each other and merge, and the convection columns can develop into cumulus clouds of thunderstorm dimensions generating lightning, thunder, and rain (Vogl 1974).

Grassland fires consume above-ground plant biomass and raise local temperature. The amount of biomass consumed and fire intensity (energy output per metre of fire front) depends on three main factors (1) weather conditions including air temperature, humidity, and wind speed; (2) topography, including slope and aspect; and (3) the kind, amount, and disposition of fuel including the mass of the burning material, its chemical composition, dryness, and aeration (Daubenmire 1968). Fire intensity represents the rate of heat release from a linear segment of the perimeter of a fire and is the product of the amount of fuel consumed, heat yield of the fuel, and the rate of spread. Heat yield of the fuel depends on the way that the fire burns, but also varies among fuel types. For example, heat yields of some Australian grasses burned when at 10% moisture content were: *Phalaris tuberosa* 13 700–13 900 kJ kg^{-1}, *Themeda australis* 14 500–14 900 kJ kg^{-1}, *Eriachne* spp. 15 200–18 500 kJ kg^{-1}, and *Sorghum intrans* 16 900–17 600 kJ kg^{-1} (Cheney and Sullivan 1997). Fire intensity varies from $c.10$ kW m^{-1} when a back fire burns light fuel, up to 60 000 kW m^{-1} in a fast-moving wildfire.

Low humidity and dry fuel contribute to ready ignition and rapid spread of fires. As relative humidity increases, a higher wind speed is required to sustain and carry a grass fire. Models to predict the spread of grassland fires include wind speed as a major variable (Cheney et al. 1998). Topography is important because grassland fires move more rapidly upslope than on level ground, and move even more slowly downslope. Aspect affects fire behaviour and intensity through its effect on microclimate: north-facing slopes, for example, are generally cooler receiving less direct sunlight than south-facing slopes in the northern hemisphere (the opposite is true in the southern hemisphere).

The greater the biomass of fuel, the hotter the fire. However, fire temperature varies depending on the conditions at the time. Temperatures of grassland fires in North American tallgrass prairie were higher in head fires (with the wind) than in back fires (against the wind), were higher in vegetation that had not burned for several years compared with in vegetation that had burned more recently and had a lower fuel load, and were higher in lowlands compared with uplands (Fig. 9.2). The length of time that the surface temperature was raised accompanying fire in annually burned *Themeda triandra* grassland in Australia was shorter than in grasslands burnt at 4–7 year intervals (surface temperatures >100 °C were <1 min compared with 2–3 min, respectively) because of differences in fuel load (more fuel with longer time since the last fire) (Morgan 2004). Spatial heterogeneity of the vegetation and hence the fuel load leads to a correspondingly heterogeneous fire and fire temperature regime. Animal disturbances (see §9.3) can alter the fuel load at a local scale and thereby affect fire intensity.

The temperature of a grassland fire generally ranges from c.95 °C to 720 °C (DeBano et al. 1998), although temperatures up to 900 °C were recorded in 3 m high grass and pygmy bamboo stands in Thailand (Stott 1986). The release of up to 20 000 kW m^{-2} has been reported from a fast-moving (6.4 m s^{-1}) grass fire in Australia (Noble 1991). The temperature reached during a fire influences seed germination, vegetative resprouting, soil microorganisms, and soil nutrient loss (Hobbs et al. 1984). In general, grassland fires burn cooler than forest fires (DeBano et al. 1998), and the gradient of soil temperature increase with depth is extremely steep as only the uppermost few centimetres of soil are heated (Fig. 9.3). Heating of the soil lasts for only a few minutes and the underground plant organs are well protected against fire. At the scale of individual plants, fire temperature is influenced by

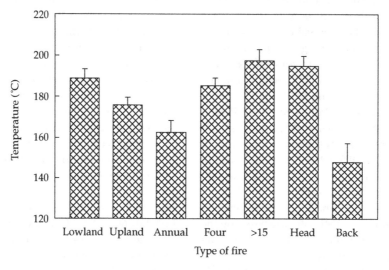

Figure 9.2 Contrasting surface temperatures in tallgrass prairie fires at Konza Prairie, KS (mean ± SE) according to fire history (Annual, annual burning; Four, 4-year burning; > 15, unburned for at least 15 years), and by fire type (Head, headfire; Back, backfire). Reproduced with permission from Gibson et al. (1990b).

the physiognomy of the vegetation. Temperatures exceeded 500 °C among the upper leaves of tussocks of *Calamagrostis* spp. during the burning of Ecuadorian grass *páramo*, whereas in the dense leaf bases temperatures were <65 °C (Ramsay and Oxley 1996).

9.2.3 Effects of fire on grasslands

There are several authoritative reviews that summarize the large literature on the effects of fire on grasslands (Collins and Wallace 1990; Daubenmire 1968; Kucera 1981; Risser *et al.* 1981; Smith and Owensby 1972; Vogl 1974; Wright and Bailey 1982). A brief summary of the main findings from these sources follows, with observations from additional studies noted where appropriate. For just about all the effects of fire that are described, the opposite effect has been reported on occasion; this is especially true with respect to the timing or season of burning (Biondini *et al.* 1989; Hover and Bragg 1981; Howe 1995).

In addition to raising the soil temperature and consuming living biomass and litter, grassland fires allow increased light levels to reach the soil surface until vegetation grows back. A dark layer of ash is deposited on the soil surface that decreases the **albedo** (reflectivity). The lower albedo of burned areas compared with unburned areas increases the absorption of solar radiation, thereby warming the soil surface for sometimes several months until vegetation cover develops. For example, mid-afternoon soil temperatures were 2.2–9.7 °C higher on spring burned compared with unburned areas in a Missouri, USA tallgrass prairie (Kucera and Ehrenreich 1962). The open, exposed soil following a fire is susceptible to increased wind and water erosion, especially in arid areas. Soil moisture can be lower in burned than in unburned areas because of the lack of an insulating litter layer and the warmer soil early on, and later because of high transpiration rates of the highly productive vegetation. Nutrients can be lost through volatilization (notably nitrogen and sulfur) or as ash is blown

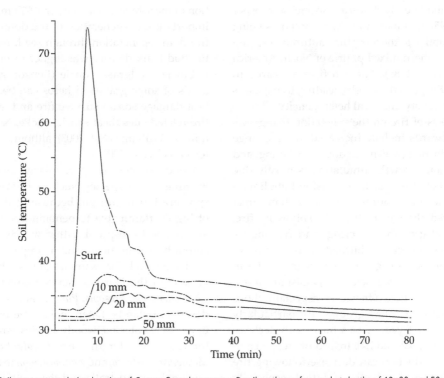

Figure 9.3 Soil temperature during burning of Campo Cerrado savanna, Brazil on the surface and at depths of 10, 20, and 50 mm. With kind permission of Springer Science and Business Media from Coutinho (1982).

away. Conversely, mineralization and nitrogen fixation rates can increase soil in burned grasslands (see §7.2) leading to an increase in soil pH, calcium, magnesium, and potassium. Experiments have shown that the response of the vegetation to fire is the result of complex interactions, with surface light, soil temperature, and nitrogen being of particular importance (Hulbert 1988).

Effects on the vegetation
The season of burning and the phenological stage of the vegetation are important in the response to fire damage. Perennials are most susceptible to burning after growth has started in the season and reserves have been translocated from storage organs. For example, spring burns in the North American tallgrass prairie burn back cool-season grasses, e.g. *Poa pratensis,* which will not grow back again that season. The warm-season grasses are protected as they have not started growth by then, and their subsequent growth is stimulated by the fire. Summer fires, by contrast, may retard the growth of the large, late-flowering C_4 grasses and allow a guild of early-flowering species to prosper (Howe 1995). Summer burns were noted as being more disruptive to the forbs than autumn or spring burns in northern mixed prairie of North America (Biondini *et al.* 1989). Autumn fires appeared to stimulate the growth of forbs, leading to increases in alpha-diversity and local heterogeneity.

The effects of fire on the vegetation that grows back afterwards include increased gas exchange rates, production (which can double), tillering, and reproduction. Growth stimulation is mostly due to removal of litter mulch, decreasing light limitation, and warming of the soil, rather than direct fire-induced changes to the soil. Following fire, precocity of growth, flowering, and fruiting by 1–3 weeks can occur on burned areas because of the warmer soil. Foliage can remain green later in the season too, and precocity can persist into a second season. Production increases, especially above ground, but also below ground, are particularly marked the first year after a burn with declines thereafter (Figs 7.3 and 7.4). In arid or steppe environments, production can decrease following fire if individual grass plants are killed, especially in dry years. *Bouteloua dactyloides, B. gracilis* and *Pascopyrum smithii* took three growing seasons to recover fully from a fire in North American shortgrass steppe (Launchbaugh 1964). In some cases the effect of burning on production is highly site specific, e.g. burning *Molinea cearulea* grassland in Great Britain can increase or decrease production, or have no effect, depending on local conditions and grazing intensity (Todd *et al.* 2000).

The ability of the vegetation to recover after a fire in grassland depends in part on the growth form and location of perennating buds. Many grassland taxa are hemicryptophytes or geophytes (Raunkiaer 1934) with their perennating buds at or below the soil surface. In grasses, leaf meristems can be protected from fire damage when located below ground and protected by closely packed leaf sheaths: for example, buds of *Pseudoroegneria spicata* are located c.1.5 cm below ground.

Enhanced flowering and seed production is observed in some grassland species after fire, but there is much variation among species and between seasons and habitats (Glenn-Lewin *et al.* 1990). The underground bud bank (below-ground population of meristems *sensu* Harper 1977) may be more important for regeneration of the dominants after fire than regeneration through seed. For example, the bud bank density was higher in burned than unburned tallgrass prairie (Benson *et al.* 2004). Seeds of some grassland herbs can be tolerant to heat damage so can survive fire and may require fire-related cues (heat shock, smoke, NO_3^-) to germinate (Williams *et al.* 2003), although not always (Overbeck *et al.* 2005b).

Woody species are particularly susceptible to burning; some species, such as cedars *Juniperus* spp., are killed outright because of the presence of highly flammable terpenoids (see §4.3.2) and are unable to resprout. Other woody plants can sprout back vigorously and a single fire may not be detrimental: for example, *Ulmus, Quercus,* and *Populus* in the North American Great Plains, or *Chrysothamnus, Prosopis, Purshia,* and *Tetradymia* in desert grasslands. Latent epicormic buds on the upper stems of some woody species can also allow for recovery after fire, e.g. *Eucalyptus, Eugenia, Melaleuca, Tristania,* and *Xanthostemon* in the tropical savannahs of Australia and the south-west Pacific (Gillison 1983). Generally, however, fire retards woody invasion of mesic and moist grasslands. The absence of fire or the suppression of the natural fire regime allows a rapid increase in woody plants in areas where there is an adequate seed source. An increase in woody cover in grasslands and savannahs has been observed worldwide after fire suppression. For example, North American tallgrass prairie left unburned for 32 years experienced a 34% increase in woody plant cover including *Ulmus, Juniperus, Quercus, Rhus,* and *Symphoricarpus* (Bragg and Hulbert 1976). Conversion of mesic grassland to shrubland over large areas of the US Midwest has been attributed to fire suppression following habitat fragmentation and alteration to the natural grazing regime (Fig. 9.4) (Briggs *et al.* 2005). At this point, these grasslands have transitioned to a new savannah-like state in which intermediate fire frequencies promote the growth of shrubs. Historically, it is likely that grassland–forest ecotones and the extent of savannahs fluctuated across the landscape in response to regional climatic variation, particularly drought, and the availability of fuel to support fire (Anderson and Brown 1986).

As a result of differential susceptibility to burning and the altered competitive environment after the burn, the plant communities of burned grassland can be substantially different from those of unburned or infrequently burned grassland. The most obvious effect of fire on grasslands is the elimination or at least reduction in the abundance of woody plants. The herbaceous flora varies under different burning regimes, as grasses are generally more abundant and forbs (non-grass herbaceous plants) are less abundant with increasing frequency of fire. Grasses frequently increase in abundance and occurrence with frequent burning, e.g. *Andropogon gerardii* in the North American tallgrass prairie and *Themeda triandra* in African grasslands. Some non-grasses respond favourably to increasing fire frequency, e.g. the lowgrowing shrub *Amorpha canescens* in tallgrass prairie (Gibson and Hulbert 1987). Annual plants can be particularly susceptible to fire, and their abundance may be closely related to the fire cycle (Gibson 1988b). Some annuals can be regarded as 'phoenix' plants, as germination and establishment may be restricted to the open soils and full sunlight conditions of post-burn environments.

Diversity generally decreases after fires as susceptible species are eliminated, with further declines in diversity with increasing fire frequency (Collins *et al.* 1995). In North American tallgrass prairie, frequent fires reduce community heterogeneity, i.e. the prairies become less patchy as the cover of the dominant grasses increases and total richness and diversity declines (Collins 1992). With decreasing fire frequency, community heterogeneity increases until the invasion and establishment of woody species in infrequently burned prairie again reduces diversity and heterogeneity. The community dynamics of post-fire vegetation development depends on the ability of the subordinate species to survive and re-establish within the matrix of fire-adapted dominants. Chance survivorship in unburned or lightly burned patches and random colonization of gaps are important in determining the resulting community (Ramsay and Oxley 1996). In tall tussock *Chinochloa rigida* grasslands of New Zealand, post-fire community structure closely followed Peet's (1992) 'gradient-in-time' and 'competitive-sorting' models (Gitay and Wilson 1995). Thus, species that established by chance early on after a fire were subject to competitive sorting into microhabitats as the community

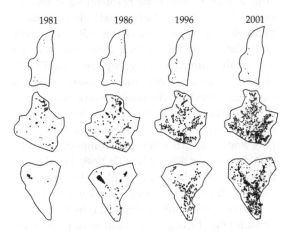

Figure 9.4 The change in woody vegetation over 20 years on three watersheds in North American tallgrass prairie at the Konza Prairie Biological Station subjected to three different burning treatments: burned annually (top series), burned every 4 years (middle series), and burned once in 20 years (bottom series). Reproduced with permission from Briggs *et al.* (2005). Copyright American Institute of Biological Sciences.

developed through time. As a result, both positive and negative interspecific associations develop and become more prevalent with time since the fire. By contrast, species richness and species turnover increased the year after fire in southern Brazilian subtropical grassland (Overbeck et al. 2005a). In this subtropical grassland, fire reduces cover of the dominants allowing rapid colonization by subordinate species, thus enhancing species richness for 2–3 years before the former dominants increase in abundance again.

9.2.4 Invasive exotics and grassland fire regimes

A number of invasive exotic species (see §1.3) have been implicated in altering fuel loads, thereby altering fire behaviour, intensity, severity, seasonality, and frequency in grasslands (Brooks et al. 2004; D'Antonio and Vitousek 1992). Invasion of the exotic grasses establish a positive-feedback, self-perpetuating **grass–fire cycle** (D'Antonio and Vitousek 1992). The grass–fire cycle occurs when an exotic grass invades a grassland and increases the abundance of fine fuels, which in turn increases fire frequency, area, and intensity, leading to a decline in native species including trees and shrubs, thus facilitating further invasion by the exotic grass. The altered fire regime caused by the presence of the exotic can significantly affect grassland community and ecosystem structure. For example, the annual grass Bromus tectorum (cheatgrass) occurs in $>40 \times 10^6$ ha in western North America in areas originally dominated by perennial grasses and shrubs. Bromus tectorum increases fuel continuity as it fills in interstices of shrubs, increasing fire intensity and frequency from which the native shrub-steppe species cannot recover. The altered habitat is a threat to other members of the system, especially some of the consumers including sage grouse Centrocercus urophasianus, black-tailed jackrabbit Lepus californicus and their predators, golden eagles Aquila chrysaetos, and prairie falcons Falco mexicanus. Another exotic grass, Taeniatherum caput-medusae, is subsequently invading portions of the range of B. tectorum, and is also highly combustible and fire-enhancing (Mutch and Philpot 1970). The African grass Andropogon gayanus was originally introduced as a pasture grass to Australia in 1931, but is now spreading throughout Australia's savannahs. Fuel load in A. gayanus-invaded savannah is seven times higher than normal and supports fires that are eight times more intense (Rossiter et al. 2003). Chromolaena odorata is an Asian vine invading South African savannahs (see Table 1.8) that leads to increased vertical fuel continuity. In Australia, invasive exotic grasses including the African perennial grass Pennisetum polystachyon are replacing the native Sorghum intrans in the Kakadu National Park. Pennisetum polystachyon alters the fire regime as it produces more litter and carries more intense fires than the native species (D'Antonio and Vitousek 1992). Tree invasion of grassland leads to increased fire severity, as fires can now be crown fires and not just ground fires. Conversely, the Asian tree Sapium sebiferum has highly decomposable leaf litter and so its presence reduces fuel loads, preventing fires that would otherwise keep woody species out of North American coastal prairies (Barrilleaux and Grace 2000).

9.3 Herbivory

Herbivores are organisms that eat only plants. In this chapter, they are divided very generally into large and small vertebrate herbivores and invertebrates. Herbivores disturb grasslands both above and below ground, consuming $\geq 50\%$ of ANPP and $\geq 20\%$ of BNPP (Detling 1988). Invertebrates and small mammals as a group consume 10–15% of ANPP, and large mammals consume more. In many grasslands, herbivores enhance biodiversity through alterations to the balance between local colonization of species from the regional species pool and local extinction and competitive exclusion dynamics (Olff and Ritchie 1998). Local colonization can be enhanced through higher propagule input by dispersal on hooves, seeds attached to fur or feathers, or in dung. Enhanced colonization can be facilitated through the creation of gaps in the grassland sward following disturbances such as digging and trampling. Local extinction by herbivores in grassland can be brought about through deposition of urine and faeces, intense selective grazing, and aggregated soil disturbances following digging, trampling of paths, and wallowing

(Olff and Ritchie 1998). The importance of canopy gaps and soil disturbances has been investigated through experimental studies in which various types of artificial disturbances were created and monitored. These studies confirm the importance of local colonization and extinction dynamics and emphasize their interaction with larger-scale disturbances such as fire (Bullock *et al.* 1995; Rogers and Hartnett 2001a, 2001b; Umbanhowar 1995). For example, vegetative regrowth was observed to be the dominant recolonization mechanism on soil disturbances created to mimic pocket gophers in North American tallgrass prairie, with colonization from the seed bank playing only a minor role (Rogers and Hartnett 2001a). Graminoid colonization on these disturbances was higher in burned than in unburned prairie, whereas forb colonization was higher in unburned prairie (Rogers and Hartnett 2001b). These and other studies indicate that the effect of herbivory on grasslands varies in different environments and across environmental gradients (Table 9.1).

9.3.1 Large vertebrate herbivores

All natural grasslands are grazed to some extent by one or more species of large vetebrate herbivore. Historically, these large herbivores were native animals which are postulated to have co-evolved with the grasses (Chapter 2) and helped shape the structure and composition of grassland ecosystems. Nowadays, and for more than 5000 years, most grassland, especially managed pasture, has also been or is instead grazed by domestic grazers (see §10.1.2). Native grazers in North America include bison *Bos bison*, elk, wild horses (reintroduced by the Spanish), deer, and a suite of many other large herbivores that are now extinct there. The herbivore fauna of the East African grasslands and savannahs are the richest and most spectacular of any on Earth and include migratory wildebeest *Connochaetes taurinus*, elend *Taurortragus oryx*, gazelle *Gazella granti* and *G. thompsoni*, Bubal hartebeest *Alcelaphus buselaphus*, and Burchell's zebra *Equus burchelli*, with non-migratory herbivores including African buffalo *Synerus caffer*, topi *Damaliscus korrigum*, elephant *Loxodonta africana*, black rhinoceros *Diceros bicornis*, bush pig *Potamochoerus porcus*, duikers *Cephalophus* spp. and *Sylvicarpa grimmia*, waterbucks *Kobus* spp., bushbuck *Tragelaphus scroptus*, and hippopotamus *Hippopotamus amphibus* (see §6.1.1 and §8.2.1 for a description of the Serengeti) (Cumming 1982; Herlocker *et al.* 1993). In the steppes of China, large native herbivores include

Table 9.1 Hypothesized effects of herbivory on grassland plant diversity in different grassland environments

Precipitation	Soil	Major limiting resource[a]	Herbivore characteristics[b]	Effects on plant diversity				Net effects
				Through extinction[c]		Through colonization[d]		
				Large herbivores	Small herbivores	Large herbivores	Small herbivores	
Dry	Infertile	Water/nutrients	Rare, small	− −	−	−	+	0/−
Dry	Fertile	Water	Abundant, diverse[e]	− −	− −	+	+	−
Wet	Infertile	Nutrients/light	Intermediately abundant, large	+ +	−	+	+	+ +
Wet	Fertile	Light	Abundant, diverse[e]	+	−	+	+	+/−

[a] Resource competed for by ungrazed plants.
[b] Herbivores can be rare, intermediately abundant, or abundant, and herbivore community is dominated by small or large herbivores or both (diverse).
[c] Key: +, diversity increased because extinction is reduced; −, diversity decreased because extinction rates are higher.
[d] Key: +, diversity increased because local colonization is enhanced; −, diversity decreased because local colonization is reduced.
[e] A diverse herbivore assemblage depends on the presence of large herbivores to facilitate smaller herbivores; otherwise herbivores will be rare.
With permission from Olff and Ritchie (1998).

Mongolian gazelle *Procapra gutturosa* and goitered gazelle *Gazella subgutturosa*; Mongolian wild horse *Equus przewalksi* and Bactrian camel *Camelus bactrianus* were also once widespread but are now near extinction (Ting-Cheng 1993). The large herbivores of South American savannahs are ungulates and include three species of tapirs *Tapirus* spp., three peccaries (*Catagonus wagneri, Dicotyles tajacu, Tayassu peccari*), four camelids (*Lama* spp., *Vicugna vicugna*), and eleven species of deer (*Blastocerus dichotomus, Hippocamelus* spp., *Mazama* spp., *Pudu* spp., *Odocoileus virginianus, Ozotocerus bezoarticus*) (Ojasti 1983). The Pampas deer *O. bezoarticus celer* was the principal large native herbivore of the South American Pampas, restricted to dry open grasslands, but is now almost extinct and restricted to a few wildlife reserves (Soriano 1992). Native mammalian herbivores are lacking in Australia and New Zealand, but introduced herbivores such as red deer *Cervus elaphus,* chamois *Rupicapra rupicapra,* and Himalayan thar *Hemitragus jemlahicus* are widespread through mountainous areas of New Zealand (Mark 1993). Marsupial mammals (Family Macropodidae) are the dominant large native herbivores in Australia and include >50 species ranging in size from the small rat-kangaroo *Hypsiprymnodon moschatus* (c.454 g) to the red kangaroo *Macropus rufa* (>82 kg), although there are also large populations of introduced herbivores including feral horses *Equus caballus,* camel *Camelus dromedarius,* goats *Capra hircus,* and several species of deer (Frith 1970). Large non-marsupial mammals became extinct in Australia in the late Pleistocene about 10 000 years ago, probably as a result of overhunting (Johnson and Prideaux 2004). Large herbivore communities can be diverse as in the East Africa Serengeti fauna and as exemplified by the following list of ungulates in three types of Indian grassland (Singh and Gupta 1993):

- **alluvial grasslands:** hog deer *Axis porcinus*, wild water buffalo *Bubalus bubalis*, swamp deer *Cervus duvauceli*, Indian rhinoceros *Rhinocerus unicornis*
- **acid grasslands:** black buck *Antilope cervicapra*, wild ass *Equus hemonius khur*, chinkara *Gazella gazelle*
- **high-altitude grasslands:** ibex *Capra ibex*, Tibetan wild ass *Equus hemonius kiang*, tahr *Hemitragus jemlahicus*, goral *Nemorhaedus goral*, argali *Ovis ammon*, urial *Ovis orientalis*, Tibetan gazelle *Procapra picticaudata*, bharal *Pseudois nayaur*.

Compared with native herbivore communities, domestic herbivore communities are relatively simple, with low diversity, generally composed of one species of commercial interest per grassland plus any other native animals that are able to subsist without bothering the land managers too much. For example, although cattle are the domestic large herbivore of commercial interest in North American ranges, elk and deer can be common. In grassland across the world, domestic herbivores are primarily either cattle or sheep, but include horses, llamas, camels, and goats. Regardless of native or domestic status, there is a strong positive correlation between vertebrate biomass and annual precipitation, and between vertebrate biomass and annual above-ground plant biomass (Dyer *et al.* 1982).

Grazing effects on grassland
As a disturbance, grazing by large herbivores changes grassland composition and structure. While climate determines grassland composition and structure at the regional scale, the effect of grazing is important at scales from that of the landscape down. Grazers selectively remove plant materials and plant organs (**defoliation**), principally leaves and stems, and in doing so alter canopy structure, trample the vegetation and compact and/or disturb the soil, and concentrate nutrients in patches through urination and defecation (Fig. 9.5). By removing the living parts of plants and affecting growth rates and fitness, grazers shift the balance between plant species. The major compositional shifts exhibited under some grazing regimes probably result from a combination of local extinction of non-tolerant species and germination and establishment of new species in the sward. Experiments with species characteristic of southern mixed prairie in the USA suggested that replacement of mid-seral species with late-seral species under grazing pressure resulted from a greater competitive ability and herbivory tolerance of the latter species compared with the former (Anderson and Briske 1995). Some grazers are highly selective, e.g. bison

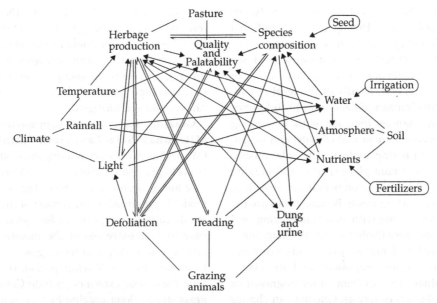

Figure 9.5 Schematic diagram of the principal interactions between grazing animals and grazed grassland. Soil and climatic factors are included to show how some of the interactions between plants and animals are mediated by, or influenced by, the abiotic environment. Other disturbances, such as fire, impact the interactions shown here. Reproduced with permission from Snaydon (1981).

and sheep, whereas others are more cosmopolitan, e.g. goats. For example, the unpalatable annual *Aristida bipartita* increased under heavy grazing by cattle in African savannah during drought years, whereas, the palatable perennials *Heteropogon contortus* and *Themeda triandra* showed extensive mortality (O'Connor 1994).

Grazing is not uniform in space or time; even with a single pasture there are areas of heavy use and areas of disuse. **Grazing lawns** are areas or patches of vegetation that are established and maintained by grazers through intensive utilization. These areas have a reduced canopy height, stimulating the activation of tillers and producing a prostrate, dense canopy (McNaughton 1984). Maintained by continuous grazing, grazing lawns are enriched by urine and dung resulting in high nitrogen mineralization rates and plants with lower average leaf age and higher digestibility (see §4.2), crude protein, and sodium than surrounding grasslands. Grazing lawns occur in areas where herds of grazing herbivores are common. They were first described in the Serengeti grasslands of Africa and have also been described in response to bison in North America (see below), and to kob *Kobus kob* and hippo *Hippopotamus amphibius* in West African savannah (Verweij et al. 2006). The grazing lawns maintained by ungulates in Terai grasslands of Nepal were described as patches of shortgrass/forb communities with higher diversity than the surrounding *Imperata cylindrica*-dominated phantas grasslands (Karki et al. 2000). By providing high-quality forage, grazing lawns contribute significantly to the nutrition of herbivores (Verweij et al. 2006).

Plant communities can be characterized by their relationship to the grazing regime (see Chapter 10). For example, in the western Rift Valley of Uganda, distinctive grassland zones can be recognized according to distance from permanent water from where hippopotami emerge at night to graze (Lock 1972). At small scales, spatial heterogeneity within communities is generated and sustained under grazing by the response of species to preferential grazing at different intensities, and small-scale disturbances generated by herbivores such as hoof prints, pawing, wallowing, and dung and urine patches (Bakker et al. 1983).

Comparing the effects of grazing from studies across the globe shows that species compositional changes affected by grazing are a function primarily of ANPP (itself reflective of moisture level, see Chapter 7), secondly of the evolutionary history of grazing at a site, and, thirdly of the level of consumption (Milchunas and Lauenroth 1993). Species composition changes increase with increasing ANPP (i.e. with a more humid climate) and longer, more intense grazing history. The relationship between the dominant plant species and grazing is more sensitive to environmental variables (e.g. location) than grazing inself. Bunchgrasses are the most likely to decrease with increased grazing, and perennials are more likely to decrease than annuals. Ungrazed and grazed grasslands are more dissimilar in species composition and abundance of the dominant species than either ungrazed or grazed shrublands or forests. Grazing can change grasslands to shrublands.

The difference in ANPP between grazed and ungrazed grassland is least in grasslands with the longest evolutionary history of grazing, with the greatest reduction in production with grazing when ANPP is high, and secondarily when consumption was high or treatment is long. Grazing can increase ANPP (the herbivore optimization hypothesis, below). Species compositional changes in response to grazing are fast, ANPP changes intermediate, and changes to the soil nutrient pool slow. There is a loose relationship between changes in species composition and changes in ANPP with grazing, but no relationship between changes in species composition and changes in root mass, soil carbon, and soil nitrogen.

Grazing effects are summarized in the **Milchunas–Sala–Lauenroth (MSL) model** (Milchunas *et al.* 1988) relating diversity in grasslands to grazing intensity along gradients of moisture and evolutionary history (Fig. 9.6). The MSL model is presented in the context of the intermediate disturbance hypothesis (see §9.1) with reference to four extremes of the moisture gradient and evolutionary-history-of-grazing gradient. Examples used by Milchunas *et al.* (1988) to illustrate these four extremes include Colorado shortgrass steppe (long grazing history, semi-arid), the Serengeti of East Africa (long history, subhumid), Argentina Patagonian arid steppe (short history, semi-arid), and Argentina flooding Pampa (short history, subhumid). The MSL model suggests that changes in species composition and diversity with respect to grazing are relatively small or moderate in semi-arid grasslands with a long or short evolutionary history of grazing, respectively (although see Cingolani *et al.* 2003). By contrast, changes in

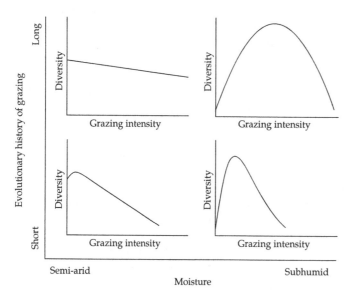

Figure 9.6 Plant diversity of grassland communities in relation to grazing intensity along gradients of moisture and evolutionary history of grazing. Reproduced with permission from Milchunas *et al.* (1988).

subhumid grasslands are relatively large no matter what the evolutionary history of grazing in the grassland is. A humped-back diversity–disturbance relationship is predicted to be most evident in subhumid grasslands with a long evolutionary history of grazing, reflecting the past selection for grazing tolerance and canopy dominance. In the absence of grazing, tall species predominate, suppressing diversity; under moderate grazing there is a mosaic with shortgrasses in heavily grazed patches and tallgrasses in ungrazed patches, thus allowing higher diversity; and under heavy grazing just a few species dominate the canopy, decreasing diversity. The ample soil moisture under subhumid conditions allows for sufficient production for tallgrasses to become an important component of the community, whereas under more arid conditions the communities are predominantely shortgrasses, thereby disallowing the mosaic of patches that drives up diversity under more mesic, grazed conditions.

Field experiments have supported the recognition that large herbivores increase plant diversity under conditions of high productivity (Bakker *et al.* 2006). Mechanistically, this increase in diversity is due to enhanced colonization of new species because of lowered light limitations on recruitment in productive grassland. Grazing opens up establishment gaps in an otherwise closed sward, facilitating colonization of subdominants (Noy-Meir *et al.* 1989). The humped-back relationship between plant diversity and productivity under grazing by large vertebrate herbivores (Fig. 9.6) does not appear to apply to the grazing effects of small vertebrate herbivores (Bakker *et al.* 2006).

The MSL model does not adequately recognize the potential existence of alternate stable vegetation states with different diversity values that can occur after vegetation change in response to grazing has passed a threshold trigger point. By contrast, the multiple equilibrium, **state-and-transition** or **state-and-threshold (S-T) model** (Laycock 1991; Westoby *et al.* 1989) developed for range management (see §10.2.2) implies irreversible transitions and alternate equilibria. An expansion of the MSL model incorporates the existence of multiple alternate vegetation states irreversibly attained in response to grazing grassland with short evolutionary history (Cingolani *et al.* 2005). In grassland with a short evolutionary history of grazing, the plants have been selected to invest resources in either sustained growth under highly productive conditions or tolerance to low resource availability under conditions of low production that may not necessarily include pre-adaptation to grazing. The introduction of grazing to these systems can cause local extinction of grazing-susceptible plants, leading to a decrease in local diversity. Intensive grazing, particularly in low-productivity systems, can lead to other ecosystem changes such as exposure of the topsoil to erosion. Together these changes can trigger irreversible changes in vegetation composition, disallowing recovery to the original state even if grazing ceases. For example, long-term heavy grazing of California's Mediterranean grasslands leads to irreversible transitions from perennial to annual grassland as the native species (e.g. *Nassella pulchra*) are replaced by invading species (e.g. *Bromus mollis, Taeniatherum caput-medusae*) (George *et al.* 1992).

The **herbivore optimization hypothesis** (Dyer *et al.* 1982) predicts that an intermediate level of grazing can have a positive effect on ANPP or plant fitness (Fig. 9.7). The hypothesis is controversial (Belsky *et al.* 1993; Crawley 1987; Detling 1988; McNaughton 1979), as under this scenario grazers could increase their own harvest through

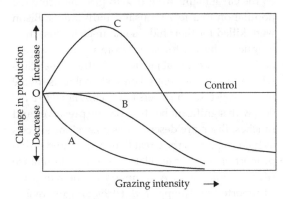

Figure 9.7 Predictions of how primary production may be affected by grazing. With increased grazing intensity primary production may (A) decrease, (B) be relatively unaffected under light grazing before decreasing under heavy grazing, or (C) show a maximum under some optimal level of grazing intensity (i.e. the herbivore or grazing optimization hypothesis). With kind permission of Springer Science and Business Media from Detling (1988).

repeated grazing (see earlier discussion of grazing lawns). Nevertheless, the herbivore optimization hypothesis is supported by the data summarized by Milchunas and Lauenroth (1993) for grasslands when evolutionary history is long and productivity low. In these situations it is postulated that grazing removes and breaks up the layers of mulch which would otherwise shade new growth. The hypothesis is also supported when there are large losses of a limiting nutrient during detrital recycling or when herbivores act to bring limiting nutrients from outside the ecosystem (i.e. offsetting loss) (de Mazancourt et al. 1998). Studies on goose herbivory are supportive of the hypothesis as biomass and tiller production of *Festuca rubra* indicated that carrying capacity was higher in grazed than in ungrazed areas (Graaf et al. 2005).

North American bison
The bison *Bos bison* was, until European settlement, the dominant herbivore across the Great Plains grasslands of North America. Other herbivores that occur at a lower density and have less of an effect on the system include elk *Cervus canadensis*, feral horses *Equus caballus* (reintroduced by the Spanish), antelope *Antilocapra americana*, and deer *Odocoileus virginianus*. Before European settlement, bison numbers are estimated to have been $30–60 \times 10^6$. Wholesale slaughter of bison occurred on the Great Plains from 1830 to 1880, reducing the population to a few thousand individuals. Bison were killed for their hide, sometimes for just their tongue or horns, for sport from train windows, and in an effort to subjugate the Native Americans whose culture was centred on sustainable use of the bison. Now bison numbers have rebounded to c.150 000, with significant herds in prairie preserves and ranches. Bison are described as a keystone species as their impact on the Great Plains grassland is disproportionately large relative to their abundance (Knapp et al. 1999). A summary of their impacts as reported by Knapp et al. (1999, and references therein) from studies conducted in tallgrass prairie in Kansas is described below.

Bison graze grasses preferentially and avoid forbs and woody species (<10% of diet). Forbs left ungrazed become surrounded by grazed grasses; *Ambrosia psilistachya* and *Vernonia baldwinii* are examples of two forbs specifically avoided by the bison (Fahnestock and Knapp 1994). This foraging pattern increases local biodiversity as it allows forbs to exist in bison-grazed grassland even with frequent fire. There are two patterns of bison grazing: (1) distinct grazing patches of $20–50 \, m^2$, (2) grazing lawns $>400 \, m^2$. Bison revisit these areas throughout the growing season, leading to repeated defoliation of patches and individual plants. There are sharp boundaries among grazed and ungrazed patches. Grazed grasses exhibit postgrazing enhancement in photosynthetic rates due to increased light levels and reduced water stress in grazed patches. Grazed tissues have enhanced nitrogen levels as nutrients are translocated from the roots. Bison exhibit a marked preference for burned sites rather than unburned sites, especially moist lowland areas late in the summer. As grazed patches become more dominated by forbs, so bison locate new grass-rich patches. Hence, over time the grazing patches migrate across the landscape.

Non-grazing activities of bison include deposition of dung and urine, trampling, and wallowing, all of which increase spatial heterogeneity of the prairie. Urine patches lead to locally high soil nitrogen, with plants growing in these patches having higher tissue nitrogen. Bison are attracted to these patches because of the more nutritious forage; as a result there is lower grass cover on urine patches, due to grazing (Day and Detling 1990). Wallowing is an activity in which bison paw the ground and roll in the exposed soil. This activity produces depressions in the soil, 3–5 m in diameter and 10–30 cm deep, that the bison will return to repeatedly. These denuded patches have compacted soil, retain water (although they become very dry in summer droughts), and have unique vegetation with a higher diversity than the surrounding vegetation. Active wallows may be colonized by ruderals including a number of annuals, whereas abandoned wallows are colonized by perennials. Wallows do not burn easily, as the fuel load is usually low. These unique patches have been found in prairie that has not be grazed for >100 years (Gibson 1989). Bison carcasses (adults can weigh >800 kg) return very high (toxic) levels of nutrients to local patches, resulting in a $4–6 \, m^2$ denuded zone that is colonized by early successional species.

Bison are not equivalent to cattle. Although both are generalist herbivores there are important differences in their foraging patterns, with bison eating more grasses and cattle eating more forbs and browse species. Bison graze for less time than cattle and spend more time in non-feeding activities such as wallowing (which cattle do not do) and horning. Bison are more often found in open grassland, whereas cattle prefer wooded areas, given the choice. In the winter, the bison is able to use its large head as a snowplough; by sweeping its head from side to side it can brush away snow to expose forage. Cattle do not do this.

The overall assessment is that bison affect the tallgrass prairie disproportionately in relation to their abundance. They affect the ecosystem at various scales, and in conjunction with other disturbances, particularly fire, enhance grassland biodiversity (Collins et al. 1998a).

9.3.2 Small vertebrate herbivores

There are a large number of small herbivorous mammals, including rodents and lagomorphs, that affect the structure and composition of grassland. The rodents (Order Rodentia) important in grassland include mice, voles, shrews, gophers, and prairie dogs. Lagomorphs (Order Lagomorpha) include hares and rabbits (Family Leporidea) and pikas (Family Ochotonidae). A number of carnivores such as members of the Family Mustelidae, subfamilies Melinae (Eurasian badgers), Mellivornae (ratel or honey badger), Taxideinae (American badger) affect grassland through their burrowing and grubbing activities. The Australian grassland fauna includes a wide diversity of murid rodents (Family Muridae), as well as small marsupial mammals such as the herbivorous wombats (*Vombatus ursinus, Lasiorhinus latifrons,* and *L. krefftii*) which primarily eat monocots (Green 2005; Wells 1989). Rabbits and hares have been introduced from Europe to many parts of the world, including Australia, where they have a large effect on the native fauna and vegetation of rangeland.

The effects of small herbivores on grassland include the following (De Vos 1969):

- On soil, burrowing brings loose soil to the surface and can honeycomb the subsurface with a network of passages. These activities aerate the soil, altering water-holding capacity, and incorporate humus. Exposing the soil can accelerate erosion. Urine and faeces enrich the soil.
- The production of mounds is characteristic of pocket gophers, ground squirrels, and prairie dogs. The flat surface of the soil is broken up, and subsoil is brought up and deposited on the surface. Mounds often occur at the entrance of burrows where they can affect microclimate and alter patterns of wind erosion. Mounds tend to be drier than the surrounding soil, hillocky, and giving a mottled aspect to landscape.
- Direct effects on vegetation include removing forage, selective foraging, removing seed (especially detrimental to annuals), cutting roots (e.g. by pocket gophers), and disseminating and redistributing seed. These activities can accelerate exotic species or shrub invasion.

As with other agents of disturbance, small herbivores interact with each other and with their environment to enhance grassland heterogeneity. For example, several important species of grassland interact with each other. Bison are attracted to the nutritious grass that grows on colonies of black-tailed prairie dog *Cynomys ludovidianus* in South Dakota, USA (see below), and in desertified grassland of northern Mexico where these prairie dogs and banner-tailed kangaroo rats *Dipodomys spectabilis* co-occur, they each have unique but additive effects on local biodiversity (Davidson and Lightfoot 2006).

The diet and activities of small herbivores in grassland vary considerably. Some are specialist herbivores focusing on particular plant groups feeding either exclusively above ground (e.g. rabbits) or below (e.g. gophers). Physical disturbance to grassland can arise from burrowing and nesting activities. Some examples are detailed below.

Fossorial rodents
Pocket gophers (Geomyidae), common in arid and semi-arid regions in North America and other fossorial rodents, e.g. molerats (Bathyergidae, Spalacidae) in Africa and Asia, and tuco-tucos and cururos (Ctenomyidae) in South America, construct extensive underground burrow systems within which they establish feeding gardens to graze

roots. Excess soil is deposited in abandoned tunnels and outside tunnel entrances where it smothers vegetation. Fossorial herbivores are considered 'ecosystem engineers' because of the extensive manner in which they alter resource availability for other organisms and modify, maintain, or create habitats (Huntly and Reichman 1994; Jones et al. 1997; Reichman and Seabloom 2002). The tunnels can underlie 7.5% of the ground with the mounds covering an additional 5–8%.

The grazing of plant roots can reduce performance of individual species; up to 30% of subterranean primary productivity can be consumed, approaching the energy flow of larger aboveground herbivores. Gophers are selective feeders showing a preference for tap-rooted plants, especially nitrogen-rich legumes. Mounds are gaps in the grassland community offering an opportunity for germination and seedling establishment, and reduced competition for plants growing at the edge. The biomass over burrows and mounds is generally less than in the surrounding prairie, with an enhanced biomass wave at the edge (Reichman et al. 1993). In tallgrass prairie, perennial graminoids are the dominant species colonizing gopher mounds, preempting colonization by forbs. In this system, the effects of gopher mounds on local species richness and production are transitory (<3 years) (Rogers et al. 2001). By contrast, the vegetation colonizing gopher mounds in California serpentine annual grassland depended to a large extent on the time of year in which the mound was formed, with short-lived species characterizing spring-formed mounds (Hobbs and Mooney 1985). Generally, the grassland dominants were poor colonizers of gopher mounds in this system because of limited seed dispersal within the closed vegetation. Rather, species with tall flowering stems were those best able to disperse on to the gopher mounds.

Fossorial herbivores have been implicated in the formation of **mima-mounds** in North and South America and Africa. Mima-mounds are areas of raised topography with a distinct flora and enhanced productivity. The soil of mima-mounds is characteristically higher in nutrients, organic matter, and water-holding capacity than surrounding intermound areas (Huntly and Reichman 1994).

For example, mima-mounds in Kenyan tropical savannahs are attributed to the activity of rhizomyid molerats *Tachyoryctes splendens* and have a lower cover of grasses and higher cover of bare ground, forbs, and shrubs than off-mound areas (Cox and Gakahu 1985). The dominant grass on the mounds was *Cynodon dactylon* compared with *Themeda triandra* in the savannah.

American badger Taxidea taxus
These animals establish underground dens, and during excavation deposit large mounds of soil on to the existing vegetation outside the entrance to the den. They also excavate mounds of soil while foraging for ground squirrels *Spermophilus tridecemlineatus*. These local disturbances enhance the spatial heterogeneity of the tallgrass prairie in which badgers are found. The vegetation of the mounds comprise a guild of fugitive species including several wind-dispersed biennials, with life histories intermediate between r- and K-selected species (see §6.4.1) adapted to the environmental conditions of the badger mounds (open space and high soil moisture) (Platt 1975). The fugitive guild are maintained in the prairie system through an inverse relationship between immigration among badger mounds and interspecific exploitive competition on mounds (Platt and Weis 1985). Some of the prairie dominants also colonize the badger mounds, including *Andropogon gerardii*, with maximum growth occurring later in the season than the fugitive species.

Banner-tail kangaroo rat Dipodomys spectabilis
The burrowing and mound-creating activities of this animal in shortgrass steppe–desert grassland ecotones leads to lower plant cover and higher diversity on mounds than on adjacent off-mound areas. The mounds (1 m high and 1.5–4.5 m in diameter) which cap subterranean chambers and tunnel systems had drier soil with higher nitrogen content, and were dominated by forbs, shrubs, and succulents in grasslands otherwise dominated by the grasses *Bouteloua gracilis* or *B. eriopoda* (Fields et al. 1999).

Prairie dogs Cynomys spp.
The prairie dog is considered by some authors as a keystone species (see above for a description

as bison as a keystone species) of the North American grasslands because of the large multi-scale effects it has (Davidson and Lightfoot 2006; Kotliar 2000; Natasha et al. 1999). It is a large herbivorous rodent ($c.1$ kg as adult) that historically occupied 40×10^6 ha of the US Great Plains, i.e. >20% of the natural shortgrass and mixed prairies. Populations of prairie dogs were reduced to <2% of their historic levels by habitat loss, plague in the western part of their range, and eradication programmes established because of reduced livestock weight gains in pastures occupied by prairie dogs (Derner et al. 2006). Eradication of prairie dogs coincided with tree/shrub encroachment and a loss of grassland/savannah habitat (Weltzin et al. 1997). Individual colonies fluctuate widely in size over time, from tens to hundreds of individuals on deep soils with gentle slopes (<7%) with densities of 10–55 animals ha^{-1}. Within colonies there can be 50–300 burrows ha^{-1} each with two entrances, 1–3 m deep, 15 m in length, and 10–13 cm in diameter. Burrowing mixes 200–225 kg soil per burrow system with much of the soil deposited as mounds 1–2 m in diameter at burrow entrances. There is continuous, intense burrowing and grazing by prairie dogs in their colonies. Studies from Wind Cave National Park, South Dakota, USA (Whicker and Detling 1988) indicate that prairie dogs forage above ground throughout the year, clipping and grazing plants to maintain a turf with a canopy less than 5–10 cm in height compared with nearby uncolonized areas which generally have a turf of 20–30 cm. As a result of generations of grazing, distinct 'dwarf' ecotypes of plants occur on prairie dog towns. The grazing by the prairie dogs alters the competitive balance among species producing a mosaic of community types within the colonies, characteristically composed of shortgrass species and annual forbs. Local diversity is altered by this activity with diversity peaking in areas subject to moderate impact or occupied for intermediate periods of time (cf. the intermediate disturbance hypothesis: see §9.1) (Archer et al. 1987; Coppock et al. 1983). On prairie dog colonies 60–80% of ANPP is consumed or wasted, a high amount compared with the 5–30% consumption rates of large herbivores. The biomass of the plants is 1/3–2/3 that of plants growing on adjacent uncolonized areas, with the standing live:dead biomass ratio 2–4 higher on colonies. Plants growing on colonies have a higher shoot nitrogen content than plants growing off colonies because of the large number of younger leaves and nitrogen translocation from the roots. Defoliation above ground also reduces below-ground biomass, leading to up to 40% lower root production. Bison, elk, and pronghorn preferentially graze the grassland associated with prairie dog colonies, especially areas of young grass.

Prairie vole Microtus ochrogaster
The prairie vole exhibits 3–5 year cyclic patterns of abundance in tallgrass prairie with its spring abundance influenced by the vegetation and positively related to ANPP from the previous season (Kaufman et al. 1998). The voles maintain runways above ground through the grass that are $c.5$ cm wide and pressed into the litter layer and soil. Grass leaves that grow into the trails are trimmed away by the voles, with fresh cuttings left along the pathway. The runways fan out from nests that are 12–25 cm in diameter and 30 cm below the soil surface (Reichman 1987). The vegetation above the nests can have different species composition from the adjacent prairie. For example, significantly higher frequencies of *Physalis pumila*, lower frequencies of *Symphyotrichum ericoides* and *Ruellia humilis*, and a lower cover of *Dicanthelium oligosanthes* characterized prairie vole nests in Kansas, USA tallgrass prairie compared with adjacent reference areas (Gibson 1989). In sagebrush rangelands of Utah and Nevada, USA, longtailed voles *Microtus longicaudus* girdle brush and when at high enough population densities can improve the range for cattle (Frischknecht and Baker 1972).

Effects of small-mammal herbivory
Although the small-scale effect of small mammals such as voles on individual plants is readily observed (examples above and Gibson 1989), the net effect on grassland community composition and structure is equivocal. A 4 year experiment in mature tallgrass prairie revealed few significant effects of *Microtus ochrogaster* except that richness of C_4 grasses and some annual forbs were reduced by the small mammals (Gibson et al. 1990a). By contrast, dramatic effects of small mammals on

grassland vegetation have been observed in young grasslands. For example, in 3 year old experimental prairie plantings, herbivory by the vole *Microtus pennsylvanicus* reduced densities of legumes and grasses, allowing an increase of unpalatable species such as the forbs *Echinacea purpurea* and *Rudbeckia hirta*, resulting in an overall decline in species richness of 19%. The composition of the vegetation of these young assembling prairie communities reflected the competitive release of unpalatable (to voles) species, illustrating the 'ghost of herbivory past' (Howe et al. 2002). These sorts of effects may, however, be transient as after 4 years diversities converged in tallgrass prairie plantings in Wisconsin, USA to which voles had access and plantings from which voles were excluded (Howe and Lane 2004). Exclusion of voles over 54 months in a young tall fescue *Schedonorus phoenix* pasture increased the frequency of *Neotyphodium ceonophialum* endophyte-infected *S. phoenix* individuals. Endophyte-infected *S. phoenix* had increased competitive vigour and biomass leading to a decrease in the abundance of forbs (Clay et al. 2005). In desert shrublands, dominance shifts from forbs to grasses were observed precipitating a change to a grassland community when the seed herbivore *Dipodomys* spp. (kangaroo rat) was excluded for 12 years. Specifically, there was a 20-fold increase in a tall perennial (*Eragrostis lehmanniana*) and 3-fold increase in annual grass (*Aristida adscensionis*) density (Brown and Heske 1990). Overall, it appears that the herbivory by small mammals as discussed here can influence successional pathways through selective foraging which can eliminate or reduce the abundance of palatable species, allowing competitive release of non-preferred species. As a result, community fate is affected. The influence of these herbivores on established grassland communities is less obvious.

9.3.3 Invertebrate herbivores

From the biblical plagues of grasshoppers to a localized outbreak, invertebrates remove 5–15% of ANPP above ground and, primarily thanks to nematodes, 6–40% of BNPP below ground in grasslands (Detling 1988). Overall, grasshoppers and termites are considered the dominant native invertebrate herbivores in grassland and savannahs worldwide and their effects are considered in some detail here. However, despite the clearly significant effects on production, the effects of many invertebrate groups on grassland composition and structure are poorly understood, especially the role of below-ground invertebrate herbivores and detritivores (consumers of dead organic matter) (Andersen and Lonsdale 1990; Whiles and Charlton 2006). Meta-analyses and comparisons among studies have yielded equivocal results in comparing the effects of different kinds of invertebrates or effects among different types of herbaceous-dominated ecosystems, including grasslands (Coupe and Cahill 2003; Schadler et al. 2003). For example, in a species-poor acidic grassland in the UK (dominants included *Festuca rubra*, *Trifolium repens*, and *Agrostis capillaris*), invertebrate and mollusc herbivory affected the abundance of some plant species, but they had less of an effect than herbivory by rabbits or interspecific competition from the dominant monocots (del-Val and Crawley 2005).

Herbivory and disturbance by invertebrates differs from that by vertebrates. Invertebrate damage is more diverse in type and spatial distribution: it can be chronic, generalized across all plant tissues, or highly specific and localized (Kotanen and Rosenthal 2000). Invertebrates can also spread disease, such as the Barley Yellow Dwarf virus that is spread among grasses through the saliva of cereal aphids (Grafton et al. 1982; Gray et al. 1991). Highly abundant and aggregated invertebrates are, however, similar to vertebrates in terms of the large-scale damage to plants for which they can be responsible. Grasses posses morphological and physiological features that confer tolerance to insect herbivory (Tscharntke and Greiler 1995). High levels of silica in grass cells make cells tough and difficult to chew (see §4.4). However, grasses are generally less well protected by anti-herbivory secondary compounds than forbs, except for the presence of alkaloids in endophyte-infected grasses that deter aphid grazing (see §4.3). It is postulated that grasses and insects co-evolved along with grazing mammals during the early rise of grassland communities (Chapter 2).

In the North American tallgrass prairie several arthropod herbivores are represented including the mandibulate folivores (Orthoptera, Lepidoptera, and some coleopterans), hemipteriod sap-feeders, and insects that live within plant tissues such as gallmakers. These arthropod herbivores do not control primary production but affect grassland ecosystem components, processes and function in several direct and indirect ways (Fig. 9.8). Many arthropod groups can have subtle effects, e.g. grasshoppers (Orthoptera) are primarily folivores with their most significant impact being on the grassland forbs, the most diverse group of plants (Whiles and Charlton 2006), except during outbreak periods (see below). Grasshopper feeding varies markedly in time and space and may remove $c.1$–5% of total ANPP. Some grasshoppers are generalists, feeding on a wide range of grasses and forbs, whereas others are **monophagous** or **oligophagous**, feeding on a restricted or single group of species.

Much invertebrate damage is sustained at the individual plant level, affecting plant growth rate, plant form, or plant fecundity. Many specialist invertebrates are limited in their distribution to a single species, leading to a heterogeneous spatial distribution based on the distribution of their host plant. For example, adults of the seed-feeding bug *Lygaeus kalmii* (Hemiptera: Lygaeidae) occur exclusively on the seed heads of milkweed *Asclepias viridis* in North American prairie. Hence, herbivory from *L. kalmii* is limited to milkweeds and depends on the distribution of reproductive host plants with respect to their neighbours (Evans 1983). Similarly, the cynipid gall wasp *Antistrophus silphii* (Hymenoptera: Cynipidae) oviposits into the apical meristem of the perennial tallgrass prairie forb *Silphium integrifolium*, inducing the formation of an abnormal spherical growth (the gall, 1–4 cm diameter) on the stem. Gall formation causes allocation changes in the plants, leading to reduced or inhibited shoot growth and leaf production, and delayed and reduced reproduction (Fay and Hartnett 1991).

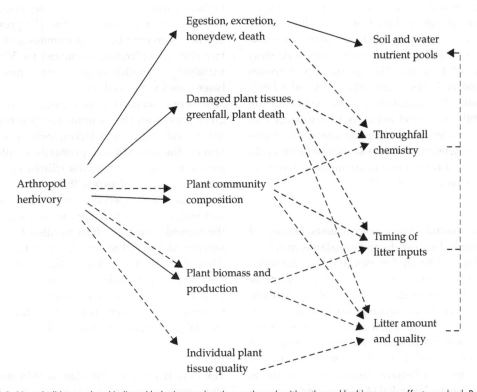

Figure 9.8 Direct (solid arrows) and indirect (dashed arrows) pathways through with arthropod herbivores can affect grassland. Reproduced with permission from Whiles and Charlton (2006) in *Annual Review of Entomology*, Volume 51. © 2006 by Annual Reviews.

Below-ground herbivores and detritivores affect root dynamics and rhizosphere nutrient cycling. Some below-ground herbivores such as large-bodied cicadas redistribute materials during adult emergence.

Mass outbreaks of invertebrates illustrate the most obvious effects of these organisms as part of the disturbance regime of grasslands. Grasshoppers are considered the most destructive invertebrates in most grassland. In the western USA, 21–23% of available forage can be eaten by grasshoppers and damage can exceed foliage production by a factor of 3 (Tscharntke and Greiler 1995). In some regions a single swarming grasshopper species is responsible for the damage (e.g. *Schistocerca gregaria* in south-west Asia into Africa or *Nomadacris septemfasciata* in southern Africa); in other areas several species can be responsible for an outbreak with the assemblage changing region to region and year to year (e.g. 12 or more grasshopper species in the western USA) (Watts et al. 1982). Reports of outbreaks by other invertebrates in grasslands include grass bugs (*Labops* spp., *Irbisia* spp., *Leptopterna* spp.), range caterpillars *Hemileuca olivia*, army worms *Spodoptera* spp., sod webworms or grass worms *Crambus* spp., mormon crickets *Anabrus simplex*, white grubs *Phyllophaga* spp., termites (Isoptera), and ants (Tscharntke and Greiler 1995). Such outbreaks can reduce grass biomass, allowing competitive release of forbs. As a disturbance that varies considerably in time and space, insect outbreaks may, because of their severity when they do occur, play a central role in community organization (Carson and Root 2000).

Termites

Termite mounds are a conspicuous feature of many tropical savannahs and grasslands in Africa, Australia, and South America. Termites are social insects (Order Isoptera) that build large nests inside or outside trees, underground, or above ground. Termite mounds are above-ground nests that can be large, up to 5 m in height at low densities <5 ha^{-1} (*Macrotermes* spp. in Africa, *Nasutitermes triodiae* in Australia) or smaller at higher densities up to several hundred per hectare (*Trinervitermes* in Africa, *Tumulitermes* in Australia) (Andersen and Lonsdale 1990). Densities of termites can range from 1000 to 10 000 m^{-2}, comprising a total biomass of 5–50 g m^{-2}. Most termites are detritivores predominantly feeding on dead grass, others feed on live grass or fungi. The mound-building and other activities redistribute soil materials from lower levels of the soil horizon (Josens 1983). As a result of these activities termites substantially affect the heterogeneity of the soil and the vegetation. Living termitaria are usually free of vegetation, creating an open patch in the grassland; after abandonment the mounds are colonized by vegetation. In open savannahs, trees and shrubs may be restricted to abandoned termite mounds (De Vos 1969). In the Serengeti National Park, Tanzania, termite activity converted patches of alkaline, shortgrass soils (dominated by *Digitatis scalarum*) into mid-grass soils as abandoned termite mounds were colonized by clonal mid-grass species (e.g. *Pennisetum mezianum* and *Themeda triandra*) (Belsky 1983, 1988). Soil under termite mounds was more homogeneous than in surrounding undisturbed shortgrass areas. The soil under abandoned termite mounds had faster water infiltration rates, higher concentrations of potassium, magnesium, and calcium, and lower sodium. The presence or absence of mound-building termites and subsurface Na+ concentration accounted for 55% of the variability in models of grassland compositional heterogeneity. In north-eastern Australia, termitaria were associated with a green 'halo' of dark green grasses and herbs immediately surrounding the mounds (Spain and McIvor 1988). The vegetation on the mounds was primarily small-seeded annual grasses (e.g. *Digitaria ciliaria*) and dicotyledons (e.g. *Sida spinosa*) with an increase in perennial grasses (especially *Heteropogon contortus*) and sedges *Fimbristylis* spp. with distance from the mound. The halo effect resulted from higher nutrient status of the soils close to the mounds. Also, plants growing close to the mounds extend their root systems below the mound and into the mound bases. The more abundant plants close to the mounds were more nutritious, attracting preferential grazing by domestic stock.

Ants

The ants (Family Formicidae, within the Order Hymenoptera) are another large, speciose (c.15 000 species) group of social insects whose nests (anthills or mounds), like those of termites, significantly affect grassland soil and vegetation. Omnivorous ants in meadows can consume up to 3% of NPP per season (Folgarait 1998). Grassland systems with abundant anthills are referred to as 'antscapes' (Kovář et al. 2001) because of the landscape-wide importance of this disturbance. Anthills can be tens of centimetres in height and diameter and may be regularly spaced across large areas. Some ant mounds grow in size each year, and anthill size has been used to estimate the date since last ploughing of chalk grasslands in the UK (King 1981). Ant mounds can have higher or lower soil nutrients, lower bulk density, and less compact soil than surrounding undisturbed areas as a result of mound-building activities of the ants and chemical enrichment from ant excreta, storage of insect components, and stimulation of ammonifying bacteria (Dostál et al. 2005). The mycorrhizal flora associated with anthills can be particularly rich because of the deposition of fungal spores, which can facilitate plant establishment. The vegetation on anthills can be distinct from the surrounding grassland and they are often dominated by annuals while active; these become displaced by perennials as the anthills age and are abandoned (King 1977a, 1977b; Wali and Kannowski 1975). For example, anthills in a Colorado, USA montane meadow otherwise dominated by *Poa interior* were characterized by *Bromus polyanthus* and *Achillea millefolium* (Culver and Beattie 1983). Ant-dispersed plants (myrmecochores, e.g. *Claytonia lanceolata, Delphinium nelsoni, Mertensia fusiformis,* and *Viola nuttallii*) possessing an eliasome were common in the grassland patches immediately surrounding anthills. An **eliasome** is fleshy oil-rich tissue on the seed of some plants that is an attractive food source for ants. The seed of these plants are carried to the mound by the worker ants and discarded intact after removal of the eliasome. In montane grassland in Slovakia, a number of annuals and several perennials (e.g. *Agrostis capillaris, Dianthus deltoides, Polystichum commune* agg., *Thymus pulegioides,* and *Veronica officinalis*) showed a preference for growing on anthills (Kovář et al. 2001). In tallgrass prairie in Kansas, USA, the perennial grass *Schizachyrium scoparius* was more frequent on anthills than in areas of surrounding grassland (Gibson 1989).

9.4 Drought

Regional precipitation patterns determine the large-scale distribution and characteristics of grasslands (Chapter 8). The amount and variability of annual precipitation is strongly related to grassland productivity (Chapter 7, Fig. 7.4, and Nippert et al. 2006), although the strength of precipitation–productivity relationships varies among grasslands (e.g. between North American and South African C_4 grasslands, Knapp et al. 2006). As generally arid systems (Chapter 1), grasslands are often characterized by frequent drought such as the widely known and reported 1930s droughts in the US Great Plains (see below). Although drought can be considered a rather normal disturbance in grasslands, sustained or extreme droughts can have dramatic and far-reaching effects on these ecosystems. The influences of drought on grassland communities are summarized in Table 9.2.

The North American drought of the early 1930s had a dramatic effect on the prairies of the Midwest. This 'dustbowl' drought began in 1931, becoming most severe in 1934. Nevertheless, the 1930s drought was not the most severe on record; many before 1600 were of longer duration and more extensive. Paleoclimatic records show that major 1930s-magnitude droughts occur in the US Great Plains about once every 50 years, with more severe

Table 9.2 Influences of drought on grassland communities

Changes in the physical environment and resource base
Removal or depression of population size and geographic extent of most species
Competitive release of some native species and explosion in their populations
Invasion of the grassland by ruderals and other fugitive species
Shifts in phonological events, especially flowering time and reduction in flowering period
Shifts in productivity and fecundity of some species
Changes in demographic and genetic structure of populations
Changes in species composition of communities and reduction in their species diversity
More frequent and perhaps stronger competitive encounters between populations of some species

---

With permission from Bazzaz and Parrish (1982). Copyright © 1982 by the University of Oklahoma Press.

decadal-length droughts once every 500 years (Woodhouse and Overpeck 1998). Droughts in the US Great Plains are due to alterations to large-scale atmospheric circulation patterns disrupting the usual flow of moist tropical air masses from the Gulf of Mexico in the summer. The large-scale atmospheric circulation patterns are themselves driven by sea surface temperatures in the Gulf of Mexico and conditions in the Atlantic and Pacific oceans (including patterns related to the El Niño–Southern Oscillation) (Woodhouse and Overpeck 1998). The 1930s drought is well known because of the accompanying large-scale soil erosion that led to the 'dust bowl' conditions resulting from poor agricultural practices (primarily 75 years of overgrazing and the ploughing up of over 2.1×10^6 ha of virgin prairie between 1925 and 1930). The 1930s drought covered much of the US Great Plains, with the dustbowl itself centred in the southern plains of the Midwest, principally large areas of Texas, New Mexico, Colorado, Oklahoma, and Kansas. Dust storms consisted of large amounts of airborne dust—topsoil—being carried up into the air by strong winds and sweeping across the landscape. In Amarillo, Texas there were an average of nine dust storms per month during January–April from 1933 to 1939, each lasting an average of 10 hours with visibility often reduced to zero, depositing drifts up to 7 m in height (Lockeretz 1978). One newspaper headline from 1935 read 'The day the southern plains went east' (Floyd 1983). The dust storms and the drought led to staggering economic losses and a breakdown of the agricultural economy and society, displacing many farmers and their families (Howarth 1984). The era is memorialized in novels such as John Stienbeck's 1939 *Grapes of Wrath* and folk music such as Woody Guthrie's 1940 album *Dust Bowl Ballads*.

The ecological effects of the 1930s drought were reported by Weaver following extensive studies in Lincoln, Nebraska (Weaver 1954; Weaver and Albertson 1936) and illustrate the effects of drought outlined in Table 9.2. Droughts in the region occurred in 1931 and 1933, with the major drought in 1934. That year started with an unusually warm winter of little snowfall, followed by a dry spring, with high, dust-laden winds. Severe drought occurred in May through mid-August which was exacerbated by a heat wave in June–July with average daily temperatures >38 °C. Soil moisture was lost first from the upper layers of the soil, with water infiltration rates being decreased and run-off increasing substantially. As a result of the severe water deficit, foliage height was only 12–18 cm rather than the usual 23–36 cm. Rooting depth of the most stressed plants increased and the roots branched more profusely, at least initially. Plants with shallow roots suffered first during the drought, especially those growing on hilltops and xeric slopes. Plants growing in lowlands showed symptoms of drought by August.

The drought caused widespread losses of plants—up to 95% on level ground in some southwesterly areas—producing open areas and bare soil. In some areas, a mosaic of native grasses was able to persist. Nevertheless, all native species suffered, especially the most shallow-rooted species such as *Schizachryium scoparium*, *Koeleria macrantha*, *Hesperostipa spartea*, and *Poa pratensis*. *Andropogon gerardii* with its deep roots suffered the least, but it still failed to grow over considerable areas. Small grasses and forbs were lost from the ground layer, e.g. *Antennaria neglecta*, *P. pratensis*, *Dichanthelium oligosanthes var. scribnerianum* and *D. wilcoxianum*, *Sisyrinchium campestre*, and *Viola* spp. Competitive release following the death of many dominant species allowed some prairie grasses to spread such as *Pascopyrum smithii* which had previously been limited to sparse, thin soils, roadsides, clay pan soils, and disturbances. Because of its early spring growth and spreading rhizomes, *P. smithii* became dominant in many areas during the drought, especially areas where *A. gerardii* or *S. scoparium* had died. Drought-adapted *Bouteloua dactyloides* and *B. gracilis* spread in mixed prairie areas as the taller competitors died. The annual grass *Vulpia octoflora* became locally dominant in some areas. The herbaceous forb *Symphyotrichum ericoides* became very important in most areas, except where disturbed. The annual *Erigeron strigosus* also spread widely where it seeded thickly to a density of 20–30 plants dm^{-2}. Some ruderal species, such as *Lepidium densiflorum* which produced huge amounts of seed, *Conyza canadensis* (eastern half of the region), *Bromus commutatus* (western part), and *B. tectorum* (eastward) increased and became abundant in different areas. The stress of surviving under drought conditions altered the phenology of many plants, inducing early and shortened flowering. Some deep-rooted forbs flowered in great abundance, e.g. *Oenothera, Psoralea, Rosa*.

The drought continued until 1941, when rains and humid air returned. Overall, *S. scoparium* had suffered the most, with a 95% loss in some places. By 1941 it was ranked sixth in abundance in the prairie, mostly due to losses in 1934 with areas where it was formerly dominant invaded by *Hesperostipa spartea, Sporobolus heterolepis, Bouteloua gracilis,* and *Pascopyrum smithii*. Although reduced in abundance, *Andropogon gerardii* was never entirely lost because of its deep roots, and it came back vigorously in 1938 after moderate rainfall. The total basal area in the prairie was reduced to 1/3–1/2 of previous levels, with the lower layers of grass and forbs lost from the sward. There was a general destruction of the characteristic plant associations and a replacement with more xeric ones, especially a replacement of true prairie (*A. gerardii–Schizachryium scoparium* dominated) with mixed prairie (*P. smithi–B. gracillis–B. dactyloides* dominated) over 7 years (1934–1941) across a 160 km wide area of central Kansas, eastern Nebraska, and eastern South Dakota.

After the drought, there was a gradual recovery of the vegetation as weedy forbs, annual grasses, and weeds were replaced by perennial grasses. Seedling recruitment was observed to be very high when the rains returned, e.g. *Bouteloua gracilis* seedlings were very dense. The crowns and rhizomes of grasses that had not died were rejuvenated from long drought-induced deep dormancy and sprouted new shoots, despite their often dead roots. Bare soil was repopulated by 3–4 years after the end of the drought. Lowland communities recovered better and quicker than upland communities. Before the drought, three major communities of the true prairie were recognized, and two *Schizachryium scoparium* communities were the top-ranked community types. Afterwards, eight communities were recognized, including three that were newly developed: *Pascopyrum smithii, B. gracilis,* and mixed prairie types. The two pre-drought *Schizachryium scoparium* communities were modified into three (relic *Andropogon gerardii–S. scoparium, A. gerardii,* and mixed grasses), and the rank order of abundance of the communities had also changed (see Table 9.3). Some of the effects were long-lasting; even 12 years after the drought ended, mixed prairie still persisted in areas it had invaded.

The detailed studies of the US Midwest during the 1930s drought summarized above (and see Coupland 1958; Robertson 1939; Tomanek and Hulett 1970), along with similar records from drought in grassland elsewhere across the globe (e.g. Milton and Dean 2004; Walker *et al.* 1987) help us to understand the sensitivity of grasslands to climate variability. Grasslands are non-equilibrium systems in which periods of drought are the norm (Illius and O'Connor 1999). Current research is investigating the effects of potential climatic shifts, including changes in interannual variability in precipitation in the context of global environmental change (Fay *et al.* 2000, 2003). Under future global change scenarios, temporal variability in precipitation may be as important as mean precipitation in determining the structure and function of grasslands, although these effects are poorly known (Collins *et al.* 1998b). The episodic nature of precipitation in many semi-arid grasslands means that grasslands may exist in one or more alternate stable states determined by past precipitation history. Changes in precipitation patterns, either a decrease or an increase in precipitation frequency or amount, may be sufficient stimulus to alter the trajectory of ecosystem structure and function in seemingly irreversible

Table 9.3 Major communities of upland true prairie after the great drought of the 1930s

Community	Rank
Mixed prairie (N)	1
Pascopyrum smithii (N)	2
Hesperostipa spartea (M)	3
Relic *Andropogon gerardii–Schizachryium scoparium* (M)	4
Bouteloua gracilis (N)	5
Andropogon gerardii (M)	6
Mixed grasses (M)	7
Sporobolus heterolepis (M)	8

M, old or modified types; N, newly developed types.
Data from Weaver (1954).

ways (Knapp *et al.* 2002; Potts *et al.* 2006). Climatic variability, particularly periods of low or variable precipitation (i.e. drought), interact with fire and grazing, the other two main forcing variables in grassland, to determine ecosystem characteristics of grassland at all hierarchical levels, from effects on individual organisms through community composition to ecosystem function (e.g. see discussion of the transient maxima hypothesis, §7.1). The response of different grasslands to temporal patterns of precipitation varies across the globe, but comparative studies suggest a paramount role of precipitation availability in determining grassland structure and function despite sometimes different and divergent evolutionary histories (Knapp *et al.* 2006).

CHAPTER 10

Management and restoration

> Restoration provides...an acid test because each time we undertake restoration we are seeing whether, in the light of our knowledge, we can recreate ecosystems that function, and function properly.
>
> (Bradshaw 1987)

This chapter deals with applied grassland ecology and discusses means of managing grasslands for human needs and conservation. The discussion in §10.1 focuses on pastoralism as an ancient method of rangeland utilization, and the use of management tools such as grazing, fire, fertilizers, herbicides, and pesticides. Rangeland assessment (§10.2) is important for determining the suitability of grasslands for various uses and to allow the development of management plans. Where grassland has been destroyed or degraded, restoration is necessary and is described in §10.3 where the focus is on management and restoration of natural or semi-natural grassland. Management of amenity and turf grassland, e.g. golf courses and other sports facilities, parks, and roadsides, largely through mowing, weeding, irrigation and herbicide application, is outside the scope of this book and is covered elsewhere (Aldous 1999; Brown 2005; Rorison and Hunt 1980)

10.1 Management techniques and goals

Depending on the desired use, grasslands are managed for one of several goals, including production of forage for raising domestic and wild animals, production of biofuels, or conservation of natural areas. Grassland management and conservation objectives can be focused either on the system as a whole, on specific plant functional types (e.g. tussock grasses, Díaz *et al.* 2002), or for particular species (e.g. to promote rare species or species of high forage value, or reduce undesirable or invasive species). Many natural areas require management intervention because natural disturbance regimes are disrupted or no longer in place (Chapters 1 and 9).

10.1.1 Pastoralists and communal grazing

Pastoralism is the raising of livestock on natural pasture unimproved by human intervention (Salzman 2004). Although the grasses and shrubs in these pastures are not planted or tended by people, these rangelands are often the result of human intervention such as forest clearance and may be maintained by fire or grazing.

Grasslands and other ecosystems used by pastoralists include savannahs, shrublands, and woodlands, and were first used by hunter–gatherer societies. These hunter–gatherers depended entirely on the natural resources of their environment to provide for all their needs. About 11 000 years ago, subsistence pastoral systems became established as isolated groups of people started to domesticate animals and plants (Grice and Hodgkinson 2002). A **subsistence, nomadic,** or **transhumant pastoral** system is one in which the people subsist and centre their culture on livestock, which they move daily or seasonally across the landscape in search of fresh pasture and water. There are still subsistence pastoral societies today, although their numbers are increasingly threatened by commercial pastoralism (below), privatization and forced settlement, conventional agriculture, habitat fragmentation, and shrinking population numbers as young people move to urban centres in search of work. For example, the Quashqai people of Iran traditionally herded flocks of goats and sheep from summer highland pastures near Shiraz to lowland winter pastures near the Persian Gulf 480 km to the south. In recent times, many pastoral groups maintain their lifestyle through increased market involvement in response to adverse socio-economic and environmental conditions (e.g. desertification). For example, the Fulani people of West Africa are traditionally nomadic pastoralists herding cattle, goats, and sheep. In the Ferlo region of northern Senegal the Fulani have become semi-sedentary and diversified their activities following the drilling of boreholes to create wells since the 1950s and a series of severe droughts in the 1970s. Their activities now include small-scale cultivation and commercial pastoralism (below) with a change in the composition of herds from predominantly cattle to include more small ruminants such as sheep (Adriansen 2006). Small villages tend to develop around the wells, and this has led to agricultural encroachment from within, so to speak, although most Fulani still live in the bush. There is a fee for using the wells and the formerly communal Ferlo region is divided into resource management units called pastoral units (*unités pastorals*).

Commercial pastoralists began exploiting pastures during recent historical times. These pastoralists have a stronger reliance on goods and services from outside of the rangelands they occupy than subsistence pastoralists, although they are still reliant on domestic livestock. They market their products to outside peoples. Commercial pastoralists focus more on the buying and selling of livestock including the slaughtered products of meats and hides than subsistence pastoralists who rely more on live animal products, such as milk, hair, blood, manure, and livestock offspring. With both pastoralist strategies the people involved typically raise more than one type of livestock, e.g. the raising of large stock, such as cattle or camel, along with small stock such as sheep, goats, and donkeys in many African pastoral societies (Salzman 2004).

Although an ancient practice, pastoralism exists as a specialized system of production and land management adapted for marginal, usually arid, environments. Even today pastoralism is important as it is a consumption system supporting 100–200 × 10^6 people globally. It is an important part of the economy in several developing countries with a high percentage of grassland land cover, and accounts for 35% of the African agricultural gross domestic product (GDP) (Scoones 1996), reaching 78% of the agricultural GDP in Senegal, 80% in Sudan, and 84% in Niger (Hatfield and Davies 2006). Production per hectare from pastoral systems (in cash, energy, and protein terms) equals or exceeds returns from ranching in most comparisons (Scoones 1996). Pastoralism is an important and integrated part of the culture and economy of societies and indigenous people worldwide and makes efficient and productive use of marginal rangeland.

Subsistence pastoralists and early commercial pastoralists utilize **communal rangelands** that are managed through controlled access as open range in which individual free-ranging livestock are the property of several different owners. Communal rangelands are not individually owned and have the potential to suffer from a **'tragedy of the commons'** (Hardin 1968) in which each of the local livestock owners allows more than an equitable number of stock to graze the range, leading to overstocking, reduced forage availability, and, eventually, range degradation. Under this scenario, it is to the benefit of an individual herdsman to add livestock once the carrying capacity of the commons (pasture) is reached; even though adding more livestock causes a decline in yield, the loss is spread across all herdsmen. Even though the individuals who added the livestock causing the carrying capacity to be exceeded will gain, the livestock is not shared among the herders. For example, on communal rangeland in the Eastern Cape, South Africa, cattle grazed continuously at a very high stocking rate of <2 ha per animal unit (AU; see §10.1.2) compared to an adjacent intensively managed commercial farm grazed according to a short-duration, high intensity, multi-camp grazing regime of about 12 ha/AU (Fabricius *et al.* 2004). The communal rangeland was severely degraded, with low vegetation cover. In a semi-arid area of Namibia, there was severe bush encroachment and reduced perennial plant diversity on communal rangeland than on commercial rangeland (the latter with lower stocking rate and privately owned and fenced); however, in an arid area differences between these types of grazing strategy were minor, reflecting an over-riding effect of abiotic factors on environmental quality (Ward *et al.* 2000). Communal grassland in the South African veld had a higher proportion of short-lived, perennial grasses of poor grazing value than commercially grazed land (O'Connor *et al.* 2003). In contrast, grazed communal rangeland in Highland Sourveld grassland of South Africa supported a higher richness of graminoid species and total species richness than comparable lightly grazed conservation grassland, suggesting that the effects of communal grazing was no worse, and perhaps better, than the effects of open-range grazing in the conservation grassland (O'Connor 2005); hence in this situation, no tragedy of commons.

The 'tragedy of the commons' theory has been used to legitimize nationalization and takeover of traditional pastoral rangeland in many parts of Africa. The result has been an undermining of the customary and traditional land tenure regimes among herders, leading to many environmental, economic, institutional, and social problems (Lane and Moorehead 1996; Reid *et al.* 2005). It is argued that a more appropriate role of governments, development agencies, and other outside groups in pastoral societies is to strengthen traditional institutions through the provision of information and legal support, public interest services (health and education), an appropriate macro-economic framework, protection for weaker groups, and a minimum of technical support (to allow locally generated technical innovation) (Swift 1996).

10.1.2 Grazing

Grassland management using grazing for forage and livestock production requires maintenance of a stocking rate (production goal) sustainable for livestock (e.g. cattle, sheep, horses) in the particular environment and plant community. The land manager is seeking to maximize conversion of solar energy to marketable animal product while maintaining the grassland ecosystem. The benefit of grazing is that a material humans cannot eat (grass) is converted into a product they can eat—meat, blood, or milk. Part of the difficulty encountered by the land manager in grazing systems is that conversion of solar energy into primary production is frequently only about 1%, of which less than 20% of above-ground annual primary production is utilized by herbivores with only about 10% of ingested energy converted into animal gain (Briske and Heitschmidt 1991). Grasses co-evolved with grazers (see §2.4); nevertheless, overstocking leads to loss of plants when they are unable to replace the leaf area, meristems, and energy reserves that are removed through grazing. Semi-arid and arid grasslands are particularly susceptible as continual overstocking leads to a loss of production, land degradation and desertification. **Overgrazing** occurs when management decisions (e.g. overstocking or inappropriate seasons of grazing) limit livestock production per unit area because plants of high nutritive value become limited in the amount of solar energy that they can convert to primary production (Briske and Heitschmidt 1991). **Undergrazing** occurs when management decisions limit livestock production per unit land area because of under-utilization of species of high nutritive value. There is thus an 'ecological dilemma', that overstocking leads to degradation while undergrazing allows for higher levels of primary production that is unused. The herbivore optimization hypothesis (see §9.3.1) suggests that an optimum, intermediate level of grazing can maximize primary production and hence stocking rate. Any change in quantity and/ or quality of available forage that facilitates sustained livestock production is referred to as **range improvement** (Heitschmidt and Taylor 1991).

The Society for Range Management (1998) defines the following terms:

- **Animal unit** (AU): One 1000 lb (450 kg) cow, either dry or with calf up to 6 months old, or the equivalent, consuming about 26 lb (12 kg) of forage on an oven-dry basis. Other classes and kinds of animals can be related to this standard, e.g. a bull 1.25 AU, a horse 1.2 AU, a sheep 0.20 AU.
- **Animal unit month** (AUM): The amount of oven-dry forage (forage demand) required by 1 AU for a standardized period of 30 animal-unit-days.
- **Stocking density** (SD): The relationship between number of animals and the specific unit of land being grazed at any one point in time. For example, 15 cows on 10 ha = 1.5 AU ha^{-1}.

- **Stocking rate** (SR): The relationship between number of animals and the grazing management unit utilized over a specified time period. SR may be expressed as animal units/unit area of land (i.e. SD) integrated over time, expressed as AUM per unit area.

On this basis, and with knowledge of rangeland carrying capacity, seasonal tolerances of plants to grazing, and condition (see §10.2), estimates of optimum stocking rates and stocking seasons can be obtained. Individual animal weight gain declines with increased stocking rate as competition for patchily distributed nutritious forage increases. Similarly, individual animal gain increases as stocking rate declines. The **critical stocking rate** is the stocking rate at which the decline in individual animal performance begins. Seasonal, year-to-year, and spatial variation among forage species and among sites means that range carrying capacity and condition can be highly variable complicating the accuracy of estimating stocking rates.

A **grazing system** is considered as 'a specialization of grazing management which defines the periods of grazing and non-grazing'(Society for Range Management 1998). Non-grazing here refers to deferment of pastures or management units from grazing. There are at least six basic types of grazing system (Heitschmidt and Taylor 1991; Tainton 1981b):

- continuous (season-long and dormant season) grazing
- deferred rotation (DR)
- rest rotation (RR)
- high intensity–low frequency (HILF)
- short duration (SD)
- zero grazing.

In continuous (season-long) grazing, livestock are allowed access to the range when the forage is ready to graze at the start of the season, and remain on the range until the end of the growing season. Stocking density may be varied through the season and individuals may be removed and replaced from time to time. True continuous grazing never removes the livestock from the land regardless of growing season, whereas season-long grazing occurs only during the growing season. Dormant season grazing is a form of continuous grazing that occurs in locations where the herbaceous vegetation becomes dormant and grazing only occurs during this dormant period.

Deferred, rest, high intensity-low frequency, and short duration are all forms of rotational grazing. These grazing systems include a period of deferment (also termed rest or absence by some authors) during which animals are taken off the range sometime during the season to allow the plants to recover within the current growing season or during the following one (Fig. 10.1). The goal of altering or rotating grazing periods with rest periods is to enhance livestock production by improving or stabilizing the quantity or quality of forage production. Rotational grazing systems allow for a greater control of grazing frequency and intensity than continuous or season-long grazing. Zero grazing is a grazing system where forage is cut and fed green to animals that do not have direct access to grazing that part of the range.

With a rotational grazing system, stocking density can be manipulated to achieve either high-utilization grazing (HUG) or high-production (or performance) grazing (HPG). With a HUG strategy, stocking density is high enough to ensure that all plants are moderately to intensively defoliated during the grazing period, whereas with a HPG strategy stocking density is low enough so that only the most nutritious (preferred) plants are defoliated and then only at light to moderate intensities (Heitschmidt and Taylor 1991). In arid regions HPG is preferred, to allow time for sward recovery under conditions that are likely to include drought (Tainton 1981b). By contrast, HUG is argued to be more appropriate for fire/grazing adapted grasslands of humid and productive regions with infrequent summer drought where sward recovery is assured between successive grazing periods.

Various grazers show distinct preferences for different types of forage. Generally, cattle and bison prefer grass (bison exclusively so), sheep prefer grass and forbs, goats prefer browse and grass. These differences in diet preference reflect variations in forage quality among plants (§4.2) and the manner in which these animals feed. **Bulk or roughage grazers** include cattle, bison, and cape buffalo, which graze by wrapping their tongue

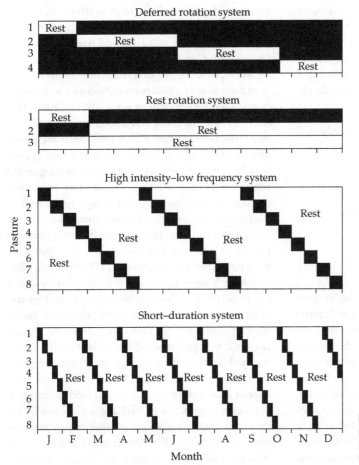

Figure 10.1 Conceptual model showing graze–rest periods for four types of grazing systems for livestock production. Reproduced with permission from Heitschmidt and Taylor (1991).

around clumps of plants, and breaking the foliage on the clump loose with a short jerking motion of the head, before drawing it into their mouth (Huston and Pinchak 1991). By contrast, **concentrate selectors** including deer and dik-dik have soft muzzles, agile tongues, and pliable, often split lips, allowing them to carefully select plants or plant parts (e.g. single leaves, leaf tips, fruits, seeds) that are high in protein and other soluble fractions and low in cell-wall fractions. **Intermediate feeders** such as goats and, to a lesser extent, domestic sheep, have a variable diet characterized by frequent compositional changes in forage. As a result of the ability of different animals to use different parts of the range, mixed grazing systems are utilized in many parts of the world.

The somewhat formulaic system of range management described above represents the common traditional approach in the rangelands of the USA, Australia, and New Zealand, and the veld of South Africa (Stoddart *et al.* 1975; Tainton 1981b) It is now being challenged as non-equilibrium based models are increasingly utilized (see §10.2). These systems also have limited utility for the communally managed rangelands common in Africa and Asia (§10.1.1).

10.1.3 Fire

Fire is a natural phenomenon in most grasslands (Chapter 9). Hence, all native species are capable of surviving certain fire regimes—some better than others. Fire in the form of prescribed burning is also a powerful tool in grassland management (Vogl 1979). Specific objectives are necessary to use prescribed fire effectively. These objectives depend on the grassland type, its condition, and management objectives, and usually include the following:

- control of woody plants and noxious weeds
- stimulation of herbage/forage growth and reproduction (including seed yield)
- soil exposure and seedbed preparation
- removal of accumulated inedible forage
- wildfire hazard reduction or elimination
- habitat conservation
- maintenance of ecological diversity.

For example, conservation of the hummock (spinifex) grassland (dominants include *Triodia* and *Plectrachne* spp., see Plate 9) habitat of the western hare-wallaby (*Lagorchestes hirsutus*) in the Tanami desert of central Australia is best obtained through periodic burning of vegetation, which favours a mosaic vegetation structure (Hodgkinson *et al.* 1984; Orr and Holmes 1984). More recently, an additional benefit of prescribed burning of grassland is the control of certain invasive exotic species, e.g. yellow starthistle *Centaurea solstitialis* L., in California grassland (DiTomaso *et al.* 1999). Some invasives can be effectively controlled through the use of fire, whereas others are adapted to fire (e.g. *Rosa bracteata*), and some enhance fire intensity and frequency (e.g. *Bromus tectorum*) (Brooks and Pyke 2001; Grace *et al.* 2001b). Care has to be taken, as some fire prescriptions can enhance the invasiveness of exotic species and firebreaks can act as dispersal corridors (Keeley 2006).

Prescribed fire was used extensively by native peoples in many arid and semi-arid grasslands the world over (e.g. in Australian hummock grasslands and savannahs, and in the US Great Plains). In South Africa, early Portuguese explorers called the county 'Terra de Fume' because of the pall of smoke frequently encountered through the interior of the country (Tainton 1981a). In the North American prairies, European settlement and a general aversion to burning disrupted the existing fire regimes leading in many areas to invasion by woody species, growth of forage of low nutritive value, or both. As European settlers started to use these areas as rangeland for raising cattle and sheep in the 1800s, most viewed fire as a destructive element and suppressed it where possible. However, in some regions fire was used to remove or reduce woody plant competition and to increase herbaceous forage. It was not until the 1960s that fire was widely recognized as a useful management tool (Wright and Bailey 1982). Burning annually or every few years in some regions is now widely incorporated as part of the management regime (Anderson 2006). Traditional grassland management of upland and rough grassland in Europe also frequently includes annual burning to reduce accumulated litter or mulch, particularly in moor and heathland areas and in areas that have been ungrazed or undergrazed (Duffey *et al.* 1974). When such areas have been regularly burned they are described as having been 'swaled'. The use of prescribed burning for management of South African veld has been controversial as the various vegetation types react differently to fire. While the fynbos, fire climax grasslands, and savannah (bushveld) require frequent fire, the karroo areas appear less well adapted; prescribed burning is not a general practice in such areas (Tainton 1981c). Prescribed burning is rare in other grasslands too, such as the Mitchell (*Astrebla* spp.) grasslands of Australia (Plate 8), because in these grasslands fire generally reduces forage yields and control of woody regrowth is usually unnecessary (Orr and Holmes 1984).

The timing of burning is important, as it affects which species are affected by the fire. Species may be susceptible to burning in one season but adapted to being burnt in another. In England the prescribed burning of rough grassland in the Cotswolds and in the Peak District of Derbyshire is usually in early February or early March before growth of the desired species and when the litter is usually dry and able to carry the fire (Duffey *et al.* 1974). Similarly, prescribed burning of North American tallgrass prairie is usually in early spring (late March–early May) before growth of the desired warm-season grasses has commenced but after the first growth of the cool season species. In these grasslands, species such as *Poa pratensis* (a cool-season grass) are not able to recover from a spring fire, many forbs have not yet started to grow above ground or are able to resprout, and the late spring emergence and growth of the warm-season grasses is encouraged by the warm soil and open, litter-free conditions that follow fire (Chapter 9 and Gibson and Hulbert 1987; Gibson et al. 1993). Although growing-season summer burns may enhance diversity and control woody species and cool-season invaders, insufficient research has been conducted to determine whether or not fires at this time of the season should be part of a recommended fire management regime (Anderson 2005). Late-season autumn fires in the North American prairies may be difficult to set because of the cool temperatures, may be difficult to control and of high intensity because of the high flammability of the senescing plant material, and can destroy winter food and shelter for wildlife (Pauly 2005). Nevertheless, autumn burns present less risk to breeding and nesting animals than spring burns. In Illinois, most of the fires set by Native Americans occurred in the autumn (Roger Anderson, personal communication).

Conducting a prescribed burn
Competent prescribed burns require a great deal of experience, preparation, organization and planning, appropriate equipment and personnel, and possession of an appropriate qualification or licence from a fire-school (Fig. 10.2). In many areas, submission and approval of a burn plan is required. For example, the US-based Nature Conservancy which is involved with prescribed burning >101 000 ha yr^{-1} requires a Site Fire Management Plan that outlines the ecological and technical information needed to justify a prescribed burn and a Prescribed Burn Unit plan, a field document for each burn that is being planned (Seamon 2007). Burn plans should

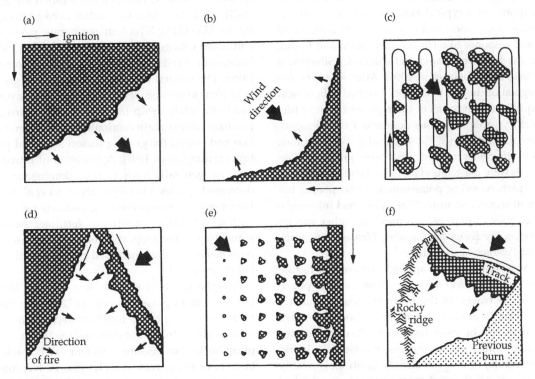

Figure 10.2 Methods used to ignite and control prescribed fire: (a), headfire; (b), backfire; (c), spot ground ignition; (d), drawing fires together; (e), grid aerial ignition; (f), mosaic burning using natural firebreaks. Reproduced with permission from Hodgkinson et al. (1984). © CSIRO 1984.

include a detailed description of the area to be burned (vegetation and fuel characteristics), management goals and objectives, fuel and weather prescriptions, fire and smoke behaviour objectives, crew and equipment lists, maps, post-burn activities, and emergency plans.

10.1.4 Fertilizers

The application of fertilizer is common in semi-natural and managed pastures and is undertaken to raise production and increase representation of the most productive and palatable species, and decrease the abundance of less desirable species ('weeds'). Such an 'improvement' in the production and composition of grassland is referred to as **agricultural intensification**. Before the advent of inorganic fertilizers, traditional methods applied farmyard manure, fish guano, potash, or lime (the latter to control soil acidity). Indeed, the long-running Park Grass Experiment (PGE) in England (see §6.2.2) was established in 1856 to assess the response of a typical meadow to traditional fertilizer application and haying. The botanical and yield response to fertilizer application and liming in the PGE were rapid and dramatic (Silvertown 1980a; Silvertown et al. 2006). After 40 years, the vegetation in the PGE plots reached a botanical equilibrium in terms of the representation of life-history groups: grasses dominated the nitrogen-fertilized plots (c.90%, especially *Festuca rubra*, *Agrostis capillaris*, and *Alopecurus pratensis*), legumes were most abundant (30%, *Lotus corniculatus*) on plots receiving potassium and phosphorus but no nitrogen, and unfertilized plots had intermediate ratios of grasses, legumes, and other species (especially *Leontodon hispidus*, *Plantago lanceolata*, and *Poterium sanguisorba*).

Most fertilizers are applied to correct or reduce deficiencies in soil nitrogen, phosphorus, or potassium. Nitrogen is the nutrient most frequently limiting in grassland and is the most important constituent of most fertilizers. Phosphorus can also be limiting in some grasslands (see §7.2.3), and potassium, even if not already limiting, is readily leached. All three of these nutrients can be lost from grassland systems as a result of vegetation removal by grazing. Plant response to fertilization is dependent on weather conditions, soil fertility, structure, temperature and moisture, botanical species composition of the sward, and management (e.g. whether the sward is cut or grazed). As an example of agricultural application rates, grazed grasslands in Northern Ireland received $109 \pm 3.1\,kg\,N\,ha^{-1}$, $9.0 \pm 0.3\,kg\,P\,ha^{-1}$, and $21 \pm 0.7\,kg\,P\,ha^{-1}$ in 2000, with the largest applications being in dairy farms compared with cattle, sheep, or tillage systems (Coulter et al. 2002).

Commonly, the percentage of nitrogen in fertilizer for various chemical formulations is anhydrous ammonia (82%), urea (46%), NH_4NO_3 (33.5%), $(NH_4)_2SO_4$ (21%), NO_3^- solutions (28–32%) (Barker and Collins 2003). Ammonia nitrogen in fertilizer increases soil acidity and can require liming to correct the pH (as it does in the PGE experiment discussed above). The presence of legumes in a grassland sward increases soil nitrogen because of nitrogen fixation effectively fertilizing the grassland by as much as $44\,g\,N\,m^2\,yr^{-1}$ (see §7.2.2). Yield increases following nitrogen fertilization are generally 4.5–23 kg forage dry matter per kg nitrogen up to 375–$452\,kg\,N\,ha^{-1}$ in temperate pastures; additional nitrogen application can decrease yield because of stand thinning and increased competition. The general form of the response of grassland yield to nitrogen fertilizer is that of an inverse quadratic relationship (Morrison 1987). Nitrogen fertilizer can maintain continuity of forage production and extend the grazing season in grazed pastures (Humphreys 1997). Application of nitrogen fertilizer generally leads to grass dominance and decreased species diversity (Titlyanova et al. 1990). For example, temperate-zone grasslands fertilized with 40–$650\,kg\,ha^{-1}$ of nitrogen fertilizer showed a doubling in the proportion of some grasses (e.g. *Arrhenathrum elatius* increased from 17 to 35% in Polish grassland).

Phosphorus is the second most common limiting nutrient in grassland soils (§7.2.3). The need to provide phosphorus fertilizer depends on the underlying bedrock, but even soils with adequate phosphorus can become deficient after grazing, haying, or silage removal. Phosphorus fertilizers are of two main kinds: water-soluble superphosphate (usually monocalcium phosphate $Ca(H_2PO_4)$) or, less commonly, water-insoluble rock phosphate (fluorapatite $Ca_{10}(PO_4)_6F_2$ or hydroxyapatite $Ca_{10}(PO_4)_6(OH)_2$). Commercial phosphorus fertilizers of different formulations include the following amounts of P_2O_5: superphosphate (20%), triple superphosphate (45%), rock phosphate (53%). Other sources include diammonium phosphate (53% P_2O_5, 21% nitrogen), monoammonium phosphate (48% P_2O_5, 11% nitrogen), and ammonium phosphate sulfate (20% P_2O_5, 16% nitrogen). The objective of adding phosphorus fertilizers is usually to raise the phosphorus content of the soil to enhance yield. In South African pastures, for example, raising the phosphorus content of clayey or sandy soils by $1\,\mu g\,g^{-1}$ requires respectively 5.0 or $6.5\,kg\,ha^{-1}$ of superphosphate (de V. Booysen 1981). Grasslands in which the herbage has been regularly cut and removed are generally the most responsive to phosphorus fertilizer application, as are slightly acid or neutral soils where the added phosphorus can remain soluble (Whitehead 1966).

Potassium is usually applied as potash (originally ash from coal furnaces) which is potassium carbonate (K_2CO_3), or sometimes as potassium chloride (KCl, muriate of potash, 60% K_2O) or potassium oxide (K_2O). Since the concentration of potassium in plant tissues is comparatively high (see Chapter 4), the loss of potassium from grassland after grazing, haying, silage making, or leaching can be higher than that of other cations. Rates of potassium application in intensively managed grassland can be in excess of $400\,kg\,K\,ha^{-1}\,yr^{-1}$ (Whitehead 1966); however, most of the applied potassium is either taken up directly by plants or lost through leaching. In South African pastures, for example, $2\,kg\,ha^{-1}$ of potash are required to raise the soil potassium content by $1\,\mu g\,g^{-1}$ for obtaining adequate yields (de V. Booysen 1981).

Other nutrients are included in fertilizer applications when necessary, for example calcium as limestone to reduce soil acidity, and magnesium in dolomitic limestone or fertilizers such as magnesium sulfate or potash-magnesia.

10.1.5 Mowing and haying

Mowing is a common management practice in many grasslands, especially in European meadows mown for hay, and of course in many grasslands where it may be the only practical management approach in the absence of grazers or fire. Mowing is used to encourage growth of grasses, reduce the growth of forbs, keep out woody plants, maintain swards at an acceptable height, and provide hay. The frequency and height of mowing is based on the specific management objectives. Mowing is not a surrogate for grazing as it is non-selective, with everything above the height of the mowing blade or scythe being severed. Mowing can also lead to significant nutrient loss from the grassland if the clippings are removed, as they are in haying or silage making.

In western Europe, mowing meadows is a traditional method of grassland management, with written records dating back to a grant by King Hlothere of Kent (England) dated 679 AD for an estate with meadows. Before that, the Romans had scythes and probably used them on meadows. The importance of meadows is embodied in many place names, especially those derived or including the Old English *mæd* derived from the verb *māwan*, to mow, e.g. Runnymede (Rackham 1986).

Haying involves cutting the meadow at 5–10 cm and raking the crop into swaths or windrows to cure before baling or stacking. Alternatively, the cut forage is dispersed across the entire field for curing, a process known as **tedding**. The goal is to minimize dry matter losses and reduced protein digestibility in the hay by maintaining moisture levels low enough to avoid microbial growth and heating (Collins and Owens 2003).

Silage is cut forage preserved, sometimes for months or years, in anaerobic conditions which encourage fermentation of sugars to organic acids. Like hay, cut silage may be windrowed in the field and allowed to wilt before harvesting.

Much like the effects of disturbance (Chapter 9), mowing affects the botanical composition of grassland through alterations to the competitive environment of the plants (Fig. 10.3). As noted above, mowing favours grasses in particular and keeps woody plants in check. Frequent mowing, with its repeated defoliation, produces a low, high-light canopy allowing the growth and spread of low, prostrate-growing forbs. Species able to complete flowering and seed dispersal before an annual mowing or hay cutting are also favoured. As a

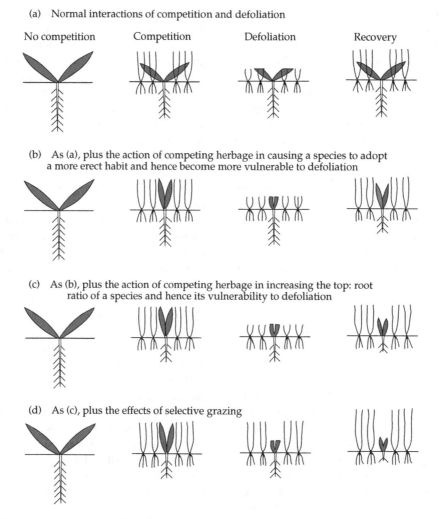

Figure 10.3 Diagrammatic representation of the effects of defoliation by mowing or selective grazing on competitive relationships in a pasture. Redrawn with permission from Norman (1960).

result, hay meadows are often sites of high botanical diversity and high conservation value. For example, upland hay meadows in the UK conforming to the Mesotrophic grassland *Anthoxanthum odoratum–Geranium sylvaticum* vegetation type are species-rich, hosting several rare and scarce species such as the endangered *Alchemilla monticola, A. subcrenata, A. wichurae,* and *Crepis molis.* Traditionally, these hay meadows were managed by applying lime and farmyard manure to offset the annual mid-July mowing and removal of hay (Jefferson 2005). Sheep are grazed on the meadows in the early spring and sheep or cattle are grazed late in the summer. More recently, these meadows have been under threat and have suffered species loss following agricultural improvement through application of inorganic fertilizers and herbicides, drainage, ploughing and reseeding to low-diversity permanent pasture for silage production.

Although not a part of the natural disturbance regime of grasslands, mowing has become an important management tool to control woody plant encroachment in many natural grasslands. For example, in North American prairie, mowing followed by hay removal is a common management practice. In this system, spring mowing increases the abundance of C_4 grasses, while summer mowing increases C_3 species abundance (Hover and Bragg 1981). An increase in the abundance of exotic species often occurs in mowed grassland, e.g. the exotic grasses *Bromus inermis* and *Bothriochloa bladhii* increased in mowed compared with unmowed tallgrass prairie on Konza Prairie, Kansas, USA (Gibson et al. 1993); see also Lunt (1991) for a comparable Australian example. Conversely, mowing can be used to control invasives in some grassland. For example, mowing and biomass removal over 5 years shifted a coastal prairie in California dominated by exotic annual grass to a mixed exotic/native forb assemblage (Maron and Jefferies 1991). Similarly, annual mowing was found to be an effective management prescription to reduce the abundance of exotic perennial grasses (e.g. *Arrhenatherum elatius*) in a native *Danthonia californica–Festuca roemeri* prairie in western Oregon, USA (Wilson and Clark 2001).

Mowing can affect the abundance and performance of endangered species. For example, for *Asclepias meadii* (Mead's milkweed, a plant on the US federal list of threatened species) the density of vegetative ramets was higher but its reproduction was lower in annually mowed meadows than in former meadows (Bowles et al. 1998). Mowing can be used as a management treatment to keep open grassland habitat, thereby helping to maintain the habitat necessary for rare grassland species. In these cases, mowing should be performed after the species of concern has reproduced and shed its seed (Eisto et al. 2000).

10.1.6 Herbicides

Herbicides can be applied to control invasion of exotics or other undesirable plants, although as a management tool they have to be used with care to avoid killing plants of conservation value (Solecki 2005). For example, the herbicides atrazine and 2,4-D have been used to maintain grass dominance by reducing forbs and cool-season grasses, especially annuals, and selectively controlling the abundance of *Poa pratensis* and *Bromus inermis*, in established tallgrass prairie (Engle et al. 1993; Gillen et al. 1987; Mitchell et al. 1996). Picloram, clopyralid, and a mixture of clopyralid and 2,4-D were found to be effective in controlling the exotic *Centaurea maculosa* in *Festuca altaica–F. idahoensis–Pseudoroegneria spicata* grassland in Montana, USA (Rice et al. 1997). Few and only transitory negative effects on diversity of the grassland were observed after herbicide application, and the herbicides had high efficacy on the target plant. The effectiveness of herbicides depends on the application rate, and season and timing of application in relation to other management treatments such as burning and fertilizers. Some herbicides are relatively selective (e.g. 2,4-D targets broadleaved plants) whereas others are non-selective and kill most plants (e.g. the glyphosates). Some herbicides, such as atrazine, remain active in the soil for some time, but others, such as glyphosate, are rapidly inactivated in the soil and do not constitute a long-term problem. Gene flow through pollen and seed dispersal from herbicide-resistant biotypes from agriculture or cultivation into natural habitats has recently been documented in *Agrostis stolonifera* and is a growing conservation concern (Zapiola et al. 2008).

10.2 Range assessment

It is important to determine the current status of grassland as a basis for developing management decisions. This need was first recognized in the 1900s in US rangeland where estimates of stocking rate were required. An early survey method to estimate forage availability was developed by James Jardine in 1910, based on plant cover estimates (National Research Council 1994). Jardine's method, and others like it at the time, were largely reconnaissance surveys (Humphrey 1947). In this section, the approaches that have been used in the past to assess rangeland (§10.2.1) are described along with modern ideas to include non-equilibrium concepts into rangeland assessment (§10.2.2), and methodology for assessing rangeland health (§10.2.3).

Although the search for a single, universal method of range assessment may be appealing, rangelands are diverse and have intricate and highly dynamic interactions between biotic and abiotic components. Ultimately, rangeland assessment should consider the management objectives and be tailored to reliably measure change while recognizing the ecological, cultural, economic, and political factors that impact rangelands (Friedel *et al.* 2000).

10.2.1 The range condition concept

Formalized by E.J. Dyksterhuis (1949), **range condition analysis** (also called **range ecological condition**) relates the current state of the vegetation to a perceived, historical climax (natural potential) community and to the plant's responses to herbivory for the site (Society for Range Management 1998). Range condition analysis is based on the successional theory of Clements which supposes that plant succession is an orderly, predictable process with vegetation composition moving unidirectionally to a theoretical endpoint, the **climax community** (Clements 1936). Knowledge of range condition allows the land manager to adjust stocking rates and develop range management plans. Assumptions of this approach include the following: the condition of the vegetation is a good predictor of animal production; that more intense grazing pressure will cause the site to regress to an earlier, lower-quality, successional stage; and those sites close to or at the climax condition are the most productive and suitable for livestock.

Range condition analysis was used extensively by the US Department of Agriculture (USDA) Natural Resources Conservation Service (NRCS; Soil Conservation Service (SCS) until 1994) from the 1940s to the 1980s as the preferred method of assessing the status of rangeland in the USA (Joyce 1993; Pendleton 1989). Calculating the similarity index of a site in comparison to the historic climax plant community is still currently a part of the NRCS basic policy and procedures for resource conservation on non-federal grazing lands (Grazing Lands Technology Institute 2003) even though the use of state-and-transition dynamics (see §10.2.2) is a part of describing rangeland ecological sites (Stringham *et al.* 2003). Other federal agencies have used different methods of assessing range condition (reviewed in West 2003). From the 1950s, the Bureau of Land Management (BLM) used the Deming Two-Phase Method to place sites into one of five range condition classes based on the estimated condition of the forage stand and the site-soil mantle (Wagner 1989). In 1977 the BLM modified the SCS range condition approach with the Soil-Vegetation Inventory Method (SVIM) which included more intense sampling, stratification of areas, data documentation, and data computation methodology. However, this approach was short-lived, largely because of high survey costs and procedural shortcomings (West 2003). In 1982, the BLM adopted the Range Site Inventory procedure in which ecological status ratings based on the percentage composition of the plant community at a site were compared with the potential vegetation of the site taking into account associated soil and climatic data. Current BLM practice is to determine rangeland health (see §10.2.3) rather than range condition (Pellant *et al.* 2005). The US Forest Service assigns ecological status ratings based on estimated departure of the present community from the potential natural community in terms of the forage condition of the range (Moir 1989). As noted below, range condition is a specific assessment of rangeland that is not necessarily correlated with rangeland health.

Calculating range condition
Quantities of the species present as a percentage of those expected in the climax vegetation are used to describe range condition according to four range condition classes that represent stages in secondary succession towards, or regression from, the presumed historic climax community (excellent condition: 75–100%, good: 50–75%, fair: 25–50%, poor 0–25%). Stocking density can then be calculated on the basis of the range condition class of a site (Stoddart *et al.* 1975). To determine range condition class, plant species abundance is determined using ocular estimates of plant abundance or direct measurements of forage volume or biomass ('t Mannetje and Jones 2000) with individual species being characterized in response to grazing as increasers, decreasers or invaders as defined by the Society for Range Management (1998):

- **increaser:** for a given plant community, those species that increase in amount as a result of specific abiotic/biotic influence or management practice

- **decreaser:** for a given plant community, those species that decrease in amount as a result of specific abiotic/biotic influence or management practice
- **invader:** plant species that were absent in undisturbed portions of the original vegetation of a specific range site and will invade or increase following disturbance or continued heavy grazing.

The range condition for a site is calculated by determining a similarity index as the sum of the climax portion for each species or type of species present (Table 10.1, Fig. 10.4). To do so, the observed proportion of each species is compared with the expected proportion under undisturbed climax conditions. For example, if a decreaser is expected to contribute 80% to the climax composition and 10% is observed, the full observed amount (10%) is retained as the contribution of that species in calculating the range condition index for a site. For increasers, no more than the expected climax proportion is retained in calculating range condition. For example, if a particular increaser is expected to contribute 5% but 10% is observed, only the expected 5% is retained. Invaders are not expected to be present in the climax condition and so none of their observed cover is retained. A completed range condition worksheet for a clay upland pasture in Kansas is shown in Fig. 10.5.

Rangeland trend is the direction of change in condition or state of rangeland (Friedel et al. 2000) and is measured by comparing condition changes over time following a monitoring programme, often using series of permanent evaluation sites. The goal of determining rangeland trend is often to monitor vegetation changes under different stocking regimes.

10.2.2 State-and-transition models

The incorporation of non-equilibrium viewpoints in ecological thinking represented a shift in understanding (Blumler 1996; Briske et al. 2003). Widespread ecological views that were held in the past included the plant community concept, succession as a ubiquitous and linear progress to a climax community, the existence of a 'balance of nature,' the nature as undisturbed vs human disturbed contrast, and the view that human impacts on nature were linear and degradational. The range condition concept based on the Clementsian equilibrium view was criticized on the following grounds (Risser 1989): (1) the ambiguity of establishing climax vegetation, (2) observations of unknown successional or retrogressional stages, (3) the occurrence of prolonged subclimax stages and differing communities of supposedly similar

Table 10.1 Calculation of range condition from composition of the cover (%) from a pasture near San Angelo, Texas, USA. Calculated similarity index = 30, giving a range condition class of Fair (see Figs 10.4 and 10.5, and see text for details).

Species or group	Contribution to climax composition[a]	Observed cover	Climax portion retained for calculation
Bouteloua curtipendula (D)	80	10	10
Aristida purpurea (I)	5	10	5
Bouteloua rigidiseta (I)	5	5	5
Forb increasers	10	5	5
Woody increasers	5	20	5
Erioneuron pilosum (invader)	0	15	0
Annuals (invaders)	0	35	0
Total		100	Similarity index = 30

[a] Maximum allowable composition under undisturbed conditions.
D, decreaser; I, increaser.
With permission from Dyksterhuis (1949).

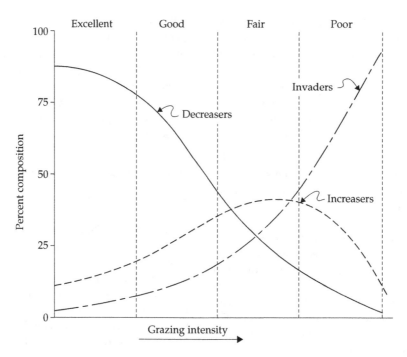

Figure 10.4 Relationship between relative proportion of decreasers, increasers, and invaders and range condition under increasing levels of grazing intensity. With permission from Stoddart et al. (1975). © McGraw-Hill Companies, Inc.

successional stage, (4) changes in range condition due to climate independent of management practices, (5) the realization that climax vegetation may not be the most productive condition for some sites, (6) herbaceous climax vegetation in forested rangeland may not be the most desirable, (7) severely disturbed sites may not be able to support climax vegetation, (8) some introduced species may actually be desirable forage species, (9) soil conditions are not included in the rating scheme, (10) species composition does not accurately predict animal performance, and (11) detecting vegetation change is often logistically difficult and time consuming.

Modern views are contrary to those of the range condition concept and are characterized by a non-equilibrium perspective, with nature seen as chaotic subject to pervasive disturbance (Chapters 6 and 9). Succession represents a shift in species groups as species respond individualistically to changing biotic and abiotic conditions through time.

It became apparent that the range condition approach (as employed by the Soil Conservation Service) and the ecological status approach (employed by the Bureau of Land Management and the Forest Service) were not underpinned by modern ecological theory (Risser 1989). As a result of changing ecological perspectives, these approaches have been phased out of in favour of measuring rangeland health (see §10.2.3) with the acceptance of state-and-transition models and ecological thresholds as underlying theory (National Research Council 1994).

State-and-transition models proposed by Westoby et al. (1989) exemplify the viewpoint that vegetation change in rangelands is not arranged along a single, linear grazing intensity to climax axis. Rather, there is a broad spectrum of multivariate vegetation dynamics in which several ecological processes are important including episodic drought, favourable precipitation, fire regimes, and soil erosion as well as grazing. Multiple stable states are thus possible in response to the processes operating at a particular site (Fig. 10.6). A 'state' is a recognizable, resistant, and resilient complex of the interaction

USDA–SCS KS–ECS–11
WORKSHEET FOR DETERMINING RANGE CONDITION 3/83
AND
EVALUATING FORAGE PREFERENCES AND USE

Range Site _Clay upland_ Location _No. 1 of 4_
Pasture No. _____ Name _DC POl of EC/ARG_
Date _6/1/87_ Conservationist _J.F.C_

A. Species Potential & Presetn	B. % in Climax	C. Present Composition lbs.	C. Present Composition %	D. % Climax Present	E. 1/ Interpretation – Plant Pregerence or Use By: Kind of Animal			
XXXXXXXXXXXXXXXXXXXXXXXXX	XXXXX	XXXXXXX	XXXXXXX	XXXXXX				
Andropogon gerardii	40		65	40				
Schizachyrium scoparium	25		T	T				
Sorghastrum nutans	30 / 15		2	2				
Panicum virgatum	15		5	5				
Sporobolus asper	5		12	5				
Panicum scribnerianum			1					
Bouteloua dactyloides			1					
Pascopyrum smithii	5		2	5				
Koeleria macranta			T					
Carex sp.			1					
Trifolium repens			1					
Ambrosia psilostachya			1					
Tragopogon dubius			T					
Symphyotrichum ericoides			1					
Achillea millefolium var. occidentalis			T					
Dalea multiflora			T					
Artemesia ludoviciana			T					
Asclepias sp.			T					
Dalea purpurea			T					
Psoralidium tenuiflorum			2					
Erigeron strigosus			T					
Oenothera speciosa			T					
Vernonia baldwinii			T					
Amorpha canescens	T		T					
TOTAL				57				

F. Range Trend _0_
  Up Static Down
  +   0   –
G. Canopy % _99_

H. Est. Initial Stocking Rate

Range Condition
Range Condition is the % of Present Plants that are Climax – Excellent 76–100; Good 51–75; Fair 26–50; Poor 0–25.

I. TOTAL ESTIMATED YIELD IN CLIMAX _____

1/ Interpretations:
  * – Key Grazing Plant
  H – High Grazing Preference
  M – Medium Grazing Preference
  L – Low or No Grazing Preference
  C – Used for Cover
  F – Used for Food
  N – Used for Nesting

Figure 10.5 Completed United States Department of Agriculture worksheet showing range condition for a clay upland pasture in Kansas tallgrass prairie.

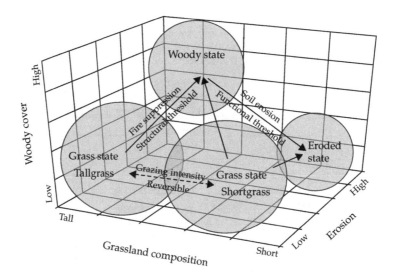

Figure 10.6 Conceptual model of the multiple stable state concept illustrating stable states (spheres) and transitions (arrows) that may occur on an individual site. The three-dimensional model illustrates the alternative states that the site can support through time. The Range Condition Concept model is represented by the single grazing intensity transition. Non-reversible transitions are depicted by solid arrows, reversible transitions by dashed arrows. Reproduced with permission from Briske et al. (2005).

between the soil base and vegetation components, albeit encompassing a certain amount of variation in time and space (Stringham et al. 2003; Westoby et al. 1989). Multiple steady states exist for many rangeland vegetation types and they may not respond to changes in grazing regime, or even removal of grazing (Laycock 1991). 'Thresholds' are boundaries in space and time among the multiple stable communities that can occupy a site and can be categorized as pattern, process, or degradation thresholds that themselves can be used to define two types of classification (prevention or restoration) thresholds depending on the emphasis on different ecosystem indicators (Bestelmeyer 2006). **Pattern thresholds** reflect changes in the distribution of features of the rangeland such as grass cover or bare patch site, **process thresholds** emphasize changes in the rates of certain processes such as erosion or dispersal (that can drive patterns), and **degradation thresholds** identify when a habitat becomes unsuitable for the species that characterize a particular state.

Negative feedbacks enforce ecological stability, reinforcing resilience and recovery following disturbance (Briske et al. 2005, 2006). Positive feedbacks degrade ecosystem resilience, and promote conversion to alternate stable states (Briske et al. 2006). The switch from negative to positive feedbacks contributes to occurrence of a threshold (Fig. 10.7). 'Transitions' between states are trajectories between states and are either short-lived or persistent (Stringham et al. 2003). A site can persist in a certain state pending the necessary conditions allowing it to transition a threshold to another state. The 'ball-and-cup' analogy is used to illustrate state-and-transition dynamics with the ball representing the system or plant community and the cup the present stable state. A large amount of disturbance is required to move the system out of the cup or trough over a threshold to one stable state to another (George et al. 1992).

Recognition of the state-and-threshold model necessitated alternative procedures to evaluate rangeland and a need to interpret continuous reversible and discontinuous non-reversible vegetation dynamics. Managers may manage for a desired community although they are not explicitly required to, but this community is not adopted as the standard for comparison. The desired community should be one found in the reference state and falls within the range of variation of

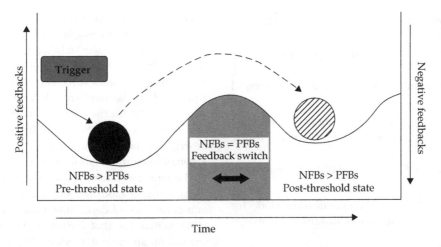

Figure 10.7 Schematic illustrating threshold occurrence as a feedback switch mechanism (shaded region). Thresholds represent the point at which feedbacks switch from a dominance of negative feedbacks (NFB) that maintain resilience of the pre-threshold state (solid ball) to a dominance of positive feedbacks (PFB) that decrease resilience of the pre-threshold state and enable an alternative post-threshold state (cross-hatched ball) to occupy the site. The feedback switch determines the degree of discontinuity associated with threshold initiation. Triggers represent events that initiate the feedback switch to begin threshold progression. Reproduced with permission permission from Briske et al. (2006).

the composition expected in the many dynamic communities found within the reference state (as defined by Stringham et al 2003).

The invasion of California grasslands by yellowstar thistle *Centaurea solstitialis*, exotic forbs, and annual grasses provides an example of an application of the state-and-transition model. These invaded grasslands represent a stable state of lower diversity compared to previous native plant dominated grasslands that existed before invasion by *C. solstitialis* (Kyser and DiTomaso 2002). Prescribed burning, herbicide application, or revegetation was unsuccessful in allowing the establishment and maintenance of an alternate stable state of higher diversity and lower populations of *C. solstitialis*. In these grasslands, land managers need to identify the set of conditions that would allow the *C. solstitialis*-dominated stable state to transition across the threshold to something more like the original highly diverse native grassland.

10.2.3 Rangeland health

Rangeland health is the degree to which the integrity of the soil, vegetation, water, and air, as well as the ecological processes of the rangeland ecosystem, are balanced and sustained. It is evaluated on the basis of three interrelated attributes: soil/site stability, hydrologic function, and biotic integrity (Pellant et al. 2005). Rangeland is evaluated on the basis of perceived degree of departure of each attribute from a reference condition.

Programmes for assessing rangeland health are currently being developed and since 2001 the **Sustainable Rangeland Roundtable** (SSR) with representatives from more than 75 organizations worldwide has been working to develop criteria and indicators for standardized inventory, monitoring, and reporting of rangelands that can be broadly accepted by the rangeland community (http://sustainablerangelands.warnercnr.colostate.edu/). The SSR has identified 64 core indicators categorized under 5 criteria (examples listed below are from the list of 64 indicators):

- conservation and maintenance of soil and water resources on rangelands
  - **example indicator:** area and percent of rangeland with accelerated soil erosion by water or wind.
- conservation and maintenance of plant and animal resources on rangelands.

– **example indicator:** rangeland area by plant community.
- maintenance of productive capacity on rangelands.
  – **example indicator:** rangeland above-ground phytomass.
- maintenance and enhancement of multiple social and economic benefits to present and future generations.
  – **example indicator:** value of forage harvested from rangeland by livestock.
- legal, institutional, and economic frameworks for rangeland conservation and sustainable management.
  – **example indicator:** measuring and monitoring. extent to which agencies, institutions, and organizations devote resources to measuring and monitoring changes in the condition of rangelands.

The SSR is currently developing methodology to measure and evaluate the full range of indicators. However, although coming up with a list of suitable assessment criteria may not be too difficult, in many cases the data are not currently being collected to allow an assessment or new methodology may to be developed. The Heinz Center attempted to use 14 indicators to assess the status of grasslands and shrublands in the USA, but found that partial or complete data were available for only 6 indicators, i.e. area of grasslands and shrublands, land use, number and duration of dry periods in streams and rivers, at-risk species, population trends in invasive and non-native birds, and cattle production (Heinz Centre 2002 (2005 update)).

Two methods of assessing rangeland health are described here, the **interpreting indicators of rangeland health (IIRH)** technique used in the USA and **landscape function analysis (LFA)** used in Australia. In both countries, these methods are used in conjunction with monitoring programmes used to assess rangeland trend. Monitoring programmes involve the regular, repeated collection of vegetation, soils, and other ecosystem data at a variety of temporal and spatial scales with the latter ranging from individual plant-scaled data (e.g. seedlings) up to continent scaled satellite-based data (e.g. NOAA and Landsat satellite imagery) (Friedel *et al.* 2000).

The IIRH technique involves rating a site according to 17 indicators designed to assess 3 ecosystem attributes (soil and site stability, hydrologic function, and biotic integrity) (Pellant *et al.* 2005; Pyke *et al.* 2002). Indicators at the site are observed and compared to the narrative describing the variation expected for the indicator on that site found in a reference sheet (Fig. 10.8). If they match the description then the indicator is given a rating of a none-to-slight deviation from what is expected at the site. Deviations beyond that are derived using either a modified site-specific description or a generic set of descriptions found in an evaluation matrix for that indicator. The ratings are entered on an evaluation sheet (Fig. 10.9). The reference sheet shows the state where the functional capacities represented by soil/site stability, hydrologic function, and biotic integrity are considered to be performing at a near-optimum level under the natural disturbance regime; this may include the traditionally viewed climax state, but should also incorporate all communities found in the reference state with reversible pathways. The reference sheet is thus prepared showing the state of each of the 17 indicators in the reference state. An indicator summary is then prepared showing the frequency distribution of indicators for each of the rangeland health attributes and allows development of an overall rating for each attribute for the site (bottom of Fig. 10.9, Table 10.2). The IIRH is not for monitoring or determining trend, or for making grazing or management decisions; it is a moment-in-time assessment that can provide early warnings of resource problems in rangeland and identify areas that are potentially about to cross a threshold. It should not be used solely to generate regional or national assessments of rangeland health. The IIRH is part of extensive monitoring programmes (Herrick *et al.* 2005a, 2005b) and is also used in Mexico and Mongolia. Similar approaches are being developed in Canada (David Pyke, personal communication).

Landscape function analysis (Tongway and Hindley 2004, http://www.cse.csiro.au/research/efa/index.htm) has been developed in Australia as a method to assess rangeland health. Formal methods of rangeland health assessment in Australia began in the 1970s with different methods adopted

Reference Sheet (Standard Example)

Author(s)/participant(s): Winnemucca Class Participants (May 12-15, 2005)

Contact for lead author: _____ **Reference site used?** Yes

Date: 5/11/05 **MLRA:** 024XY **Ecological Site:** Loamy 8-10" PZ, 024XY005NV. This *must* be verified based on soils and climate (see Ecological Site Description). Current plant community cannot be used to identify the ecological site.

Composition (indicators 10 and 12) based on: X Annual Production, __Foliar Cover, __Biomass

Indicators. For each indicator, describe the potential for the site. Where possible, (1) use numbers, (2) include expected range of values for above- and below-average years for **each** community and natural disturbance regimes within the reference state, when appropriate and (3) cite data. Continue descriptions on separate sheet.

1. **Number and extent of rills:** Minimal on slopes less than 10% and increasing slightly as slopes increase up to 50%. Rills spaced 15-50 feet apart when present on slopes of 10-50%. After wildfires, high levels of natural herbivory or extended drought, or combinations of these disturbances, rills may double in numbers on slopes from 10-50% after high intensity summer thunderstorms.

2. **Presence of water flow patterns:** Generally up to 20 feet apart and short (less than 10 feet long) with numerous obstructions that alter the water flow path. On slopes of 10-50%, flow patterns increase in number and length. Flow pattern length and numbers may double after wildfires, high levels of natural herbivory, extended drought, or combinations of these disturbances if high intensity summer thunderstorms occur.

3. **Number and height of erosional pedestals or terracettes:** Plant or rock pedestals and terracettes are almost always in flow patterns. Wind caused pedestals are rare and only would be on the site after wildfires, high levels of natural herbivory, extended drought, or combinations of these disturbances. Pedestals of Sandberg bluegrass on pedestals outside water flow patterns are generally caused by frost heaving, not erosion. Pedestals and terracettes would be particularly apparent on 10-50% slopes, especially immediately after high intensity summer thunderstorms.

4. **Bare ground from Ecological Site Description or other studies (rock, litter, standing dead, lichen, moss, plant canopy are not bare ground):** 10-20% or less bare ground with bare patches less than 10% of the evaluation area occurring as intercanopy patches larger than 2 feet in diameter (intercanopy patches can include areas that are not bare ground). Most large patches can include areas that are not bare ground. Within this range, lower slopes are expected to have less bare ground than steeper slopes. Upper end of precip range (10") will also have less bare ground. Canopy gaps generally less than 12 inches in diameter in the intervals between natural disturbance events. Bare ground would be expected to increase to 80% or more the first year following wildfire but to decrease to prefire levels within 2-5 years depending on climate and other disturbances. Multi-year droughts can also cause bare ground to increase to 30%.

5. **Number of gullies and erosion associated with gullies:** Gullies are rare and would only be present when a high intensity summer thunderstorm occurs after wildfires, with high levels of natural herbivory, extended drought, or combinations of these disturbances.

6. **Extent of wind scoured, blowouts and/or depositional areas:** Wind erosion is minimal. Moderate wind erosion can occur when disturbances such as severe wildfires, high levels of natural herbivory, extended drought, or combinations of these disturbances. After rain events, exposed soil surfaces form a physical crust that tends to reduce wind erosion.

7. **Amount of litter movement (describe size and distance expected to travel):** Litter movement consists primarily of redistribution of fine litter (herbaceous plant material) in flow patterns for distances of 1-3 feet on 2-15% slopes, 4-6 feet on 15-30% slopes, and 7-10 feet on 30-50% slopes. After wildfires, high levels of natural herbivory, extended drought, or combinations of these disturbances, size of litter and distance litter moves can increase with coarse woody litter and fine litter moving up to 10' (2-15% slope); 25' (15-30% slope); 100' (30-50% slope).

8. **Soil surface (top few mm) resistance to erosion (stability values are averages – most sites will show a range of values):** Values of 4.5-5.5 under canopies and in intercanopy spaces.

9. **Soil surface structure and SOM content (include type and strength of structure, and A-horizon color and thickness):** Surface layer is light brown and 6-7 inches thick with moderate granular structure. Loss of several millimeters of soil may occur immediately after a high intensity wildfire, high levels of natural herbivory, extended drought, or combinations of these disturbances.

10. **Effect of plant community composition (relative proportion of different functional groups) and spatial distribution on infiltration and runoff:** Perennial plants and especially sagebrush capture snow, increasing soil water availability in the spring. High bunchgrass density increases infiltration by improving soil structure and slowing runoff. Loss of sagebrush after a high intensity wildfire reduces snow accumulation in the winter, reducing the depth of soil water recharge negatively affecting growth and production of deep rooted forbs and perennial grasses. This reduced soil water recharge is part of the site dynamics if exotics or other management actions don't delay the succession back to a sagebrush-grass plant community.

Figure 10.8 *continues*

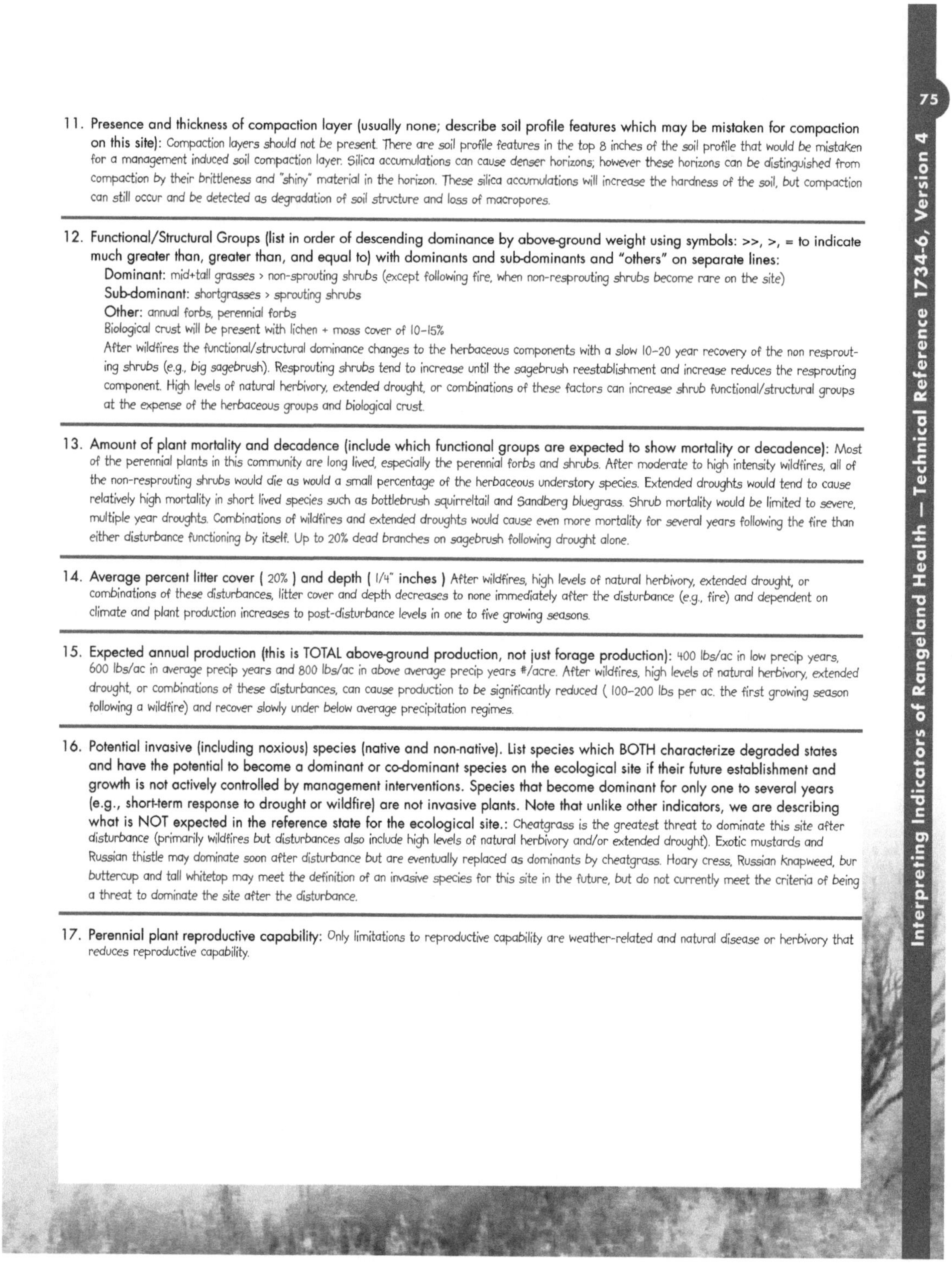

11. **Presence and thickness of compaction layer** (usually none; describe soil profile features which may be mistaken for compaction on this site): Compaction layers should not be present. There are soil profile features in the top 8 inches of the soil profile that would be mistaken for a management induced soil compaction layer. Silica accumulations can cause denser horizons; however these horizons can be distinguished from compaction by their brittleness and "shiny" material in the horizon. These silica accumulations will increase the hardness of the soil, but compaction can still occur and be detected as degradation of soil structure and loss of macropores.

12. **Functional/Structural Groups** (list in order of descending dominance by above-ground weight using symbols: >>, >, = to indicate much greater than, greater than, and equal to) with dominants and sub-dominants and "others" on separate lines:
    Dominant: mid+tall grasses > non-sprouting shrubs (except following fire, when non-resprouting shrubs become rare on the site)
    Sub-dominant: shortgrasses > sprouting shrubs
    Other: annual forbs, perennial forbs
    Biological crust will be present with lichen + moss cover of 10–15%.
    After wildfires the functional/structural dominance changes to the herbaceous components with a slow 10–20 year recovery of the non resprouting shrubs (e.g., big sagebrush). Resprouting shrubs tend to increase until the sagebrush reestablishment and increase reduces the resprouting component. High levels of natural herbivory, extended drought, or combinations of these factors can increase shrub functional/structural groups at the expense of the herbaceous groups and biological crust.

13. **Amount of plant mortality and decadence** (include which functional groups are expected to show mortality or decadence): Most of the perennial plants in this community are long lived, especially the perennial forbs and shrubs. After moderate to high intensity wildfires, all of the non-resprouting shrubs would die as would a small percentage of the herbaceous understory species. Extended droughts would tend to cause relatively high mortality in short lived species such as bottlebrush squirreltail and Sandberg bluegrass. Shrub mortality would be limited to severe, multiple year droughts. Combinations of wildfires and extended droughts would cause even more mortality for several years following the fire than either disturbance functioning by itself. Up to 20% dead branches on sagebrush following drought alone.

14. **Average percent litter cover** (20%) **and depth** (1/4" inches) After wildfires, high levels of natural herbivory, extended drought, or combinations of these disturbances, litter cover and depth decreases to none immediately after the disturbance (e.g., fire) and dependent on climate and plant production increases to post-disturbance levels in one to five growing seasons.

15. **Expected annual production** (this is TOTAL above-ground production, not just forage production): 400 lbs/ac in low precip years, 600 lbs/ac in average precip years and 800 lbs/ac in above average precip years #/acre. After wildfires, high levels of natural herbivory, extended drought, or combinations of these disturbances, can cause production to be significantly reduced (100–200 lbs per ac. the first growing season following a wildfire) and recover slowly under below average precipitation regimes.

16. **Potential invasive (including noxious) species** (native and non-native). List species which BOTH characterize degraded states and have the potential to become a dominant or co-dominant species on the ecological site if their future establishment and growth is not actively controlled by management interventions. Species that become dominant for only one to several years (e.g., short-term response to drought or wildfire) are not invasive plants. Note that unlike other indicators, we are describing what is NOT expected in the reference state for the ecological site.: Cheatgrass is the greatest threat to dominate this site after disturbance (primarily wildfires but disturbances also include high levels of natural herbivory and/or extended drought). Exotic mustards and Russian thistle may dominate soon after disturbance but are eventually replaced as dominants by cheatgrass. Hoary cress, Russian knapweed, bur buttercup and tall whitetop may meet the definition of an invasive species for this site in the future, but do not currently meet the criteria of being a threat to dominate the site after the disturbance.

17. **Perennial plant reproductive capability:** Only limitations to reproductive capability are weather-related and natural disease or herbivory that reduces reproductive capability.

Figure 10.8 Interpreting indicators of rangeland health (IIRH) reference sheet. Reproduced with permission from Pellant *et al.* (2005).

Evaluation Sheet (Example) (Front)

Aerial Photo: _____

Management Unit: Allotment I, pasture I State: NM Office: Las Cruces Range/Ecol. Site Code: 042XB999NM
(Allotment or pasture)

Ecological Site Name: Limy Soil Map Unit/Component Name: Nickel gravelly fine sandy loam

Observers: Joe Smith, Jose Garcia, and Thaddeus Jones Date: June 10, 2002

Location (description): Limy site two miles north of windmill in S.E. pasture

T. 11 S R. 23 W or _____ N. Lat. Or UTM E _____ m Position by GPS? Y / N No
UTM Zone ___, Datum ___
Sec. 12, NE 1/4 _____ W. Long. N _____ m Photos taken? Y / N Yes

Size of evaluation area: Evaluation area is approximately 3 ac. and represents entire ecological site in this pasture

Composition (Indicators 10 and 12) based on: __ Annual Production, _X_ Cover Produced During Current Year or __ Biomass

Soil/site verification:
Range/Ecol. Site Descr., Soil Surv., and/or Ecol. Ref. Area: Evaluation Area:
Surface texture grfsl, grlfs, gl Surface texture gfsl
Depth: very shallow __, shallow __, moderate __, deep X Depth: very shallow __, shallow __, moderate __, deep X
Type and depth of diagnostic horizons: Type and depth of diagnostic horizons:
1. Calcic horizon w/in 20" 3. _____ 1. Calcic horizon at 15" 3. _____
2. _____ 4. _____ 2. _____ 4. _____
Surf. Efferv.: none __, v. slight __, slight __, strong X, violent __ Surf. Efferv.: none __, v. slight __, slight __, strong X, violent __

Parent material Alluvium Slope 0-5 % Elevation 4100 ft. Topographic position toeslope Aspect south

Average annual precipitation 8-12 inches Seasonal distribution Summer thunderstorms dominate

Recent weather (last 2 years) (1) drought ____, (2) normal X, or (3) wet ____.

Wildlife use, livestock use (intensity and season of allotted use), and recent disturbances:
Wildlife use is dominated by pronghorn antelope in the winter. Livestock use was extremely heavy yearlong during 1900-1930. Last 50 years, livestock use has been cow/calf moderate yearlong use.

Off-site influences on evaluation area:
None

Criteria used to select this particular evaluation area as REPRESENTATIVE (specific info. and factors considered; degree of "representativeness")
Area is located near a pasture key area. It is located in the center of the ecological site and represents the typical amount of livestock, wildlife and recreational uses on this area. This ecological site dominates this pasture. The area is 3/4 of a mile from the closest water source.

Other remarks (continue on back if necessary)

Reference: (1) Reference Sheet: Limy SD—42B ; Author: J. Christensen ; Creation Date: 03/23/2002
or (2) Other (e.g., name and date of ecological site description; locations of ecological reference area(s)) Limy Ecological Site
042XB999NM, June 2001

Figure 10.9 *continues*

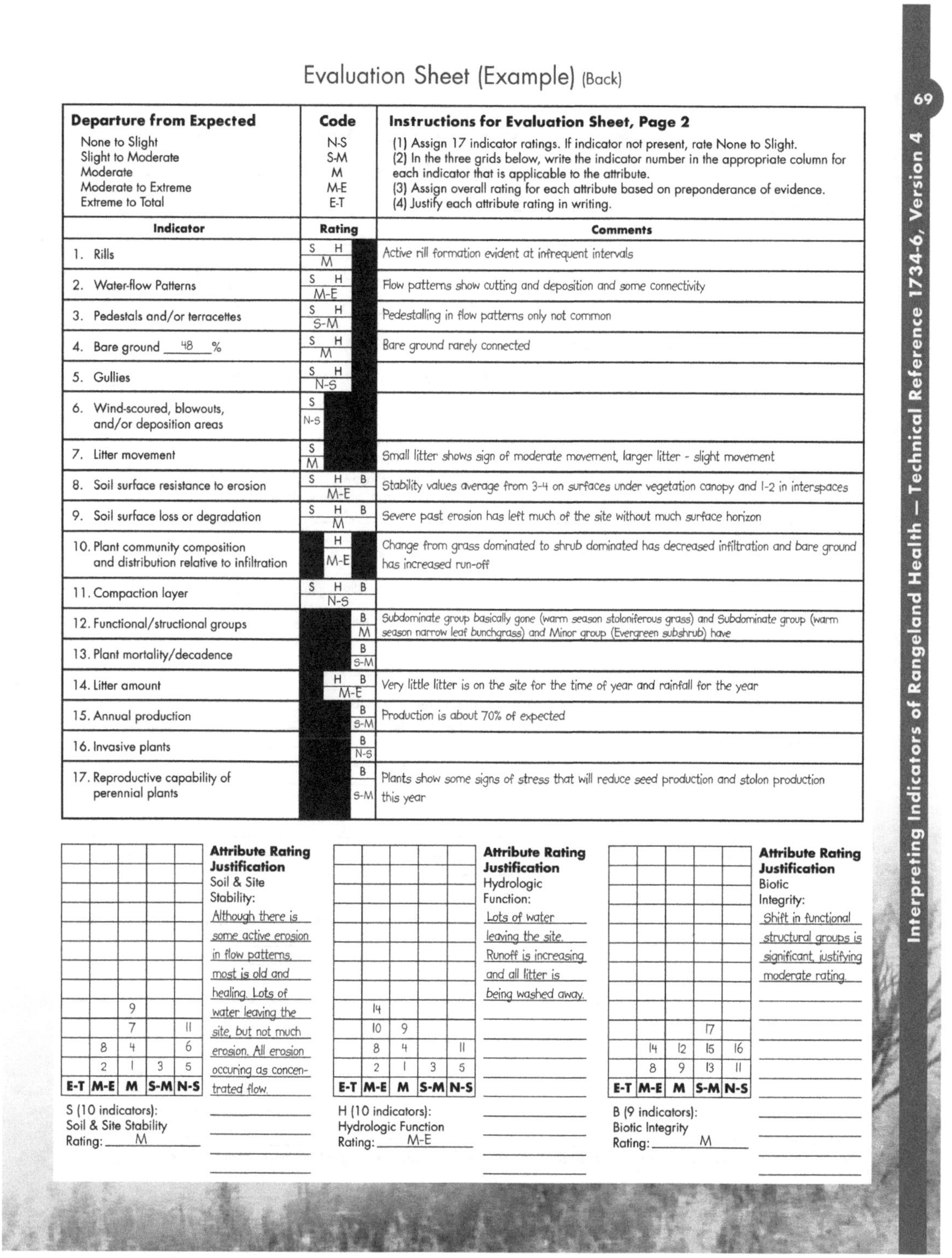

Interpreting Indicators of Rangeland Health — Technical Reference 1734-6, Version 4   69

Evaluation Sheet (Example) (Back)

Departure from Expected	Code
None to Slight	N-S
Slight to Moderate	S-M
Moderate	M
Moderate to Extreme	M-E
Extreme to Total	E-T

Instructions for Evaluation Sheet, Page 2

(1) Assign 17 indicator ratings. If indicator not present, rate None to Slight.
(2) In the three grids below, write the indicator number in the appropriate column for each indicator that is applicable to the attribute.
(3) Assign overall rating for each attribute based on preponderance of evidence.
(4) Justify each attribute rating in writing.

Indicator	Rating	Comments
1. Rills	S H — M	*Active rill formation evident at infrequent intervals*
2. Water-flow Patterns	S H — M-E	*Flow patterns show cutting and deposition and some connectivity*
3. Pedestals and/or terracettes	S H — S-M	*Pedestalling in flow patterns only not common*
4. Bare ground 48 %	S H — M	*Bare ground rarely connected*
5. Gullies	S H — N-S	
6. Wind-scoured, blowouts, and/or deposition areas	S — N-S	
7. Litter movement	S — M	*Small litter shows sign of moderate movement, larger litter - slight movement*
8. Soil surface resistance to erosion	S H B — M-E	*Stability values average from 3-4 on surfaces under vegetation canopy and 1-2 in interspaces*
9. Soil surface loss or degradation	S H B — M	*Severe past erosion has left much of the site without much surface horizon*
10. Plant community composition and distribution relative to infiltration	H — M-E	*Change from grass dominated to shrub dominated has decreased infiltration and bare ground has increased run-off*
11. Compaction layer	S H B — N-S	
12. Functional/structional groups	B — M	*Subdominate group basically gone (warm season stoloniferous grass) and Subdominate group (warm season narrow leaf bunchgrass) and Minor group (Evergreen subshrub) have*
13. Plant mortality/decadence	B — S-M	
14. Litter amount	H B — M-E	*Very little litter is on the site for the time of year and rainfall for the year*
15. Annual production	B — S-M	*Production is about 70% of expected*
16. Invasive plants	B — N-S	
17. Reproductive capability of perennial plants	B — S-M	*Plants show some signs of stress that will reduce seed production and stolon production this year*

E-T	M-E	M	S-M	N-S
		9		
		7		11
	8	4		6
	2	1	3	5

S (10 indicators): Soil & Site Stability Rating: M

Attribute Rating Justification Soil & Site Stability: *Although there is some active erosion in flow patterns, most is old and healing. Lots of water leaving the site, but not much erosion. All erosion occuring as concentrated flow.*

E-T	M-E	M	S-M	N-S
	14			
	10	9		
	8	4		11
	2	1	3	5

H (10 indicators): Hydrologic Function Rating: M-E

Attribute Rating Justification Hydrologic Function: *Lots of water leaving the site. Runoff is increasing and all litter is being washed away.*

E-T	M-E	M	S-M	N-S
			17	
	14	12	15	16
	8	9	13	11

B (9 indicators): Biotic Integrity Rating: M

Attribute Rating Justification Biotic Integrity: *Shift in functional structural groups is significant, justifying moderate rating.*

Figure 10.9 Interpreting indicators of rangeland health (IIRH). Evaluation sheet (a) front, and (b) back italicized text indicates the evaluator's comments. Reproduced with permission from Pellant *et al.* (2005).

Table 10.2 (a) Completed indicator summary, Part 3 of the IIRH Evaluation Summary Worksheet, using the information in Fig. 10.9 to show the frequency distribution of indicators for each of the rangeland health attributes and (b) the attribute summary that gives the evaluator's judgment regarding the overall rating for each attribute at the site. Italicized text indicates the evaluator's comments regarding the attribute summary.

Rangeland health attributes	Extreme	Moderate to extreme	Moderate	Slight to moderate	None to slight	Σ
S—Soil/site stability (Indicators 1–6, 8, 9 & 11)			✓✓✓✓	✓✓✓✓	✓	9
W—Hydrologic function (Indicators 1–5, 7–11 & 14)		✓	✓✓✓✓✓	✓✓✓✓	✓	11
B—Biotic integrity (Indicators 8–9 &111–17)	✓✓✓	✓✓	✓✓✓		✓	9

(b) Attribute summary—Check the category that best fits the 'preponderance of evidence' for each of the 3 attributes relative to the distribution of indicator ratings in the preceding Indicator Summary table

Attribute	Extreme	Moderate to extreme	Moderate	Slight to moderate	None to slight
Soil/site stability Rationale: *Interspaces all show signs of erosion*	☐	☐	☒	☐	☐
Hydrologic function Rationale: *Water appears to be moving on the surface and low infiltration on the site*	☐	☐	☒	☐	☐
Biotic integrity Rationale: *Only invasive plants indicator was rated higher than Moderate*	☐	☒	☐	☐	☐

With permission from Pyke *et al.* (2002).

by each state (Wilson 1989). Generally, rangeland condition was based on a single parameter, the density of shrubs.

The LFA method is made up of three modules: a conceptual framework, field methodology, and interpretational framework. The conceptual framework is a systems-based, rangeland function depending on a trigger-transfer-reserve-pulse (TTRP) framework representing sequences of ecosystem processes and feedback loops (Fig. 10.10). Field methodology at the monitoring site involves line transects established across the landscape orientated in the direction of resource flow, usually downslope. Soil and vegetation data are collected along the sampling transects from patches, where resources tend to accumulate, and interpatches, where resources tend to be mobilized and transported away. The interpretative framework involves combining the soil surface data to derive indices of stability, infiltration, and nutrient cycling. The index values are interpreted in a landscape context by generating a response surface that relates functional status with stress and disturbance (Fig. 10.11). Sites from throughout the range of conditions as well as 'typical' sites are required to develop the response surfaces with the slope of the curve representing the 'robustness' or 'fragility' of the system.

The LFA forms part of the Australian Collaborative Rangeland Information System (ACRIS), a partnership between the Commonwealth and New South Wales, Queensland, the Northern Territory, Western Australia, and South Australia. ACRIS has used results from the Western Australian Rangeland Monitoring System (WARMS) maintained by the Department of Agriculture and Food as a pilot project to determine how existing data could be accessed and reported in a useful form for the whole of Australia (Watson 2006). WARMS includes a set of about 1600 permanent sites (numbers vary from one year to the next) in which attributes of perennial vegetation and landscape function (i.e. LFA) are measured and used as indicators of change in the rangeland (Holm *et al.* 1987). In arid rangelands of Western Australia, WARMS

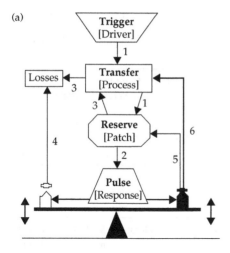

Ref	Processes involved
1	- Run-on - Storage/capture - Deposition - Saltation capture
2	- Plant germination, growth - Uptake processes - Nutrient mineralisation
3	- Run-off into streams - Sheet erosion out of stream - Rill flow and erosion - Wind erosion out of system
4	- Herb array - Harvesting - Fire - Deep drainage
5	- Seed pool replenishment - Organic matter cycling/decomposition processes - Harvest/concentration by soil micro-fauna
6	- Physical obstruction/absorption processes

Figure 10.10 Trigger–transfer–reserve–pulse (TTRP framework representing sequences of ecosystem processes and feedback loops underlying the Landscape Function Analysis system for monitoring rangeland function: (a) TTRP framework, (b) some of the processes operating at different locations. Reproduced with permission from Tongway and Hindley (2004).

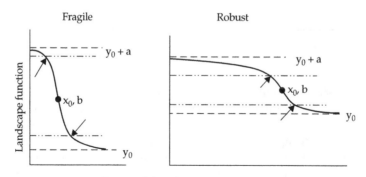

Figure 10.11 Landscape function analysis: response curves for fragile and robust landscapes relating changes in landscape to gradients in applied stress/disturbance regimes. Arrows represent critical thresholds. b = slope factor between the upper and lower asymptotes, an indicator of fragility/robustness. The upper asymptote represents the biogeochemical potential of the site limited by climate and parent material whereas the lower asymptote represents the lower limit of function under the existing land use. Reproduced with permission from Tongway and Hindley (2004).

sites were assessed using LFA indicating positive change in functionally intact areas of rangeland (Pringle et al. 2006). In contrast, aerial reconnaissance surveys that included smaller parts of the landscape indicated widespread range degradation and desiccation. The apparent contradiction in assessment arising from the two methods reflects the differing spatial focus of the two approaches with the WARMS sites being positioned to avoid historically degraded areas.

10.2.4 Adaptive management

Traditional management utilizes one or more of the methods described in §10.1 in an attempt to change a grassland from one state to another (e.g. from a shrubland to a grassland) or maintain a grassland in a desired state. By contrast, **adaptive management (AM)** is a dynamic, best-practice decision-making process in which different management alternatives are tested on the landscape (Holling 1978; Morghan *et al.* 2006; Schreiber *et al.* 2004). The goal of AM is to gain knowledge about a system by integrating research, design, management, and monitoring techniques. Sometimes called 'learning by doing', AM views management actions as experiments to test assumptions of the system, allowing changes in management strategies to be embraced as new information and enhanced understanding develops. Some of the key premises in AM include the involvement of stakeholders in decision-making, the incorporation of risk and uncertainty into a dynamic management strategy (i.e. the idea of adapt and learn), implementing several 'best bet' management strategies across an area rather than imposing a single management regime, and comprehensive documentation of management strategies and rigorous assessment of outcomes through monitoring (Lunt and Morgan 1999). For example, the implementation of adaptive management trails for restoring degraded *Themeda triandra* grassland in southern Australia was proposed (Lunt 2003). Trials included comparing the survival and growth of desired indicator species in plots that were subject to either reserve-wide management or species specific management. The AM plan was to compare the outcome of the two management options after 5 years of treatment and monitoring and use the outcome to guide future management.

The principles of AM are being used increasingly in the management (§10.1) and restoration (§10.3) of grass- and rangeland. For example, AM has been used to seek optimum management solutions to ensure steady-state productivity of goat herds belonging to pastoralists (§10.1) in northern Kenya (Hary 2004). As these groups of formerly nomadic pastoralists become increasingly sedentary, alternatives to herd mobility were sought to make up seasonality in forage production. Over 4 years AM included six different mating system treatments to determine the optimal season to allow breeding to maximize biological productivity of the pastoral goat flocks.

An adaptive management model was developed to quantitatively evaluate alternate management strategies for countering increased densities of shrubs in semi-arid pastoral rangeland of Australia (Noble and Walker 2006). The model included input from researchers, landholders, and agency personnel describing the relationship between woody weed invasion, control options, property economics, and management constraints. In keeping with the principles of AM, one of the goals of the study was 'to measure the outcome of particular management strategies, and quickly and efficiently to adapt them where necessary for future management.'

10.3 Restoration

As discussed in Chapter 1, grasslands the world over have suffered catastrophic losses in area, habitat quality, and biodiversity. To preserve and enhance the goods and services offered by grasslands, it is important that lost habitat is recovered and that damaged or degraded grasslands are rehabilitated and restored. In this section the objectives and issues related to ecological restoration of grassland are discussed.

10.3.1 Defining restoration

Ecological restoration is 'the process of assisting the recovery of an ecosystem that has been degraded, damaged, or destroyed' (SER 2004). The Society for Ecological Restoration further notes that restoration is 'an intentional activity that initiates or accelerates the recovery of an ecosystem with respect to its health, integrity and sustainability.'

Any restoration needs targets. These are objectives and goals for the restoration and can be individuals or groups of desired species, desired communities, or vegetation types, or desired parameters of ecosystem function (see reference states in §10.2.2). These targets must reflect the hierarchy of scale in ecological systems from population through community to ecosystem scales. Targets should be set early on in the restoration, preferably before starting. Monitoring is needed to evaluate progress towards reaching targets. Some targets are reached more easily and quickly than others. 'Means' are management activities undertaken in order to make progress towards targets, e.g. establish a grazing unit, impoverish the soil (if previously fertilized), practice reduced intensive farming, or restore natural processes (Bakker *et al.* 2000). Ideally, targets are set with respect to a known reference point, such as nearby high-quality grassland remnants. However, in many cases reference states are unavailable, remnants may be different from the original habitat, or the target conditions may be incompletely known. For example, the only remnant prairies near the prairie restorations at Nachusa Grasslands, Illinois, USA are grazed upland areas occurring on a quite different substrate from the post-agricultural lowlands that are being restored (Taft *et al.* 2006). The structure and composition of target communities can be controversial, as questions about the most suitable pre-disturbance conditions are raised. In North American prairie restoration, it is common to try to restore to pre-European settlement conditions, i.e. pre-1830s. However, it is clear that the Native Americans had an influence on the composition of their environment, often through deliberate or accidental burning, making it difficult to establish an appropriate and uncontroversial restoration target. In this case, the reference state(s) should incorporate at a minimum both burned and unburned communities if using the state-and-transition concepts defined by Stringham *et al.* (2003). Furthermore, quantitative data on the condition of the North American prairies before European settlement are lacking and even the earliest records are often incomplete, or from areas that were disturbed or may not have been suitable for agricultural development in the first place, and are, therefore, unrepresentative of the rest of the prairie (e.g. unploughed areas on rocky soils such as those described above for Nachusa Grasslands) (Allison 2004).

Restoration has been described as the 'acid test' for ecology (Bradshaw 1987). Many ecological concepts are important for the practice of restoration ecology, and at the same time as more research is undertaken, the results of restoration ecology research increasingly inform ecology (Table 10.3). As shown in Table 10.3, restorations offer a unique opportunity to test fundamental ecological concepts such as issues related to the multiple constraints imposed on the assembly of plant communities.

For example, experimental restoration plantings of North American prairie species indicated that diversity after 6–7 growing seasons was strongly and positively related to the diversity of the initial seed mixture (Piper and Pimm 2002). Moreover, reseeding experiments in these plantings showed that dominant species had essentially one chance to become established in the system and benefited most from inclusion in the initial mix. A related study showed that the establishment of a dominant species, such as *Panicum virgatum* in tallgrass prairie restoration, can mask a wide range of environmental heterogeneity and determine overall community composition (Baer *et al.* 2005). The findings from studies such as these inform our understanding of biodiversity and community assembly as related to succession.

10.3.2 From minimal intervention to large-scale reconstruction

Some grassland restorations of degraded or agriculturally improved grassland require little more than mechanical removal of undesirable (often woody) species to allow a restoration of biodiversity and a desired botanical composition to develop. For example, Simpson's Barrens in southern Illinois, USA was restored in the mid-1980s after local botanists noticed prairie plants growing along the roadside of a forested area. Intrigued, they noted that the original government surveyor's reports from the 1830s used the term 'barrens' to describe much of the area but contained no information about the herbaceous ground layers. Early aerial photographs from 1938 showed an open landscape of open-grown 'wolf trees' (Stritch 1990). Hand thinning of woody species and initiation of a prescribed burning programme led to a rapid and spectacular restoration of a prairie community. It is believed that most of the prairie species that returned so rapidly were growing under the forest canopy as depauperate remnant individuals (Mohlenbrock 1993).

Other restorations represent re-establishment of grassland in areas where little if anything of the original system remains, such as in former arable fields. For example, the 7695 ha Midewin National tallgrass prairie 72 km south-west of Chicago, USA

Table 10.3 Established ecological concepts that are generally understood by restoration practitioners. Some of these are deeply embedded in the knowledge base of restorationists (and agronomists); others are in the process of being incorporated into restoration practice

1	Competition: (plant) species compete for resources, and competition increases with decreasing distance between individuals and with decreasing resource abundance
2	Niches: species have physiological and biotic limits that restrict where they can thrive. Species selection and reference communities need to match local conditions
3	Succession: in many ecosystems, communities tend to recover naturally from natural and anthropogenic disturbances following the removal of these disturbances. Restoration often consists of assisting or accelerating this process. In some cases, restoration activities may need to repair underlying damage (soils) before secondary succession can begin
4	Recruitment limitation: the limiting stage for the establishment of individuals of many species is often early in life, and assistance at this stage (such as irrigation or protection from competitors and herbivores) can greatly increase the success of planted individuals
5	Facilitation: the presence of some plant species (guilds) enhances natural regeneration. These include N-fixers and overstorey plants, including shade plantings and brush piles
6	Mutualisms: mycorrhizae, seed dispersers and pollinators are understood to have useful and even critical roles in plant regeneration
7	Herbivory/predation: seed predators and herbivores often limit regeneration of natural and planted populations
8	Disturbance: disturbance at a variety of spatial and temporal scales is a natural, and even essential, component of many communities. The restoration of disturbance regimes may be critical
9	Island biogeography: larger and more connected reserves maintain more species, and facilitate colonizations, including invasions
10	Ecosystem function: nutrient and energy fluxes are essential components of ecosystem function and stability at a range of spatial and temporal scales
11	Ecotypes: populations are adapted to local conditions, at a variety of spatial and temporal scales. Matching ecotypes to local conditions increases restoration success
12	Genetic diversity: all else being equal, populations with more genetic diversity should have greater evolutionary potential and long-term prospects than genetically depauperate populations

Reproduced with permission from Young et al. (2005).

was established in 1996 on the site of a decommissioned US Army arsenal and ammunition plant. Few remnants of original upland prairie remained on the site (<3% of the area) and a large-scale restoration effort was planned, which included establishment of native seed gardens to provide the propagules necessary to reseed areas of the disturbed land back to prairie (Midewin National Tallgrass Prairie 2002). A restoration of this magnitude (in terms of work to be done as well as spatial extent) involves more than restoration of the biotic prairie ecosystem, but participation and acceptance of the change from a agricultural/developed landscape to the more natural grassland landscape with goods and services (Chapter 1) may not be fully appreciated by local landowners and the public (Davenport et al. 2007; Stewart et al. 2004). Although it will probably be decades before prairie grassland once again covers the landscape of the Midewin, the site is already yielding important findings regarding the population ecology of some prairie species in the few remnant areas that remain; for example, on the reproductive biology of the threatened *Agalinis auriculata* (Mulvaney et al. 2004).

Between the somewhat minimal approach of cutting out a few trees and prescribed burning and the extensive reestablishment of grassland in massively disturbed areas, as described above, lies a whole continuum of requirements for a grassland restoration programme. For example, restoration of Dutch calcareous grassland has been necessary on fertilized sites, abandoned grasslands, and former arable fields (Willems 2001). Similarly, restoration of species-rich grasslands in France required a variety of approaches depending on the extent of degradation of individual sites which included formerly grazed land, sites subject to extensive agricultural production, and sites degraded by intensification, ski tracks, or civil engineering installations (e.g. dams) (Muller et al. 1998). Each type of site offers a different challenge and requires a different approach. Nevertheless, several phases of activity are common to all grassland restorations (Willems 2001):

- **Pre-restoration phase:** data on land-use history are gathered and restoration objectives established.
- **Initial restoration phase:** prior non-conservation land use strategies are halted or undone, allowing germination and establishment of desired species from the seed bank (if present), or desired species are seeded or planted into the site. Appropriate ecological engineering may be required on severely degraded sites.
- **Consolidation phase:** appropriate management strategies for the restored grassland are put in place.
- **Long-term conservation strategy development:** necessary to ensure that the sites are not negatively affected by outside disturbance and genetic erosion of endangered plant populations.

10.3.3 Restoration issues and decisions

Grassland restoration can range from upgrading an existing degraded site to establishment of a grassland *de novo* on a site where grassland species have been completely lost. In restorations that involve more than a change in management strategy, there are several important decisions and issues that need to be considered. These include (Kline and Howell 1987; Packard and Mutel 2005):

- site preparation
- choice of an appropriate seed mix
- choice of appropriate seed provenances and use of local ecotypes
- raising of seedlings for planting (if necessary)
- timing of introduction of desired species (e.g. is it best to establish grasses first and introduce forbs later?)
- planting pattern (e.g. establishment of forb-rich patches)
- optimum management strategies (e.g. when or how often to burn).

Site preparation is important as the condition of the site, particularly the soil, can constrain and determine the outcome of restoration programme.

Depending on site history, removal of woody plants or exotics may be necessary through mechanical means, herbicide, or fire. If the restoration is to be started from seed or seedlings, disking or ploughing may be necessary.

A reduction of soil nutrients on ex-arable land or improved grasslands is necessary to reduce production and allow diversity to increase (Bakker and van Diggelen 2006; Hutchings and Stewart 2002), and can be achieved by biomass removal (clippings or hay), carbon addition (see Chapter 7), burning (see §10.1.3), topsoil stripping, or cultivation to promote leaching. Grazing can break up swards, increasing heterogeneity (see §9.3), and provide colonization opportunities, and can help lower soil fertility if animals are removed at night. Reducing soil nutrient levels may not, however, encourage native species where soil organic matter has been lost in long-cultivated fields; in these cases fertilization may be necessary (Wilson 2002).

Mycorrhizae are ubiquitous and important in grasslands (see §5.2.3) and their levels can be low in agricultural fields. Adding mycorrhizal inoculum can help restorations, especially in heavily disturbed soils (Huxman *et al.* 1998; Smith *et al.* 1998).

The use of local ecotypes is particularly important as it is becoming increasingly clear that the choice of local ecotypes (see §5.3.1) affects competitiveness of that species vs the rest of the species mixture (Gustafson *et al.* 2005). Use of a highly competitive cultivar of one of the dominant grasses has the potential to limit diversity of the other species in a restoration. It has long been an rule of thumb in grassland restorations that local ecotypes should be used wherever possible (Schramm 1970, 1992), but empirical data to support this rule have been lacking until recently. The use of molecular methods has allowed the genetic basis underlying the differential competitiveness of different seed sources to be determined (Falk *et al.* 2001; Gustafson *et al.* 2002; Lesica and Allendorf 1999).

Many grassland restorations reflect the work of grassroots volunteer organizations with well-intentioned individuals, local land managers, architects, and property owners who just want to bring back lost ecosystems. For example, the US prairie restoration movement has over the last 50+ years restored thousands of hectares of grassland.

The restoration efforts of these groups tend to concentrate on restoring the botanical richness of the desired grasslands. Although the value in terms of biodiversity of these restored grasslands is clear, this approach has been criticized because of the build-up of a 'folklore' of restoration knowledge, a frequent lack of empirical research to back up claims of successful methodology, and a lack of involvement of the human and cultural dimensions into the restored landscapes. Prairie restoration has been described as an art rather than a science (Schramm 1992), and likened to 'ecological gardening' (Allison 2004). In part to remedy such criticism, the Society for Ecological Restoration International was formed in 1987 with the subsequent publication of two professional, peer-reviewed journals: The first publication was *Restoration and Management Notes*, which became *Ecological Restoration* in 1987; *Restoration Ecology* was established later (1993) as a peer-reviewed publication that focused more on scientific studies related to restoration. *Restoration and Management Notes* was established as a forum for restorationists to exchange ideas. It has morphed into a more scientific peer-reviewed journal, *Ecological Restoration*. There has been a dramatic increase in the publication of articles on restoration ecology, much of it based in grasslands, with their relative contribution accounting for $c.4\%$ of all ecology papers by 2004 (Young *et al.* 2005).

Although many, if not most, grassland restorations reflect the efforts of individuals or small groups, there are a number of government-supported programmes that offer incentives to land owners to restore habitat, including grassland (see papers in Part 1 Restoration policy and infrastructure, in Perrow and Davy 2002). The US Conservation Reserve Program and Grassland Reserve Program and the European agri-environment schemes are examples described here. These and other similar schemes offer tangible incentives to increase the extent of grassland cover, and they encourage the planting of native species and restoration to fully functioning native species grassland.

The US Conservation Reserve Program (CRP) and Grassland Reserve Program (GRP) are voluntary incentive programmes administered since their creation through federal legislation in 1985 and 2002, respectively, by the USDA NRCP. The CRP provides technical and financial assistance to eligible farmers and ranchers to address soil, water, and related natural resource concerns (e.g. soil erosion, water quality, and wildlife habitat problems) on their lands in an environmentally beneficial and cost-effective manner. The goal of the CRP is to reduce soil erosion and convert highly erodible lands from crop production by providing perennial vegetation cover that can include planting of native grasses. Farmers receive an annual rental payment for the term of a 10–15 year contract. Fifty per cent cost sharing is provided to establish permanent cover with native perennial grasses. By the end of fiscal year 2006, 14 570 018 ha were enrolled in active contracts in the CRP, with most of the area in the Great Plains region being converted to perennial grasslands.

The GRP is a voluntary programme specifically to protect, preserve, and restore grasslands, including rangeland, pastureland, and shrublands, and protect these lands from conversion to cropland while maintaining them as grazing lands. The programme places an emphasis on working grazing operations. Landowners receive an annual rental payment on the property for as long as it is enrolled in the programme and as long as they convert the land to vegetative cover that can include native grasses. Existing grasslands can be enrolled as (1) permanent easement, (2) 30-year easement, or (3) under a rental agreement of 10, 15, or 20 years. Payment depends on market value of the property and the type of arrangement entered into. Alternatively, landowners can enter into a cost-share restoration agreement and receive up to 90% of the costs on land never cultivated or up to 75% on lands previously cropped. In 2004, 114 663 ha were enrolled in GRP including 31 654 ha of native grassland. Participants must implement and maintain an approved conservation plan that will lead to improved grassland management, enhanced infiltration, reduced soil erosion, increased carbon sequestration, and reduced water run-off. Future development and cropping uses of enrolled land must be limited, while the land owner retains the right to conduct common grazing practices (i.e. mowing, hay production, harvesting for seed). In essence, the GRP is an incentive plan that pays private landowners to be good stewards of their land.

Neither the CRP nor the GRP is intended as prairie restoration *per se*, as the establishment of native biodiversity is not a goal of either programme. Nevertheless, the characteristics of grasslands restored under these programmes provide a measure of the extent to which ecosystem characteristics can be re-established following the planting of perennial grasses in former cropland. A study of a chronosequence of CRP grasslands in Nebraska, USA planted to native grasses indicated recovery of several key ecosystem functions (primary productivity, cover of native grasses, below-ground biomass, and the labile carbon pool) to levels comparable with native prairie by 12 years (Baer *et al.* 2002). Other functions (i.e. soil structure, total soil carbon, and microbial biomass) showed an increase towards the steady-state conditions of native prairie but had not yet reached that point after 12 years. Total soil nitrogen processes showed little or no recovery with age, in part due to a lack of legumes in the planted species pools. Thus, the recovery of soil carbon pools in CRP restorations was faster than in unmanaged, fallow cropland, but dominance by C_4 grasses and a near absence of forbs is a stark contrast to the high diversity of forbs that characterizes native prairie. Similarly, in Kansas, CRP grasslands out of cultivation for an average of 15 years seeded with the same native warm-season grasses as in the Nebraska study, plus some native forbs (*Cassia chamaecrista*, *Helianthus maximiliani*, *Desmanthus illinoiensis*, and *Dalea purpureum*), were more similar to cool-season hayfields (seeded with the non-natives *Bromus inermis* or *Schedonorus phoenix*) than they were to native warm-season prairies. In particular, the CRP grasslands had lower soil organic matter and nitrogen and higher soil pH, bulk density, and clay than the native prairies (Murphy *et al.* 2006). These two studies, along with others, indicate that the CRP and GRP programmes are important in restoring grassland in the US Great Plains. This restored grassland has recovered some of the ecosystem function of native prairie. Nevertheless, the levels of biodiversity do not approach native prairie, in large part because of a lack of forb diversity in the planting mixes.

In Europe, agri-environment schemes (AES) are national (or local) schemes that pay farmers to farm in an environmentally sensitive way. They form part of the European Union's Common Agricultural Policy, with specific objectives that differ among EU member countries. Farmers enrolled in an AES are paid to modify management practices for environmental benefits that include reduced nutrient and pesticide emissions, protection of biodiversity, restoration of landscapes, and prevention of rural depopulation. Since 1994, €24.3 billion has been spent on AES in the EU (Kleijn and Sutherland 2003).

Although there are problems with evaluation and monitoring of AES, and their effectiveness has been debated (Whittingham 2007), grassland diversity can be improved under these schemes. Although not a grassland restoration scheme as such, much of the land included in AES is grassland pasture subject to fertilizer input and seeded to low-diversity pastures. For example, in Switzerland the objectives of AES were to enhance biodiversity in extensively used hay meadows through a prescription of no fertilizer use, pesticide application for problem weeds only, and haying only once per year. After 6 years, AES in Switzerland showed enhanced plant and bee diversity (Kleijn *et al.* 2006) . In the UK, 6 m wide grass field margin strips provide wildlife habitats, corridors, and buffers. The AES prescription for these field margins has been for natural regeneration or sowing of grass/forb mixtures, haying once per year, and no pesticides except for controlling problem weeds. Results from UK field margin AES showed enhanced plant and orthoptera diversity after 3.5 years.

The DEFRA Environmental Stewardship scheme in the UK (http://www.defra.gov.uk/erdp/schemes/es/) is an AES established in 2005 that includes encouragement of grassland buffer strips around field margins, and the maintenance and restoration of species-rich semi-natural grassland. Rewards are offered to land managers for delivery of a wider range of biodiversity benefits on targeted sites, such as those with existing high-priority environmental features. Entry-level enrollment in the scheme provides the landowner with £30 ha^{-1} of land enrolled, with amounts rising depending on the agreement; for example, payments rise to £280 ha⁻¹ under the Higher Level Stewardship Scheme for sowing seed mixes containing a range of forbs. There are several positive economic incentives for enrolling land in the Environmental Stewardship scheme (see discussion in Bullock et al. 2007) including reports of increased hay production and farmer income on recreated species-rich grassland that was formerly under intensive agriculture.

10.3.4 Case study: the Curtis Prairie

The Curtis Prairie in Madison, Wisconsin, USA is the oldest prairie restoration in the world (Plate 14). It is part of the c.486 ha University of Wisconsin Arboretum that was dedicated in 1934 by Aldo Leopold to recreate the natural habitats of Wisconsin (Blewett and Cottam 1984). Most of the former cornfields and overgrazed woodlands of the arboretum have been restored to forest, but 38 ha have been restored to prairie. At the suggestion of Aldo Leopold, a prairie restoration was established on 24 ha of old farm soil that had been tilled for row crops from 1836 to 1920. Originally oak savannah, at the time of acquisition by the University of Wisconsin in 1932–33, the area consisted of fields dominated by bluegrass (*Poa pratensis* and *P. compressa*). Adjacent to the restoration area is 1.2 ha of undisturbed wet-mesic prairie that provides a remnant for comparative purposes. There is also a smaller 14 ha prairie restoration, the Henry Greene Prairie, which was established later than the Curtis Prairie (1945–52).

Curtis Prairie, as it came to be called later on, started with a small prairie restoration experiment in 1935 by Norman Fassett. In 1937 Theodore Sperry was hired to direct and supervise the main restoration work across the entire area. Much of the labour was provided through the Civilian Conservation Corps, a federal government labor programme. Forty-two species were planted in large blocks using transplanted prairie sod, hay, and seed from remnant prairies around the state. The transplanted prairie sods were each 10 or 20 cm in diameter and were placed out in the bluegrass matrix. However, although selected to obtain a single desired species, each sod contained other species that soon became established too. Of 198 of the plantings that were relocated in 1982, 55% contained the originally planted species (Sperry 1994).

In 1940 John Curtis joined the University of Wisconsin arboretum as Director of Plant Research, an appointment split with Aldo Leopold who was Director of Animal Research. Virtually nothing was known about prairie restoration at that time. The early trial-and-error efforts led John Curtis to recognize that seed required cold stratification for germination and that use of transplanted prairie sod was too expensive given the large scale of the arboretum prairie restoration being undertaken. Other research by Curtis and his students investigated techniques for prairie restoration including the use of prescribed fire to eliminate *Poa* spp. sod and competition, fire and raking as soil preparation for *Andropogon gerardii* and *Schizachrium scoparium*, benefits of fire on flower production, and effects of clipping, burning, and competition on seedling establishment and survival (Howell and Stearns 1993).

Curtis was particularly interested in the role of fire. The first major burn of the prairie restoration was undertaken in 1950 under his direction. The outcome of this and subsequent burns was the demonstration that fire is an effective management tool favouring prairie grasses at the expense of cool-season grasses and other weeds (Anderson 1973; Curtis and Partch 1948). Following the initiation of a prescribed biennial burning schedule in 1950, a second major planting programme of 156 species was undertaken through 1957 into select areas under the direction of Arboretum botanist David Archbald. A Friends of the Arboretum group was established in 1962, the year after Curtis died, and the first meeting dedicated the prairie to Curtis.

Since 1946, the Curtis Prairie has been surveyed using a series of about 1300 1 m² quadrats regularly spaced in the corners of 15.2 m blocks in a grid along a permanent baseline. The 1946 and 1956 surveys were of the planted species only, but in subsequent surveys all plants were recorded. More than 300 prairie natives and several dozen weedy exotics had been recorded from the Curtis Prairie, more species per unit area than any other area in Wisconsin. Analyses of the survey data through 1961 showed that the composition of the restored prairie reflected local soil moisture conditions (ranging from very dry to very wet), that although the restored prairie was becoming similar to native prairie it retained more weed species than native prairie, but that the latter were decreasing in importance (Cottam and Wilson 1966). The frequency of the dominant prairie grasses (*Andropogon gerardii, Schizachyrium scoparium, Sorghastrum nutans*) and forbs (e.g. *Solidago* spp., *Ratibida pinnata, Eryngium yuccifolium*) had increased, and weedy species including *Pastinaca sativa* and *Poa pratensis* had decreased (Anderson 1973).

Recent research shows that after 65 years the Curtis Prairie is functioning similarly to the adjacent prairie remnant in terms of several ecosystem parameters, i.e. species richness per unit area, ANPP, peak leaf area index, and total root biomass. However, average annual soil respiration and net ecosystem production (NEP) was higher in the restored prairie and soil surface carbon was lower compared with the remnant. High variability in NEP values made it impossible to determine whether the Curtis Prairie was acting as a carbon sink, as it should (see Chapter 7), or as a carbon source (Kucharik *et al.* 2006).

What lessons have we learned from the oldest prairie restoration? It is clear that we still have much more to learn about restoration and can expect the 'unexpected to be expected' (Cottam 1987). Life history of individual species governs success and patterns of community development and although the introduction of many prairie species is not too difficult, the removal of undesirable weeds is much more of a problem. Time is an important ingredient of success and with good management can help with weed problems and hasten development of a rich prairie.

References

Aarssen LW and Turkington R (1985a). Biotic specialization between neighbouring genotypes in *Lolium perenne* and *Trifolium repens* from a permanent pasture. *Journal of Ecology*, 73, 605–614.

Aarssen LW and Turkington R (1985b). Within-species diversity in natural populations of *Holcus lanatus*, *Lolium perenne* and *Trifolium repens* from four different-age pastures. *Journal of Ecology*, 73, 869–886.

Abdul-Wahab A and Rice E (1967). Plant inhibition by Johnson grass and its possible significance in old-field succession. *Bulletin of the Torrey Botanical Club*, 94, 486–497.

Abrams MD (1988). Effects of burning regime on buried seed banks and canopy coverage in a Kansas tallgrass prairie. *Southwestern Naturalist*, 33, 65–70.

Acton DF (1992). Grassland soils. In RT Coupland, ed. *Ecosystems of the world: natural grasslands: introduction and western hemisphere*, Vol. 8A, pp. 25–54. Elsevier, Amsterdam.

Adams DE, Perkins WE, and Estes JR (1981). Pollination systems in *Paspalum dilatatum* Poir. (Poaceae): an example of insect pollination in a temperate grass. *American Journal of Botany*, 68, 389–394.

Adler PB (2004). Neutral models fail to reproduce observed species-area and species-time relationships in Kansas grassland. *Ecology*, 85, 1265–1272.

Adriansen HK (2006). Continuity and change in pastoral livelihoods of Senegalese Fulani. *Agriculture and Human Values*, 23, 215–229.

Aerts R and Berendse F (1988). The effect of increased nutrient availability on vegetation dynamics in wet heathlands. *Vegetatio*, 76, 63–69.

Aerts R, de Caluwe H, and Beltman B (2003). Plant community mediated vs. nutritional controls on litter decomposition rates in grasslands. *Ecology*, 84, 3198–3208.

Agarie S, Agata W, Uchida H, Kubota F, and Kaufman PB (1996). Function of silica bodies in the epidermal system of rice (*Oryza sativa* L): Testing the window hypothesis. *Journal of Experimental Botany*, 47, 655–660.

Aguiar MR and Sala OE (1994). Competition, facilitation, seed distribution and the origin of patches in a Patagonian steppe. *Oikos*, 70, 26–34.

Aguiar MR, Sorianao A, and Sala OA (1992). Competition and facilitation in the recruitment of seedlings in Patagonian steppe. *Functional Ecology*, 6, 66–70.

Albertson FW (1937). Ecology of mixed prairie in west central Kansas. *Ecological Monographs*, 7, 481–547.

Albertson FW and Tomanek GW (1965). Vegetation changes during a 30-year period in grassland communities near Hays, Kansas. *Ecology*, 46, 714–720.

Aldous DE, ed. (1999). *International turf management handbook*. CRC Press, Boca Raton, FL.

Al-Hiyaly SA, McNeilly T, and Bradshaw, AD (1988). The effects of zinc contamination from electricity pylons—evolution in a replicated situation. *New Phytologist*, 110, 571–580.

Allen TFH and Starr TB (1982). *Hierarchy perspectives for ecological complexity*. University of Chicago Press, Chicago.

Allen TFH and Wyleto EP (1983). A hierarchical model for the complexity of plant communities. *Journal of Theoretical Biology*, 101, 529–540.

Allison GW (1999). The implications of experimental design for biodiversity manipulations. *American Naturalist*, 153, 26–45.

Allison S (2004). What *do* we mean when we talk about ecological restoration? *Ecological Restoration*, 22, 281–286.

Al-Mufti MM, Sydes CL, Furness SB, Grime JP, and Band SR (1977). A quantitative analysis of shoot phenology and dominance in herbaceous vegetation. *Journal of Ecology*, 65, 759–791.

Altesor A, Di Landro E, May H, and Ezucurra E (1998). Long-term species change in a Uruguayan grassland. *Journal of Vegetation Science*, 9, 173–180.

Altesor A, Oesterheld M, Leoni E, Lezama F, and Rodríguez C (2005). Effect of grazing on community structure and productivity of a Uruguayan grassland. *Plant Ecology*, 179, 83–91.

Altesor A, Piñeiro G, Lezama F, Jackson RB, Sarasola M, and Paruelo JM (2006). Ecosystem changes associated with grazing in subhumid South American grasslands. *Journal of Vegetation Science*, 17, 323–332.

Amboseli Baboon Research Project (2001). Day in the life of a baboon. http://www.princeton.edu/~baboon/day_in_life.html. Accessed 17 March 2008.

Ambus P (2005). Relationship between gross nitrogen cycling and nitrous oxide emission in grass-clover pasture. *Nutrient Cycling in Agroecosystems*, 72, 189–199.

Amthor JS (1995). Terrestrial higher-plant response to increasing atmospheric CO_2 in relation to the global carbon cycle. *Global Change Biology*, 1, 243–274.

Anand M and Orloci L (1997). Chaotic dynamics in a multispecies community. *Environmental and Ecological Statistics*, 4, 337–344.

Anders W (1999). The effect of systemic rusts and smuts on clonal plants in natural systems. *Plant Ecology*, 141, 93–97.

Andersen AN and Lonsdale WM (1990). Herbivory by insects in Australian tropical savannas: a review. *Journal of Biogeography*, 17, 433–444.

Anderson RC (1973). The use of fire as a management tool on the Curtis Prairie. *Proceedings of the 12th Annual Tall Timbers Fire Ecology Conference (1972)*, 23–35.

Anderson RC (1991). *Illinois prairies: a historical perspective*. Illinois Natural History Survey Bulletin, Springfield, IL.

Anderson RC (2005). Summer fires. In S Packard and CF Mutel, eds. *The tallgrass restoration handbook*, pp. 245–249. Island Press, Washington, DC.

Anderson RC (2006). Evolution and origin of the Central Grassland of North America: climate, fire, and mammalian grazers. *Journal of the Torrey Botanical Society*, 133, 626–647.

Anderson RC and Brown LE (1986). Stability and instability in plant communities following fire. *American Journal of Botany*, 73, 364–368.

Anderson RC and Schelfhout S (1980). Phenological patterns among tallgrass prairie plants and their implications for pollinator competition. *American Midland Naturalist*, 104, 253–263.

Anderson RC, Hetrick BAD, and Wilson GWT (1994). Mycorrhizal dependence of *Andropogon gerardii* and *Schizachyrium scoparium* in two prairie soils. *American Midland Naturalist*, 132, 366–376.

Anderson TM, Metzger KL, and McNaughton SJ (2007). Multi-scale analysis of plant species richness in Serengeti grasslands. *Journal of Biogeography*, 34, 313–323.

Anderson VJ and Briske DD (1995). Herbivore-induced species replacement in grasslands: is it driven by herbivory tolerance or avoidance? *Ecological Applications*, 5, 1014–1024.

Anker PJ (2000). Holism and ecological racism: the history of South African human ecology. Online abstracts, Available http://depts.washington.edu/hssexec/annual/abstractsp1.html. Accessed 17 January 2008. In History of Science Society Annual Meeting, Vancouver.

Antonovics JA, Bradshaw AD, and Turner RG (1971). Heavy metal tolerance in plants. *Advances in Ecological Research*, 7, 1–85.

Anttila CK, Daehler CC, Rank NE, and Strong DR (1998). Greater male fitness of a rare invader (*Spartina alterniflora*, Poaceae) threatens a common native (*Spartina foliosa*) with hybridization. *American Journal of Botany*, 85, 1597–1601.

Apel P (1994). Evolution of the C_4 photosynthetic pathway: a physiologists' point of view. *Photosynthetica*, 30, 495–502.

Archbold S, Bond WJ, Stock WD, and Fairbanks DHK (2005). Shaping the landscape: fire-grazer interactions in an African savanna. *Ecological Applications*, 15, 96–109.

Archer S (1984). The distribution of photosynthetic pathway types on a mixed-grass prairie hillside. *American Midland Naturalist*, 111, 138–142.

Archer S, Garrett MG, and Detling JK (1987). Rates of vegetation change associated with prairie dog (*Cynomys ludovicianus*) grazing in North American mixed-grass prairie. *Vegetatio*, 72, 159–166.

Ash HJ, Gemmell RP, and Bradshaw AD (1994). The introduction of native plant species on industrial waste heaps: a test of immigration and other factors affecting primary succession. *Journal of Applied Ecology*, 31, 74–84.

Australian National Botanic Gardens (1998). Aboriginal plant use in south-eastern Australia. Common Reed, *Phragmites australis*. http://www.anbg.gov.au/aborig.s.e.aust/phragmites-australis.html. Accessed 10 January 2008.

Avdulov NP (1931). Karyo-systematische untersuchungen der familie Gramineen. *Bulletin of Applied Botany Plant Breeding Supplement*, 44, 1–428.

Axelrod DI (1985). Rise of the grassland biome, central North America. *Botanical Review*, 51, 163–201.

Baer SG, Kitchen J, Blair JM, and Rice CW (2002). Changes in ecosystem structure and function along a chronosequence of restored grasslands. *Ecological Applications*, 12, 1688–1701.

Baer SG, Blair JM, Collins SL, and Knapp AK (2003). Soil resources regulate productivity and diversity in newly established tallgrass prairie. *Ecology*, 84, 724–735.

Baer SG, Collins SL, Blair JM, Knapp AK, and Fiedler AK (2005). Soil heterogeneity effects on Tallgrass prairie community heterogeneity: an application of ecological theory to restoration ecology. *Restoration Ecology*, 13, 413–424.

Bai Y, Han X, Wu J, Chen Z, and Linghao L (2004). Ecosystem stability and compensatory effects in the Inner Mongolian grassland. *Nature*, 431, 181–184.

Bailey C (1997). Rhizosheath occurrence in South African grasses. *South African Journal of Botany*, 63, 484–490.

Bailey RG (1996). *Ecosystem geography*. Springer-Verlag, New York.

Bailey RG (1998). *Ecoregions: the ecosystem geography of the oceans and climates*. Springer-Verlag, New York.

Baker AJM (1987). Metal tolerance. *New Phytologist*, 106, 93–111.

Bakker ES, Ritchie ME, Olff H, Milchunas DG, and Knops JMH (2006). Herbivore impact on grassland plant diversity depends on habitat productivity and herbivore size. *Ecology Letters*, 9, 780–788.

Bakker JP and van Diggelen R (2006). Restoration of dry grasslands and heathlands. In J Van Andel and J Aronson, eds. *Restoration ecology: the new frontier*, pp. 95–110. Blackwell, Oxford.

Bakker JP, de Leeuw J, and van Wieren SE (1983). Micropatterns in grassland vegetation created and sustained by sheep grazing. *Vegetatio*, 55, 153–163.

Bakker JP, Grootjans AP, Hermy M, and Poschlod P (2000). How to define targets for ecological restoration? *Applied Vegetation Science*, 3, 3–7.

Balasko JA and Nelson CJ (2003). Grasses for northern areas. In RF Barnes, CJ Nelson, M Collins and KJ Moore, eds. *Forages: an introduction to grassland agriculture*, Vol. 1, pp. 125–148. Iowa State Unviversity Press, Ames, IA.

Baldini JI and Baldini VLD (2005). History on the biological nitrogen fixation research in graminaceous plants: special emphasis on the Brazilian experience. *Anais da Academia Brasileira de Ciências*, 77, 549–579.

Ball DM, Pedersen JF, and Lacefield GD (1993). The tall-fescue endophyte. *American Scientist*, 81, 370–371.

Balsberg P and Anna M (1995). Growth, radicle and root hair development of *Deschampsia flexuosa* (L.) Trin. seedlings in relation to soil acidity. *Plant and Soil*, 175, 125–132.

Barbehenn RV, Chen Z, Karowe DN, and Spickards A (2004). C_3 grasses have higher nutritional quality than C_4 grasses under ambient and elevated atmospheric CO_2. *Global Change Biology*, 10, 1565–1575.

Bardgett RD, Streeter TC, and Bol R (2003). Soil microbes compete effectively with plants for organic nitrogen inputs to temperate grasslands. *Ecology*, 84, 1277–1287.

Bardgett RD, Smith RS, Shiel RS, Peacock S, Simkin JM, Quirk H, and Hobbs PJ (2006). Parasitic plants indirectly regulate below-ground properties in grassland ecosystems. *Nature*, 439, 969–972.

Barker DJ and Collins M (2003). Forage fertilization and nutrient management. In RF Barnes, CJ Nelson, M Collins and KJ Moore, eds. *Forages: an introduction to grassland agriculture*, Vol. 1, pp. 263–293. Iowa State University Press, Ames, IA.

Barkworth ME, Capels KM, Long S, and Piep MB, eds. (2003). *Flora of North America north of Mexico: Volume 25 Magnoliophyta: Commelinidae (in part): Poaceae, part 2*. Vol. 25. Oxford University Press, New York.

Barkworth ME, Capels KM, Long S, Anderton LK, and Piep MB, eds. (2007). *Flora of North America north of Mexico: Volume 24 Magnoliophyta: Commelinidae (in part): Poaceae, part 1*. Vol. 24. Oxford University Press, New York.

Barnes JI (2001). Economic returns and allocation of resources in the wildlife sector of Botswana. *South African Journal of Wildlife Research*, 31, 141–153.

Barnes JI, Schier C, and van Rooy G (1999). Tourists' willingness to pay for wildlife viewing and wildlife conservation in Namibia. *South African Journal of Wildlife Research*, 29, 101–11.

Barnes PW, Tieszen LT, and Ode DJ (1983). Distribution, production, and diversity of C3- and C4-dominated communities in a mixed prairie. *Canadian Journal of Botany*, 61, 741–751.

Barnes RF and Nelson CJ (2003). Forages and grasslands in a changing world. In RF Barnes, CJ Nelson, M Collins and KJ Moore, eds. *Forages: an introduction to grassland agriculture*, Vol. 1, pp. 3–24. Iowa State University Press, Ames, IA.

Barnes RF, Nelson CJ, Collins M, and Moore KJ, eds. (2003). *Forages: an introduction to grassland agriculture*. Vol. 1. Iowa State University Press, Ames, IA.

Barrilleaux TC and Grace JB (2000). Growth and invasive potential of *Sapium sebiferum* (Euphorbiaceae) within the coastal prairie region: the effects of soil and moisture regime. *American Journal of Botany*, 87, 1099–1106.

Bascompte J and Rodríguez MA (2000). Self-disturbance as a source of spatiotemporal heterogeneity: the case of tallgrass prairie. *Journal of Theoretical Biology*, 204, 153–164.

Baskin CC and Baskin JM (1998). Ecology of seed dormancy and germination in grasses. In GP Cheplick, ed. *Population biology of grasses*, pp. 30–83. Cambridge University Press, Cambridge.

Bassman JH (2004). Ecosystem consequences of enhanced solar ultraviolet radiation: secondary plant metabolites as mediators of multiple trophic interactions in terrestrial plant communities. *Photochemistry and Photobiology*, 79, 382–398.

Baxter DR and Fales SL (1994). Plant environment and quality. In GC Fahey, ed. *Forage quality, evaluation, and utilization*, pp. 155–199. American Society of Agronomy, Inc., Crop Science Society of America, Inc., Soil Science Society of America, Inc., Madison, WI.

Bazzaz FA and Parrish JAD (1982). Organization of grassland communities. In JR Estes, RJ Tyrl and JN Brunken, eds. *Grasses and grassland communities: systematics and ecology*, pp. 233–254. University of Oklahoma Press, Norman, OK.

Becker DA and Crockett JJ (1976). Nitrogen fixation in some prairie legumes. *American Midland Naturalist*, 96, 133–143.

Beetle AA (1980). Vivipary, proliferation, and phyllody in grasses. *Journal of Range Management*, 33, 256–261.

Begon M, Townsend CR, and Harper JL (2006). *Ecology: from individuals to ecosystems*, 4th edition. Blackwell, Oxford.

Bekker RM, Verweij GL, Smith REN, Reine R, Bakker JP, and Schneider S (1997). Soil seed banks in European grasslands: does land use affect regeneration perspectives? *Journal of Applied Ecology*, 34, 1293–1310.

Bell DT and Muller CH (1973). Dominance of California annual grasslands by *Brassica nigra*. *American Midland Naturalist*, 90, 277–299.

Bell TJ and Quinn JA (1985). Relative importance of chasmogamously and cleistogamously derived seeds of *Dichanthelium clandestinum* (L.) Gould. *Botanical Gazette*, 146, 252–258.

Belsky AJ (1983). Small-scale pattern in grassland communities in the Serengeti National Park. *Vegetatio*, 55, 141–151.

Belsky AJ (1988). Regional influences on small-scale vegetational heterogeneity within grasslands in the Serengeti National Park, Tanzania. *Vegetatio*, 74, 3–10.

Belsky AJ, Carson WP, Jensen CL, and Fox GA (1993). Overcompensation by plants: herbivore optimization or red herring? *Evolutionary Ecology*, 7, 109–121.

Benson EJ, Hartnett DC, and Mann KH (2004). Belowground bud banks and meristem limitation in tallgrass prairie plant populations. *American Journal of Botany*, 91, 416–421.

Bentham G (1878). *Flora Australiensis*, 7, 449–670.

Bentham G (1881). Notes on Gramineae. *Journal of the Linnean Society. Botany*, 19, 14–134.

Bentham G and Hooker JD (1883). *Genera Plantarum*, vol. 3(2). Lovell Reeve, London.

Bentivenga SP and Hetrick BAD (1991). Relationship between mycorrhizal activity, burning, and plant productivity in tallgrass prairie. *Canadian Journal of Botany*, 69, 2597–2602.

Bertness MD and Callaway R (1994). Positive interactions in communities. *Trends in Ecology and Evolution*, 9, 191–193.

Bestelmeyer BT (2006). Threshold concepts and their use in rangeland management and restoration: the good, the bad, and the insidious. *Restoration Ecology*, 14, 325–329.

Bever JD (1994). Feedback between plants and their soil communities in an old field community. *Ecology*, 75, 1965–1977.

Bews JW (1918). *The grasses and grasslands of South Africa*. P. Davis & Sons, Pietermaritzburg.

Bews JW (1927). Studies in the ecological evolution of the angiosperms. *New Phytologist*, 26, 1–21, 65–84, 129–148, 209–248, 273–294.

Bews JW (1929). *The world's grasses: their differentiation, distribution, economics and ecology*. 1979 reissue. Russel & Russel, New York.

Biddiscome EF (1987). The productivity of Mediterranean and semi-arid grasslands. In RW Snaydon, ed. *Ecosystems of the world: managed grasslands: analytical studies*, Vol. 17B, pp. 19–27. Elsevier, Amsterdam.

Bierzychudek P and Eckhart V (1988). Spatial segregation of the sexes of dioecious plants. *American Naturalist*, 132, 34–43.

Biondini ME, Steuter AA, and Grygiel CE (1989). Seasonal fire effects on the diversity patterns, spatial distribution and community structure of forbs in the Northern Mixed Prairie, USA. *Vegetatio*, 85, 21–31.

Bisigato AJ and Bertiller MB (2004). Seedling recruitment of perennial grasses in degraded areas of the Patagonian Monte. *Journal of Range Management*, 57, 191–196.

Blackstock TH, Rimes CA, Stevens DP, Jefferson RG, Robertson HJ, Mackintosh J, and Hopkins JJ (1999). The extent of semi-natural grassland communities in lowland England and Wales: a review of conservation surveys 1978–96. *Grass and Forage Science*, 54, 1–18.

Blair JM (1997). Fire, N availability, and plant response in grasslands: a test of the transient maxima hypothesis. *Ecology*, 78, 2359–2368.

Blair JM, Seastedt TR, Rice CW, and Ramundo RA (1998). Terrestrial nutrient cycling in tallgrass prairie. In AK Knapp, JM Briggs, DC Hartnett and SL Collins, eds. *Grassland dynamics*, pp. 222–243. Oxford University Press, New York.

Blake AK (1935). Viability and germination of seeds and early life history of prairie plants. *Ecological Monographs*, 5, 405–460.

Blank RR and Young JA (1992). Influence of matric potential and substrate characteristics on germination of Nezpar Indian ricegrass. *Journal of Range Management*, 45, 205–209.

Blažková D (1993). Phytosociological study of grassland vegetation in North Korea. *Folia Geobotanica et Phytotaxonomica*, 28, 247–260.

Blewett TJ and Cottam G (1984). History of the University of Wisconsin Arboretum prairies. *Transactions of the Wisconsin Academy of Science, Arts, and Letters*, 72, 130–144.

Blumler MA (1996). Ecology, evolutionary theory and agricultural origins. In DR Harris, ed. *The origins and spread of agriculture and pastoralism in Eurasia*, pp. 25–50. Smithsonian Institution Press, Washington, DC.

Bobbink R, Hornung M, and Roelofs JGM (1998). The effects of air-borne nitrogen pollutants on species diversity in natural and semi-natural European vegetation. *Journal of Ecology*, 86, 717–738.

Bochert JR (1950). The climate of the central North American grassland. *Annals of the Association of American Geographers*, 40, 1–39.

Bock JH and Bock CE (1995). The challenges of grassland conservation. In A Joern and KH Keeler, eds. *The changing prairie: North American grasslands*, pp. 199–222. Oxford University Press, New York.

Boe A, Bortnem R, and Kephart KD (2000). Quantitative description of the phytomers of big bluestem. *Crop Science*, 40, 737–741.

Böhm W (1979). *Methods of studying root systems*. Springer-Verlag, Berlin.

Bond WJ, Midgley GF, and Woodward FI (2003). What controls South African vegetation—climate or fire? *South African Journal of Botany*, 69, 79–91.

Bond WJ, Woodward FI, and Midgley GF (2005). The global distribution of ecosystems in a world without fire. *New Phytologist*, 165, 525–538.

Bone E and Farres A (2001). Trends and rates of microevolution in plants. *Genetica*, 112–113, 165–182.

Boonman JG and Mikhalev SS (2005). The Russian steppe. In JM Suttie, SG Reynolds and C Batello, eds. *Grasslands of the world*, pp. 381–416. Food and Agriculture Organization of the United Nations, Rome.

Booth MS, Stark JM, and Rastetter E (2005). Controls on nitrogen cycling in terrestrial ecosystems: a synthetic analysis of literature data. *Ecological Monographs*, 75, 139–157.

Booth WE (1941). Revetation of abandoned fields in Kansas and Oklahoma. *American Journal of Botany*, 28, 415–422.

Borchert JR (1950). The climate of the central North American grassland. *Annals of the Association of American Geographers*, 40, 1–39.

Bossuyt B and Hermy M (2003). The potential of soil seedbanks in the ecological restoration of grassland and heathland communities. *Belgium Journal of Botany*, 136, 23–34.

Bouchereau A, Guenot P, and Lather F (2000). Analysis of amines in plant materials. *Journal of Chromotography B—Analytical Technologies in the Biomedical and Life Sciences*, 747, 49–67.

Bourlière F and Hadley M (1983). Present-day savannas: an overview. In F Bourliére, ed. *Ecosystems of the world 13: Tropical savannas*, pp. 1–17. Elsevier, Amsterdam.

Bowles ML, McBride JL, and Betz RF (1998). Management and restoration ecology of the federal threatened Mead's milkweed, *Asclepias meadii* (Asclepiadaceae). *Annals of the Missouri Botanical Garden*, 85, 110–125.

Bradshaw AD (1952). Populations of *Agrostis tenuis* resistant to lead and zinc poisoning. *Nature*, 169, 1098.

Bradshaw AD (1987). Restoration: an acid test for ecology. In WR Jordan III, ME Gilpin and JD Aber, eds. *Restoration ecology: a synthetic approach to ecological research*, pp. 23–30. Cambridge University Press, Cambridge.

Bragg TB and Hulbert LC (1976). Woody plant invasion of unburned Kansas bluestem prairie. *Journal of Range Management*, 29, 19–24.

Bransby DI (1981). Forage Quality. In NM Tainton, ed. *Veld and pasture management in South Africa*, pp. 175–214. Shuter & Shooter, Pietermaritzburg.

Branson FA (1956). Quantitative effects of clipping treatments on five range grasses. *Journal of Range Management*, 9, 86–88.

Breckle S-W (2002). *Walter's vegetation of the earth*, 4th edn. Springer-Verlag, Berlin.

Bredenkamp GJ, Spada F, and Kazmierczak E (2002). On the origin of northern and southern hemisphere grasslands. *Plant Ecology*, 163, 209–229.

Breymeyer AI (1987–1990). *Managed grasslands: A. regional studies*. Elsevier, Amsterdam.

Briggs D and Walters SM (1984). *Plant variation and evolution*. 2nd edn. Cambridge University Press, Cambridge.

Briggs JM, Nellis MD, Turner CL, Henebry GM, and Su H (1998). A landscape perspective of patterns and processes in tallgrass prairie. In AK Knapp, JM Briggs, DC Hartnett and SL Collins, eds. *Grassland dynamics*, pp. 265–279. Oxford University Press, New York.

Briggs JM, Knapp AK, Blair JM, Heisler JL, Hoch GA, Lett MS, and McCarron JK (2005). An ecosystem in transition: causes and consequences of the conversion of mesic grassland to shrubland. *BioScience*, 55, 243–254.

Briske DD and Anderson VJ (1992). Competitive ability of the bunchgrass *Schizachyrium scoparium* as affected by grazing history and defoliation. *Vegetatio*, 103, 41–49.

Briske DD and Derner JD (1998). Clonal biology of ceaspitose grasses. In GP Cheplick, ed. *Population biology of grasses*, pp. 106–135. Cambridge University Press, Cambridge.

Briske DD and Heitschmidt RK (1991). An ecological perspective. In RK Heitschmidt and JW Stuth, eds. *Grazing management: an ecological perspective*, pp. Chapter 1. Timber Press, Portland, OR.

Briske DD, Fuhlendorf SD, and Smeins FE (2003). Vegetation dynamics on rangelands: a critique of the current paradigms. *Journal of Applied Ecology*, 40, 601–614.

Briske DD, Fuhlendorf SD, and Smeins FE (2005). State-and-transition models, thresholds, and rangeland health: a synthesis of ecological concepts and perspectives. *Rangeland Ecology and Management*, 58, 1–10.

Briske DD, Fuhlendorf SD, and Smeins FE (2006). A unified framework for assessment and application of ecological thresholds. *Rangeland Ecology and Management*, 59, 225–236.

Brooker RW, Maestre FT, Callaway RM et al. (2008). Facilitation in plant communities: the past, the present, and the future. *Journal of Ecology*, 96, 18–34.

Brooks ML and Pyke DA (2001). Invasive plants and fire in the deserts of North America. In *Proceedings of the Invasive Species Workshop: the role of fire in the control and spread of invasive species*. Fire Conference 2000: the First National Congress on Fire Ecology, Prevention, and Management. Miscellaneous Publication No. 11 (eds KEM Galley and TP Wilson), pp. 1–14. Tall Timbers Research Station, Tallahassee, FL.

Brooks ML, D'Antonio CM, Richardson DM et al. (2004). Effects of invasive alien plants on fire regimes. *BioScience*, 54, 677–688.

Brown JH (1984). On the relationship between abundance and distribution of species. *American Naturalist*, 124, 255–279.

Brown JH and Heske EJ (1990). Control of a desert-grassland transition by a keystone rodent guild. *Science*, 250, 1705–1707.

Brown R (1810). *Prodromus florae Novae Hollandiae et insulae Van-Diemen*, vol 1. J. Johnson, London.

Brown S (2005). *Sports turf and amenity grassland management*. Crowood Press, Marlborough, Wilts.

Brown SA (1981). Courmarins. In EE Conn, ed. *The biochemistry of plants: a comprehensive treaty*, Vol. 7 *Secondary plant products*, pp. 269–300. Academic Press, New York.

Brown WV (1958). Leaf anatomy in grass systematics. *Botanical Gazette*, 119, 170–178.

Brown WV (1974). Another cytological difference among the Kranz subfamilies of the Gramineae. *Bulletin of the Torrey Botanical Club*, 101, 120–124.

Brown WV (1975). Variations in anatomy, associations, and origins of Kranz tissue. *American Journal of Botany*, 62, 395–402.

Brown WV and Emery WHP (1958). Apomixis in the Gramineae: Panicoideae. *American Journal of Botany*, 45, 253–263.

Brundrett M (1991). Mycorrhizas in natural ecosystems. *Advances in Ecological Research*, 21, 171–311.

Brunken JN and Estes JR (1975). Cytological and morphological variation in *Panicum virgatum* L. *Southwestern Naturalist*, 19, 379–385.

Bruno JFS, Stachowicz JJ, and Bertness, M.D. (2003). Inclusion of facilitation into ecological theory. *Trends in Ecology and Evolution*, 18, 119–125.

Bruun HH (2000). Patterns of species richness in dry grassland patches in an agricultural landscape. *Ecography*, 23, 641–650.

Buckner RC and Bush LP (1979). *Tall fescue*. American Society of Agronomy, Madison, WI.

Buffington LC and Herbel CH (1965). Vegetational changes on a semidesert grassland range. *Ecological Monographs*, 35, 139–164.

Bullock JM, Clear Hill B, Silvertown J, and Sutton M (1995). Gap colonization as a source of grassland community change: effects of gap size and grazing on the rate and mode of colonization by different species. *Oikos*, 72, 273–282.

Bullock JM, Pywell RF, and Walker KJ (2007). Long-term enhancement of agricultural production by restoration of biodiversity. *Journal of Applied Ecology*, 44, 6–12.

Burdon JJ, Thrall PH, and Ericson AL (2006). The current and future dynamics of disease in plant communities. *Annual Review of Phytopathology*, 44, 19–39.

Burkart A (1975). Evolution of grasses and grasslands in South America. *Taxon*, 24, 53–66.

Burke IC, Lauenroth WK, and Coffin DP (1995). Soil organic matter recovery in semiarid grasslands: implications for the conservation reserve program. *Ecological Applications*, 5, 793–801.

Butler JL and Briske DD (1988). Population structure and tiller demography of the bunchgrass *Schizachyrium scoparium* in response to herbivory. *Oikos*, 51, 306–312.

Buxton DR, Russell JR, and Wedin WF (1987). Structural neutral sugars in legume and grass stems in relation to digestibility. *Crop Science*, 27, 1279–1285.

Cable DR (1971). Growth and development of arizona cottontop (*Trichachne californica* [Benth.] Chase). *Botanical Gazette*, 132, 119–145.

Cai H, Hudson EA, Mann P et al. (2004). Growth-inhibitory and cell cycle-arresting properties of the rice bran constituent tricin in human-derived breast cancer cells in vitro and in nude mice *in vivo*. *British Journal of Cancer*, 91, 1364–1371.

Callaway RM and Aschehoug ET (2000). Invasive plants versus their new and old neighbors: a mechanism for exotic invasion. *Science*, 290, 521–523.

Callaway RM and Walker, L.R. (1997). Competition and facilitation: a synthetic approach to interactions in plant communities. *Ecology*, 78, 1958–1965.

Campbell BD and Stafford Smith DM (2000). A synthesis of recent global change research on pasture and rangeland production: reduced uncertainties and their management implications. *Agriculture, Ecosystems & Environment*, 82, 39.

Campbell CS and Kellogg EA (1986). Sister group relationships of the Poaceae. In TR Soderstrom, KW Hilu, CS Campbell and ME Barkworth, eds. *Grass systematics and evolution*, pp. 217–224. Smithsonian Institution Press, Washington, DC.

Campbell CS, Quinn JA, Cheplick GP, and Bell TJ (1983). Cleistogamy in grasses. *Annual Review of Ecology and Systematics*, 14, 411–441.

Canales J, Trevisan MC, Silva JF, and Caswell H (1994). A demographic study of an annual grass (*Andropogon brevifolius* Schwarz) in burnt and unburnt savanna. *Acta Oecologia*, 15, 261–273.

Carey PD, Fitter AH, and Watkinson AR (1992). A field study using the fungicide benomyl to investigate the effect of mycorrhizal fungi on plant fitness. *Oecologia*, 90, 550–555.

Carpenter JR (1940). The grassland biome. *Ecological Monographs*, 10, 619–684.

Carson WP and Root RB (2000). Herbivory and plant species coexistence: community regulation by an outbreaking phytophagous insect. *Ecological Monographs*, 70–99, 73.

Casler MD (2005). Ecotypic variation among switchgrass populations from the northern USA. *Crop Science*, 45, 388–398.

Cerling TE, Wang Y, and Quade J (1993). Expansion of C_4 ecosystems as an indicator of global ecological change in the late Miocene. *Nature*, 361, 344–345.

Cerling TE, Harris JM, MacFadden BJ *et al.* (1997). Global vegetation change through the Miocene/Pliocene boundary. *Nature*, 389, 153–159.

Cerling TE, Ehleringer JR, and Harris JM (1998). Carbon dioxide starvation, the development of C_4 ecosystems, and mammalian evolution. *Philosophical Transactions of the Royal Society, London, Series B*, 353, 159–171.

Cerling TE, Harris JM, and Leakey MG (1999). Browsing and grazing in elephants: the isotope record of modern and fossil proboscideans. *Oecologia*, 120, 364–374.

Chaffey NJ (1994). Structure and function of the membraneous grass ligule: a comparative study. *Botanical Journal of the Linnean Society*, 116, 53–69.

Chaffey N (2000). Physiological anatomy and function of the membranous grass ligule. *New Phytologist*, 146, 5–21.

Chaneton EJ, Lemcoff, JH, and Lavado RS (1996). Nitrogen and phosphorus cycling in grazed and ungrazed plots in a temperate subhumid grassland in Argentina. *Journal of Applied Ecology*, 33, 291–302.

Chaneton EJ, Perelman SB, Omacini M, and León JC (2002). Grazing, environmental heterogeneity, and alien plant invasions in temperate Pampa grasslands. *Biological Invasions*, 4, 7–24.

Chaneton EJ, Perelman SB, and Leon RJC (2005). Floristic heterogeneity of Flooding Pampa grasslands: a multi-scale analysis. *Plant Biosystems*, 139, 245–254.

Chapman GP (1996). *The biology of grasses*. CAB International, Wallingford, Oxon.

Chapman GP and Peat WE (1992). *Introduction to the grasses*. CAB International, Wallingford, Oxon.

Chase A (1908). Notes on cleistogamy in grasses. *Botanical Gazette*, 45, 135–6.

Chase A (1918). Axillary cleistogenes in some American grasses. *American Journal of Botany*, 5, 254–258.

Chatterton NJ, Harrison PA, Bennett JH, and Asay KH (1989). Carbohydrate partitioning in 185 accessions of Gramineae grown under warm and cool temperatures. *Journal of Plant Physiology*, 134, 169–179.

Chavannes E (1941). Written records of forest succession. *Scientific Monthly*, 53, 76–80.

Cheney NP, Gould JS, and Catchpole WR (1998). Prediction of fire spread in grasslands. *International Journal of Wildland Fire*, 8, 1–13.

Cheney P and Sullivan A (1997). *Grassfires: fuel, weather and fire behaviour*. CSIRO, Collingwood, Australia.

Cheng D-L, Wang G-X, Chen B-M, and Wei X-P (2006). Positive interactions: crucial organizers in a plant community. *Journal of Integrative Plant Biology*, 48, 128–136.

Cheplick GP (1997). Effects of endophytic fungi on the phenotypic plasticity of *Lolium perenne* (Poaceae). *American Journal of Botany*, 84, 34–40.

Cheplick GP, ed. (1998a). *Population biology of grasses*. Cambridge University Press, Cambridge.

Cheplick GP (1998b). Seed dispersal and seedling establishment in grass populations. In GP Cheplick, ed. *Population biology of grasses*, pp. 84–105. Cambridge University Press, Cambridge.

Cheplick GP and Cho R (2003). Interactive effects of fungal endophyte infection and host genotype on growth and storage in *Lolium perenne*. *New Phytologist*, 158, 183–191.

Cheplick GP and Grandstaff K (1997). Effects of sand burial on purple sandgrass (*Triplasis purpurea*): the significance of seed heteromorphism. *Plant Ecology*, 133, 79–89.

Cheplick GP and Quinn JA (1983). The shift in aerial/subterranean fruit ratio in *Amphicarpum purshii*: causes and significance. *Oecologia*, 57, 374–379.

Cheplick GP and Quinn JA (1986). Self-fertilization in *Amphicarpum purshii*: its influence on fitness and variation of progeny from aerial panicles. *American Midland Naturalist*, 116, 394–402.

Cheplick GP and Quinn JA (1987). The role of seed depth, litter, and fire in the seedling establishment of amphicarpic peanutgrass (*Ampicarpum purshii*). *Oecologia*, 73, 459–464.

Cheplick GP, Perera A, and Koulouris K (2000). Effect of drought on the growth of *Lolium perenne* genotypes with and without fungal endophytes. *Functional Ecology*, 14, 657–667.

Cherney JH, Cherney DJR, and Bruulsema TW (1998). Potassium management. In JH Cherney and DJR Cherney, eds. *Grass for dairy cattle*, pp. 137–160. CABI International, Wallingford, Oxon.

Chiy PC and Phillips CJC (1999). Sodium fertilizer application to pasture. 8. Turnover and defoliation of leaf tissue. *Grass & Forage Science*, 54, 297–311.

Cibils AF and Borreli PR (2005). Grasslands of Patgonia. In JM Suttie, SG Reynolds and C Batello, eds. *Grasslands of the world*, pp. 121–170. Food and Agriculture Organization of the United Nations, Rome.

Ciepiela AP and Sempruch C (1999). Effect of L-3,4-dihydroxyphenylalanine, ornithine and gamma-aminobutyric acid on winter wheat resistance to grain aphid. *Journal of Applied Entomology*, 123, 285–288.

Cingolani AM, Cabido MR, Renison D, and Solís Neffa V (2003). Combined effect of environment and grazing on vegetation structure in Argentine granite grasslands. *Journal of Vegetation Science*, 14, 223–232.

Cingolani AM, Noy-Meir I, and Diaz S (2005). Grazing effects on rangeland diversity: a synthesis of contemporary models. *Ecological Applications*, 15, 757–773.

Clark DL and Wilson MV (2003). Post-dispersal seed fates of four prairie species. *American Journal of Botany*, 90, 730–735.

Clark EA (2001). Diversity and stability in humid temperate pastures. In PG Tow and A Lazenby, eds. *Competition and succession in pastures*, pp. 103–118. CAB Publishing, Wallingford, Oxon.

Clark FE and Woodmansee RG (1992). Nutrient cycling. In RT Coupland, ed. *Ecosystems of the world: natural grasslands: introduction and western hemisphere*, Vol. 8A, pp. 137–146. Elsevier, Amsterdam.

Clark LG and Fisher JB (1986). Vegetative morphology of grasses: shoots and roots. In TR Soderstrom, KW Hilu, CS Campbell and ME Barkworth, eds. *Grass systematics and evolution*, pp. 37–45. Smithsonian Institution Press, Washington, DC.

Clark LG, Zhang W, and Wendel JF (1995). A phylogeny of the grass family (Poaceae) based on *ndhf* sequence data. *Systematic Botany*, 20, 436–460.

Clark LG, Kobayashi M, Mathews S, Spangler RE, and Kellogg EA (2000). The Puelioideae, a new subfamily of Poaceae. *Systematic Botany*, 25, 181–187.

Clarke S and French K (2005). Germination response to heat and smoke of 22 Poaceae species from grassy woodlands. *Australian Journal of Botany*, 53, 445–454.

Clay K (1987). Effects of fungal endophytes on the seed and seedling biology of *Lolium perenne* and *Festuca arundinacea*. *Oecologia*, 73, 358–362.

Clay K (1990a). Comparative demography of three graminoids infected by systemic, clavicipitaceous fungi. *Ecology*, 71, 558–570.

Clay K (1990b). Fungal endophytes of grasses. *Annual Review of Ecology and Systematics*, 21, 275–297.

Clay K (1994). The potential role of endophytes in ecosystems. In CW Bacon and JF White, eds. *Biotechnology of endophytic fungi of grasses*, pp. 73–86. CRC Press, Boca Raton, FL.

Clay K (1997). Fungal endophytes, herbivores, and the structure of grassland communities. In A Gange, C. and VK Brown, eds. *Multitrophic interactions in terrestrial systems*, pp. 151–170. Blackwell Science, Oxford.

Clay K and Holah J (1999). Fungal endophyte symbiosis and plant diversity in successional fields. *Science*, 285, 1742–1744.

Clay K and Schardl C (2002). Evolutionary origins and ecological consequences of endophyte symbiosis with grasses. *American Naturalist*, 160, supplement, S99-S127.

Clay K, Cheplick GP, and Marks S (1989). Impact of the fungus *Balansia henningsiana* on *Panicum agrostoides*: frequency of infection, plant growth and reproduction, and resistance to pests. *Oecologia*, 80, 374–380.

Clay K, Marks S, and Cheplick GP (1993). Effects of insect herbivory and fungal endophyte infection on competitive interactions among grasses. *Ecology*, 74, 1767–1777.

Clay K, Holah J, and Rudgers JA (2005). Herbivores cause a rapid increase in heredity symbiosis and alter plant community composition. *Proceedings of the National Academy of Science, USA*, 102, 12465–12470.

Clayton WD and Renvoize SA (1986). *Genera Graminum: grasses of the world*. HMSO, London.

Clayton WD and Renvoize SA (1992). A system of classification for the grasses. In GP Chapman, ed. *Grass evolution and domestication*, pp. 338–353. Cambridge University Press, Cambridge.

Clements FE (1936). Nature and structure of the climax. *Journal of Ecology*, 24, 252–284.

Clements FE, Weaver JE, and Hanson HC (1929). *Plant competition: an analysis of community functions*. Carnegie Institution, Washington DC.

Cleveland CC, Townsend AR, Schimel DS, Fisher H, Howarth RW, Hedin LO, Perakis SS, Latty EF, Von

Fischer JC, Elseroad A, and Wasson MF (1999). Global patterns of terrestrial biological nitrogen (N_2) fixation in natural ecosystems. *Global Biogeochemical Cycles*, 13, 623–646.

Clifford HT (1986). Spikelet and flora morphology. In TR Soderstrom, KW Hilu, CS Campbell and ME Barkworth, eds. *Grass systematics and evolution*, pp. 21–30. Smithsonian Institution Press, Washington, DC.

Cochrane V and Press MC (1997). Geographical distribution and aspects of the ecology of the hemiparasitic angiosperm *Striga asiatica* (L.) Kuntze: a herbarium study. *Journal of Tropical Ecology*, 13, 371–380.

Cofinas M and Creighton C, eds. (2001). *Australian native vegetation assessment*. National Land and Water Resources Audit, Land & Water Australia, Canberra.

Cole CV, Innis GS, and Stewart JWB (1977). Simulation of phosphorus cycling in semiarid grasslands. *Ecology*, 58, 1–15.

Collins M and Fritz JO (2003). Forage quality. In RF Barnes, CJ Nelson, M Collins and KJ Moore, eds. *Forages: an introduction to grassland agriculture*, Vol. 1, pp. 363–390. Iowa State University Press, Ames, IA.

Collins M and Owens VN (2003). Preservation of forage as hay and silage. In RF Barnes, CJ Nelson, M Collins and KJ Moore, eds. *Forages: an introduction to grassland agriculture*, Vol. 1, pp. 443–472. Iowa State University Press, Ames, IA.

Collins SL (1990). Patterns of community structure during succession in tallgrass prairie. *Bulletin of the Torrey Botanical Club*, 117, 397–408.

Collins SL (1992). Fire frequency and community heterogeneity in tallgrass prairie vegetation. *Ecology*, 73, 2001–2006.

Collins SL and Adams DE (1983). Succession in grasslands: thirty-two years of change in a central Oklahoma tallgrass prairie. *Vegetatio*, 51, 181–190.

Collins SL and Barber SC (1985). Effects of disturbance on diversity in mixed-grass prairie. *Vegetatio*, 64, 87–94.

Collins SL and Glenn SM (1988). Disturbance and community structure in North American prairies. In HJ During, MJA Werger and JH Willems, eds. *Diversity and pattern in plant communities*, pp. 131–143. SPB Academic, The Hague.

Collins SL and Glenn SM (1990). A hierarchical analysis of species abundance patterns in grassland vegetation. *American Naturalist*, 176, 233–237.

Collins SL and Glenn SM (1991). Importance of spatial and temporal dynamics in species regional abundance and distribution. *Ecology*, 72, 654–664.

Collins SL and Glenn SM (1997). Intermediate disturbance and its relationship to within- and between patch dynamics. *New Zealand Journal of Ecology*, 21, 103–110.

Collins SL and Steinauer EM (1998). Disturbance, diversity, and species interactions in tallgrass prairie. In AK Knapp, JM Briggs, DC Hartnett and SL Collins, eds. *Grassland dynamics: long-term ecological research in tallgrass prairie*, pp. 140–156. Oxford University Press, New York.

Collins SL and Wallace LL, eds. (1990). *Fire in North American tallgrass prairies*, pp. 175. University of Oklahoma Press, Norman, OK.

Collins SL, Bradford JA, and Sims PL (1987). Succession and fluctuation in *Artemisia* dominated grassland. *Vegetatio*, 73, 89–99.

Collins SL, Glenn SM, and Gibson DJ (1995). Experimental analysis of intermediate disturbance and initial floristic composition: decoupling cause and effect. *Ecology*, 76, 486–492.

Collins SL, Knapp AK, Briggs JM, Blair JM, and Steinauer EM (1998a). Modulation of diversity by grazing and mowing in native tallgrass prairie. *Science*, 280, 745–747.

Collins SL, Knapp AK, Hartnett DC, and Briggs JM (1998b). The dynamic tallgrass prairie: synthesis and research opportunities. In AK Knapp, JM Briggs, DC Hartnett and SL Collins, eds. *Grassland dynamics: long-term ecological research in tallgrass prairie*, pp. 301–315. Oxford University Press, New York.

Connell JH (1978). Diversity in tropical rainforests and coral reefs. *Science*, 199, 1302–1310.

Conner HE (1986). Reproductive biology in grasses. In TR Soderstrom, KW Hilu, CS Campbell and ME Barkworth, eds. *Grass systematics and evolution*, pp. 117–132. Smithsonian Institution Press, Washington, DC.

Connor HE (1956). Interspecific hybrids in New Zealand *Agropyron*. *Evolution*, 10, 415–420.

Connor HE, Anton AM, and Astegiano ME (2000). Dioecism in grasses in Argentina. In SWL Jacobs and J Everett, eds. *Grasses: systematics and evolution*, pp. 287–293. CSIRO, Collingwood, Australia.

Coppock DL, Detling JK, Ellis JE, and Dyer MI (1983). Plant-herbivore interactions in a North American mixed-grass prairie. *Oecologia*, 56, 1–9.

Cornet B (2002). Upper Cretaceous facies, fossil plants, amber, insects and dinosaur bones, Sayreville, New Jersey. Available http://www.sunstar-solutions.com/sunstar/Sayreville/Kfacies.htm. Accessed 17 March 2008.

Costanza R and Farber S (2002). Introduction to the special issue on the dynamics and value of ecosystem services: integrating economic and ecological perspectives. *Ecological Economics*, 41, 367–373.

Costanza R, dArge R, deGroot R et al. (1997). The value of the world's ecosystem services and natural capital. *Nature*, 387, 253–260.

Cottam G (1987). Community dynamics on an artificial prairie. In WR Jordan III, ME Gilpin and JD Aber, eds. *Restoration ecology: a synthetic approach to ecological research*, pp. 257–270. Cambridge University Press, Cambridge.

Cottam G and Wilson HC (1966). Community dynamics on an artificial prairie. *Ecology*, 47, 88–96.

Couch HB (1973). *Diseases of turfgrasses*. 2nd edn. Robert E. Krieger, Huntington, NY.

Coughenour MB (1985). Graminoid responses to grazing by large herbivores: adaptations, exaptations, and interacting processes. *Annals of the Missouri Botanical Garden*, 72, 852–863.

Coulter BS, Murphy WE, Culleton N, Finnerty E, and Connolly L (2002). *A survey of fertilizer use in 2000 for grassland and arable crops*. Teagasc, Johnstown Castle Research Centre, Wexford.

Counce PA, Keisling TC, and Mitchell AJ (2000). A uniform, objective, and adaptive system for expressing rice development. *Crop Science*, 40, 436–443.

Coupe MD and Cahill J, J.F. (2003). Effects of insects on primary production in temperate herbaceous communities: a meta-analysis. *Ecological Entomology*, 28, 511–521.

Coupland RT (1958). The effects of fluctuations in weather upon the grasslands of the Great Plains. *Botanical Review*, 24, 271–317.

Coupland RT (1961). A reconsideration of grassland classification in the Northern Great Plains of North America. *Journal of Ecology*, 49, 135–167.

Coupland RT, ed. (1992a). *Ecosystems of the world: natural grasslands: introduction and western hemisphere*. Vol. 8A. Elsevier, Amsterdam.

Coupland RT (1992b). Mixed prairie. In RT Coupland, ed. *Ecosystems of the world: natural grasslands: introduction and western hemisphere*, Vol. 8A, pp. 151–182. Elsevier, Amsterdam.

Coupland RT (1992c). Overview of South American grasslands. In RT Coupland, ed. *Ecosystems of the world: natural grasslands: introduction and western hemisphere*, Vol. 8A, pp. 363–366. Elsevier, Amsterdam.

Coupland RT, ed. (1993a). *Ecosystems of the world: natural grasslands: eastern hemisphere and résumé*. Elsevier, Amsterdam.

Coupland RT (1993b). Review. In RT Coupland, ed. *Ecosystems of the world: eastern hemisphere and résumé*, Vol. 8B, pp. 471–482. Elsevier, Amsterdam.

Coutinho LM (1982). Ecological effects of fire in Brazilian cerrado. In BJ Huntley and BH Walker, eds. *Ecology of tropical savannas*, pp. 272–291. Springer-Verlag, Berlin.

Cowles HC (1899). The ecological relations of the vegetation on the sand dunes of Lake Michigan. *Botanical Gazette*, 27, 95–177, 167–202, 281–308, 361–391.

Cowling RM and Holmes PM (1992). Endemism and speciation in a lowland flora from the Cape Floristic Region. *Biological Journal of the Linnaean Society*, 47, 367–383.

Cox GW and Gakahu G (1985). Mima mound microtopography and vegetation pattern in Kenyan savannas. *Journal of Tropical Ecology*, 1, 23–36.

Cox RM and Hutchinson TC (1979). Metal co-tolerances in the grass *Deschampsia cespitosa*. *Nature*, 279, 231–233.

Craine JM (2005). Reconciling plant strategy theories of Grime and Tilman. *Journal of Ecology*, 93, 1041–1052.

Craine JM and Reich PB (2001). Elevated CO_2 and nitrogen supply alter leaf longevity of grassland species. *New Phytologist*, 150, 397–403.

Crawley MJ (1987). Benevolent herbivores? *Trends in Ecology & Evolution*, 2, 167–168.

Crepet WL and Feldman GD (1991). The earliest remains of grasses in the fossil record. *American Journal of Botany*, 78, 1010–1014.

Crider FJ (1955). *Root-growth stoppage*. Report No. 1102, United States Department of Agriculture, Washington, DC.

Cronquist A (1981). *An integrated system of classification of flowering plants*. Columbia University Press, New York.

Culver DC and Beattie AJ (1983). Effects of ant mounds on soil chemistry and vegetation patterns in a Colarado montane meadow. *Ecology*, 64, 485–492.

Cumming DHM (1982). The influence of large herbivores on savanna structure in Africa. In BJ Huntley and BH Walker, eds. *Ecology of Tropical Savannas*, pp. 217–245. Springer-Verlag, Berlin.

Curtis JT and Partch ML (1948). Effect of fire on the competition between blue grass and certain prairie plants. *American Midland Naturalist*, 39, 437–443.

Czapik R (2000). Apomixis in monocotyledons. In SWL Jacobs and J Everett, eds. *Grasses: systematics and evolution*, pp. 316–321. CSIRO, Collingwood, Australia.

D'Angela E, Facelli JM, and Jacobo E (1988). The role of the permanent soil seed bank in early stages of a post-agricultural succession in the Inland Pampa, Argentina. *Plant Ecology*, 74, 39–45.

D'Antonio CM and Vitousek PM (1992). Biological invasions by exotic grasses, the grass/fire cycle, and global change. *Annual Review of Ecology and Systematics*, 23, 63–87.

Dahlgren RMT, Clifford HT, and Yeo PF (1985). *The families of monocotyledons*. Springer-Verlag, Berlin.

Darwin CR (1845). *Journal of researches into the natural history and geology of the countries visited during the voyage of H.M.S. 'Beagle' round the world*. Ward, Lock & Co., London.

Darwin CR (1877). *The different forms of flowers on plants of the same species*. D. Appleton & Company, New York.

Daubenmire R (1968). Ecology of fire in grasslands. *Advances in Ecological Research*, 5, 209–266.

Davenport MA, Leahy JE, Anderson DH, and Jakes PJ (2007). Building trust in natural resource management within local communities: a case study of the Midewin National Tallgrass Prairie. *Environmental Management*, 39, 353–368.

Davidson AD and Lightfoot DC (2006). Keystone rodent interactions: prairie dogs and kangaroo rats structure the biotic composition of a desertified grassland. *Ecography*, 29, 755–765.

Davidson EA and Verchot LV (2000). Testing the hole-in-the-pipe model of nitric and nitrous oxide emissions from soils using the TRAGNET databse. *Global Biogeochemical Cycles*, 14, 1035–1043.

Davidson EA, Keller M, Erickson HE, Verchot LV, and Veldkamp E (2000). Testing a conceptual model of soil emissions of nitrous and nitric oxides. *BioScience*, 50, 667–680.

Davidson IA and Robson MJ (1986). Effect of temperature and nitrogen supply on the growth of perennial ryegrass and white clover. 2. A comparison of monocultures and mixed swards. *Annals of Botany*, 57, 709–719.

Davies A (2001). Competition between grasses and legumes in established pastures. In PG Tow and A Lazenby, eds. *Competition and succession in pastures*, pp. 63–83. CAB Publishing, Wallingford, Oxon.

Davies DM, Graves JD, Elias CO, and Williams PJ (1997). The impact of *Rhinanthus* spp. on sward productivity and composition: implications for the restoration of species-rich grasslands. *Biological Conservation*, 82, 98–93.

Davis AS, Dixon PM, and Liebman M (2004). Using matrix models to determine cropping system effects on annual weed demography. *Ecological Applications*, 14, 655–668.

Davis JI and Soreng RJ (2007). A preliminary phylogenetic analysis of the grass subfamily Pooideae (Poaceae), with attention to structural features of the plastic and nuclear genomes, including an intron loss in GBSSI. *Aliso*, 23, 335–348.

Davis MA, Thompson K, and Grime JP (2005). Invasibility: the local mechanism driving community assembly and species diversity. *Ecography*, 28, 696–704.

Day TA and Detling JK (1990). Grassland patch dynamics and herbivore grazing preference following urine deposition. *Ecology*, 71, 180–188.

de Kroon H and Visser EJW, eds. (2003). *Root ecology*. Springer-Verlag, Berlin.

de Kroon H, Plaisier A, van Groenendael JM, and Caswell H (1986). Elasticity as a measure of the relative contribution of demographic parameters to population growth rate. *Ecology*, 67, 1427–1431.

de Mazancourt C, Loreau M, and Abbadie L (1998). Grazing optimization and nutrient cycling: when do herbivores enhance plant production? *Ecology*, 79, 2242–2252.

de V. Booysen P (1981). Fertilization of sown pastures. In NM Tainton, ed. *Veld and pasture management in South Africa*, pp. 122–151. Shuter & Shooter, Pietermaritzburg.

De Vos A (1969). Ecological conditions affecting the production of wild herbivorous mammals on grasslands. *Advances in Ecological Research*, 6, 139–183.

De Wet JMJ (1986). Hybridization and polyploidy in the Poaceae. In TR Soderstrom, KW Hilu, CS Campbell and ME Barkworth, eds. *Grass systematics and evolution*, pp. 188–194. Smithsonian Institution Press, Washington, DC.

de Wit CT (1960). On competition. *Verslagen Can Landouskundige Onderzoekingen*, 66, 1–82.

de Wit MP, Crookes DJ, and Van Wilgen BW (2001). Conflicts of interest in environmental management: estimating the costs and benefits of a tree invasion. *Biological Invasions*, 3, 167–178.

DeBano LF, Neary DG, and Ffolliott PF (1998). *Fire's effects on ecosystems*. John Wiley & Sons, New York.

Defossé GE, Robberecht R, and Bertillier MB (1997a). Effects of topography, soil moisture, wind and grazing on *Festuca* seedlings in a Patagonian grassland. *Journal of Vegetation Science*, 8, 677–684.

Defossé GE, Robberecht R, and Bertillier MB (1997b). Seedling dynamics of *Festuca* spp in a grassland of Patagonia, Argentina, as affected by competition, microsites, and grazing. *Journal of Range Management*, 50, 73–79.

del-Val E and Crawley MJ (2005). What limits herb biomass in grasslands: competition or herbivory? *Oecologia*, 142, 202–211.

Deregibus VA, Sanchez RA, Casal JJ, and Trlica MJ (1985). Tillering responses to enrichment of red light beneath the canopy in a humid natural grassland. *Journal of Applied Ecology*, 22, 199–206.

Derner JD, Detling JK, and Antolin MF (2006). Are livestock weight gains affected by black-tailed prairie dogs? *Frontiers in Ecology and the Environment*, 4, 459–464.

Deshmukh I (1985). Decomposition of grasses in Nairobi National Park, Kenya. *Oecologia*, 67, 147–149.

Detling JK (1988). Grassland and savannas: regulation of energy flow and nutrient cycling by herbivores. In LR

Pomeroy and JJ Alberts, eds. *Concepts of ecosystem ecology*, pp. 131–148. Springer-Verlag, New York.

Dewey DR (1969). Synthetic hybrids of *Agropyron caespitosum* × *Agropyron spicatum*, *Agropyron caninum*, and *Agropyron yezoense*. *Botanical Gazette*, 130, 110–116.

Dewey DR (1972). Genome analysis of South American *Elymus patagonicus* and its hybrids with two North American and two Asian *Agropyron* species. *Botanical Gazette*, 133, 436–443.

Dewey DR (1975). The origin of *Agropyron smithii*. *American Journal of Botany*, 62, 524–530.

Diamond DD and Smeins FE (1988). Gradient analysis of remnant True and Upper Coastal prairie grasslands of North America. *Canadian Journal of Botany*, 66, 2152–2161.

Díaz S, Briske DD, and McIntyre S (2002). Range management and plant functional types. In AC Grice and KC Hodgkinson, eds. *Global rangelands: progress and prospects*, pp. 81–100. CABI Publishing, Wallingford, Oxon.

Dickinson CE and Dodd JL (1976). Phenological pattern in the shortgrass prairie. *American Midland Naturalist*, 96, 367–378.

Dijkstra FA, Hobbie SE, and Reich PB (2006). Soil processes affected by sixteen grassland species grown under different environmental conditions. *Soil Science Society of America Journal*, 70, 770–777.

DiTomaso JM, Kyser GB, and Hastings MS (1999). Prescribed burning for control of yellow starthistle (*Centaurea solstitialis*) and enhanced native plant diversity. *Weed Science*, 47, 233–242.

Dix RL (1964). A history of biotic and climatic changes within the North American grassland. In DJ Crisp, ed. *Grazing in terrestrial and marine environments*, pp. 71–90. Blackwell Scientific, Oxford.

Dodd JD (1983). Grassland associations in North America. In FW Gould and MR Shaw, eds. *Grass systematics*. 2nd edn. pp. 343–357. Texas A&M Press, College Station, TX.

Donald CM (1970). Temperate pasture species. In RM Moore, ed. *Australian grasslands*, pp. 303–320. Australian National University Press, Canberra.

Dormaar JF (1992). Decomposition as a process in natural grasslands. In RT Coupland, ed. *Ecosystems of the world: natural grasslands: introduction and western hemisphere*, Vol. 8A, pp. 121–136. Elsevier, Amsterdam.

Dostál P, Březnová M, Kozlíčková V, Herben T, and Kovář P (2005). Ant-induced soil modification and its effect on plant below-ground biomass. *Pedobiologia*, 49, 127–137.

Drenovsky RE and Batten KM (2006). Invasion by *Aegilops triuncialis* (barb goatgrass) slows carbon and nutrient cycling in a serpentine grassland. *Biological Invasions*, 9, 107–166.

Driessen P and Deckers J, eds. (2001). *Lecture notes on the major soils of the world*, pp 307. Food and Agriculture Organization of the United Nations, Rome.

Duffey E, Morris MG, Sheail J, Ward LK, Wells DA, and Wells TCA (1974). *Grassland ecology and wildlife management*. Chapman & Hall, London.

Dukes JS (2001). Biodiversity and invasibility in grassland microcosms. *Oecologia*, 126, 563.

Dunn J and Diesburg K (2004). *Turf management in the transition zone*. John Wiley & Sons, Hoboken, NJ.

Dunnett NP, Willis AJ, Hunt R, and Grime JP (1998). A 38-year study of relations between weather and vegetation dynamics in road verges near Bibury, Gloucestershire. *Journal of Ecology*, 86, 610–623.

Dvořák J, Luo M-C, and Yang Z-L (1998). Genetic evidence on the origin of *Triticum aestivum* L. In AB Damania, J Valkoun, G Willcox and CO Qualset, eds. *The origins of agriculture and crop domestication*, Ch 10. International Center for Agricultural Research in the Dry Areas, Aleppo, Syria.

Dyer AR and Rice KJ (1997). Intraspecific and diffuse competition: the response of *Nassella pulchra* in a California grassland. *Ecological Applications*, 7, 484–492.

Dyer MI, Detling JK, Coleman DC, and Hilbert DW (1982). The role of herbivores in grasslands. In JR Estes, RJ Tyrl and JN Brunken, eds. *Grasses and grassland communities: systematics and ecology*, pp. 255–295. University of Oklahoma Press, Norman, OK.

Dyksterhuis EJ (1949). Condition and management of rangeland based on quantitative ecology. *Journal of Range Management*, 2, 105–115.

Easton HS, Mackey AD, and Lee J (1997). Genetic variation for macro- and micro-nutrient concentration in perennial ryegrass (*Lolium perenne* L.). *Australian Journal of Agricultural Research*, 48, 657–666.

Edwards GR and Crawley MJ (1999). Herbivores, seed banks and seedling recruitment in mesic grassland. *Journal of Ecology*, 87, 423–435.

Edwards PJ and Tainton NM (1990). Managed grasslands in South Africa. In AI Breymeyer, ed. *Managed grasslands: regional studies: ecosystems of the world 17A*, pp. 99–128. Elsevier, Amsterdam.

Ehleringer JR and Monson RK (1993). Evolutionary and ecological aspects of photosynthetic pathway variation. *Annual Reviews of Ecology and Systematics*, 24, 411–439.

Ehrenfeld JG (1990). Dynamics and processes of barrier island vegetation. *Reviews in Aquatic Science*, 2, 437–480.

Einhellig FA and Souza IF (1992). Phytotoxicity of sorgoleone found in grain-sorghum root exudates. *Journal of Chemical Ecology*, 18, 1–11.

Eisele KA, Schimel DS, Kapustka LA, and Parton WJ (1989). Effects of available P-ratio and N-P-ratio on non-symbiotic dinitrogen fixation in tallgrass prairie soils. *Oecologia*, 79, 471–474.

Eissenstat DM and Caldwell MM (1988). Competitive ability is linked to rates of water extraction. *Oecologia*, 75, 1–7.

Eisto A-K, Kuitunen M, Lammi A, Saari V, Suhonen J, Syrjasuo S, and Tikka PM (2000). Population persistence and offspring fitness in the rare bellflower *Campanula cervicaria* in relation to population size and habitat quality. *Conservation Biology*, 14, 1413–1421.

Ellison L (1954). Subalpine vegetation of the Wasatch Plateau, Utah. *Ecological Monographs*, 24, 89–184.

Ellstrand NC (2001). When transgenes wander, should we worry? *Plant Physiology*, 125, 1543–1545.

Ellstrand NC, Prentice HC, and Hancock JF (1999). Gene flow and introgression from domesticated plants into their wild relatives. *Annual Review of Ecology and Systematics*, 30, 539–563.

Elmqvist T and Cox PA (1996). The evolution of vivipary in flowering plants. *Oikos*, 77, 3–9.

Elton CS (1958). *The ecology of invasion by plant and animals.* Methuen, London.

Emmett BA (2007). Nitrogen saturation of terrestrial ecosystems: some recent findings and their implications for our conceptual framework. *Water, Air, & Soil Pollution: Focus*, 7, 99–109.

Engle DM, Stritzke JF, Bidwell TG, and Claypool PL (1993). Late-summer fire and follow-up herbicide treatments in tallgrass prairie. *Journal of Range Management*, 46, 542–547.

Engler A (1892). *Syllabus der voelesungen über specielle und medicinisch-pharmaceutisch botanik.* Gebrüder Borntrager, Berlin.

Eppley SM (2001). Gender-specific selection during early life history stages in the dioecious grass *Distichlis spicata. Ecology*, 82, 2022–2031.

Epstein HE (1994). The anomaly of silica in plant biology. *Proceedings of the National Academy of Science, USA*, 91, 11–17.

Epstein HE (1999). Silicon. *Annual Review of Plant Physiology and Plant Molecular Biology*, 50, 641–664.

Epstein HE, Burke IC, and Lauenroth WK (2002a). Regional patterns of decomposition and primary production rates in the U.S. Great Plains. *Ecology*, 83, 320–327.

Epstein HE, Gill RA, Paruelo JM, Lauenroth WK, Jia GJ, and Burke IC (2002b). The relative abundance of three plant functional types in temperate grasslands and shrublands of North and South America: effects of projected climate change. *Journal of Biogeography*, 29, 875–888.

Escribano-Bailón M, Santos-Buelga C, and Rivas-Gonzalo JC (2004). Anthocyanins in cereals. *Journal of Chromotography A*, 1054, 129–141.

Evans EW (1983). The influence of neighboring hosts on colonization of prairie milkweeds by a seed-feeding bug. *Ecology*, 64, 648–653.

Evans EW, Briggs JM, Finck EJ, Gibson DJ, James SW, Kaufman DW, and Seastedt TR (1989). Is fire a disturbance in grasslands? In TB Bragg and J Stubbendieck, eds. *Proceedings of the eleventh North American prairie conference. prairie pioneers: ecology, history and culture*, pp. 159–161. University of Nebraska, Lincoln, NE.

Evans LT (1964). Reproduction. In C Barnard, ed. *Grasses and grasslands*, pp. 126–153. Macmillan & Co, London.

Evans MW (1946). *The grasses: their growth and development.* Ohio Agricultural Experiment Station, Wooster, OH.

Eyre FH (1980). *Forest cover types of the United States and Canada.* Society of American Foresters, Washington, DC.

Faber-Langendoen D, ed. (2001). *Plant communities of the Midwest: classification in an ecological context.* Association for Biodiversity Information, Arlington, VA.

Faber-Langendoen D, Tart D, Gray A *et al.* (2008 (in prep)). *Guidelines for an integrated physiognomic—floristic approach to vegetation classification.* Hierarchy Revisions Working Group, Federal Geographic Data Committee, Vegetation Subcommittee, Washington, DC.

Fabricius C, Palmer AR, and Burger M (2004). Landscape diversity in a conservation area and commercial and communal rangeland in xeric succulent thicket, South Africa. *Landscape Ecology*, 17, 531–537.

Facelli JM, Leon RJC, and Deregibus VA (1989). Community structure in grazed and ungrazed grassland sites in the Flooding Pampa, Argentina. *American Midland Naturalist*, 121, 125–133.

Faeth SH and Sullivan TJ (2003). Mutualistic asexual endophytes in a native grass are usually parasitic. *American Naturalist*, 161, 310–325.

Faeth SH, Helander ML, and Saikkonen KT (2004). Asexual *Neotyphodium* endophytes in a native grass reduce competitive abilities. *Ecology Letters*, 7, 304–313.

Fahnestock JT and Knapp AK (1994). Plant responses to selective grazing by bison: interactions between light, herbivory and water stress. *Vegetatio*, 115, 123–131.

Falk DA, Krapp EE, and Guerrant EO (2001). *An introduction to restoration genetics.* Available: http://www.ser.org/pdf/SER_restoration_genetics.pdf Accessed 14 January 2008. Society for Ecological Restoration.

Farber S, Costanza R, Childers DL *et al.* (2006). Linking ecology and economics for ecosystem management. *BioScience*, 56, 121–133.

Fargione J, Brown CS, and Tilman D (2003). Community assembly and invasion: an experimental test of neutral versus niche processes. *Proceedings of the National Academy of Sciences, USA*, 100, 8916–8920.

Fay PA and Hartnett DC (1991). Constraints on growth and allocation patterns of *Silphium integrifolium* (Asteraceae) caused by a cynipid gall wasp. *Oecologia*, 88, 243–250.

Fay PA, Carlisle JD, Knapp AK, Blair JM, and Collins SL (2000). Altering rainfall timing and quantity in a mesic grassland ecosystem: design and performance of rainfall manipulation shelters. *Ecosystems*, 3, 308–309.

Fay PA, Carlisle JD, Knapp AK, Blair JM, and Collins SL (2003). Productivity responses to altered rainfall patterns in a C_4-dominated grassland. *Oecologia*, 137, 245–251.

Fenner M and Thompson K (2005). *The ecology of seeds.* Cambridge University Press, Cambridge.

Ferreira VLP, Yotsuyanagi K, and Carvalho CRL (1995). Elimination of cyanogenic compounds from bamboo shoots *Dendrocalamus giganteus* Munro. *Tropical Science*, 35, 342–346.

Fester T, Maier W, and Strack D (1999). Accumulation of secondary compounds in barley and wheat roots in response to inoculation with an arbuscular mycorrhizal fungus and co-inoculation with rhizosphere bacteria. *Mycorrhiza*, 8, 241–246.

Fields MJ, Coffin DP, and Gosz JR (1999). Burrowing activities of kangaroo rats and patterns in plant species dominance at a shortgrass steppe-desert grassland ecotone. *Journal of Vegetation Science*, 10, 123–130.

Firestone MK and Davidson EA (1989). Microbiological basis of NO and N_2O production and consumption in soil. In MO Andreae and DS Schmel, eds. *Exchange of trace gases between terrestrial ecosystems and the atmosphere*, pp. 7–21. John Wiley & Sons, New York.

Fischer Walter LE, Hartnett DC, Hetrick BAD, and Schwab AP (1996). Interspecific nutrient transfer in a tallgrass prairie plant community. *American Journal of Botany*, 83, 180–184.

Fitter AH (2005). Presidential Address. Darkness visible: reflections on underground ecology. *Journal of Ecology*, 93, 231–243.

Fleming GA (1963). Distribution of major and trace elements in some common pasture species. *Journal of the Science of Food and Agriculture*, 14, 203–208.

Flint CL (1859). *Grasses and forage plants.* Phillips, Sampson & Co, Boston, MA.

Floate MJS (1987). Nitrogen cycling in managed grasslands. In RW Snaydon, ed. *Ecosystems of the world: managed grasslands: analytical studies*, Vol. 17B, pp. 163–172. Elsevier, Amsterdam.

Flora of Australia (2002). *Volume 43, Poaceae 1: introduction and atlas.* ABRS/CSIRO, Melbourne.

Flora of Australia (2005). *Volume 44B, Poaceae 3.* ABRS/CSIRO, Melbourne.

Flora of China Editorial Committee (2006). *Flora of China, Volume (22), Poaceae.* Missouri Botanical Garden Press, Saint Louis, MO.

Floyd J (1983). The day the southern plains went west. In *Saint Louis Globe Democrat*, pp. 1–12, Saint Louis, MO.

Folgarait PJ (1998). Ant biodiversity and its relationship to ecosystem functioning: a review. *Biodiversity and Conservation*, 7, 1221–1244.

Fone AL (1989). A comparative demographic study of annual and perennial *Hypochoeris* (Asteraceae). *Journal of Ecology*, 77, 495–508.

Forage and Grazing Terminology Committee (1992). Terminology for grazing lands and grazing animals. *Journal of Production Agriculture*, 5, 191–201.

Fossen T, Slimestad R, Øvstedal DO, and Andersen ØM (2002). Anthocyanins of grasses. *Biochemical Systematics and Ecology*, 30, 855–864.

Foster BL (2002). Competition, facilitation, and the distribution of *Schizachrium scoparium* along a topographic-productivity gradient. *Ecoscience*, 9, 355–363.

Foster BL and Tilman D (2003). Seed limitation and the regulation of community structure in oak savanna grassland. *Journal of Ecology*, 91, 999–1007.

Foster BL, Dickson TL, Murphy CA, Karel IS, and Smith VH (2004). Propagule pools mediate community assembly and diversity-ecosystem regulation along a grassland productivity gradient. *Journal of Ecology*, 92, 435–449.

Fox JF and Harrison AT (1981). Habitat assortment of sexes and water balance in a dioecious grass. *Oecologia*, 49, 233–235.

Franco M and Silvertown J (2004). A comparative demography of plants based upon elasticities of vital rates. *Ecology*, 85, 531–538.

Franks SJ (2003). Facilitation in multiple life-history stages: evidence for nucleated succession in coastal dunes. *Plant Ecology*, 168, 1–11.

Franzén D and Erikkson O (2001). Small-scale patterns of species richness in Swedish semi-natural grasslands: the effects of community species pools. *Ecography*, 24, 505–510.

Freckleton RP and Watkinson AR (2002). Large-scale spatial dynamics of plants: metapopulations, regional ensembles and patchy populations. *Journal of Ecology*, 90, 419–434.

Freckleton RP and Watkinson AR (2003). Are all plant populations metapopulations? *Journal of Ecology*, 91, 321–324.

Friedel MH, Laycock WA, and Bastin GN (2000). Assessing rangeland condition and trend. In L 't Mannetje and RM Jones, eds. *Field and laboratory methods for grassland and animal production research*, pp. 227–262. CABI Publishing, Wallingord, Oxon.

Frischknecht NC and Baker MF (1972). Voles can improve sagebrush rangelands. *Journal of Range Management*, 25, 466–468.

Frith HJ (1970). The herbivorous wild animals. In RM Moore, ed. *Australian grasslands*, pp. 74–83. Australian National University Press, Canberra.

Fritsche F and Kaltz O (2000). Is the *Prunella* (Lamiaceae) hybrid zone structured by an environmental gradient? Evidence from a reciprocal transplant experiment. *American Journal of Botany*, 87, 995–1003.

Fry J and Huang B (2004). *Applied turfgrass science and physiology*. John Wiley & Sons, Hoboken, NJ.

Fuhlendorf SD and Smeins FE (1996). Spatial scale influence on longterm temporal patterns of a semi-arid grassland. *Landscape Ecology*, 11, 107–113.

Fukui Y and Doskey PV (2000). Identification of non-methane organic compound emissions from grassland vegetation. *Atmospheric Environment*, 34, 2947–2956.

Fuller RM (1987). The changing extent and conservation interest of lowland grasslands in England and Wales: review of grassland surveys 1930–84. *Biological Conservation*, 40, 281–300.

Furniss PR, Ferrar P, Morris JW, and Bezuidenhout JJ (1982). A model of savanna litter decomposition. *Ecological Modelling*, 17, 33–50.

Fynn RWS and O'Connor TG (2005). Determinants of community organization of a South African mesic grassland. *Journal of Vegetation Science*, 16, 93–102.

Gale GW (1955). *John William Bews: a memoir*. University of Natal Press, Pietermaritzburg.

Galloway JN, Dentener FJ, Capone DG et al. (2004). Nitrogen cycles: past, present, and future. *Biogeochemistry*, 70, 153–226.

Gandolfo MA, Nixon KC, and Crepet WL (2002). Triuridaceae fossil flowers from the Upper Cretaceous of New Jersey. *American Journal of Botany*, 89, 1940–1957.

Garcia-Guzman G and Burdon JJ (1997). Impact of the flower smut *Ustilago cynodontis* (Ustilaginaceae) on the performance of the clonal grass *Cynodon dactylon* (Gramineae). *American Journal of Botany*, 84, 1565–1571.

Garcia-Guzman G, Burdon JJ, Ash JE, and Cunningham RB (1996). Regional and local patterns in the spatial distribution of the flower- infecting smut fungus *Sporisorium amphilophis* in natural populations of its host *Bothriochloa macra*. *New Phytologist*, 132, 459–469.

Garrett KA, Dendy SP, Frank EE, Rouse MN, and Travers SE (2006a). Climate change effects on plant disease: genomes to ecosystems. *Annual Review of Phytopathology*, 44, 489–509.

Garrett KA, Hulbert SH, Leach JE, and Travers SE (2006b). Ecological genomics and epidemiology. *European Journal of Plant Pathology*, 115, 35–51.

Gartside DW and McNeilly T (1974). Genetic studies in heavy metal tolerant plants I. Genetics of zinc tolerance in *Anthoxanthum odoratum*. *Heredity*, 32, 287–297.

Garwood EA (1967). Seasonal variation in appearance and growth of grass roots. *Journal of the British Grassland Society*, 22, 121–130.

Gatsuk LE, Smirnova OV, Vorontzova LI, Zaugolnova LB, and Zhukova LA (1980). Age states of plants of various growth forms: a review. *Journal of Ecology*, 68, 675–696.

Gaulthier DA and Wiken E (1998). The Great Plains of North America. *Parks*, 8, 9–20.

George MR, Brown JR, and Clawson WJ (1992). Application of nonequilibrium ecology to management of Mediterranean grasslands. *Journal of Range Management*, 45, 436–440.

Ghermandi L, Guthmann N, and Bran D (2004). Early post-fire succession in northwestern Patagonia grasslands. *Journal of Vegetation Science*, 15, 67–76.

Gianoli E and Niemeyer HM (1998). DIBOA in wild Poaceae: sources of resistance to the Russian wheat aphid (*Diuraphis noxiai*) and the greenbug (*Schizaphis graminum*). *Euphytica*, 102, 317–321.

Gibson CC and Watkinson AR (1989). The host range and selectivity of a parasitic plant: *Rhinanthus minor* L. *Oecologia*, 401–406.

Gibson DJ (1988a). The maintenance of plant and soil heterogeneity in dune grassland. *Journal of Ecology*, 76, 497–508.

Gibson DJ (1988b). Regeneration and fluctuation of tallgrass prairie vegetation in response to burning frequency. *Bulletin of the Torrey Botanical Club*, 115, 1–12.

Gibson DJ (1988c). The relationship of sheep grazing and soil heterogeneity to plant spatial patterns in dune grassland. *Journal of Ecology*, 76, 233–252.

Gibson DJ (1989). Effects of animal disturbance on tallgrass prairie vegetation. *American Midland Naturalist*, 121, 144–154.

Gibson DJ (1998). Review of *Population biology of grasses*, ed. G.P. Cheplick. *Ecology*, 79, 2968–2969.

Gibson DJ (2002). *Methods in comparative plant population ecology*. Oxford University Press, Oxford.

Gibson DJ and Hetrick BAD (1988). Topographic and fire effects on endomycorrhizae species composition on tallgrass prairie. *Mycologia*, 80, 433–451.

Gibson DJ and Hulbert LC (1987). Effects of fire, topography and year-to-year climatic variation on species composition in tallgrass prairie. *Vegetatio*, 72, 175–185.

Gibson DJ and Newman JA (2001). *Festuca arundinacea* Schreber (*F. elatior* subsp. *arundinacea* (Schreber) Hackel). *Journal of Ecology*, 89, 304–324.

Gibson DJ and Risser PG (1982). Evidence for the absence of ecotypic development in *Andropogon virginicus* (L.) on metalliferous mine wastes. *New Phytologist*, 92, 589–599.

Gibson DJ and Taylor I (2003). Performance of *Festuca arundinacea* Schreb. (Poaceae) populations in England. *Watsonia*, 24, 413–426.

Gibson DJ, Freeman CC, and Hulbert LC (1990a). Effects of small mammal and invertebrate herbivory on plant species richness and abundance in tallgrass prairie. *Oecologia*, 84, 169–175.

Gibson DJ, Hartnett DC, and Merrill GLS (1990b). Fire temperature heterogeneity in contrasting fire prone habitats: Kansas tallgrass prairie and Florida sandhill. *Bulletin of the Torrey Botanical Club*, 117, 349–356.

Gibson DJ, Seastedt TR, and Briggs JM (1993). Management practices in tallgrass prairie: large- and small-scale experimental effects on species composition. *Journal of Applied Ecology*, 30, 247–255.

Gibson DJ, Ely JS, Looney PB, and Gibson PT (1995). Effects of inundation from the storm surge of Hurricane Andrew upon primary succession on dredge spoil. *Journal of Coastal Research*, 21 Special Issue, 208–216.

Gibson DJ, Ely JS, and Collins SL (1999). The core-satellite species hypothesis provides a theoretical basis for Grime's classification of dominant, subordinate, and transient species. *Journal of Ecology*, 87, 1064–1067.

Gibson DJ, Spyreas G, and Benedict J (2002). Life history of *Microstegium vimineum* (Poaceae), an invasive grass in southern Illinois. *Journal of the Torrey Botanical Society*, 129, 207–219.

Gibson DJ, Middleton BA, Foster K, Honu YAK, Hoyer EW, and Mathis M (2005). Species frequency dynamics in an old-field succession: effects of disturbance, fertilization and scale. *Journal of Vegetation Science*, 16, 415–422.

Gicquiaud L, Hennion F, and Esnault MA (2002). Physiological comparisons among four related Bromus species with varying ecological amplitude: polyamine and aromatic amine composition in response to salt spray and drought. *Plant Biology*, 746–753.

Gill RA and Burke IC (2002). Influence of soil depth on the decomposition of *Bouteloua gracilis* roots in the shortgrass steppe. *Plant and Soil*, 241, 233–242.

Gill RA and Jackson RB (2000). Global patterns of root turnover for terrestrial ecosystems. *New Phytologist*, 147, 13–31.

Gill RA, Kelly RH, Parton WJ et al. (2002). Using simple environmental variables to estimate below-ground productivity in grasslands. *Global Ecology & Biogeography*, 11, 79–86.

Gillen RL, Rollins D, and Stritzke JF (1987). Atrazine, spring burning, and nitrogen for improvement of tallgrass prairie. *Journal of Range Management*, 40, 444–447.

Gillingham AG (1987). Phosphorus cycling in managed grasslands. In RW Snaydon, ed. *Ecosystems of the world: managed grasslands: analytical studies*, Vol. 17B, pp. 173–180. Elsevier, Amsterdam.

Gillison AN (1983). Tropical savannas of Australia and the southwest pacific. In F Bourliére, ed. *Ecosystems of the world 13: tropical savannas*, pp. 183–243. Elsevier Scientific Publishing Company, Amsterdam.

Gillison AN (1992). Overview of the grasslands of Oceania. In RT Coupland, ed. *Natural grasslands: introduction and western hemisphere*, Vol. 8A, pp. 303–313. Elsevier, Amsterdam.

Gillison AN (1993). Grasslands of the south-west Pacific. In RT Coupland, ed. *Natural grasslands: eastern hemisphere and résumé*, Vol. 8B, pp. 435–470. Elsevier, Amsterdam.

Gitay H and Wilson JB (1995). Post-fire changes in community structure of tall tussock grasslands: a test of alternative models of succession. *Journal of Ecology*, 83, 775–782.

Gleason HA (1917). The structure and development of the plant association. *Bulletin of the Torrey Botanical Club*, 44, 463–481.

Gleason HA (1922). The vegetation history of the Middle West. *Annals of the Association of American Geographers*, 12, 39–85.

Gleason HA (1926). The individualistic concept of the plant association. *American Midland Naturalist*, 21, 92–110.

Glenn SM, Collins SL, and Gibson DJ (1992). Disturbances in tallgrass prairie: local and regional effects on community heterogeneity. *Landscape Ecology*, 7, 243–251.

Glenn-Lewin DC, Johnson LA, Jurik TW, Kosek A, Leoschke M, and Rosburg T (1990). Fire in central North American grasslands: vegetative reproduction, seed germination, and seedling establishment. In SL Collins and LL Wallace, eds. *Fire in North American tallgrass prairies*, pp. 28–45. University of Oklahoma Press, Norman, OK.

Godley EJ (1965). The ecology of the Subantarctic islands of New Zealand: notes on the vegetation of the Auckland Islands. *New Zealand Journal of Ecology*, 12, 57–63.

Godt MJW and Hamrick JL (1998). Allozyme diversity in the grasses. In GP Cheplick, ed. *Population biology of grasses*, pp. 11–29. Cambridge University Press, New York.

Goldberg DE (1990). Components of resource competition in plant communities. In JB Grace and D Tilman, eds. *Perspectives on plant competition*, pp. 27–50. Academic Press, San Diego, CA.

Goodland R and Pollard R (1973). The Brazilian cerrado vegetation: a fertility gradient. *Journal of Ecology*, 61, 219–224.

Gotelli NJ and Simberloff D (1987). The distribution and abundance of tallgrass prairie plants: a test of the core-satellite hypothesis. *American Naturalist*, 130, 18–35.

Gould FW (1955). An approach to the study of grasses, the 'Tribal Triangle'. *Journal of Range Management*, 8, 17–19.

Gould FW and Shaw RB (1983). *Grass systematics*. Texas A & M University Press, College Station, TX.

GPWG (2000). Grass Phylogeny Working Group. Phylogeny and subfamilial classification of the grasses (Poaceae). Available: http://www.virtualherbarium.org/grass/gpwg/default.htm. Accessed 6 February 2004.

GPWG (2001). Grass Phylogeny Working Group. Phylogeny and subfamilial classification of the grasses (Poaceae). *Annals of the Missouri Botanical Garden*, 88, 373–457.

Graaf AJ, Stahl J, and Bakker JP (2005). Compensatory growth of *Festuca rubra* after grazing: can migratory herbivores increase their own harvest during staging? *Functional Ecology*, 19, 961–969.

Grace J (1990). On the relationship between plant traits and competitive ability. In JB Grace and D Tilman, eds. *Perspectives on plant competition*, pp. 51–66. Academic Press, San Diego, CA.

Grace J (1999). The factors controlling species diversity in herbaceous plant communities: an assessment. *Perspectives in Plant Ecology, Evolution and Systematics*, 2, 1–28.

Grace J, Meir P, and Malhi Y (2001a). Keeping track of carbon flows between biosphere and atmosphere. In MC Press, NJ Huntly and S Levin, eds. *Ecology: achievement and challenge*, pp. 249–269. Blackwell Science, Oxford.

Grace JB, Smith MD, Grace SL, Collins SL, and Stohlgren TJ (2001b). Interactions between fire and invasive plants in temperate grasslands of North America. In *Proceedings of the Invasive Species Workshop: the role of fire in the control and spread of invasive species*. Fire Conference 2000: the First National Congress on Fire Ecology, Prevention, and Management. Miscellaneous Publication No. 11 (eds KEM Galley and TP Wilson), pp. 40–65. Tall Timbers Research Station, Tallahassee, FL.

Grafton KF, Poehlman JM, and Sechler DT (1982). Tall fescue as a natural host and aphid vectors of barley yellow dwarf virus in Missouri. *Plant Disease*, 66, 318–320.

Gray SM, Power AG, Smith DM, Seamon AJ, and Altman NS (1991). Aphid transmission of barley yellow dwarf virus: acquisition access periods and virus concentration requirements. *Phytopathology*, 81, 539–545.

Grazing Lands Technology Institute (2003). *National range and pasture handbook*. United States Department of Agriculture, Natural Resources Conservation Service, Fort Worth, TX.

Green JO (1990). The distribution and management of grasslands in the British Isles. In AI Breymeyer, ed. *Managed grasslands: regional studies. ecosystems of the world 17A*, pp. 15–36. Elsevier, Amsterdam.

Green K (2005). Winter home range and foraging of common wombats (*Vombatus ursinus*) in patchily burnt subalpine areas of the Snowy Mountains, Australia. *Wildlife Research*, 32, 525–529.

Greig-Smith P (1983). *Quantitative plant ecology*. 3rd edn. Blackwell Scientific, Oxford.

Greipsson S, El-Mayas H, and Ahokas H (2004). Variation in populations of the coastal dune building grass *Leymus arenarius* in Iceland revealed by endospermal prolamins. *Journal of Coastal Conservation*, 10, 101–108.

Grice AC and Hodgkinson KC (2002). Challenges for rangeland people. In AC Grice and KC Hodgkinson, eds. *Global rangelands: progress and prospects*, pp. 1–9. CAB International, Wallingford, Oxon.

Grime JP (1979). *Plant strategies and vegetation processes*. John Wiley & Sons, Chichester.

Grime JP (1998). Benefits of plant diversity to ecosystems: immediate, filter and founder effects. *Journal of Ecology*, 86, 902–910.

Grime JP (2007). Plant strategy theories: a comment on Craine (2005). *Journal of Ecology*, 95, 227–230.

Grime JP and Hillier SH (2000). The contribution of seedling regeneration to the structure and dynamics of plant communities, ecosystems and larger units of the landscape. In M Fenner, ed. *Seeds: the ecology of regeneration in plant communities*. 2nd edn, pp. 361–374. CABI Publishing, Wallingford, Oxon.

Grime JP, Hodgson JG, and Hunt R (1988). *Comparative plant ecology*. Unwin, Hyman, London.

Gross KL, Mittelbach GG, and Reynolds HL (2005). Grassland invasibility and diversity: responses to nutrients, seed input, and disturbance. *Ecology*, 86, 476–486.

Grossman DH, Faber-Langendoen D, Weakley AS *et al.* (1998). *International classification of ecological communities: terrestrial vegetation of the United States*. Volume 1. *The National Vegetation Classification Scheme: development, status, and applications*. Nature Conservancy, Arlington, VA.

Gustafson DJ, Gibson DJ, and Nickrent NL (1999). Random amplified polymorphic DNA variation among remnant big bluestem (*Andropogon gerardii* Vitman) populations from Arkansas' Grand Prairie. *Molecular Ecology*, 8, 1693–1701.

Gustafson DJ, Gibson DJ, and Nickrent NL (2001). Characterizing three restored *Andropogon gerardii* Vitman (big bluestem) populations established with Illinois and non-Illinois seed: established plants and their offspring. In *17th North American Prairie Conference. Seeds for the future; roots of the past* (eds NP Bernstein and LJ Ostrander), pp. 118–124. North Iowa Area Community College, Mason City, IA.

Gustafson DJ, Gibson DJ, and Nickrent DL (2002). Genetic diversity and competitive abilities of *Dalea purpurea* (Fabaceae) from remnant and restored grasslands. *International Journal of Plant Science*, 163, 979–990.

Gustafson DJ, Gibson DJ, and Nickrent DL (2004). Conservation genetics of two co-dominant grass species in an endangered grassland ecosystem. *Journal of Applied Ecology*, 41, 389–397.

Gustafson DJ, Gibson DJ, and Nickrent DL (2005). Empirical support for the use of local seed sources in prairie restoration. *Native Plants Journal*, 6, 25–28.

Haahtela K, Wartiovaara T, Sundman V, and Skujins J (1981). Root-associated N_2 fixation (acetylene reduction) by Enterobacteriaceae and Azospirillum strains in cold-climate spodosols. *Applied and Environmental Microbiology*, 41, 203–206.

Hanski I (1982). Dynamics of regional distribution: the core and satellite species hypothesis. *Oikos*, 38, 210–221.

Hanski I (1999). *Metapopulation ecology*. Oxford University Press, New York.

Hanski I and Gyllenberg M (1993). Two general metapopulation models and the core-satellite species hypothesis. *American Naturalist*, 142, 17–41.

Harberd DJ (1961). Observations on population structure and longevity in *Festuca rubra* L. *New Phytologist*, 60, 184–206.

Harborne JB (1967). *Comparative biochemistry of the flavonoids*. Academic Press, London.

Harborne JB (1977). *Introduction to ecological biochemistry*. Academic Press, London.

Harborne JB and Williams CA (1986). Flavonoids patterns of grasses. In TR Soderstrom, KW Hilu, CS Campbell and ME Barkworth, eds. *Grass systematics and evolution*, pp. 107–113. Smithsonian Institution Press, Washington, DC.

Hardin G (1968). The tragedy of the commons. *Science*, 162, 1243–1248.

Harlan JR (1956). *Theory and dynamics of grassland agriculture*. Van Nostrand, Princeton, NJ.

Harniss RO and Murray RB (1973). 30 years of vegetal change following burning of sagebrush-grass range. *Journal of Range Management*, 26, 322–325.

Harper JL (1977). *Population biology of plants*. Academic Press, London.

Harper JL (1978). Plant relations in pastures. In JR Wilson, ed. *Plant relations in pastures*, pp. 3–16. CSIRO, East Melbourne, Australia.

Harpole SW (2006). Resource-ratio theory and the control of invasive plants. *Plant and Soil*, 280, 23–27.

Harpole WS and Tilman D (2006). Non-neutral patterns of species abundance in grassland communities. *Ecology Letters*, 9, 15–23.

Hartley W (1958). Studies on the origin, evolution, and distribution of the Gramineae I. the tribe Andropogoneae. *Australian Journal of Botany*, 6, 115–128.

Hartley W (1961). Studies on the origin, evolution, and distribution of the Gramineae: the genus *Poa*. *Australian Journal of Botany*, 9, 152–161.

Hartnett DC (1993). Regulation of clonal growth and dynamics of *Panicum virgatum* (Poaceae) in tallgrass prairie: effects of neighbour removal and nutrient addition. *American Journal of Botany*, 80, 1114–1120.

Hartnett DC and Fay PA (1998). Plant populations: patterns and processes. In AK Knapp, JM Briggs, DC Hartnett and SL Collins, eds. *Grassland dynamics*, pp. 81–100. Oxford University Press, New York.

Hartnett DC and Keeler KH (1995). Population processes. In A Joern and KH Keeler, eds. *The changing prairie: North American grasslands*, pp. 82–99. Oxford University Press, New York.

Hartnett DC and Wilson GWT (1999). Mycorrhizae influence plant community structure and diversity in tallgrass prairie. *Ecology*, 80, 1187–1195.

Hartnett DC, Hetrick BAD, Wilson GWT, and Gibson DJ (1993). Mycorrhizal influence on intra- and interspecific neighbour interactions among co-occurring prairie grasses. *Journal of Ecology*, 81, 787–795.

Hartnett DC, Samensus RJ, Fischer LE, and Hetrick BAD (1994). Plant demographic responses to mycorrhizal symbiosis in tallgrass prairie. *Oecologia*, 99, 21–26.

Hary I (2004). Assessing the effect of controlled seasonal breeding on steady-state productivity of pastoral goat herds in northern Kenya. *Agricultural Systems*, 81, 153–175.

Hatch MD and Slack CR (1966). Photosynthesis by sugarcane leaves. *Biochemical Journal*, 101, 103–111.

Hatfield R and Davies J (2006). *Global review of the economics of pastoralism*. IUCN (World Conservation Union), Nairobi.

Hattersley PW (1986). Variations in photosynthetic pathway. In TR Soderstrom, KW Hilu, CS Campbell and ME Barkworth, eds. *Grass systematics and evolution*, pp. 49–64. Smithsonian Institution Press, Washington, DC.

Heady HF, Bartolome JW, Pitt MD, Savelle GD, and Stroud MC (1992). California prairie. In RT Coupland, ed. *Natural grasslands: introduction and western hemisphere*, Vol. 8A, pp. 313–335. Elsevier, Amsterdam.

Heaton EA, Clifton-Brown J, Voigt TB, Jones MB, and Long SP (2004). Miscanthus for renewable energy generation: European Union experience and projections for Illinois. *Mitigation and Adaptation Strategies for Global Change*, 9, 433–451.

Hector A, Dobson K, Minns A, Bazeley-White E, and Hartley Lawton J (2001). Community diversity and invasion resistance: an experimental test in a grassland ecosystem and a review of comparable studies. *Ecological Research*, 16, 819.

Hector A, Schmid B, Beierkuhnlein C et al. (1999). Plant diversity and productivity experiments in European grasslands. *Science*, 286, 1123–1127.

Heinz Center (2002; 2005 update) The H. John Heinz III Center for Science Economics and the Environment. *The state of the nation's ecosystems: measuring the lands, waters, and living resources of the United States*. Cambridge University Press, New York.

Heitschmidt RK and Taylor CAJ (1991). Livestock production. In RK Heitschmidt and JW Stuth, eds. *Grazing managment: an ecological perspective*, pp. 161–178. Timber Press, Portland, OR.

Helgadóttir Á and Snaydon RW (1985). Competitive interactions between populations of *Poa pratensis* and *Agrostis tenuis* from ecologically-contrasting environments. *Journal of Applied Ecology*, 22, 525–537.

Helm A, Hanski I, and Partel M (2006). Slow response of plant species richness to habitat loss and fragmentation. *Ecology Letters*, 9, 72–77.

Hendon BC and Briske DD (1997). Demographic evaluation of a herbivory-sensitive perennial bunchgrass: does it possess an Achilles heel? *Oikos*, 80, 8–17.

Hendrickson JR and Briske DD (1997). Axillary bud banks of two semiarid perennial grasses: occurrence, longevity, and contribution to population persistence. *Oecologia*, 110, 584–591.

Henry HAL and Jefferies RL (2003). Plant amino acid uptake, soluble N turnover and microbial N capture in soils of a grazed Arctic salt marsh. *Journal of Ecology*, 91, 627–636.

Henwood WD (1998a). Editorial—the worlds temperate grasslands: a beleaguered biome. *Parks*, 8, 1–2.

Henwood WD (1998b). An overview of protected areas in the temperate grasslands biome. *Parks*, 8, 3–8.

Herendeen RA and Wildermuth T (2002). Resource-based sustainability indicators: Chase County, Kansas, as example. *Ecological Economics*, 42, 243–257.

Herlocker DJ, Dirschl HJ, and Frame G (1993). Grasslands of East Africa. In RT Coupland, ed. *Ecosystems of the world: eastern hemisphere and résumé*, Vol. 8B, pp. 221–264. Elsevier, Amsterdam.

Herrick JE, Van Zee JW, Havstad KM, Burkhett LM, and Whiford WG (2005a). *Monitoring manual for grassland, shrubland and savanna ecosytems. Volume 1. Quick start*. USDA-ARS Jornada Experimental Range, Las Cruces, NM.

Herrick JE, Van Zee JW, Havstad KM, Burkhett LM, and Whiford WG (2005b). *Monitoring manual for grassland, shrubland and savanna ecosytems. Volume 2. Design, supplementary methods and interpretation*. USDA-ARS Jornada Experimental Range, Las Cruces, NM.

Heslop-Harrison J and Heslop-Harrison Y (1986). Pollen-stigma interaction in grasses. In TR Soderstrom, KW Hilu, CS Campbell and ME Barkworth, eds. *Grass systematics and evolution*, pp. 133–142. Smithsonian Institution Press, Washington, DC.

Hetrick BAD, Kitt DG, and Wilson GWT (1986). The influence of phosphorus fertilizer, drought, fungal species, and nonsterile soil on the mycorrhizal growth response in tallgrass prairie. *Canadian Journal of Botany*, 64, 1199–1203.

Hetrick BAD, Kitt DG, and Wilson GWT (1988). Mycorrhizal dependence and growth habit of warm-season and cool-season tallgrass prairie plants. *Canadian Journal of Botany*, 66, 1376–1380.

Hetrick BAD, Hartnett DC, Wilson GWT, and Gibson DJ (1994). Effects of mycorrhizae, phosphorus availability, and plant density on yield relationships among competing tallgrass prairie grasses. *Canadian Journal of Botany*, 72, 168–176.

Heywood VH, ed. (1978). *Flowering plants of the world*. Mayflower Books, New York.

Hierro JL and Callaway RM (2003). Allelopathy and exotic plant invasion. *Plant and Soil*, 256, 29–39.

Higgins KF (1984). Lightning fires in North Dakota grasslands and in pine-savanna lands of South Dakota and Montana. *Journal of Range Management*, 37, 100–103.

Hilton JR (1984). The influence of temperature and moisture status on the photoinhibition of seed germination in *Bromus sterilis* L. by the far-red absorbing form of Phytochrome. *New Phytologist*, 97, 369–374.

Hitchcock AS and Chase A (1950). *Manual of the grasses of the United States*. 2nd edn. Miscellaneous Publication No. 200, Department of Agriculture, Washington, DC.

Hnatiuk RJ (1993). Grasslands of the sub-Antarctic islands. In RT Coupland, ed. *Natural grasslands: eastern hemisphere and résumé*, Vol. 8B, pp. 411–434. Elsevier, Amsterdam.

Hobbs RJ and Mooney HA (1985). Community and population dynamics of serpentine grassland annuals in relation to gopher disturbance. *Oecologia*, 67, 342–351.

Hobbs RJ, Currall JE, and Gimingham CH (1984). The use of 'thermocolor' pyrometers in the study of heath fire behaviour. *Journal of Ecology*, 72, 241–250.

Hodgkinson KC, Harrington GN, Griffin GF, Noble JC, and Young MD (1984). Management of vegetation with fire. In GN Harrington, AD Wilson and MD Young, eds. *Management of Australia's rangelands*, pp. 141–156. CSIRO, East Melbourne, Australia.

Hodgson J (1979). Nomenclature and definitions in grazing studies. *Grass and Forage Science*, 34, 11–18.

Hodkinson TR, Salamin N, Chase MW, Bouchenak-Khelladi Y, Renvoize SA, and Savolainen V (2007a). Large trees, supertrees, and diversification of the grass family. *Aliso*, 23, 248–258.

Hodkinson TR, Savolainen V, Jacobs SWL, Bouchenak-Khelladi Y, Kinney MS, and Salamin N (2007b). Supersizing: progress in documenting and understanding grass species richness. In TR Hodkinson and JAN Parnell, eds. *Reconstructing the tree of life: taxonomy and systematics of species rich taxa*, pp. 279–298. CRC Press, Boca Raton, FL.

Hoekstra JM, Boucher TM, Ricketts TH, and Roberts C (2005). Confronting a biome crisis: global disparities of habitat loss and protection. *Ecology Letters*, 8, 23–29.

Høiland K and Oftedal P (1980). Lead-tolerance in *Deschampia flexuosa* from a naturally lead polluted area in S. Norway. *Oikos*, 34, 168–172.

Holdaway RJ and Sparrow AD (2006). Assembly rules operating along a primary riverbed-grassland successional sequence. *Journal of Ecology*, 94, 1092–1102.

Holling CS (1978). *Adaptive environmental assessment and management*. John Wiley & Son, Chichester.

Holm AM, Burnside DG, and Mitchell AA (1987). The development of a system for monitoring trend in range condition in the arid shrublands of Western Australia. *Rangeland Journal*, 9, 14–20.

Holmgren M, Scheffer M, and Huston MA (1997). The interplay of facilitation and competition in plant communities. *Ecology*, 78, 1966–1975.

Honey M (1999). *Ecotourism and sustainable development: who owns paradise?* Island Press, Washington DC.

Hover EI and Bragg TB (1981). Effect of season of burning and mowing on an Eastern Nebraska *Stipa–Andropogon* prairie. *American Midland Naturalist*, 105, 13–18.

Howarth W (1984). The Okies: beyond the dust bowl. *National Geographic*, 166, 322–349.

Howe HF (1995). Succession and fire season in experimental prairie plantings. *Ecology*, 76, 1917–1925.

Howe HF and Lane D (2004). Vole-driven succession in experimental wet prairie restorations. *Ecological Applications*, 14, 1295–1305.

Howe HF, Brown JS, and Zorn-Arnold B (2002). A rodent plague on prairie diversity. *Ecology Letters*, 5, 30–36.

Howell E and Stearns F (1993). The preservation, management, and restoration of Wisconsin Plant Communities: the influence of John Curtis and his students. In JS Fralish, RP McIntosh and OL Loucks, eds. *John Curtis: fifty years of Wisconsin plant ecology*, pp. 57–66. Wisconsin Academy of Sciences, Art & Letters, Madison, WI.

Hu Z and Zhang D (2003). China's pasture resources. In JM Suttie and SG Reynolds, eds. *Transhumant grazing systems in temperate Asia*. Food and Agriculture Organization of the United Nations, Rome.

Huang BR and Liu XZ (2003). Summer root decline: production and mortality for four cultivars of creeping bentgrass. *Crop Science*, 43, 258–265.

Huang S-Q, Yang H-F, Lu I, and Takahashi Y (2002). Honeybee-assisted wind pollination in bamboo *Phyllostachys nidularia* (Bambusoideae: Poaceae)? *Botanical Journal of the Linnean Society*, 138, 1–7.

Hubbard CE (1954). *Grasses*. Penguin, Harmondsworth.

Hubbard CE (1984). *Grasses*. 3rd edn. Penguin, Harmondsworth.

Hubbell SP (2001). *The unified neutral theory of biodiversity and biogeography*. Princeton University Press, Princeton, NJ.

Hudson EA, Dinh PA, Kokubun T, Simmonds MSJ, and Gescher A (2000). Characterization of potentially chemopreventive phenols in extracts of brown rice that inhibit the growth of human breast and colon cancer cells. *Cancer Epidemiology Biomarkers & Prevention*, 9, 1163–1170.

Hufford KM and Mazer SJ (2003). Plant ecotypes: genetic differentiation in an age of ecological restoration. *Trends in Ecology and Evolution*, 18, 147–155.

Hui D and Jackson RB (2005). Geographical and interannual variability in biomass partitioning in grassland ecosystems: a synthesis of field data. *New Phytologist*, 169, 85–93.

Hulbert LC (1955). Ecological studies of *Bromus tectorum* and other annual bromegrasses. *Ecological Monographs*, 25, 181–213.

Hulbert LC (1988). Causes of fire effects in tallgrass prairie. *Ecology*, 69, 46–58.

Humphrey RR (1947). Range forage evaluation by the range condition method. *Journal of Forestry*, 45, 10–16.

Humphrey RR and Mehrhoff LA (1958). Vegetation changes on a southern Arizona grassland range. *Ecology*, 39, 720–726.

Humphreys LR (1997). *The evolving science of grassland improvement*. Cambridge University Press, Cambridge.

Hunt HW (1977). A simulation model for decomposition in grasslands. *Ecology*, 58, 469–484.

Hunt HW, Coleman DC, Ingham ER *et al.* (1987). The detrital food web in a shortgrass prairie. *Biology and Fertility of Soils*, 3, 57–68.

Hunt MG and Newman JA (2005). Reduced herbivore resistance from a novel grass–endophyte association. *Journal of Applied Ecology*, 42, 762–769.

Hunt R, Hodgson JG, Thompson K, Bungener P, Dunnett NP, and Askew AP (2004). A new practical tool for deriving a functional signature for herbaceous vegetation. *Applied Vegetation Science*, 7, 163–170.

Huntley BJ and Walker BH, eds. (1982). *Ecology of tropical savannas*. Springer-Verlag, Berlin.

Huntly NJ and Reichman OJ (1994). Effects of subterranean mammalian herbivores on vegetation. *Journal of Mammalogy*, 75, 852–859.

Huston JE and Pinchak WE (1991). Range Animal Nutrition. In RK Heitschmidt and JW Stuth, eds. *Grazing management: an ecological perspective*, Chapter 2. Timber Press, Portland, OR.

Huston MA (1979). A general hypothesis of species diversity. *American Naturalist*, 113, 81–101.

Hutchings MJ and Booth KD (1996). Studies on the feasibility of recreating chalk grassland on ex-arable land. I. The potential roles of the seedbank and seed rain. *Journal of Applied Ecology*, 33, 1171–1181.

Hutchings MJ and Stewart AJA (2002). Calcareous grasslands. In MR Perrow and AJ Davy, eds. *Handbook of ecological restoration. Vol. 2. Restoration in practice*, pp. 419–443. Cambridge University Press, Cambridge.

Huxman TE, Hamerlynck EP, Jordan DN, Salsman KJ, and Smith SD (1998). The effects of parental CO_2 environment on seed quality and subsequent seedling performance in *Bromus rubens*. *Oecologia*, 114, 202–208.

Illinois Department of Energy and Natural Resources (1994). *The changing Illinois environment: critical trends*. Illinois Department of Energy and Natural Resources, Springfield, IL.

Illius AW and O'Connor TG (1999). On the relevance of nonequilibrium concepts to arid and semiarid grazing systems. *Ecological Applications*, 9, 798–813.

Iltis HH (2000). Homeotic sexual translocations and the origin of maize (*Zea mays*, Poaceae): a new look at an old problem. *Economic Botany*, 54, 7–42.

Ingalls JJ (1948). In praise of blue grass: 1872 address. In The Yearbook Committee, ed. *Grass: the yearbook of agriculture 1948*, pp. 6–8. U.S. Department of Agriculture, Washington, DC.

IPCC (2007). *Climate change 2007: IPCC 4th assessment report*. Intergovernmental Panel on Climate Change, Geneva, Switzerland.

Ishikawa Y and Kanke T (2000). Role of graminea in the feeding deterrence of barley against the migratory locust, *Locusta migratoria* (Orthoptera: Acrididae). *Applied Entomology and Zoology*, 35, 251–256.

Ito I (1990). Managed grassland in Japan. In AI Breymeyer, ed. *Managed grasslands: regional studies. ecosystems of the world 17A*, pp. 129–148. Elsevier, Amsterdam.

Iturralde-Vinent MA and MacPhee RDE (1996). Age and paleogeographical origin of Dominican amber. *Science*, 273, 1850–1852.

IUCN/SSC Invasive Species Specialist Group (2003). Global invasive species database. Available: http://www.issg.org/database/welcome/. Accessed 17 March 2008.

IUCN-WCPA (2000). Proceedings. In *Seminar on the protection and conservation of grasslands in east Asia*, pp. 75. IUCN-WCPA, Ulaanbaatar, Mongolia.

Jacobs BF, Kingston JD, and Jacobs LL (1999). The origin of grass-dominated ecosystems. *Annals of the Missouri Botanical Garden*, 86, 590–643.

Jameson DA (1963). Responses of individual plants to harvesting. *Botanical Review*, 29, 532–594.

Janis CM, Damuth J, and Theodor JM (2004). The species richness of Miocene browsers, and implications for habitat type and primary productivity in the North American grassland biome. *Paleogeography, Paleoclimatology, Palaeoecology*, 207, 371–398.

Janssen T and Bremer K (2004). The age of major monocot groups inferred from 800+ rbcL sequences. *Botanical Journal of the Linnean Society*, 146, 385–398.

Janssens F, Peeters A, Tallowin JRB, Bakker JP, Bekker RM, Fillat F, and Oomes MJM (1998). Relationship between soil chemical factors and grassland diversity. *Plant and Soil*, 202, 69.

Jean JP and Keith C (2003). Infection by the systemic fungus *Epichloë glyceriae* alters clonal growth of its grass host, *Glyceria striata*. *Proceedings of the Royal Society B, Biological Sciences*, 270, 1585–1591.

Jefferson RG (2005). The conservation management of upland hay meadows in Britain: a review. *Grass and Forage Science*, 60, 322–331.

Jennersten O (1988). Pollination in *Dianthus deltoides* (Caryophyllaceae): effects of habitat fragmentation on visitation and seed set. *Conservation Biology*, 2, 359–366.

Joern A (1995). The entangled bank: species interactions in the structure and functioning of grasslands. In A

Joern and KH Keeler, eds. *The changing prairie: North American grasslands*, pp. 100–127. Oxford University Press, New York.

Johnson CN and Prideaux GJ (2004). Extinctions of herbivorous mammals in the late Pleistocene of Australia in relation to their feeding ecology: no evidence for environmental change as cause of extinction. *Austral Ecology*, 29, 553–557.

Johnson FL, Gibson DJ, and Risser PG (1982). Revegetation of unreclaimed coal strip-mines in Oklahoma. *Journal of Applied Ecology*, 19, 453–463.

Jones CG, Lawton JH, and Shachak M (1997). Positive and negative effects of organisms as physical ecosystem engineers. *Ecology*, 78, 1946–1957.

Jones M (1985). Modular demography and form in silver birch. In J White, ed. *Studies on plant demography: a festschrift for John L. Harper*, pp. 223–237. Academic Press, London.

Jones MB and Woodmansee RG (1979). Biogeochemical cycling in annual grassland ecosystems. *Botanical Review*, 45, 111–144.

Jones TA (2005). Genetic principles and the use of native seeds—just the FAQs, please, just the FAQs. *Native Plants Journal*, 6, 14–24.

Josens G (1983). The soil fauna of tropical savannas, III. The termites. In F Bourliére, ed. *Ecosystems of the world 13: tropical savannas*, pp. 505–524. Elsevier, Amsterdam.

Joshi J, Matthies D, and Schmid B (2000). Root hemiparasites and plant diversity in experimental grassland communities. *Journal of Ecology*, 88, 634–644.

Josse C, Navarro G, Comer P et al. (2003). *Ecological systems of Latin America and the Caribbean: a working classification of terrestrial systems*. NatureServe, Arlington, VA.

Jowett D (1964). Population studies on lead-tolerant *Agrostis tenuis*. *Evolution*, 18, 70–81.

Joyce LA (1993). The life cycle of the range condition concept. *Journal of Range Management*, 46, 132–138.

Kaiser P (1983). The role of soil micro-organisms in savanna ecosystems. In F Bourliére, ed. *Ecosystems of the world 13: tropical savannas*, pp. 541–557. Elsevier, Amsterdam.

Kalamees R and Zobel M (1998). Soil seed bank composition in different successional stages of a species rich wooded meadow in Laelatu, western Estonia. *Acta Oecologica*, 19, 175–180.

Kalisz S and McPeek MA (1992). Demography of an age-structured annual: resampled projection matrices, elasticity analyses, and seed bank effects. *Ecology*, 73, 1082–1093.

Kapadia ZJ and Gould FW (1964). Biosystematic studies in the *Bouteloua curtipendula* complex. IV. Dynamics of variation in *B. curtipendula* var. *caespitosa*. *Bulletin of the Torrey Botanical Club*, 91, 465–478.

Karataglis SS (1978). Studies on heavy metal tolerance in popuations of *Anthoxanthum odoratum*. *Deutsche Botanische Gesellschaft Berichte*, 91, 205–216.

Karki JB, Jhala YV, and Khanna PP (2000). Grazing lawns in Terai grasslands, Royal Bardia National Park, Nepal. *Biotropica*, 32, 423–429.

Kaufman DW, Kaufman GA, Fay PA, Zimmerman JL, and Evans EW (1998). Animal populations and communities. In AK Knapp, JM Briggs, DC Hartnett and SL Collins, eds. *Grassland dynamics*, pp. 113–139. Oxford University Press, New York.

Keeler KH (1990). Distribution of polyploid variation in big bluestem (*Andropogon gerardii*, Poaceae) across the tallgrass prairie region. *Genome*, 33, 95–100.

Keeler KH (1992). Local polyploid variation in the native prairie grass *Andropogon gerardii*. *American Journal of Botany*, 79, 1229–1232.

Keeler KH (1998). Population biology of intraspecific polyploidy in grasses. In GP Cheplick, ed. *Population biology of grasses*, pp. 183–208. Cambridge University Press, Cambridge.

Keeler KH and Davis GA (1999). Comparison of common cytotypes of *Andropogon gerardii* (Andropogoneae, Poaceae). *American Journal of Botany*, 86, 974–979.

Keeler KH and Kwankin B (1989). Polyploid polymorphism in grasses of the North American prairie. In JH Bock and YB Linhart, eds. *The evolutionary ecology of plants*. Westview Press, Boulder, CO.

Keeler KH, Kwankin B, Barnes PW, and Galbraith DW (1987). Polyploid polymorphism in *Andropogon gerardii*. *Genome*, 29, 374–379.

Keeley JE (2006). Fire management impacts on invasive plants in the Western United States. *Conservation Biology*, 20, 375–384.

Keeley JE and Rundel PW (2005). Fire and the Miocene expansion of C_4 grasslands. *Ecology Letters*, 8, 683–690.

Kellogg EA (2000). The grasses: a case study in macroevolution. *Annual Review of Ecology and Systematics*, 31, 217–38.

Kellogg EA (2001). Evolutionary history of the grasses. *Plant Physiology*, 125, 1198–1205.

Kellogg EA (2002). Classification of the grass family. *Flora of Australia*, 43, 19–36.

Kemp DR and King WM (2001). Plant competition in pastures—implications for management. In PG Tow and A Lazenby, eds. *Competition and succession in pastures*, pp. 85–102. CABI Publishing, Wallingford, Oxon.

Kepe T (2001). Tourism, protected areas and development in South Africa: views of visitors to Mkambati Nature

Reserve. *South African Journal of Wildlife Research*, 31, 155–159.

Kesselmeier J and Staudt M (1999). Biogenic volatile organic compounds (VOC): an overview on emission, physiology and ecology. *Journal of Atmospheric Chemistry*, 33, 23–88.

Kikuzawa K and Ackerly D (1999). Significance of leaf longevity in plants. *Plant Species Biology*, 14, 39–45.

Kikvidze Z (1996). Neighbour interaction and stability in subalpine meadow communities. *Journal of Vegetation Science*, 7, 41–44.

King TJ (1977a). The plant ecology of ant-hills in calcareous grasslands: I. Patterns of species in relation to ant-hills in southern England. *Journal of Ecology*, 65, 235–256.

King TJ (1977b). The plant ecology of ant-hills in calcareous grasslands: II. Succession on the mounds. *Journal of Ecology*, 65, 257–278.

King TJ (1981). Ant-hills and grassland history. *Journal of Biogeography*, 8, 329–334.

Kirkham FW, Mountford JO, and Wilkins RJ (1996). The effects of nitrogen, potassium and phosphorus addition on the vegetation of a Somerset peat moor under cutting management. *Journal of Applied Ecology*, 33, 1013 1029.

Kirschbaum MUF (1994). The sensitivity of C_3 photosynthesis to increasing CO_2 concentration: a theoretical analysis of its dependence on temperature and background CO_2 concentration. *Plant, Cell and Environment*, 17, 747–754.

Kitajima K and Fenner M (2000). Ecology of seedling regeneration. In M Fenner, ed. *Seeds: the ecology of regeneration in plant communities*. 2nd edn, pp. 331–360. CABI Publishing, Wallingford, Oxon.

Kitajima K and Tilman D (1996). Seed banks and seedling establishment on an experimental productivity gradient. *Oikos*, 76, 381–391.

Kleijn D and Sutherland WJ (2003). How effective are European agri-environment schemes in conserving and promoting biodiversity? *Journal of Applied Ecology*, 40, 947–969.

Kleijn D, Baquero RA, Clough Y et al. (2006). Mixed biodiversity benefits of agri-environment schemes in five European countries. *Ecology Letters*, 9, 243–254.

Kline VM and Howell EA (1987). Prairies. In WR Jordan III, ME Gilpin and JD Aber, eds. *Restoration ecology: a synthetic approach to ecological research*, pp. 75–83. Cambridge University Press, Cambridge.

Knapp AK and Seastedt T (1986). Detritus accumulation limits productivity in tallgrass prairie. *BioScience*, 36, 662–668.

Knapp AK and Seastedt TR (1998). Introduction. In AK Knapp, JM Briggs, DC Hartnett and SL Collins, eds. *Grassland dynamics*, pp. 3–18. Oxford University Press, New York.

Knapp AK, Hamerlynck EP, and Owensby CE (1993). Photosynthetic and water relations responses to elevated CO_2 in the C_4 grass *Andropogon gerardii*. *International Journal of Plant Science*, 154, 459–466.

Knapp AK, Briggs JM, Blair JM, and Turner CL (1998). Patterns and controls of aboveground net primary production in tallgrass prairie. In AK Knapp, JM Briggs, DC Hartnett and SL Collins, eds. *Grassland dynamics*, pp. 193–221. Oxford University Press, New York.

Knapp AK, Blair JM, Briggs JM et al. (1999). The keystone role of bison in North American tallgrass prairie. *BioScience*, 49, 39–50.

Knapp AK, Fay PA, Blair JM et al. (2002). Rainfall variability, carbon cycling and plant species diversity in a mesic grassland. *Science*, 298, 2202–2205.

Knapp AK, Burns CE, Fynn RWS, Kirkman KP, Morris CD, and Smith MD (2006). Convergence and contingency in production-precipitation relationships in North American and South African C_4 grasslands. *Oecologia*, 149, 456–464.

Knorr M, Frey SD, and Curtis PS (2005). Nitrogen additions and litter decomposition: a meta-analysis. *Ecology*, 86, 3252–3257.

Koide R, Li M, Lewis J, and Irby C (1988). Role of mycorrhizal infection in the growth and reproduction of wild vs. cultivated plants. *Oecologia*, 77, 537–543.

Koide RT (1991). Tansley Review No. 29: Nutrient supply, nutrient demand and plant response to mycorrhizal infection. *New Phytologist*, 117, 365–386.

Kolasa J and Rollo CD (1991). Introduction: the heterogeneity of heterogeneity: a glossary. In J Kolasa and STA Pickett, eds. *Ecological heterogeneity*, pp. 1–23. Springer-Verlag, New York.

Komarek EVS (1964). The natural history of lightning. *Proceedings of the Third Annual Tall Timbers Fire Ecology Conference*, 3, 139–184.

König G, Brunda M, Puxbaum H, Hewitt CN, and Duckham SC (1995). Relative contribution of oxygenated hydrocarbons to the total biogenic VOC emissions of selected mid-European agricultural and natural plant species. *Atmospheric Environment*, 29, 861–874.

Kotanen PM and Rosenthal JP (2000). Tolerating herbivory: does the plant care if the herbivore has a backbone? *Evolutionary Ecology*, 14, 537–549.

Kotliar NB (2000). Application of the New Keystone-Species concept to prairie dogs: how well does it work? *Conservation Biology*, 14, 1715–1721.

Kovář P, Kovářová P, Dostál P, and Herben T (2001). Vegetation of ant-hills in a mountain grassland: effects of mound history and of dominant ant species. *Plant Ecology*, 156, 215–227.

Krueger-Mangold J, Sheley R, and Engel R (2006). Can R*s predict invasion in semi-arid grasslands? *Biological Invasions*, 8, 1343–1354.

Kucera CL (1981). Grasslands and fire. In HA Mooney, TM Bonnicksen, NL Christensen, JE Lotan and WA Reiners, eds. *Fire regimes and ecosystem properties*, pp. 90–111. USDA For. Serv. Gen. Tech. Rep. WO-26.

Kucera CL (1992). Tall-grass prairie. In RT Coupland, ed. *Ecosystems of the world: natural grasslands: introduction and western hemisphere*, Vol. 8A, pp. 227–268. Elsevier, Amsterdam.

Kucera CL and Ehrenreich JH (1962). Some effects on annual burning on central Missouri prairie. *Ecology*, 43, 334–336.

Kucharik CJ, Fayram N, and Cahill KN (2006). A paired study of prairie carbon stocks, fluxes, and phenology: comparing the world's oldest prairie restoration with an adjacent remnant. *Global Change Biology*, 12, 122–139.

Küchler AW (1964). Potential natural vegetation of the conterminous United States. *American Geographical Society Special Publication*, No. 36.

Küchler AW (1974). A new vegetation map of Kansas. *Ecology*, 55, 586–604.

Kwak MM, Velterop O, and van Andel J (1998). Pollen and gene flow in fragmented habitats. *Applied Vegetation Science*, 1, 37–54.

Kyser GB and DiTomaso JM (2002). Instability in a grassland community after the control of yellow starthistle (*Centaurea solstitialis*) with prescribed burning. *Weed Science*, 50, 648–657.

Lack AJ (1982). The ecology of flowers of chalk grassland and their insect pollinators. *Journal of Ecology*, 70, 773–790.

Lambers H, Chapin FSI, and Pons TL (1998). *Plant physiological ecology*. Springer-Verlag, New York.

Lambers JHR, Harpole WS, Tilman D, Knops J, and Reich PB (2004). Mechanisms responsible for the positive diversity–productivity relationship in Minnesota grasslands. *Ecology Letters*, 7, 661–668.

Lambert MG, Renton SW, and Grant DA (1982). Nitrogen balance studies in some North Island hill pastures. In PW Gandar and DS Bertaud, eds. *Nitrogen balances in New Zealand ecosystems*, pp. 35–40. Department of Scientific and Industrial Research, Palmerston North, New Zealand.

Lamotte M and Bourliére F (1983). Energy flow and nutrient cycling in tropical savannas. In F Bourliére, ed. *Ecosystems of the world 13: tropical savannas*, pp. 583–603. Elsevier, Amsterdam.

Lane C and Moorehead R (1996). New directions in rangeland resource tenure and policy. In I Scoones, ed. *Living with uncertainty: new directions in pastoral development in Africa*, pp. 116–133. Intermediate Technology Publications, London.

Langer RHM (1972). *How grasses grow*. Edward Arnold, London.

Larson EC (1947). Photoperiodic responses of geographical strains of *Andropogon scoparius*. *Botanical Gazette*, 109, 132–149.

Lauenroth WK and Aguilera MO (1998). Plant-plant interactions in grasses and grasslands. In GP Cheplick, ed. *Population biology of grasses*, pp. 209–230. Cambridge University Press, Cambridge.

Lauenroth WK and Gill R (2003). Turnover of root systems. In H de Kroon and EJW Visser, eds. *Root ecology*, Vol. 168, pp. 61–89. Springer-Verlag, Berlin.

Lauenroth WK and Milchunas DG (1992). Short-grass steppe. In RT Coupland, ed. *Ecosystems of the world: natural grasslands: introduction and western hemisphere*, Vol. 8A, pp. 183–226. Elsevier, Amsterdam.

Lauenroth WK, Burke IC, and Gutmann MP (1999). The structure and function of ecosystems in the central and north American grassland region. *Great Plains Research*, 9, 223–259.

Launchbaugh JL (1964). Effects of early spring burning on yields of native vegetation. *Journal of Range Management*, 17, 5–6.

Lavrenko EM and Karamysheva ZV (1993). Steppes of the former Soviet Union and Mongolia. In RT Coupland, ed. *Ecosystems of the world: eastern hemisphere and résumé*, Vol. 8B, pp. 3–60. Elsevier, Amsterdam.

Lawrence WE (1945). Some ecotypic relations of *Deschampsia caespitosa*. *American Journal of Botany*, 32, 298–314.

Laycock WA (1991). Stable states and thresholds of range condition on North American rangelands: a viewpoint. *Journal of Range Management*, 44, 427–433.

Lazarides M (1979). *Micraira* F. Muell. (Poaceae, Micrairoideae). *Brunonia*, 2, 67–84.

Le Houerou HN and Hoste CH (1977). Rangeland production and annual rainfall relations in the Mediterranean Basin and in the African Sahelo-Sudanian zone. *Journal of Range Management*, 30, 181–189.

Lee HJ, Kuchel RE, and Trowbridge RF (1956). The aetiology of *Phalaris* staggers in sheep. II. The toxicity to sheep of three types of pasture containing *Phalaris tuberosa*. *Australian Journal of Agricultural Research*, 7, 333–344.

Lee JA and Harmer R (1980). Vivipary, a reproductive strategy in response to environmental stress? *Oikos*, 35, 254–265.

Lee TD, Reich PB, and Tjoelker MG (2003). Legume presence increases photosynthesis and N concentrations of co-occurring non-fixers but does not modulate their responsiveness to carbon dioxide enrichment. *Oecologia*, 137, 22–31.

Legendre P and Legendre L (1998). *Numerical ecology*, 2nd English edn. Elsevier, Amsterdam.

Leibold MA, Holyoak M, Mouquet N et al. (2004). The metacommunity concept: a framework for multi-scale community ecology. *Ecology Letters*, 7, 601–613.

Leis S, Engle DM, Leslie D, and Fehmi J (2005). Effects of short- and long-term disturbance resulting from military maneuvers on vegetation and soils in a mixed prairie area. *Environmental Management*, 36, 849.

Leistner E (1981). Biosynthesis of plant quinones. In EE Conn, ed. *The biochemistry of plants. a comprehensive treaty*, Vol. 7 Secondary plant products, pp. 403–423. Academic Press, New York.

Lejeune KD and Seastedt TR (2001). Centaurea species: the forb that won the West. *Conservation Biology*, 15, 1568–1574.

Lemus R and Lal R (2005). Bioenergy crops and carbon sequestration. *Critical Reviews in Plant Sciences*, 24, 1–21.

Leone V and Lovreglio R (2004). Conservation of Mediterranean pine woodlands: scenarios and legislative tools. *Plant Ecology*, 171, 221–235.

Lesica P and Allendorf FW (1999). Ecological genetics and the restoration of plant communities: mix or match? *Restoration Ecology*, 7, 42–50.

Li WB, Shi XH, Wang H, and Zhang FS (2004). Effects of silicon on rice leaves resistance to ultraviolet-B. *Acta Botanica Sinica*, 46, 691–697.

Licht DS (1997). *Ecology and economics of the Great Plains*. University of Nebraska Press, Lincoln, NE.

Lieth H, Berlekamp J, and Riediger S (1999). *Climate diagram world atlas*. Backhuys Publishers, Leiden.

Liu ZG and Zou XM (2002). Exotic earthworms accelerate plant litter decomposition in a Puerto Rican pasture and a wet forest. *Ecological Applications*, 12, 1406–1417.

Lock JM (1972). The effects of hippopotamus grazing on grasslands. *Journal of Ecology*, 60, 445–467.

Lockeretz W (1978). The lessons of the dust bowl. *American Scientist*, 66, 560–569.

Long SP, Ainsworth EA, Rogers A, and Ort DR (2004). Rising atmospheric carbon dioxide: plants face the future. *Annual Review of Plant Biology*, 55, 591–628.

López-Mariño A, Luis-Calabuig E, Fillat F, and Bermúdez FF (2000). Floristic composition of established vegetation and the soil seed bank in pasture communities under different traditional management regimes. *Agriculture, Ecosystems and Environment*, 78, 273–282.

Lortie CJ and Callaway RM (2006). Re-analysis of meta-analysis: support for the stress gradient hypothesis. *Journal of Ecology*, 94, 7–16.

Lortie CJ, Brooker RW, Choler P et al. (2004). Rethinking plant community theory. *Oikos*, 107, 433–438.

Loveland TR, Reed BC, Brown JF et al. (2000). Development of a global land cover characteristics database and IGBP DISCover from 1 km AVHRR data. *International Journal of Remote Sensing*, 21, 1303–1330.

Lovett Doust L (1981). Population dynamics and local specialization in a clonal perennial (*Ranunculus repens*). I. The dynamics of ramets in contrasting habitats. *Journal of Ecology*, 69, 743–755.

Low AB and Robelo AG, eds. (1995). *Vegetation of South Africa, Lesotho and Swaziland*. Department of Environmental Affairs and Tourism, Pretoria.

Ludwig JA and Reynolds JF (1988). *Statistical ecology*. Wiley Interscience, New York.

Lunt ID (1991). Management of remnant lowland grasslands and grassy woodlands for nature conservation: a review. *Victoria Naturalist*, 108, 56–66.

Lunt ID (2003). A protocol for integrated management, monitoring, and enhancement of degraded *Themeda triandra* grasslands based on plantings of indicator species. *Restoration Ecology*, 11, 223–230.

Lunt ID and Morgan JW (1999). Vegetation changes after 10 years of grazing exclusion and intermittent burning in a *Themeda triandra* (Poaceae) grassland reserve in South-eastern Australia. *Australian Journal of Botany*, 47, 537–552.

Ma JF, Miyake Y, and Takahashi E (2001). Silicon as a beneficial element for crop plants. In LE Datnoff, GH Snyder and GH Korndörfer, eds. *Silicon in agriculture*, pp. 17–39. Elsevier Science, Amsterdam.

Mabberley DJ (1987). *The plant book*. Cambridge University Press, Cambridge.

MacArthur RH and Wilson EO (1967). *The theory of island biogeography*. Princeton University Press, Princeton, NJ.

Mack RN and Pyke DA (1984). The demography of *Bromus tectorum*: the role of microclimate, grazing and disease. *Journal of Ecology*, 72, 731–748.

Macphail MK and Hill RS (2002). Paleobotany of the Poaceae. *Flora of Australia*, 43, 37–70.

Maestre FT, Bautista S, and Cortina J (2003). Positive, negative, and net effects in grass–shrub interactions in Mediterranean semiarid grasslands. *Ecology*, 84, 3186–3197.

Maestre FT, Valladares F, and Reynolds JF (2005). Is the change of plant-plant interactions with abiotic stress predictable? A meta-analysis of field results in arid environments. *Journal of Ecology*, 93, 748–757.

Maestre FT, Valladares F, and Reynolds JF (2006). The stress-gradient hypothesis does not fit all relationships between plant-plant interactionss and abiotic stress: further insights from arid environments. *Journal of Ecology*, 94, 17–22.

Magda D, Duru M, and Theau J-P (2004). Defining management rules for grasslands using weed demographic characteristics. *Weed Science*, 52, 339–345.

Maier W, Hammer K, Dammann U, Schulz B, and Strack D (1997). Accumulation of sesquiterpenoid cyclohexenone derivatives induced by an arbuscular mycorrhizal fungus in members of the Poaceae. *Planta*, 202, 36–42.

Makita A (1998). Population dynamics in the regeneration process of monocarpic dwarf bamboos, *Sasa* species. In GP Cheplick, ed. *Population biology of grasses*, pp. 313–332. Cambridge University Press, Cambridge.

Mallory-Smith C, Hendrickson P, and Mueller-Warrant G (1999). Cross-resistance of primisulfuron-resistant *Bromus tectorum* L. (downy brome) to sulfosulfuron. *Weed Science*, 47, 256–257.

Malmstrom CM, Hughes CC, Newton LA, and Stoner CJ (2005). Virus infection in remnant native bunchgrasses from invaded California grasslands. *New Phytologist*, 168, 217–230.

Mann LK (1986). Changes in soil carbon storage after cultivation. *Soil Science*, 142, 279–288.

Manning R (1995). *Grassland: the history, biology, politics, and promise of the American prairie*. Penguin, New York.

Manry DE and Knight RS (1986). Lightning density and burning frequency in South African vegetation. *Plant Ecology*, 66, 67–76.

Mark AF (1993). Indigenous grasslands of New Zealand. In RT Coupland, ed. *Ecosystems of the world: eastern hemisphere and résumé*, Vol. 8B, pp. 361–410. Elsevier, Amsterdam.

Mark RN (1981). Invasion of *Bromus tectorum* L. into western North America: an ecological chronicle. *Agro-Ecosystems*, 7, 145–165.

Maron JL and Jefferies RL (1991). Restoring enriched grasslands: effects of mowing on species richness, productivity, and nitrogen retention. *Ecological Applications*, 11, 1088–1100.

Marshall JK (1977). Biomass and production partitioning in response to environment in some North American grasslands. In JK Marshall, ed. *The belowground ecosystem*, pp. 73–84. Colorado State University, Fort Collins, CO.

Martin RE, Asner GP, Ansley RJ, and Mosier AR (2003). Effects of woody vegetation encroachment on soil nitrogen oxide emisssions in a temperate savanna. *Ecological Applications*, 13, 897–910.

Martinsen GD, Cushman JH, and Whitham TG (1990). Impact of pocket gopher disturbance on plant species diversity in a shortgrass prairie community. *Oecologia*, 83, 132.

Mathews S, Tritschler JPI, and Miyasaka SC (1998). Phosphorus management and sustainability. In JH Cherney and DJR Cherney, eds. *Grass for Dairy Cattle*, pp. 193–222. CABI International, Wallingford, Oxon.

Matsuoka Y (2005). Origin matters: lessons from the search for the wild ancestor of maize. *Breeding Science*, 55, 383–390.

Matthews JA (1992). *The ecology of recently-deglaciated terrain: a geoecological approach to glacial forelandss and primary succession.* Cambridge University Press, Cambridge.

Maybury KP, ed. (1999). *Seeing the forest and the trees: ecological classification for conservation.* Nature Conservancy, Arlington, VA.

Mayer PM, Tunnell SJ, Engle DM, Jorgensen EE, and Nunn P (2005). Invasive grass alters litter decomposition by influencing macrodetritivores. *Ecosystems*, 8, 200–209.

Mazzanti A, Lemaire G, and Gastal F (1994). The effect of nitrogen fertilization upon the herbage production of tall fescue swards continuously grazed with sheep. I. Herbage growth dynamics. *Grass and Forage Science*, 49, 111–120.

McCulley RL, Burke IC, Nelson JA, Lauenroth WK, Knapp AK, and Kelly EF (2005). Regional patterns in carbon cycling across the Great Plains of North America. *Ecosystems*, 8, 106–212.

McEwen LC (1962). Leaf longevity and crude protein content for roughleaf ricegrass in the Black Hills. *Journal of Range Management*, 15, 106.

McIntyre S, Heard KM, and Martin TG (2003). The relative importance of cattle grazing in subtropical grasslands: does it reduce or enhance plant biodiversity? *Journal of Applied Ecology*, 40, 445–457.

McIvor JG (2005). Australian grasslands. In JM Suttie, SG Reynolds and C Batello, eds. *Grasslands of the world*, pp. 343–380. Food and Agriculture Organization of the United Nations, Rome.

McLaughlin SB, De La Torre Ugarte DG, Garten Jr. CT et al. (2002). High-value renewable energy from prairie grasses. *Environmental Science and Technology*, 36, 2122–2129.

McMillan C (1956a). Nature of the plant community. I. Uniform garden and light period studies of five grass taxa in Nebraska. *Ecology*, 37, 330–340.

McMillan C (1956b). Nature of the plant community. II. Variation in flowering behavior within populations of *Andropogon scoparius*. *American Journal of Botany*, 43, 429–436.

McMillan C (1957). Nature of the plant communiy. III. Flowering behavior within two grassland communities under reciprocal transplanting. *American Journal of Botany*, 44, 144–153.

McMillan C (1959a). Nature of the plant community. V. Variation within the true prairie community-type. *American Journal of Botany*, 46, 418–424.

McMillan C (1959b). The role of ecotypic variation in the distribution of the Central Grassland of North America. *Ecological Monographs*, 29, 285–308.

McMillan C and Weiler J (1959). Cytogeography of *Panicum virgatum* in central North America. *American Journal of Botany*, 46, 590–593.

McNaughton SJ (1979). Grazing as an optimization process: grass-ungulate relationships in the Serengeti. *American Naturalist*, 113, 691–703.

McNaughton SJ (1983). Serengeti grassland ecology: the role of composite environmental factors and contingency in community organization. *Ecological Monographs*, 53, 291–320.

McNaughton SJ (1984). Grazing lawns: animals in herds, plant form, and coevolution. *American Naturalist*, 124, 863–886.

McNaughton SJ (1985). Ecology of a grazing ecosystem: the Serengeti. *Ecological Monographs*, 55, 259–294.

McNaughton SJ, Oesterheld M, Frank DA, and Williams KJ (1989). Ecosystem-level patterns of primary productivity and herbivory in terrestrial habitats. *Nature*, 341, 142–144.

McNaughton SJ, Tarrants JL, McNaughton MM, and Davis RH (1985). Silica as a defense against herbivory and a growth promotor in African grasses. *Ecology*, 66, 528–535.

Menaut JC (1983). The vegetation of African savannas. In F Bourliére, ed. *Ecosystems of the world 13: tropical savannas*, pp. 109–149. Elsevier, Amsterdam.

Metcalf CR (1960). *Anatomy of the Monocotyledons. I. Gramineae*. Oxford University Press, Oxford.

Metz M and Fütterer J (2002). Biodiversity (communications arising): suspect evidence of transgenic contamination (see editorial footnote). *Nature*, 416, 600–601.

Meyer CK, Whiles MR, and Charlton RE (2002). Life history, secondary production, and ecosystem significance of acridid grasshoppers in annually burned and unburned tallgrass prairie. *American Entomologist*, 48, 52–61.

Michalet R, Brooker RW, Cavieres LA *et al.* (2006). Do biotic interactions shape both sides of the humped-back model of species richness in plant communities? *Ecology Letters*, 9, 767–773.

Michaud R, Lehman WF, and Rumbaugh MD (1988). World distribution and historical development. In AA Hanson, DK D.K. Barnes and JRR Hill, eds. *Alfalfa and alfalfa improvement*, Vol. 29, pp. 25–91. American Society of Agronomy, Madison, WI.

Michelangeli FA, Davis JI, and Stevenson DW (2003). Phylogenetic relationships among Poaceae and related families as inferred from morphology, inversions in the plastid genome, and sequence data from the mitochondrial and plastid genomes. *American Journal of Botany*, 90, 93–106.

Midewin National Tallgrass Prairie (2002). http://www.fs.fed.us/mntp/plan/, accessed 28 May 2007.

Milberg P (1993). Seed bank and seedlings emerging after soil disturbance in a wet semi-nature grassland in Sweden. *Annales Botanici Fennici*, 30, 9–13.

Milberg P (1995). Soil seed bank after eighteen years of succession from grassland to forest. *Oikos*, 72, 3–13.

Milchunas DG and Lauenroth WK (1993). Quantitative effects of grazing on vegetation and soils over a global range of environments. *Ecological Monographs*, 63, 327–366.

Milchunas DG, Sala OE, and Lauenroth WK (1988). A generalized model of the effects of grazing by large herbivores on grassland communitiy structure. *American Naturalist*, 132, 87–106.

Miller DJ (2005). The Tibetan steppe. In JM Suttie, SG Reynolds and C Batello, eds. *Grasslands of the world*, pp. 305–342. Food and Agriculture Organization of the United Nations, Rome.

Miller TE, Burns JH, Munguia P *et al.* (2005). A critical review of twenty years' use of the resource-ratio theory. *American Naturalist*, 165, 439–448.

Milner C and Hughes RE (1968). *Methods for the measurement of the primary production of grassland*. Blackwell Scientific, Oxford.

Milton SJ and Dean WRJ (2004). Disturbance, drought and dynamics of desert dune grassland, South Africa. *Plant Ecology*, 150, 37–51.

Milton SJ, Dean WRJ, and Richardson DM (2003). Economic incentives for restoring natural capital in southern African rangelands. *Frontiers in Ecology and the Environment*, 1, 247–254.

Misra R (1983). Indian savannas. In F Bourliére, ed. *Ecosystems of the world 13: tropical savannas*, pp. 151–166. Elsevier, Amsterdam.

Mitchell CE (2003). Trophic control of grassland production and biomass by pathogens. *Ecology Letters*, 6, 147–155.

Mitchell RB, Masters RA, Waller SS, Moore KJ, and Young LJ (1996). Tallgrass prairie vegetation response to spring burning dates, fertilizer, and atrazine. *Journal of Range Management*, 49, 131–136.

Mohlenbrock RH (1993). Simpson Township Barrens, Illinois. In *Natural History*, pp. 25–27.

Moir WH (1989). History and development of site and condition criteria for range condition within the U.S. Forest Service. In WK Lauenroth and WA Laycock, eds. *Secondary succession and the evaluation of rangeland condition*, pp. 49–76. Westview Press, Boulder, CO.

Moloney KA (1986). Fine-scale spatial and temporal variation in the demography of a perennial bunchgrass. *Ecology*, 69, 1588–1598.

Montgomery RF and Askew GP (1983). Soils of tropical savannas. In F Bourliére, ed. *Ecosystems of the world 13: tropical savannas*, pp. 63–78. Elsevier, Amsterdam.

Moore I (1966). *Grass and grasslands*. Collins, London.

Moore KJ (2003). Compendium of common forages. In RF Barnes, CJ Nelson, M Collins and KJ Moore, eds. *Forages: an introduction to grassland agriculture*, Vol. 1, pp. 237–238. Iowa State Unviversity Press, Ames, IA.

Moore KJ and Moser LE (1995). Quantifying developmental morphology of perennial grasses. *Crop Science*, 35, 37–43.

Moore RM (1993). Grasslands of Australia. In RT Coupland, ed. *Natural grasslands: eastern hemisphere and résumé*, Vol. 8B, pp. 315–360. Elsevier, Amsterdam.

Moore RM, ed. (1970). *Australian grasslands*. Australian National University, Canberra.

Mora G and Pratt LM (2002). Carbon isotopic evidence from paleosols for mixed C-3/C-4 vegetation in the Bogota Basin, Colombia. *Quaternary Science Review*, 21, 985–995.

Morecroft MD, Sellers EK, and Lee JA (1994). An experimental investigation into the effects of atmospheric nitrogen deposition on two semi-natural grasslands. *Journal of Ecology*, 82, 475–483.

Morgan JW (2004). Defining grassland fire events and the response of perennial plants to annual fire in temperate grasslands of south-eastern Australia. *Plant Ecology*, 144, 127–144.

Morghan KJR, Sheley RL, and Svejcar T (2006). Successful adaptive management—the integration of research and management. *Rangeland Ecology and Management*, 59, 216–219.

Morris JW, Bezuidenhout JJ, and Furniss PR (1982). Litter decomposition. In BJ Huntley and BH Walker, eds. *Ecology of tropical savannas*, pp. 535–553. Springer-Verlag, Berlin.

Morrison J (1987). Effects of nitrogen fertilizer. In RW Snaydon, ed. *Ecosystems of the world: managed grasslands: analytical studies*, Vol. 17B, pp. 61–70. Elsevier, Amsterdam.

Mueggler WF (1972). Influence of competition on the response of bluebunch wheatgrass to clipping. *Journal of Range Management*, 25, 88–92.

Mueller-Dombois D and Ellenberg H (1974). *Aims and methods of vegetation ecology*. John Wiley & Sons, New York.

Muller CB and Krauss J (2005). Symbiosis between grasses and asexual fungal endophytes. *Current Opinion in Plant Biology*, 8, 450–456.

Muller S, Dutoit T, Alard D, and Grévilliot F (1998). Restoration and rehabilitation of species-rich grassland ecosystems in France: a review. *Restoration Ecology*, 6, 94–101.

Mulvaney CR, Molano-Flores B, and Whitman DW (2004). The reproductive biology of *Agalinis auriculata* (Michx.) Raf. (Orobanchaceae), a threatened North American Prairie inhabitant. *International Journal of Plant Sciences*, 165, 605–614.

Murphy BP and Bowman DMJS (2007). The interdependence of fire, grass, kangaroos and Australian Aborigines: a case study from central Arnhem Land, northern Australia. *Journal of Biogeography*, 34, 237–250.

Murphy CA, Foster BL, Ramspott ME, and Price KP (2006). Effects of cultivation history and current grassland management on soil quality in northeastern Kansas. *Journal of Soil and Water Conservation*, 61, 75–84.

Murphy JA, Hendricks MG, Rieke PE, Smucker AJM, and Branham BE (1994). Turfgrass root systems evaluated using the minirhizotron and video recording methods. *Agronomy Journal*, 86, 247–250.

Mutch RW and Philpot CW (1970). Relation of silica content to flameability in grasses. *Forest Science*, 16, 64–65.

Myers N and Mittermeier RA (2000). Biodiversity hotspots for conservation priorities. *Nature*, 403, 853–859.

Natasha BK, Bruce WB, April DW, and Glenn P (1999). A critical review of assumptions about the prairie dog as a keystone species. *Environmental Management*, 24, 177–192.

National Research Council (1992). *Grasslands and grassland sciences in Northern China*. National Academy Press, Washington, DC.

National Research Council (1994). Rangeland health: new methods to classify, inventory, and monitor rangelands. National Academy Press, Washington, DC.

NatureServe (2003). A working classification of terrestrial ecological systems in the coterminous United States. international terrestrial ecological systems classification. NatureServe, Arlington, VA.

NatureServe (2007a). *International ecological classification standard: terrestrial ecological classifications*. Data current as of 16 August 2005 edn. NatureServe Central Databases, Arlington, VA.

NatureServe (2007b). NatureServe Explorer: an online encyclopedia of life [web application]. Version 6.2. Available http://www.natureserve.org/explorer.

Accessed 10 January 2008. Data current as of 16 August 2005 edn. NatureServe, Arlington, VA.

Navas M-L (1998). Individual species performance and response of multispecific communities to elevated CO_2: a review. *Functional Ecology*, 12, 721–727.

Navas M-L, Ducout B, Roumet C, Richarte J, Garnier J, and Garnier E (2003). Leaf life span, dynamics and construction cost of species from Mediterranean old-fields differing in successional status. *New Phytologist*, 159, 213–218.

Neill C, Steudler PA, Garcia-Montiel DC *et al.* (2005). Rates and controls of nitrous oxide and nitric oxide emissions following conversion of forest to pasture in Rondônia. *Nutrient Cycling in Agroecosystems*, 71, 1–15.

Nelson CJ and Moser LE (1994). Plant factors affecting forage quality. In GC Fahey, ed. *Forage quality, evaluation, and utilization*, pp. 115–154. American Society of Agronomy, Inc., Crop Science Society of America, Inc., Soil Science Society of America, Inc., Madison, WI.

Newingham BA and Belnap J (2006). Direct effects of soil amendments on field emergence and growth of the invasive annual grass *Bromus tectorum* L. and the native perennial grass *Hilaria jamesii* (Torr.) Benth. *Plant and Soil*, 280, 29–40.

Newman EI (1982). *The plant community as a working mechanism*. Blackwell Scientific, Oxford.

Newman JA, Abner ML, Dado RG, Gibson DJ, and Hickman A (2003). Effects of elevated CO_2 on the tall fescue-endophytic fungus interaction: growth, photosynthesis, growth, chemical composition and digestibility. *Global Change Biology*, 9, 425–437.

Newsham KK and Watkinson AR (1998). Arbuscular mycorrhizas and the population biology of grasses. In GP Cheplick, ed. *Population biology of grasses*, pp. 286–312. Cambridge University Press, Cambridge.

Newsham KK, Fitter AH, and Watkinson AR (1995). Arbuscular mycorrhiza protect an annual grass from root pathogenic fungi in the field. *Journal of Ecology*, 83, 991–1000.

Nie D, He H, Kirkham MB, and Kanemasu ET (1992a). Photosynthesis of a C_3 grass and a C_4 grass under elevated CO_2. *Photosynthetica*, 26, 189–198.

Nie D, Kirkham MB, Ballou LK, Lawlor DJ, and Kanemasu ET (1992b). Changes in prairie vegetation under elevated carbon dioxide levels and two soil moisture regimes. *Journal of Vegetation Science*, 3, 673–678.

Nimbal CI, Yerkes CN, Weston LA, and Weller SC (1996). Herbicidal activity and site of action of the natural product sorgoleone. *Pesticide Biochemistry and Physiology*, 54, 73–83.

Nippert JB, Knapp AK, and Briggs JM (2006). Intra-annual rainfall varibility and grassland productivity: can the past predict the future? *Plant Ecology*, 184, 65–74.

Nix HA (1983). Climate of tropical savannas. In F Bourliére, ed. *Ecosystems of the world 13: tropical savannas*, pp. 37–62. Elsevier, Amsterdam.

Njoku OU, Okorie IN, Okeke EC, and Okafor JI (2004). Investigation on the phytochemical and antimicrobial properties of *Pennisetum purpureum*. *Journal of Medicinal and Aromatic Plant Sciences*, 26, 311–314.

Noble JC (1991). Behaviour of a very fast grassland wildfire on the riverrine plain of southeastern Australia. *International Journal of Wildland Fire*, 1, 189–196.

Noble JC and Walker P (2006). Integrated shrub management in semi-arid woodlands of eastern Australia: a systems-based decision support model. *Agricultural Systems*, 88, 332–359.

Noble JC, Cunningham GM, and Mulham WE (1984). Rehabilitation of degraded land. In GN Harrington, AD Wilson and MD Young, eds. *Management of Australia's rangelands*, pp. 171–186. CSIRO, East Melbourne, Australia.

Norby RJ and Lou Y (2004). Evaluating ecosystem responses to rising atmospheric CO_2 and global warming in a multi-factor world. *New Phytologist*, 162, 281–293.

Norman MJT (1960). The relationship between competition and defoliation in pasture. *Grass and Forage Science*, 15, 145–149.

Norton DA and Miller CJ (2000). Some issues and options for the conservation of native biodiversity in rural New Zealand. *Ecological Management and Restoration*, 1, 26–34.

Noy-Meir I, Gutmann MP, and Kaplan Y (1989). Responses of mediterranean grassland plants to grazing and protection. *Journal of Ecology*, 77, 290–310.

Nunes da Cunha C and Junk WJ (2004). Year-to-year changes in water level drive the invasion of *Vochysia divergens* in Pantanal grasslands. *Applied Vegetation Science*, 7, 103–110.

O'Connor TG (1983). Nitrogen balances in natural grasslands and extensively-managed grassland systems. *New Zealand Journal of Ecology*, 6, 1–18.

O'Connor TG (1993). The influence of rainfall and grazing on the demography of some African savanna grasses: a matrix modelling approach. *Journal of Applied Ecology*, 30, 119–132.

O'Connor TG (1994). Composition and population responses of an African savanna grassland to rainfall and grazing. *Journal of Applied Ecology*, 31, 155–171.

O'Connor TG (2005). Influence of land use on plant community composition and diversity in Highland Sourveld grassland in the southern Drakensberg, South Africa. *Journal of Applied Ecology*, 42, 975–988.

O'Connor TG and Everson TM (1998). Population dynamics of perennial grasses in African savanna and grassland. In GP Cheplick, ed. *Population biology of grasses*, pp. 333–365. Cambridge University Press, Cambridge.

O'Connor TG, Morris CD, and Marriott DJ (2003). Change in land use and botanical composition of KwaZulu-Natal's grasslands over the past fifty years: Acocks' sites revisited. *South African Journal of Botany*, 69, 105–115.

O'Lear HA, Seastedt TR, Briggs JM, Blair JM, and Ramundo RA (1996). Fire and topographic effects on decomposition rates and N dynamics of buried wood in tallgrass prairie. *Soil Biology and Biochemistry*, 3, 323–329.

Ogelthorpe DR and Sanderson RA (1999). An ecological-economic model for agri-environmental policy analysis. *Ecological Economics*, 28, 245–266.

Ohiagu CE and Wood TG (1979). Grass production and decomposition in Southern Guinea savanna, Nigeria. *Oecologia*, 40, 155–165.

Ojasti J (1983). Ungulates and large rodents of South America. In F Bourliére, ed. *Ecosystems of the world 13: tropical savannas*, pp. 427–439. Elsevier, Amsterdam.

Ojima DS, Parton WJ, Schimel DS, and Owensby CE (1990). Simulated effects of annual burning on prairie ecosystems. In SL Collins and LL Wallace, eds. *Fire in North American tallgrass prairie*, pp. 118–132. University of Oklahoma Press, Norman, OK.

Oksanen J (1996). Is the humped relationship between species richness and biomass an artifact due to plot size? *Journal of Ecology*, 84, 293–295.

Olff H and Ritchie ME (1998). Effects of herbivores on grassland plant diversity. *Trends in Ecology & Evolution*, 13, 261–265.

Olff H, Hoorens B, de Goede RGM, van der Putten WH, and Gleichman JM (2000). Small-scale shifting mosaics of two dominant grassland species: the possible role of soil-borne pathogens. *Oecologia*, 125, 45–54.

Oliveira JM and Pillar VD (2004). Vegetation dynamics on mosaics of Campos and Araucaria forest between 1974 and 1999 in Southern Brazil. *Community Ecology*, 5, 197–202.

Olmated CE (1944). Growth and development in range grasses. IV. Photoperiodic responses in twelve geographic strains of sideoats grama. *Botanical Gazette*, 106, 46–74.

Olmated CE (1945). Growth and development in range grasses. V. Photoperiodic responses of clonal divisions of three latitudinal strains of sideoats grama. *Botanical Gazette*, 106, 382–401.

Olsen JS (1963). Energy storage and the balance of producers and decomposers in ecological systems. *Ecology*, 44, 322–331.

Olson DM and Dinerstein E (2002). The Global 200: priority ecoregions for global conservation. *Annals of the Missouri Botanical Garden*, 89, 199–224.

Olson DM, Dinerstein E, Abell R et al. (2000). *The Global 200: a representation approach to conserving the Earth's distinctive ecoregions*. Conservation Science Program, World Wildlife Fund-US, Washington DC.

Olson DM, Dinerstein E, Wikramanayake ED et al. (2001). Terrestrial ecoregions of the world: a new map of life on earth. A new global map of terrestrial ecoregions provides an innovative tool for conserving biodiversity. *BioScience*, 51, 933–938.

Olson JS (1958). Rates of succession and soil changes on southern Lake Michigan sand dunes. *Botanical Gazette*, 119, 125–170.

Orr DM and Holmes WE (1984). Mitchell grasslands. In GN Harrington, AD Wilson and MD Young, eds. *Management of Australia's rangelands*, pp. 241–254. CSIRO, East Melbourne, Australia.

Ortiz-Garcia S, Ezcurra E, Schoel B, Acevedo F, Soberon J, and Snow AA (2005). Absence of detectable transgenes in local landraces of maize in Oaxaca, Mexico (2003–2004). *Proceedings of the National Academy of Science, USA*, 102, 12338–12343.

Osborne CP (2008). Atmosphere, ecology and evolution: what drove the Miocene expansion of C4 grasslands? *Journal of Ecology*, 96, 35–45.

Osborne CP and Beerling DJ (2006). Nature's green revolution: the remarkable evolutionary rise of C_4 plants. *Philosophical Transactions of the Royal Society of London, Series B*, 361, 173–194.

Overbeck GE, Müller SC, Pillar VD, and Pfadenhauer J (2005a). Fine-scale post-fire dynamics in southern Brazilian subtropical grassland. *Journal of Vegetation Science*, 16, 655–664.

Overbeck GE, Müller SC, Pillar VD, and Pfadenhauer J (2005b). No heat-stimulated germination found in herbaceous species from burned subtropical grassland. *Plant Ecology*, 184, 237–243.

Owensby CE, Coyne PI, Ham JM, Auen LM, and Knapp AK (1993). Biomass production in a tallgrass prairie ecosystem exposed to ambient and elevated CO_2. *Ecological Applications*, 5, 644–653.

Packard S and Mutel CF, eds. (2005). *The tallgrass restoration handbook*, 2nd edn. Island Press, Washington DC.

Pallarés OR, Berretta EJ, and Maraschin GE (2005). The south American Campos ecosystem. In JM Suttie, SG Reynolds and C Batello, eds. *Grasslands of the world*, pp. 171–219. Food and Agriculture Organization of the United Nations, Rome.

Parton WJ, Coughenour MB, Scurlock JMO et al. (1996). Global grassland ecosystem modelling: development and test of ecosystem models for grassland systems. In AI Breymeyer, DO Hall, JM Melillo and GI Agren, eds. *Global change: effects on coniferous forests and grasslands.* Wiley, Chichester.

Parton WJ, Stewart JWB, and Cole CV (1988). Dynamics of C, N, P and S in grassland soils: a model. *Biogeochemistry*, 5, 109–31.

Paruelo JM, Jobbágy EG, Sala OE, Lauenroth WK, and Burke IC (1998). Functional and structural convergence of temperate grassland and shrubland ecosystems. *Ecological Applications*, 8, 194–206.

Pastor J, Stillwell MA, and Tilman D (1987). Little bluestem litter dynamics in Minnesota old fields. *Oecologia*, 72, 327–330.

Pauly WR (2005). Conducting burns. In S Packard and CF Mutel, eds. *The tallgrass restoration handbook*, pp. 223–244. Island Press, Washington, DC.

Pausas JG and Austin MP (2001). Patterns of plant species richness in relation to different environments: an appraisal. *Journal of Vegetation Science*, 12, 153–166.

Payne F, Murray PJ, and Cliquet JB (2001). Root exudates: a pathway for short-term N transfer from clover and ryegrass. *Plant and Soil*, 229, 235–243.

Peacock E and Schauwecker T, eds. (2003). *Blackland prairies of the Gulf coastal plain.* University of Alabama Press, Tuscaloosa, AL.

Pearsall WH (1950). *Mountains and moorlands.* Collins, London.

Peco B, Ortega M, and Levassor C (1998). Similarity between seed bank and vegetation in Mediterranean grassland: a predictive model. *Journal of Vegetation Science*, 9, 815–828.

Peet RK (1992). Community structure and ecosystem function. In DC Glenn-Lewin, RK Peet and TT Veblen, eds. *Plant succession: theory and practice*, pp. 103–151. Chapman & Hall, London.

Peeters A (2004). *Wild and sown grasses.* Food and Agriculture Organization of the United Nations, Rome.

Pellant M, Shaver P, Pyke DA, and Herrick JE (2005). *Interpreting indicators of rangeland health, version 4. Technical Reference 1734-6. BLM/WO/ST-00/001+1734/REV05.* U.S. Department of the Interior, Bureau of Land Management, National Science and Technology Center, Denver, CO.

Pendleton DT (1989). Range condition as used in the Soil Conservation Service. In WK Lauenroth and WA Laycock, eds. *Secondary succession and the evaluation of rangeland condition*, pp. 17–34. Westview Press, Boulder, CO.

Pennington W (1974). *The history of British vegetation.* 2nd edn. English Universities Press, London.

Perelman SB, Leon JC, and Oesterheld M (2001). Cross-scale vegetation patterns of flooding Pampa grasslands. *Journal of Ecology*, 89, 562–577.

Pérez EM and Bulla L (2005). Llanos (NT0709). Available http://www.worldwildlife.org/wildworld/profiles/terrestrial/nt/nt0709_full.html Accessed November 9 2005. World Wildlife Fund.

Perrow MR and Davy AJ, eds. (2002). *Handbook of ecological restoration. Vol. 2. Restoration in practice.* Cambridge University Press, Cambridge.

Peters JC and Shaw MW (1996). Effect of artificial exclusion and augmentation of fungal plant pathogens on a regenerating grassland. *New Phytologist*, 134, 295–307.

Phoenix GK, Booth RE, Leake JR, Read DJ, Grime JP, and Lee JA (2003). Effects of enhanced nitrogen deposition and phosphorus limitation on nitrogen budgets of semi-natural grasslands. *Global Change Biology*, 9, 1309–1321.

Pickett STA and Cadenasso ML (2005). Vegetation dynamics. In E Van der Maarel, ed. *Vegetation ecology*, pp. 172–198. Blackwell, Oxford.

Pickett STA and White PS (1985). *The ecology of natural disturbance and patch dynamics.* Academic Press, Orlando, FL.

Pimm SL (1997). The value of everything. *Nature*, 38, 231–232.

Piper JK and Pimm SL (2002). The creation of diverse prairie-like communities. *Community Ecology*, 3, 205.

Piperno D and Hans-Dieter S (2005). Dinosaurs dined on grass. *Science*, 310, 1126–1128.

Platt WJ (1975). The colonization and formation of equilibrium plant species associations on badger disturbances in a tallgrass prairie. *Ecological Monographs*, 45, 285–305.

Platt WJ and Weis JM (1985). An experimental study of competition among fugitive prairie plants. *Ecology*, 66, 708–720.

Poinar GC and Columbus JT (1992). Adhesive grass spikelet with mammalian hair in Dominican amber—1st fossil evidence of epizoochory. *Experientia*, 48, 906–908.

Polley HW (1997). Implications of rising atmospheric carbon dioxide concentration for rangelands. *Journal of Range Management*, 50, 562–577.

Polley HW and Collins SL (1984). Relationships of vegetation and environment in buffalo wallows. *American Midland Naturalist*, 112, 178–186.

Polley HW, Johnson HB, Mayeux HS, and Tischler CR (1996). Are some of the recent changes in grassland communities a response to rising CO_2 concentrations? In C Körner and FA Bazzaz, eds. *Carbon dioxide,*

populations and communities, pp. 177–195. Academic Press, New York.
Poorter H (1993). Interspecific variation in the growth response of plants to an elevated ambient CO_2 concentration. *Vegetatio*, 104/105, 77–97.
Poorter H and Navas M-L (2003). Plant growth and competition at elevated CO_2: on winners, losers and functional groups. *New Phytologist*, 157, 175–198.
Potts DL, Huxman TE, Enquist BJ, Weltzin JF, and Williams DG (2006). Resilience and resistance of ecosystem funtional response to a precipitation pulse in a semi-arid grassland. *Journal of Ecology*, 94, 23–30.
Potvin C and Vasseur L (1997). Long-term CO_2 enrichment of a pasture community: species richness, dominance, and succession. *Ecology*, 78, 666–677.
Prasad V, Stromberg CAE, Alimohammadian H, and Sahni A (2005). Dinosaur coprolites and the early evolution of grasses and grazers. *Science*, 310, 1177–1180.
Prat H (1936). La systématique des Graminées. *Annales des Sciences Naturelles Série Botanique X*, 18, 165–258.
Pratt A (1873). *The flowering plants, grasses, sedges, and ferns of Great Britain, and their allies the club mosses, pepperworts and horsetails*. Frederick Warne & Co, London.
Prendergast HDV (1989). Geographical distribution of C_4 acid decarboxylation types and associated structural variants in native Australian C_4 grasses (Poaceae). *Australian Journal of Botany*, 37, 253–273.
Prendergast HDV, Hattersley PW, Stone NE, and Lazarides M (1986). C_4 acid decarboxylation type in Eragrostis (Poaceae): patterns of variation in chloroplast position, ultrastructure and geographical distribution. *Plant, Cell & Environment*, 9, 333–344.
Prentice HC, Lönn M, Lefkovitch LP, and Runyeon H (1995). Associations between allele frequencies in *Festuca ovina* and habitat variation in the alvar grassland son the Baltic island of Öland. *Journal of Ecology*, 83, 391–402.
Prentice HC, Lönn M, Lager H, Rosén E, and Van der Maarel E (2000). Changes in allozyme frequencies in *Festuca ovina* populations after a 9-year nutrient/water experiment. *Journal of Ecology*, 88, 331–347.
Prentice HC, Lönn M, Rosquist G, Ihse M, and Kindstrom M (2006). Gene diversity in a fragmented population of *Briza media*: grassland continuity in a landscape context. *Journal of Ecology*, 94, 87–97.
Press MC and Phoenix GK (2005). Impacts of parasitic plants on natural communities. *New Phytologist*, 166, 737–751.
Pringle HJR, Watson IW, and Tinley KL (2006). Landscape improvement, or ongoing degredation—reconciling apparent contradictions from the arid rangelands of Western Australia. *Landscape Ecology*, 21, 1267–1279.

Prychid CJ, Rudall PJ, and Gregory M (2003). Systematics and biology of silica bodies in monocotyledons. *Botanical Review*, 69, 377–440.
Putten WHVd, Dijk CV, and Peters BAM (1993). Plant-specific soil-borne diseases contribute to succession in foredune vegetation. *Nature*, 362, 53.
Pyke DA, Herrick JE, Shaver P, and Pellant M (2002). Rangeland health attributes and indicators for quantitative assessment. *Journal of Range Management*, 55, 584–597.
Pyne SJ (2001). *Fire: a brief history*. University of Washington Press, Seattle, WA.
Quin BF (1982). The influence of grazing animals on nitrogen balances. In PW Gandar and DS Bertaud, eds. *Nitrogen balances in New Zealand ecosystems*, pp. 95–102. Department of Scientific and Industrial Research, Palmerston North, New Zealand.
Quinn JA (1978). Plant ecotypes: ecological or evolutionary units? *Bulletin of the Torrey Botanical Club*, 105, 58–64.
Quinn JA (1991). Evolution of dioecy in *Buchloe dactyloides* (Gramineae): tests for sex-specific vegetative characters, ecological differences, and sexual niche-partitioning. *American Journal of Botany*, 78, 481–488.
Quinn JA (2000). Adaptive plasticity in reproduction and reproductive systems of grasses. In SWL Jacobs and J Everett, eds. *Grasses: systematics and evolution*, pp. 281–286. CSIRO, Collingwood, Australia.
Quinn JA and Ward RT (1969). Ecological differentiation in sand dropseed (*Sporobolus cryptandrus*). *Ecological Monographs*, 39, 61–78.
Quist D and Chapela IH (2001). Transgenic DNA introgressed into traditional maize landraces in Oaxaca, Mexico. *Nature*, 414, 541–543.
Rabinowitz D and Rapp JK (1985). Colonization and establishment of Missouri prairie plants on artificial soil disturbances. III. Species abundance distributions, survivorship, and rarity. *American Journal of Botany*, 72, 1635–1640.
Rabotnov TA (1955). Fluctuations of meadows. *Bjulleten Moskovskogo Obscestva Ispytatelej Prirody Otdel Biologiceskij*, 60, 9–30.
Rabotnov TA (1974). Differences between fluctuations and successions. In R Knapp, ed. *Vegetation dynamics*, pp. 19–24. Dr. W. Junk Publisher, The Hague.
Rackham O (1986). *The history of the countryside*. Phoenix, London.
Radcliffe JE and Baars JA (1987). The productivity of temperate grasslands. In RW Snaydon, ed. *Ecosystems of the world: managed grasslands: analytical studies*, Vol. 17B, pp. 7–17. Elsevier, Amsterdam.

Raghu S, Anderson RC, Daehler CC, Davis AS, Wiedenmann RN, Simberloff D, and Mack RN (2006). Adding biofuels to the invasive species fire? *Science*, 313, 1742.

Rahbek C (2005). The role of spatial scale and the perception of large-scale species-richness patterns. *Ecology Letters*, 8, 224–239.

Rajaniemi TK (2002). Why does fertilization reduce plant species diversity? Testing three competition-based hypotheses. *Journal of Ecology*, 90, 316–324.

Ramos-Neto MB and Pivello VR (2000). Lightning fires in a Brazilian savanna National Park: rethinking management strategies. *Environmental Mangament*, 26, 675–684.

Ramsay PM and Oxley ERB (1996). Fire temperatures and postfire plant community dynamics in Ecuadorian grass páramo. *Vegetatio*, 124, 129–144.

Ramsay PM and Oxley ERB (1997). The growth form composition of plant communities in the ecuadorian páramos. *Plant Ecology*, 131, 173–192.

Rapp JK and Rabinowitz D (1985). Colonization and establishment of Missouri prairie plants on artificial soil disturbances. I. Dynamics of forb and graminoid seedlings and shoots. *American Journal of Botany*, 72, 1618–1628.

Raunkiaer C (1934). *The life forms of plants and statistical plant geography*. Clarendon Press, Oxford.

Rawitscher F (1948). The water economy of the vegetation of the 'Campos Cerrados' in southern Brazil. *Journal of Ecology*, 36, 237–268.

Read TR and Bellairs SM (1999). Smoke affects the germination of native grasses of New South Wales. *Australian Journal of Botany*, 47, 563–576.

Redfearn DD and Nelson CJ (2003). Grasses for southern areas. In RF Barnes, CJ Nelson, M Collins and KJ Moore, eds. *Forages: an introduction to grassland agriculture*, Vol. 1, pp. 149–169. Iowa State Unviversity Press, Ames, IA.

Redmann RE (1992). Primary productivity. In RT Coupland, ed. *Ecosystems of the world: natural grasslands: introduction and western hemisphere*, Vol. 8A, pp. 75–94. Elsevier, Amsterdam.

Reich PB, Walters MB, and Ellsworth DS (1997). From tropics to tundra: global convergence in plant functioning. *Proceedings of the National Academy of Sciences, USA*, 94, 13730–13734.

Reichman OJ (1987). *Konza Prairie: a tallgrass natural history*. University of Kansas Press, Lawrence, KS.

Reichman OJ and Seabloom EW (2002). The role of pocket gophers as subterranean ecosystem engineers. *Trends in Ecology & Evolution*, 17, 44–49.

Reichman OJ and Smith SC (1985). Impact of pocket gopher burrows on overlying vegetation. *Journal of Mammalogy*, 66, 720–725.

Reichman OJ, Bendix JH, and Seastedt TR (1993). Distinct animal-generated edge effects in a tallgrass prairie community. *Ecology*, 74, 1281–1285.

Reid RS, Serneels S, Nyabenge M, and Hanson J (2005). The changing face of pastoral systems in grass-dominated ecosystems of eastern Africa. In JM Suttie, SG Reynolds and C Batello, eds. *Grasslands of the world*, pp. 19–76. Food and Agriculture Organization of the United Nations, Rome.

Reinhold B, Hurek T, Niemann E-G, and Fendrik I (1986). Close association of *Azospirillum* and diazotrophic rods with different root zones of Kallar Grass. *Applied and Environmental Microbiology*, 52, 520–526.

Renvoize S (2002). Grass anatomy. *Flora of Australia*, 43, 71–132.

Rice CW, Todd TC, Blair JM, Seastedt T, Ramundo RA, and Wilson GWT (1998). Belowground biology and processes. In AK Knapp, JM Briggs, DC Hartnett and SL Collins, eds. *Grassland dynamics: long-term ecological research in tallgrass prairie*, pp. 244–264. Oxford University Press, New York.

Rice EL (1984). *Allelopathy*. Academic Press, New York.

Rice KJ (1989). Impacts of seed banks on grassland community structure and population dynamics. In MA Leck, VT Parker and RL Simpson, eds. *Ecology of soil seed banks*, pp. 211–230. Academic Press, San Diego, CA.

Rice PM, Toney JC, Bedunah DJ, and Carlson CE (1997). Plant community diversity and growth form responses to herbicide applications for control of *Centaurea maculosa*. *Journal of Applied Ecology*, 34, 1397–1412.

Richards AJ (1990). The implications of reproductive versatility for the structure of grass populations. In GP Chapman, ed. *Reproductive versatility in the grasses*, pp. 131–153. Cambridge University Press, Cambridge.

Richards AJ (2003). Apomixis in flowering plants: an overview. *Philosophical Transactions of the Royal Society of London, Series B*, 358, 1085–1093.

Richmond KE and Sussman M (2003). Got silicon? The non-essential beneficial plant nutrient. *Current Opinion in Plant Biology*, 6, 268–272.

Riley RD and Vogel KP (1982). Chromosome numbers of released cultivars of switchgrass, indiangrass, big bluestem, and sand bluestem. *Crop Science*, 22, 1082–1083.

Risser PG (1969). Competitive relationships among herbaceous grassland plants. *Botanical Review*, 35, 251–284.

Risser PG (1988). Diversity in and among grasslands. In EO Wilson, ed. *Biodiversity*, pp. 176–180. National Academy Press, Washington, DC.

Risser PG (1989). Range condition analysis: past, present, and future. In WK Lauenroth and WA Laycock, eds.

Secondary succession and the evaluation of rangeland condition, pp. 143–156. Westview Press, Boulder, CO.

Risser PG and Parton WJ (1982). Ecosystem analysis of the tallgrass prairie: nitrogen cycle. *Ecology*, 63, 1342–1351.

Risser PG, Birney EC, Blocker HD, May SW, Parton WJ, and Weins JA (1981). *The true prairie ecosystem*. Hutchinson, Stroudsburg, PA.

Robertson JH (1939). A quantitative study of true-prairie vegetation after three years of extreme drought. *Ecological Monographs*, 9, 432–492.

Robertson PA and Ward RT (1970). Ecotypic differentiation in *Koeleria cristata* (L.) Pers. from Colorado and related areas. *Ecology*, 51, 1083–1087.

Robinson D, Hodge A, and Fitter A (2003). Constraints on the form and function of root systems. In H de Kroon and EJW Visser, eds. *Root ecology*, Vol. 168, pp. 1–31. Springer-Verlag, Berlin.

Rodríguez C, Leoni E, Lezama F, and Altesor A (2003). Temporal trends in species composition and plant traits in natural grasslands of Uruguay. *Journal of Vegetation Science*, 14, 433–440.

Rodwell JS, ed. (1992). *British plant communities*. Vol. 3. *Grasslands and montane communities*. Cambridge University Press, Cambridge.

Rodwell JS, ed. (2000). *British plant communities*. Vol. 5. *Maritime communities and vegetation of open habitats*. Cambridge University Press, Cambridge.

Rodwell JS, Schaminée JHJ, Mucina L, Pignatti JD, and Moss D (2002). The diversity of European vegetation. An overview of phytosociological alliances and their relationships to EUNIS habitats. EC-LNV. Report EC-LNV nr. 2002/054, Wageningen.

Rogers WE and Hartnett DC (2001a). Temporal vegetation dynamics and recolonization mechanisms on different-sized soil disturbances in tallgrass prairie. *American Journal of Botany*, 88, 1634–1642.

Rogers WE and Hartnett DC (2001b). Vegetation responses to different spatial patterns of soil disturbance in burned and unburned tallgrass prairie. *Plant Ecology*, 155, 99–109.

Rogers WE, Hartnett DC, and Elder B (2001). Effects of plains pocket gopher (*Geomys bursarius*) disturbances on tallgrass-prairie plant community structure. *American Midland Naturalist*, 145, 344–357.

Romo JT (2005). Emergence and establishment of *Agropyron desertorum* (Fisch.) (Crested Wheatgrass) seedlings in a sandhills prairie of central Saskatchewan. *Natural Areas Journal*, 25, 26–35.

Rorison IH and Hunt HW, eds. (1980). *Amenity grassland: an ecological perspective*. John Wiley & Sons, Chichester.

Rossiter NA, Setterfield SA, Douglas MM, and Hutley LB (2003). Testing the grass-fire cycle: alien grass invasion in the tropical savannas of northern Australia. *Diversity and Distributions*, 9, 169–176.

Rotsettis J, Quinn JA, and Fairbrothers DE (1972). Growth and flowering of *Danthonia sericea* populations. *Ecology*, 53, 227–234.

Rouget M (2003). Measuring conservation value at fine and broad scales: implications for a diverse and fragmented region, the Agulhas Plain. *Biological Conservation*, 112, 217–232.

Roxburgh SH and Wilson JB (2000). Stability and coexistence in a lawn community: experimental assessment of the stability of the actual community. *Oikos*, 88, 409–423.

Roy M and Ghosh B (1996). Polyamines, both common and uncommon, under heat stress in rice (*Oryza sativa*) callus. *Physiologia Plantarum*, 98, 196–200.

Rumbaugh MD (1990). Special purpose forage legumes. In J Janick and JE Simon, eds. *Advances in new crops*, pp. 183–190. Timber Press, Portland, OR.

Rumbaugh MD, Johnson DA, and Van Epps GA (1982). Forage yield and quality in a Great Basin shrub, grass, and legume pasture experiment. *Journal of Range Management*, 35, 604–609.

Rychnovská M (1993). Temperate semi-natural grasslands of Eurasia. In RT Coupland, ed. *Ecosystems of the world: eastern hemisphere and résumé*, Vol. 8B, pp. 125–166. Elsevier, Amsterdam.

Ryser P (1993). Influences of neighbouring plants on seedling establishment in limestone grassland. *Journal of Vegetation Science*, 4, 195–202.

Sackville Hamilton NR (1999). Genetic erosion issues in temperate grasslands. In Technical meeting on the methodology of the FAO world information and early warning system on plant genetic resources (eds J Serwinski and I Faberová). Food and Agriculture Organization of the United Nations, Research Institute of Crop Production, Prague, Czech Republic. Online: http://apps3.fao.org/wiews/Prague/tabcont.jsp.

Sackville Hamilton NR and Harper JL (1989). The dynamics of *Trifolium repens* in a permanent pasture. 1. The population dynamics of leaves and nodes per shoot axis. *Proceedings of the Royal Society of London, Series B Biological Sciences*, 237, 133–173.

Safford DH (1999). Brazilian Páramos I. An introduction to the physical environment and vegetation of the campos de altitude. *Journal of Biogeography*, 26, 693–712.

Sage RF (2004). The evolution of C_4 photosynthesis. *New Phytologist*, 161, 341–370.

Saggar S and Hedley CB (2001). Estimating seasonal and annual carbon inputs, and root decomposition rates in a temperate pasture following field 14C pulse-labelling. *Plant and Soil*, 236, 91–103.

Sagoff M (1997). Can we put a price on Nature's services? *Report from the Institute for Philosophy and Public Policy*, 17(3), 7–13

Sala O and Paruelo J (1997). Ecosystem services in grasslands. In G Daily, ed. *Nature's services: societal dependence on natural ecosystems*. Island Press, Washington, DC.

Salas ML and Corcuera LJ (1991). Effect of environment on gramine content in barley leaves and susceptibility to the aphid *Schizaphis graminum*. *Phytochemistry*, 30, 3237–3240.

Salzman PC (2004). *Pastoralists: equality, hierarchy, and the state*. Westview Press, Boulder, CO.

Sanchez JM, SanLeon DG, and Izco J (2001). Primary colonisation of mudflat estuaries by *Spartina maritima* (Curtis) Fernald in Northwest Spain: vegetation structure and sediment accretion. *Aquatic Botany*, 69, 15.

Sánchez-Ken J, G. and Clark LG (2000). Overview of the subfamily Centothecoideae (Poaceae). *American Journal of Botany*, 87 (6 supplement), 163 (abstract).

Sánchez-Moreiras AM, Weiss OA, and Reigosa-Roger MJ (2004). Allelopathic evidence in the Poaceae. *Botanical Review*, 69, 300–319.

Sangster AG, Hodson MJ, and Parry DW (1983). Silicon deposition and anatomical studies in the inflorescence bracts of four Phalaris species with their possible relevance to carcinogenesis. *New Phytologist*, 93, 105–122.

Sankaran M, Hanan NP, Scholes RJ et al. (2005). Determinants of woody cover in African savannas. *Nature*, 438, 846–849.

Sarmiento G (1983). The savannas of tropical america. In F Bourliére, ed. *Ecosystems of the world 13: tropical savannas*, pp. 245–288. Elsevier, Amsterdam.

Sarmiento G (1984). *The ecology of neotropical savannas*. Harvard University Press, Cambridge, MA.

Sarukhán J and Harper JL (1973). Studies on plant demography: *Ranuculus repens* L., *R. bulbosus* L. and *R. acris* L. I. Population flux and survivorship. *Journal of Ecology*, 61, 675–716.

Saunders DA, Hobbs RJ, and Margules CR (1991). Biological consequences of ecosystem fragmentation: a review. *Conservation Biology*, 5, 18–32.

Schadler M, Jung G, Auge H, and Brandl R (2003). Does the Fretwell–Oksanen model apply to invertebrates? *Oikos*, 100, 203–207.

Scheiner SM, Cox SB, Willig M, Mittelbach GG, Osenberg C, and Kaspari M (2000). Species richness, species-area curves and Simpson's paradox. *Evolutionary Ecology Research*, 2, 791–802.

Schimel DS and Bennett J (2004). Nitrogen mineralization: challenges of a changing paradigm. *Ecology*, 85, 591–602.

Schippfers P and Kropff MJ (2001). Competition for light and nitrogen among grassland species: a simulation analysis. *Functional Ecology*, 15, 155–164.

Schläpfer F, Pfisterer AB, and Schmid B (2005). Non-random species extinction and plant production: implications for ecosystem functioning. *Journal of Applied Ecology*, 42, 13–24.

Schmutz EM, Smith EL, Ogden PR et al. (1992). Desert grassland. In RT Coupland, ed. *Natural grasslands: introduction and western hemisphere*, Vol. 8A, pp. 337–362. Elsevier, Amsterdam.

Scholefield PA, Doick KJ, Herbert BMJ et al. (2004). Impact of rising CO_2 on emissions of volatile organic compounds: isoprene emission from *Phragmites australis* growing at elevated CO_2 in a natural carbon dioxide spring. *Plant, Cell & Environment*, 27, 393–401.

Scholes RJ and Hall DO (1996). The carbon budgets of tropical savannas, woodlands and grasslands. In AI Breymeyer, DO Hall, JM Melillo and GI Agren, eds. *Global change: effects on coniferous forests and grasslands*, Vol. 56. Wiley, Chichester.

Scholz H (1975). Grassland evolution in Europe. *Taxon*, 24, 81–90.

Schramm P (1970). The 'Do's and Don'ts' of prairie restoration. In P Schramm, ed. *Proceedings of a symposium on prairie and prairie restoration*, pp. 139–150. Knox College Biology Field Station Special Publication No. 3, Knox College, Galesburg, IL.

Schramm P (1992). Prairie restoration: a twenty-five year perspective on establishment and management. In DD Smith and CA Jacobs, eds. *Proceedings of the twelfth North American prairie conference: recapturing a vanishing heritage.*, pp. 169–177. University of Northern Iowa, Cedar Falls, IA.

Schreiber ESG, Bearlin AR, Nicol SJ, and Todd CR (2004). Adaptive management: a synthesis of current understanding and effective application. *Ecological Management and Restoration*, 5, 177–182.

Schüßler A, Schwarzott D, and Walker C (2001). A new fungal phylum, the Glomeromycota: phylogeny and evolution. *Mycological Research*, 105, 1413–1421.

Schwinning S and Parsons AJ (1996). Analysis of coexistence mechanisms for grasses and legumes in grazing systems. *Journal of Ecology*, 84, 799–814.

Scoones I (1996). New directions in pastoral development in Africa. In I Scoones, ed. *Living with uncertainty: new directions in pastoral development in Africa*, pp. 1–36. Intermediate Technology Publications, London.

Scott GAJ (1977). The role of fire in the creation and maintenance of savanna in the Montana of Peru. *Journal of Biogeography*, 4, 143–167.

Scurlock JMO, Johnson K, and Olson RJ (2002). Estimating net primary productivity from grassland biomass dynamics measurements. *Global Change Biology*, 8, 736–753.

Seamon P (2007). Fire management manual. Accessed 20 May 2007. Available: http://www.tncfiremanual.org. Nature Conservancy, Arlington, VA.

Seastedt T and Knapp AK (1993). Consequences of non-equilibrium resource availability across multiple time scales: the transient maxima hypothesis. *American Naturalist*, 141, 621–633.

Seastedt T, Coxwell CC, Ojima DS, and Parton WJ (1994). Controls of plant and soil carbon in a semihumid temperate grassland. *Ecological Applications*, 4, 344–353.

Seig CH, Flather CH, and McCanny S (1999). Recent biodiversity patterns in the Great Plains: implications for restoration and management. *Great Plains Research*, 9, 277–313.

SER (2004). Society for Ecological Restoration International Science & Policy Working Group. The SER international primer on ecological restoration, Version 2. Available: http://www.ser.org/content/ecological_restoration_primer.asp. Accessed 21 May 2007.Society for Ecological Restoration International, Tucson, AZ.

Sharifi MR (1983). The effects of water and nitrogen supply on the competition between three perennial meadow grasses. *Acta Oecologia Oecology Plantarum*, 4, 71–82.

Shaver GR and Billings WD (1977). Effects of daylength and temperature on root elongation in tundra graminoids. *Oecologia*, 28, 57–65.

Shea K, Roxburgh SH, and Rauschert ESJ (2004). Moving from pattern to process: coexistence mechanisms under intermediate disturbance regimes. *Ecology Letters*, 7, 491–508.

Sheail J, Wells TCE, Wells DA, and Morris MG (1974). Grasslands and their history. In E Duffey, MG Morris, J Shaeil, LK Ward, DA Wells and TCA Wells, eds. *Grassland ecology and wildlife management*, pp. 1–40. Chapman & Hall, London.

Shildrick J (1990). The use of turfgrasses in temperate humid climates. In AI Breymeyer, ed. *Managed grasslands: regional studies. Ecosystems of the world 17A*, pp. 255–300. Elsevier, Amsterdam.

Shu WS, Ye ZH, Zhang ZQ, Lan CY, and Wong MH (2005). Natural colonization of plants on five lead/zinc mine tailings in Southern China. *Restoration Ecology*, 13, 49–60.

Silva JF, Raventos J, Caswell H, and Trevisan MC (1991). Population responses to fire in a tropical savanna grass, *Andropogon semiberis*: a matrix model approach. *Journal of Ecology*, 79, 345–356.

Silvertown J (1980a). The dynamics of a grassland ecosystem: botanical equilibrium in the Park Grass Experiment. *Journal of Applied Ecology*, 17, 491–504.

Silvertown J (1980b). Leaf-canopy-induced seed dormancy in a grassland flora. *New Phytologist*, 85, 109–118.

Silvertown J (1981). Seed size, life span, and germination date as co-adapted features of plant life history. *American Naturalist*, 118, 860–864.

Silvertown JW and Lovett Doust J (1993). *Introduction to plant population biology*. 3rd edn. Blackwell Science, Oxford.

Silvertown JW and Charlesworth D (2001). *Introduction to plant population biology*. 4th edn. Blackwell Science, Oxford.

Silvertown J, Franco M, Pisanty I, and Mendoza A (1993). Comparative plant demography—relative importance of life-cycle components to the finite rate of increase in woody and herbaceous perennials. *Journal of Ecology*, 81, 465–476.

Silvertown J, Poulson P, Johnson J, Edwards GR, Biss P, Heard M, and Henman D (2006). The Park Grass Experiment 1856–2006: its contribution to ecology. *Journal of Ecology*, 94, 801–814.

Sims PL (1988). Grasslands. In MG Barbour and WD Billings, eds. *North American terrestrial vegetation*, pp. 265–286. Cambridge University Press, Cambridge.

Sims PL and Singh JS (1978a). The structure and function of ten western North American grasslands. II. Intraseasonal dynamics in primary producer compartments. *Journal of Ecology*, 66, 547–572.

Sims PL and Singh JS (1978b). The structure and function of ten western North American grasslands. III. Net primary production, turnover and efficiencies of energy capture and water use. *Journal of Ecology*, 66, 573–597.

Sims PL and Singh JS (1978c). The structure and function of ten western North American grasslands. IV. Compartmental transfers and energy flow within the ecosystem. *Journal of Ecology*, 66, 983–1009.

Sims PL, Singh JS, and Lauenroth WK (1978). The structure and function of ten western North American grasslands. I. Abiotic and vegetational characteristics. *Journal of Ecology*, 66, 251–285.

Singh JS and Gupta SR (1993). Grasslands of southern Asia. In RT Coupland, ed. *Ecosystems of the world: eastern hemisphere and résumé*, Vol. 8B, pp. 83–124. Elsevier, Amsterdam.

Skeel VA and Gibson DJ (1998). Photosynthetic rates and vegetative production of *Sorghastrum nutans* in response to competition at two strip mines and a railroad prairie. *Photosynthetica*, 35, 139–149.

Skerman PJ and Riveros F (1990). *Tropical grasses*. Food and Agriculture Organization of the United Nations, Rome.

Skerman PJ, Cameron DG, and Riveros F (1988). *Tropical forage legumes*. 2nd edn. Food and Agriculture Organization of the United Nations, Rome.

Smit AL, Bengough AG, Engels C, Van Noordwijk M, Pellerin S, and Van De Geijn SC (2000). *Root methods: a handbook*. Springer-Verlag, Berlin.

Smith DC, Nielsen EL, and Ahlgren HL (1946). Variation in ecotypes of *Poa pratensis*. *Botanical Gazette*, 108, 143–166.

Smith EF and Owensby CE (1972). Effects of fire on true prairie grasslands. *Proceedings of the Tall Timbers Fire Ecology Conference*, 12, 9–22.

Smith GS, Cornforth IS, and Herderson HV (1985). Critical leaf concentrations for deficiencies of nitrogen, potassium, phosphorus, sulphur and magnesium in perennial ryegrass. *New Phytologist*, 101, 393–409.

Smith KF, Rebertke GJ, Eagles HA, Anderson MW, and Easton HS (1999a). Genetic control of mineral concentration and yield in perennial ryegrass (*Lolium perenne* L.), with special emphasis on minerals related to grass tetany. *Australian Journal of Agricultural Research*, 50, 79–86.

Smith MD and Knapp AK (2003). Dominant species maintain ecosystem function with non-random species loss. *Ecology Letters*, 6, 509–517.

Smith MD, Hartnett DC, and Wilson GWT (1999b). Interacting influence of mycorrhizal symbiosis and competition on plant diversity in tallgrass prairie. *Oecologia*, 121, 547–582.

Smith MR, Charvat I, and Jacobson RL (1998). Arbuscular mycorrhizae promote establishment of prairie species in a tallgrass prairie restoration. *Canadian Journal of Botany*, 76, 1947–1954.

Smith RAH and Bradshaw AD (1979). The use of metal tolerant plant populations for reclamation of metalliferous wastes. *Journal of Applied Ecology*, 16, 595–612.

Snaydon RW (1981). The ecology of grazed pastures. In FHW Morley, ed. *Grazing animals*, pp. 13–31. Elsevier, Amsterdam.

Snaydon RW and Davies MS (1972). Rapid population divergence in a mosaic environment. II. Morphological variation in *Anthoxanthum odoratum*. *Evolution*, 26, 390–405.

Snaydon RW and Davies TM (1982). Rapid divergence of plant populations in response to recent changes in soil conditions. *Evolution*, 36, 289–297.

Society for Range Management (1998). *Glossary of terms used in range management*, 4th edition. Society for Range Management, Denver, CO.

Soderstrom TR and Calderón CE (1971). Insect pollination in tropical rain forest grasses. *Biotropica*, 3, 1–16.

Solecki MK (2005). Controlling invasive plants. In S Packard and CF Mutel, eds. *The tallgrass restoration handbook*. 2nd edn, pp. 251–278. Island Press, Washington, DC.

Sollenberger LE and Collins M (2003). Legumes for southern areas. In RF Barnes, CJ Nelson, M Collins and KJ Moore, eds. *Forages: an introduction to grassland agriculture*, Vol. 1, pp. 191–213. Iowa State Unviversity Press, Ames, IA.

Soreng RJ (2000). Apomixis and amphimixis comparative biogeography: a study in *Poa* (*Poaceae*). In SWL Jacobs and J Everett, eds. *Grasses: systematics and evolution*, pp. 294–306. CSIRO, Collingwood, Australia.

Soreng RJ and Davis JJ (1998). Phylogenetics and character evolution in the grass family (Poaceae): simulataneous analysis of morphological and chloroplast DNA restriction site character sets. *Botanical Review*, 64, 1–85.

Soreng RJ, Davis JI, and Volonmaa MA (2007). A phylogenetic analysis of *Poaceae* tribe *Poeae* s.l. based on morphological characters and sequence data from three plastid-encoded genes: evidence for reticulation, and a new classification for the tribe. *Kew Bulletin*, 62, 425–454.

Soriano A (1992). Río de la Plata grasslands. In RT Coupland, ed. *Natural grasslands: introduction and western hemisphere*, Vol. 8A, pp. 367–407. Elsevier, Amsterdam.

Soussana J-F and Luscher A (2007). Temperate grasslands and global atmospheric change: a review. *Grass and Forage Science*, 62, 127–134.

Spain AV and McIvor JG (1988). The nature of herbaceous vegetation associated with termitaria in north-eastern Australia. *Journal of Ecology*, 76, 181–191.

Spears JW (1994). Minerals in forages. In GC Fahey, ed. *Forage quality, evaluation, and utilization*, pp. 281–317. American Society of Agronomy, Inc., Crop Science Society of America, Inc., Soil Science Society of America, Inc., Madison, WI.

Spehn EM, Joshi J, Schmid B, Diemer M, and Körner C (2000). Above-ground resource use increases with plant species richness in experimental grassland ecosystems. *Functional Ecology*, 14, 326–337.

Spehn EM, Scherer-Lorenzen M, Schmid B et al. (2002). The role of legumes as a component of biodiversity in a cross-European study of grassland biomass nitrogen. *Oikos*, 98, 205–218.

Sperry TM (1994). The Curtis Prairie restoration, using the single-species planting method. *Natural Areas Journal*, 14, 124–127.

Sprague HG (1933). Root development of perennial grasses and its relation to soil conditions. *Soil Science*, 36, 189–209.

Spyreas G, Gibson DJ, and Basinger M (2001a). Endophyte infection levels of native and naturalized fescues in Illinois and England. *Journal of the Torrey Botanical Society*, 128, 25–34.

Spyreas G, Gibson DJ, and Middleton BA (2001b). Effects of endophyte infection in tall fescue (*Festuca arundinacea*: Poaceae) on community diversity. *International Journal of Plant Sciences*, 162, 1237–1245.

Srivastava DS and Vellend M (2005). Biodiversity-ecosystem function research: is it relevant to conservation? *Annual Review of Ecology and Systematics*, 36, 267–294.

Stace C (1991). *New flora of the British Isles*. Cambridge University Press, Cambridge.

Stapledon RG (1928). Cocksfoot grass (*Dactylis glomerata* L.): ecotypes in relation to biotic factor. *Journal of Ecology*, 16, 71–104.

Stebbins GL (1947). Types of polyploidy I. Their classification and significance. *Advances in Genetics*, 1, 403–429.

Stebbins GL (1956). Cytogenetics and evolution of the Grass family. *American Journal of Botany*, 43, 890–905.

Stebbins GL (1957). Self-fertilization and population variability in the higher plants. *American Naturalist*, 91, 337–354.

Stebbins GL (1975). The role of polyploid complexes in the evolution of North American grasslands. *Taxon*, 24, 91–106.

Stebbins GL (1982). Major trends of evolution in the Poaceae and their possible significance. In JR Estes, RJ Tyrl and JN Brunken, eds. *Grasses and grasslands: systematics and evolution*, pp. 3–36. University of Oklahoma Press, Norman, OK.

Stebbins GL (1987). Grass systematics and evolution: past, present and future. In TR Soderstrom, KW Hilu, CS Campbell and ME Barkworth, eds. *Grass systematics and evolution*, pp. 359–367. Smithsonian Institution Press, Washington, DC.

Stebbins GL and Crampton B (1961). A suggested revision of the grass genera of temperate North America. In *Recent Advances in Botany*. University of Toronto Press, Toronto.

Steffan-Dewenter I and Tscharntke T (2002). Insect communities and biotic interactions on fragmented calcareous grasslands—a mini review. *Biological Conservation*, 104, 275–284.

Stern H, de Hoedt G, and Ernst J (2007). Objective classification of Australian climates. Available http://www.bom.gov.au/climate/environ/other/koppen_explain.shtml. Accessed 12 December 2007. Commonwealth of Australia, Bureau of Meteorology, Canberra.

Stevens CJ, Dise NB, Mountford JO, and Cowing DJ (2004). Impact of nitrogen deposition on the species richness of grasslands. *Science*, 303, 1876–1879.

Stevens PF (2001 onwards). Angiosperm phylogeny website. Version 7, May 2006 [and more or less continuously updated since]. Available: http://www.mobot.org/MOBOT/research/APweb/. Accessed 9 December 2006.

Stewart WP, Liebert D, and Larkin KW (2004). Community identities as visions for landscape change. *Landscape and Urban Planning*, 69, 315–334.

Stiling P (1999). *Ecology: theories and applications*. 3rd edn. Prentice Hall, Englewood Cliffs, NJ.

Stoddart LA, Smith AD, and Box TW (1975). *Range management*. 3rd edn. McGraw-Hill, New York.

Stott P (1986). The spatial pattern of dry season fires in the savanna forests of Thailand. *Journal of Biogeography*, 13, 345–358.

Stringer C (2003). Human evolution: out of Ethiopia. *Nature*, 423, 692–695.

Stringham TK, Krueger WC, and Shaver P (2003). State and transition modeling: an ecological process approach. *Journal of Range Management*, 56, 106–113.

Stritch L (1990). Landscape-scale restoration of barrens-woodland wihin the oak-hickory forest mosaic. *Restoration and Management Notes*, 8, 73–77.

Strömberg CAE (2002). The origin and spread of grass-dominated ecosystems in the late Tertiary of North America: preliminary results concerning the evolution of hypsodonty. *Paleogeography, Paleoclimatology, Palaeoecology*, 177, 59–75.

Strömberg CAE (2004). Using phytolith assemblages to reconstruct the origin and spread of grass-dominated habitats in the great plains of North America during the late Eocene to early Miocene. *Paleogeography, Paleoclimatology, Palaeoecology*, 207, 239–275.

Strömberg CAE (2005). Decoupled taxonomic radiation and ecological expansion of open-habitat grasses in the Cenozoic of North America. *Proceedings of the National Academy of Science, USA*, 102, 11980–11984.

Suding KN, Collins SL, Gough L, Clark C, Cleland EE, Gross KL, Milchunas DG, and Pennings S (2005). Functional- and abundance-based mechanisms explain diversity loss due to N fertilization. *Proceedings of the National Academy of Science, USA*, 102, 4387–4392.

Suttie JM (2005). Grazing management in Mongolia. In JM Suttie, SG Reynolds and C Batello, eds. *Grasslands of the world*, pp. 265–304. Food and Agriculture Organization of the United Nations, Rome.

Suttie JM, Reynolds SG, and Batello C (2005). Other grasslands. In JM Suttie, SG Reynolds and C Batello, eds. *Grasslands of the world*, pp. 417–461. Food and Agriculture Organization of the United Nations, Rome.

Suzuki JI, Herben T, Krahulec F, and Hara T (1999). Size and spatial pattern of *Festuca rubra* genets in a mountain grassland: its relevance to genet establishment and dynamics. *Journal of Ecology*, 87, 942–954.

Suzuki JI, Herben T, Krahulec F, Štorchova H, and Hara T (2006). Effects of neighbourhood structure and tussock dynamics on genet demography of *Festuca rubra* in a mountain meadow. *Journal of Ecology*, 94, 66–76.

Swanson DA (1996). *Nesting ecology and nesting habitat requirements of Ohio's grassland-nesting birds: a literature review*. Ohio Fish and Wildlife Report. Division of Wildlife, Ohio Dept of Natural Resources, Columbus, OH.

Swift J (1996). Dynamic ecological systems and the administration of pastoral development. In I Scoones, ed. *Living with uncertainty: new directions in pastoral development in Africa*, pp. 153–173. Intermediate Technology Publications, London.

Swift MJ, Heal OW, and Anderson JM (1979). *Decomposition in terrestrial ecosystems*. Blackwell Scientific, Oxford.

't Mannetje L and Jones RM, eds. (2000). *Field and laboratory methods for grassland and animal production research*, pp. 447. CABI Publishing, Wallingford, Oxon.

Taft JB, Hauser C, and Robertson KR (2006). Estimating floristic integrity in tallgrass prairie. *Biological Conservation*, 131, 42–51.

Taiz L and Zeiger E (2002). *Plant physiology*. 3rd edn. Sinauer Associates, Sunderland, MA.

Tainton NM (1981a). Introduction to the concepts of development, production and stability of plant communities. In NM Tainton, ed. *Veld and pasture management in South Africa*, pp. 1–24. Shuter & Shooter, Pietermaritzburg.

Tainton NM, ed. (1981b). *Veld and pasture management in South Africa*, Shuter & Shooter, Pietermaritzburg.

Tainton NM (1981c). Veld burning. In NM Tainton, ed. *Veld and pasture management in South Africa*, pp. 363–391. Shuter & Shooter, Pietermaritzburg.

Tainton NM and Walker BH (1993). Grasslands of southern Africa. In RT Coupland, ed. *Ecosystems of the world: eastern hemisphere and résumé*, Vol. 8B, pp. 265–290. Elsevier, Amsterdam.

Tani T and Beard JB (1997). *Color atlas of turfgrass diseases: disease characteristics and control*. Ann Arbor Press, Chelsea, MI.

Tansley AG (1935). The use and abuse of vegetational concepts and terms. *Ecology*, 16, 284–307.

Tawaraya K (2003). Arbuscular mycorrhizal dependency of different plant species and cultivars. *Soil Science and Plant Nutrition*, 49, 655–668.

Terri JA and Stowe LG (1976). Climatic patterns and distribution of C_4 grasses in North America. *Oecologia*, 23, 1–12.

Teyssonneyre F, Picon-Cochard C, Falcimagne R, and Soussana JF (2002). Effects of elevated CO_2 and cutting frequency on plant community structure in a temperate grassland. *Global Change Biology*, 8, 1034–1046.

Thomas H (1980). Terminology and definitions in studies of grassland plants. *Grass and Forage Science*, 35, 13–23.

Thomasson JR, Nelson ME, and Zakrzewski RJ (1986). A fossil grass (Gramineae: Chloridoideae) from the Miocene with Kranz Anatomy. *Science*, 233, 876–878.

Thompson K and Grime JP (1979). Seasonal variation in the seed banks of herbaceous species in ten contrasting habitats. *Journal of Ecology*, 67, 893–921.

Thompson K, Bakker JP, and Bekker RM (1997). *The soil seed banks of North West Europe: methodology, density and longevity*. Cambridge University Press, Cambridge.

Thompson K, Grime JP, and Mason G (1977). Seed germination in response to diurnal fluctuations of temperature. *Nature*, 267, 147–149.

Thornley JHM (1998). *Grassland dynamics: an ecosystem simulation model*. CAB International, Wallingford, Oxon.

Thornley JHM and Cannell MGR (2000). Dynamics of mineral N availability in grassland ecosystems under increased $[CO_2]$: hypotheses evaluated using the Hurley Pasture Model. *Plant and Soil*, 224, 153–170.

Thornton B (2001). Uptake of glycine by non-mycorrhizal *Lolium perenne*. *Journal of Experimental Botany*, 52, 1315–1322.

Tilman D (1982). *Resource competition and community structure*. Princeton University Press, Princeton, NJ.

Tilman D (1987). Secondary succession and patterns of plant dominance along experimental nitrogen gradients. *Ecological Monographs*, 57, 189–214.

Tilman D (1988). *Plant strategies and the dynamics and structure of plant communities*. Princeton University Press, Princeton, NJ.

Tilman D (1990). Mechanisms of plant competition for nutrients: the elements of a predictive theory of competition. In JB Grace and D Tilman, eds. *Perspectives on plant competition*, pp. 117–141. Academic Press, San Diego, CA.

Tilman D (1996). Biodiversity: population versus ecosystem stability. *Ecology*, 77, 350–363.

Tilman D (2004). Niche tradeoffs, neutrality, and community structure: a stochastic theory of resource competition, invasion, and community assembly. *Proceedings of the National Academy of Science, USA*, 101, 10864–10861.

Tilman D (2007). Resource competition and plant traits: a response to Craine (2005). *Journal of Ecology*, 95, 231–234.

Tilman D and Downing JA (1994). Biodiversity and stability in grasslands. *Nature*, 367, 363–365.

Tilman D and Wedin D (1991a). Oscillations and chaos in the dynamics of a perennial grass. *Nature*, 353, 653–655.

Tilman D and Wedin D (1991b). Plant traits and resource reduction for five grasses growing on a nitrogen gradient. *Ecology*, 72, 685–700.

Tilman D, Hill J, and Lehman C (2006a). Carbon-negative biofuels from low-input high-diversity grassland biomass. *Science*, 314, 1598–1600.

Tilman D, Reich PB, and Knops JMH (2006b). Biodiversity and ecosystem stability in a decade-long grassland experiment. *Nature*, 441, 629–632.

Ting-Cheng Z (1993). Grasslands of China. In RT Coupland, ed. *Ecosystems of the world: eastern hemisphere and résumé*, Vol. 8B, pp. 61–82. Elsevier, Amsterdam.

Titlyanova AA, Zlotin RI, and French NR (1990). Changes in temperate-zone grasslands under the influence of man. In AI Breymeyer, ed. *Managed grasslands: regional studies. Ecosystems of the world 17A*, pp. 301–334. Elsevier, Amsterdam.

Todd PA, Phillips JDP, Putwain PD, and Marrs RH (2000). Control of *Molinia caerulea* on moorland. *Grass and Forage Science*, 55, 181–191.

Tomanek GW and Hulett GK (1970). Effects of historical droughts on grassland vegetation in the Central Great Plains. In *Pleistocene and recent environments of the central Great Plains*, pp. 203–210. University Press of Kansas, Lawrence, KS.

Tomlinson KW and O'Connor TG (2004). Control of tiller recruitment in bunchgrasses: uniting physiology and ecology. *Functional Ecology*, 18, 489–496.

Tongway D and Hindley N (2004). Landscape function analysis: a system for monitoring rangeland function. *African Journal of Range and Forage Science*, 21, 109–113.

Toole EH and Brown E (1946). Final results of the Duvel buried seed experiment. *Journal of Agricultural Research*, 72, 210–210.

Tow PG and Lazenby A (2001). Competition and succession in pastures—some concepts and questions. In PG Tow and A Lazenby, eds. *Competition and succession in pastures*, pp. 1–14. CABI Publishing, Wallingford, Oxon.

Towne EG and Knapp AK (1996). Biomass and density responses in tallgrass prairie legumes to annual fire and topographic position. *American Journal of Botany*, 83, 175–179.

Towne G and Owensby CE (1984). Long-term effects of annual burning at different dates in ungrazed Kansas tallgrass prairie. *Journal of Range Management*, 37, 392–397.

Transeau EN (1935). The prairie peninsula. *Ecology*, 16, 423–37.

Trewartha GT (1943). *An introduction to weather and climate*. McGraw-Hill, New York.

Tscharntke T and Greiler H-J (1995). Insect communities, grasses, and grasslands. *Annual Review of Entomology*, 40, 535–558.

Turgeon AJ (1985). *Turfgrass management*. Reston Publishing, Reston, VA.

Turkington R and Harper JL (1979). The growth, distribution and neighbour relationships of *Trifolium repens* in a permanent pasture.IV. Fine-scale biotic differentiation. *Journal of Ecology*, 67, 245–254.

Turlings TCJ, Tumlinson JH, and Lewis WJ (1990). Exploitation of herbivore-induced plant odors by host-seeking parasitic wasps. *Science*, 250, 1251–1253.

Turner CL, Kneisler JR, and Knapp AK (1995). Comparative gas exchange and nitrogen responses of the dominant C_4 grass *Andropogon gerardii* and five C_3 forbs to fire and topographic position in tallgrass prairie during a wet year. *International Journal of Plant Science*, 156, 216–226.

U.S. Soil Survey Staff (1975). *Soil taxonomy: a basic system of soil classification for making and interpreting soil surveys*. USDA Agricultural Handbook, 436. U.S. Government Printing Office, Washington, DC.

Úlehlová B (1992). Microorganisms. In RT Coupland, ed. *Ecosystems of the world: natural grasslands: introduction and western hemisphere*, Vol. 8A, pp. 95–119. Elsevier, Amsterdam.

Umbanhowar CEJ (1995). Revegatation of earthern mounds along a topographic-productivity gradient in a northern mixed prairie. *Journa of Vegetation Science*, 6, 637–646.

Usher G (1966). *The Wordsworth dictionary of botany*. Wordsworth, Ware, Herefordshire.

van Andel J (2005). Species interactions structuring plant communities. In E Van der Maarel, ed. *Vegetation ecology*, pp. 238–264. Blackwell, Oxford.

van Andel J, Snaydon RW, and Bakker JP, eds. (1987). *Disturbance in grasslands: causes, effects and processes*, pp 316. Kluwer, Dordrecht.

van der Maarel E (1981). Fluctuations in a coastal dune grassland due to fluctuations in rainfall: experimental evidence. *Vegetatio*, 47, 259–265.

van der Maarel E (1996). Pattern and process in the plant community: fifty years after A.S. Watt. *Journal of Vegetation Science*, 7, 19–28.

van Groenendael J, de Kroon H, Kalisz S, and Tuljapurkar S (1994). Loop analysis: evaluating life history pathways in population projection matrices. *Ecology*, 75, 2410–2415.

Van Treuren R, Bas N, Goosens PJ, Jansen J, and Van Soest LJM (2005). Genetic diversity in perennial ryegrass and white clover among old Dutch grasslands as compared to cultivars and nature reserves. *Molecular Ecology*, 14, 39–52.

Vandik V and Goldberg DE (2006). Sources of diversity in a grassland metacommunity: quantifying the contribution of dispersal to species richness. *American Naturalist*, 168, 157–167.

Vandvik V and Birks JHB (2002). Pattern and process in Norwegian upland grasslands: a functional analysis. *Journal of Vegetation Science*, 13, 123–134.

Vasseur L and Potvin C (1998). Natural pasture community response to enriched carbon dioxide atmosphere. *Plant Ecology*, 135, 31–41.

Vega E and Montaña C (2004). Spatio-temporal variation in the demography of a bunch grass in a patchy semi-arid environment. *Plant Ecology*, 175, 107–120.

Verweij RJT, Verrelst J, Loth PE, Heitkönig IMA, and Brunsting AMH (2006). Grazing lawns contribute to the subsidence of mesoherbivores on dystophic savannas. *Oikos*, 114, 108–116.

Vicari M and Bazely DR (1993). Do grasses fight back? The case for antiherbivore defences. *Trends in Ecology and Evolution*, 8, 137–141.

Vickery PJ (1972). Grazing and net primary production of a temperate grassland. *Journal of Applied Ecology*, 9, 307–314.

Vinton MA and Goergen EM (2006). Plant–soil feedbacks contribute to the persistence of *Bromus inermis* in tallgrass prairie. *Ecosystems*, V9, 967–976.

Virágh K and Gerencsér G (1988). Seed bank in the soil and its role during secondary successions induced by some herbicides in a perennial grassland community. *Acta Botanica Hungarica*, 34, 77–121.

Vitousek PM, Aber JD, Howarth RW *et al.* (1997). Technical report: human alteration of the global nitrogen cycle: sources and consequences. *Ecological Applications*, 7, 737–750.

Vitt DH (1979). The moss flora of the Auckland Islands, New Zealand, with a consideration of habitats, origins, and adaptations. *Canadian Journal of Botany*, 57, 2226–2263.

Vogel JC, Fuls A, and Danin A (1986). Geographical and environmental distribution of C_3 and C_4 grasses in the Sinai, Negev, and Judean deserts. *Oecologia*, 70, 258–265.

Vogl RJ (1974). Effects of fire on grasslands. In TT Kozlowski and CE Ahlgren, eds. *Fire and ecosystems*, pp. 139–194. Academic Press, New York.

Vogl RJ (1979). Some basic principles of grassland fire management. *Environmental Management*, 3, 51–57.

Voigt JW (1980). J E Weaver and the North American prairie: 'look carefully and look often'. In *Seventh North American Prairie Conference* (ed CL Kucera), pp. 317–321, Southwest Missouri State University, Springfield, MI.

Vujnovic K, Wein RW, and Dale MRT (2002). Predicting plant species diversity in response to disturbance magnitude in grassland remnants of central Alberta. *Canadian Journal of Botany*, 504–511.

Waddington DV, Carrow RN, and Shearman RC, eds. (1992). *Turfgrass*, pp. 805. American Society of Agronomy, Inc., Crop Science Society of America, Inc., Soil Science Society of America, Inc., Madison, Wisconsin, WI.

Wagner RE (1989). History and development of site and condition criteria in the Bureau of Land Management. In WK Lauenroth and WA Laycock, eds. *Secondary succession and the evaluation of rangeland condition*, pp. 35–48. Westview Press, Boulder, CO.

Wahren C-HA, Papst WA, and Williams RJ (1994). Long-term vegetation change in relation to cattle grazing in subalpine grassland and heathland on the Bogong High Plains: an analysis of vegetation records from 1945 to 1994. *Australian Journal of Botany*, 42, 607–639.

Waide RB, Willig MR, Steiner CF *et al.* (1999). The relationship between productivity and species richness. *Annual Review of Ecology and Systematics*, 30, 257–300.

Wali MK and Kannowski PB (1975). Prairie ant mound ecology: interrelationships of microclimate, soils and vegetation. In MK Wali, ed. *Prairie: a multiple view*. *Midwest Prairie Conference*, pp. 155–169. University of North Dakota, Grand Forks, ND.

Walker BH, Emslie RH, Owen-Smith RN, and Scholes RJ (1987). To cull or not to cull: lessons from a Southern African drought. *Journal of Applied Ecology*, 24, 381–401.

Wan CSM and Sage RF (2001). Climate and the distribution of C-4 grasses along the Atlantic and Pacific coasts of North America. *Canadian Journal of Botany*, 4, 474–486.

Wand SJE, Midgley GF, Jones MH, and Curtis PS (1999). Responses of wild C_4 and C_3 grass (Poaceae) species to elevated atmospheric CO_2 concentration: a meta-analytic test of current theories and perceptions. *Global Change Biology*, 5, 723–741.

Ward D, Ngairorue BT, Karamata J, Kapofi I, Samuels R, and Ofran Y (2000). Effects of communal pastoralism on vegetation and soil in a semi-arid and in an arid region of Namibia. In *Proceedings of the IAVS Symposium*, pp. 344–347. Opulus Press, Grangärde, Sweden.

Warren JM, Raybould AF, Ball T, Gray AJ, and Hayward MD (1988). Genetic structure in the perennial grasses *Lolium perenne* and *Agrostis curtisii*. *Heredity*, 81, 556–562.

Watkinson AR (1990). The population dynamics of *Vulpia fasiculata*: a nine-year study. *Journal of Ecology*, 78, 196–209.

Watkinson AR, Lonsdale WM, and Andrew MH (1989). Modelling the population dynamics of an annual plant *Sorghum intrans* in the wet-dry tropics. *Journal of Ecology*, 77, 162–181.

Watkinson AR, Freckleton RP, and Forrester L (2000). Population dynamics of *Vulpia ciliata*: regional, patch and local dynamics. *Journal of Ecology*, 88, 1012–1029.

Watson I (2006). Monitoring in the rangelands. Available: http://www.deh.gov.au/soe/2006/emerging/rangelands/index.html. Accessed 15 January 2008. In 2006 Australian State of the Environment Committee, Department of the Environment and Heritage, Canberra.

Watson L (1990). The grass family Poaceae. In GP Chapman, ed. *Reproductive versatility in the grasses*, pp. 1–31. Cambridge University Press, Cambridge.

Watson L and Dallwitz MJ (1988). *Grass genera of the world: interactive identification and information retrieval.* Research School of Biological Sciences, Australian National University, Canberra.

Watson L and Dallwitz MJ (1992 onwards). Grass genera of the world: descriptions, illustrations, identification, and information retrieval; including synonyms, morphology, anatomy, physiology, phytochemistry, cytology, classification, pathogens, world and local distribution, and references. Version: 7th April 2008, http://delta-intkey.com

Watt AS (1940). Studies in the ecology of Breckland. IV. The grass-heath. *Journal of Ecology*, 28, 42–70.

Watt AS (1947). Pattern and process in the plant community. *Journal of Ecology*, 35, 1–22.

Watt AS (1981a). A comparison of grazed and ungrazed grassland A in East Anglian Breckland. *The Journal of Ecology*, 69, 499–508.

Watt AS (1981b). Further obervations on the effects of excluding rabbits from grassland A in East Anglian Breckland: the patterns of change and factors affecting it (1936–1973). *Journal of Ecology*, 69, 509–536.

Watts JG, Huddleston EW, and Owens JC (1982). Rangeland entomology. *Annual Reviews of Entomology*, 27, 283–311.

Weaver JE (1919). *The ecological relations of roots.* Carnegie Institute Publication 286, Washington, DC.

Weaver JE (1920). *Root development in the grassland formation.* Carnegie Institute Publication 292, Washington, DC.

Weaver JE (1942). Competition of western wheatgrass with relict vegetation of prairie. *American Journal of Botany*, 29, 366–372.

Weaver JE (1950). Effects of different intensities of grazing on depth and quantity of roots of grasses. *Journal of Range Management*, 3, 100–113.

Weaver JE (1954). *North American prairie.* Johnsen Publishing, Lincoln, NE.

Weaver JE (1958). Classification of root systems of forbs of grassland and a consideration of their significance. *Ecology*, 39, 393–401.

Weaver JE (1961). The living network in prairie soils. *Botanical Gazette*, 123, 16–28.

Weaver JE and Albertson FW (1936). Effects of the great drought on the prairies of Iowa, Nebraska, and Kansas. *Ecology*, 17, 567–639.

Weaver JE and Albertson FW (1956). *Grasslands of the Great Plains.* Johnsen Publishing Company, Lincoln, Nebraska.

Weaver JE and Clements FE (1938). *Plant ecology.* McGraw-Hill, New York.

Weaver JE and Darland RW (1949a). Quantitative study of root systems in different soil types. *Science*, 110, 164–165.

Weaver JE and Darland RW (1949b). Soil-root relationships of certain native grasses in various soil types. *Ecological Monographs*, 19, 303–338.

Weaver JE and Fitzpatrick TJ (1932). Ecology and relative importance of the dominants of tall-grass prairie. *Botanical Gazette*, 93, 113–150.

Weaver JE and Fitzpatrick TJ (1934). The prairie. *Ecological Monographs*, 4, 113–295.

Weaver JE and Rowland NW (1952). Effects of excessive natural mulch on development, yield, and structure of native grassland. *Botanical Gazette*, 114, 1–19.

Weaver JE and Tomanek GW (1951). *Ecological studies in a midwestern range: the vegetation and effects of cattle on its composition and distribution.* University of Nebraska, Conservation and Survey Division, Lincoln, NE.

Weaver JE and Zink E (1945). Extent and longevity of the seminal roots of certain grasses. *Plant Physiology*, 20, 359–379.

Weaver JE, Houghen VH, and Weldon MD (1935). Relation of root distribution to organic matter in prairie soil. *Botanical Gazette*, 96, 389–420.

Weaver T, Payson EM, and Gustafson DL (1996). Prairie ecology—the shortgrass prairie. In FB Samson and FL Knopf, eds. *Prairie conservation: preserving North*

America's most endangered ecosystem, pp. 67–88. Island Press, Washington, DC.

Wedin D and Tilman D (1993). Competition among grasses along a nitrogen gradient: initial conditions and mechanisms of competition. *Ecological Monographs*, 63, 199–229.

Weigelt A, Bol R, and Bardgett RD (2005). Preferential uptake of soil nitrogen forms by grassland plant species. *Oecologia*, 142, 627–635.

Wells RT (1989). Vombatidae. In DW Walton and BJ Richardson, eds. *Fauna of Australia*, Vol. 1B *Mammalia*, pp. 1–25. Australian Government Publishing Service, Canberra.

Weltzin JF, Archer S, and Heitschmidt RK (1997). Small-mammal regulation of vegetation structure in a temperate savanna. *Ecology*, 78, 751–763.

Wentworth TR (1983). Distributions of C_4 plants along environmental and compositional gradients in south-eastern Arizona. *Vegetatio*, 52, 21–34.

West CP (1994). Physiology and drought tolerance of endophyte-infected grasses. In CW Bacon and JF White, eds. *Biotechnology of endophytic fungi of grasses*, pp. 87–99. CRC Press, Boca Raton, FL.

West CP and Nelson CJ (2003). Naturalized grassland ecosystems and their management. In RF Barnes, CJ Nelson, M Collins and KJ Moore, eds. *Forages: an introduction to grassland agriculture*, Vol. 1, pp. 315–337. Iowa State University Press, Ames, IA.

West HM, Fitter AH, and Watkinson AR (1993). Response of *Vulpia ciliata* ssp. *ambigua* to removal of mycorrhizal infection and to phosphate application under field conditions. *Journal of Ecology*, 81, 351–358.

West NE (2003). History of rangeland monitoring in the U.S.A. *Arid Land Research and Management*, 17, 495–545.

Westhoff V and van der Maarel E (1973). The Braun-Blanquet approach. In RH Whittaker, ed. *Ordination and classification of communities*, pp. 617–726. Dr. W. Junk Publishers, The Hague.

Westoby M, Walker BH, and Noy-Meir I (1989). Opportunistic management for rangelands not at equilibrium. *Journal of Range Management*, 42, 266–274.

Westover KM and Bever JD (2001). Mechanisms of plant species coexistence: roles of rhizosphere bacteria and root fungal pathogens. *Ecology*, 82, 3285–3294.

Whicker AD and Detling JK (1988). Ecological consequences of prairie dog disturbances. *BioScience*, 38, 778–785.

Whiles MR and Charlton RE (2006). The ecological significance of tallgrass prairie arthropods. *Annual Reviews of Entomology*, 51, 387–412.

White JF (1988). Endophyte-host associations in forage grasses. XI. A proposal concerning origin and evolution. *Mycologia*, 80, 442–446.

White PS and Pickett STA (1985). Natural disturbance and patch dynamics: an introduction. In STA Pickett and PS White, eds. *The ecology of natural disturbance and patch dynamics*, pp. 3–16. Academic Press, Orlando, FL.

White R, Murray S, and Rohweder M (2000). *Pilot analysis of global ecosystems: grassland ecosystems technical report*. World Resources Institute, Washington, DC.

Whitehead DC (1966). *Nutrient minerals in grassland herbage*. Grassland Research Institute, Commonwealth Agricultural Bureaux, Hurley, Berks.

Whitehead DC (2000). Nutrient elements in grassland: soil-plant-animal relationships. CABI Publishing, Wallingford, Oxon.

Whittaker RH (1951). A criticism of the plant association and climatic climax concepts. *Northwest Science*, 25, 17–31.

Whittaker RH (1973). Dominance types. In RH Whittaker, ed. *Ordination and classification of communities*, pp. 389–402. Dr. W. Junk Publishers, The Hague.

Whittaker RJ, Bush MB, and Richards K (1989). Plant recolonization and vegetation succession on the Krakatau islands, Indonesia. *Ecological Monographs*, 59, 59–123.

Whittingham MJ (2007). Will agri-environment schemes deliver substantial biodiversity gain, and if not why not? *Journal of Applied Ecology*, 44, 1–5.

Willems JH (2001). Problems, approaches, and results in restoration of Dutch Calcareous grassland during the last 30 years. *Restoration Ecology*, 9, 147–154.

Williams GJ and Markley JL (1973). The photosynthetic pathway type of North American shortgrass prairie species and some ecological implications. *Photosynthetica*, 7, 262–270.

Williams JR and Diebel PL (1996). The economic value of the prairie. In FB Sampson and FL Knopf, eds. *Prairie conservation*, pp. 19–38. Island Press, Washington, DC.

Williams PR, Congdon RA, Grice AC, and Clarke PJ (2003). Fire-related cues break seed dormancy of six legumes of tropical eucalypt savannas in north-eastern Australia. *Austral Ecology*, 28, 507–514.

Williams RJ (1982). The role of climate in a grassland classification. In AC Nicholson, A McLean and TE Baker, eds. *Grassland ecology and classification symposium proceedings*, pp. 41–51. British Columbia Ministry of Forests, Victoria, BC.

Wilsey BJ (2002). Clonal plants in a spatially heterogeneous environment: effects of integration on Serengeti grassland response to defoliation and urine-hits from grazing mammals. *Plant Ecology*, 159, 15–22.

Wilson AD (1989). The development of systems of assessing the condition of rangeland in Australia. In WK Lauenroth and WA Laycock, eds. *Secondary succession and the evaluation of rangeland condition*, pp. 77–102. Westview Press, Boulder, CO.

Wilson GWT and Hartnett DC (1997). Effects of mycorrhizae on plant growth and dynamics in experimental tallgrass prairie microcosms. *American Journal of Botany*, 84, 478–482.

Wilson GWT and Hartnett DC (1998). Interspecific variation in plant responses to mycorrhizal colonization in tallgrass prairie. *American Journal of Botany*, 85, 1732–1738.

Wilson JB (1994). The 'intermediate disturbance hypothesis' of species coexistence is based on patch dynamics. *New Zealand Journal of Ecology*, 18, 176–181.

Wilson JB and Roxburgh SH (1994). A demonstration of guild-based assembly rules for a plant community, and determination of intrinsic guilds. *Oikos*, 69, 267–276.

Wilson JB and Watkins AJ (1994). Guilds and assembly rules in lawn communities. *Journal of Vegetation Science*, 5, 591–600.

Wilson JB, Wells TCE, Trueman IC et al. (1996). Are there assembly rules for plant species abundance? An investigation in relation to soil resources and successional trends. *Journal of Ecology*, 84, 527–538.

Wilson MV and Clark DL (2001). Controlling invasive *Arrhenatherum elatius* and promoting native prairie grasses through mowing. *Applied Vegetation Science*, 4, 129–138.

Wilson SD (2002). Prairies. In MR Perrow and AJ Davy, eds. *Handbook of ecological restoration, volume 2 restoration in practice*, pp. 443–465. Cambridge University Press, Cambridge.

Winkler J, B. and Herbst M (2004). Do plants of a semi-natural grassland community benefit from long-term CO_2 enrichment? *Basic and Applied Ecology*, 5, 131–143.

Wolfe EC and Dear BS (2001). The population dynamics of pastures, with particular reference to southern Australia. In PG Tow and A Lazenby, eds. *Competition and succession in pastures*, pp. 119–148. CABI Publishing, Wallingford, Oxon.

Woodhouse CA and Overpeck JT (1998). 2000 years of drought variability in the Central United States. *Bulletin of the American Meteorological Society*, 79, 2693–2714.

Woodmansee RG and Duncan DA (1980). Nitrogen and phosphorus dynamics and budgets in annual grasslands. *Ecology*, 61, 893–904.

Woodward SL (2003). *Biomes of earth*. Greenwood Press, Westport, CT.

Wooton JT (1998). Effects of disturbance on species diversity: a multitrophic perspective. *American Naturalist*, 152, 803–825.

World Resources 2000–2001 (2000). *People and ecosystems: the fraying web of life*. World Resources Institute in collaboration with the United Nations Development Programme, The United Nations Environment Programme, and the World Bank, Washington, DC.

Wright HA and Bailey AW (1982). *Fire ecology*. John Wiley & Sons, New York.

Yarranton GA and Morrison RG (1974). Spatial dynamics of a primary succession: nucleation. *Journal of Ecology*, 62, 417–428.

Yaxing W and Quangong C (2001). Grassland classification and evaluation of grazing capacity in Naqu Prefecture, Tibet autonomous region, China. *New Zealand Journal of Agricultural Research*, 44, 253–258.

Young JA, Evans RA, Raguse CA, and Larson JR (1980). Germinable seeds and periodicity of germination in annual grasslands. *Hilgardia*, 49, 1–37.

Young TP, Petersen DA, and Clary JJ (2005). The ecology of restoration: historical links, emerging issues and unexplored realms. *Ecology Letters*, 8, 662–673.

Yu F, Dong M, and Bertil K (2004). Clonal integration helps *Psammochloa villosa* survive sand burial in an inland dune. *New Phytologist*, 162, 697–704.

Zapiola ML, Campbell CK, Butler MD, and Mallory-Smith CA (2008). Escape and establishment of transgenic glyphosate-resistant creeping bentgrass (*Agrostis stolonifera*) in Oregon, USA: a 4-year study. *Journal of Applied Ecology*, 45, 486–494.

Zavaleta ES and Hulvey KB (2004). Realistic species losses disproportionately reduce grassland resistance to biological invaders. *Science*, 306, 1175–1177.

Zavaleta ES, Shaw MR, Chiariello NR, Mooney HA, and Field CB (2003). Additive effects of simulated climate changes, elevated CO_2, and nitrogen deposition on grassland diversity. *Proceedings of the National Academy of Science, USA*, 100, 7650–7654.

Zeilhofer P and Schessl M (2000). Relationship between vegetation and environmental conditions in the northern Pantanal of Mato Grosso, Brazil. *Journal of Biogeography*, 27, 159–168.

Zhang W (2000). Phylogeny of the grass family (Poaceae) from *rp116* Intron sequence data. *Molecular Phylogenetics and Evolution*, 15, 135–146.

Zhenggang G, Tiangang L, and Zihe Z (2003). Classification management for grassland in Gansu

Province, China. *New Zealand Journal of Agricultural Research*, 46, 123–131.

Zhou H, Zhou L, Zhao X *et al.* (2006). Stability of alpine meadow ecosystem on the Qinghai-Tibetan Plateau. *Chinese Science Bulletin*, 51, 320–327.

Zizhi H and Degang Z (2005). China. In *Country Pasture/Forage Resource Profiles*: Available http://www.fao.org/ag/AGP/AGPC/doc/Counprof/china/china1.htm. Accessed August 19 2005 (ed S Reynolds).

Zobel M (1997). The relative role of species pools in determining plant species richness: an alternative explanation of species coexistence? *Trends in Ecology and Evolution*, 12, 266–269.

Plant index

Nomenclature for plant names is synonymised according to the USDA Plants National Database (http://plants.usda.gov/index.html) as of May, 2008, or for species not in the database, the name used in the original source. As a result, names are changed from the original source when a more up to date name is provided in the database. Some familiar plants have new names. For example, *Schedonorus phoenix* is used in the text in place of the older *Festuca arundinacea*. Older names are included in the index with a reference to the new name.

Acacia 166, 174
  A. cana 169
  A. constricta 170
  A. farnesiana 169
  A. mearnsii 12
Acaena magellanica 12
Achantherum splendens 183
Achillea lanulosa see *Achillea millefolium* var. *occidentalis*
  A. millefolium 120, 123, 207
  A. millefolium var. *occidentalis* 225
Achnatherum lettermani 172
  A. speciosum 169
Adesmia campestris 168
Aegilops geniculata 46
  A. triuncialis 151
  A. speltoides 107
Agalinus 120
  A. auriculata 237
Agropyron 14, 49, 82, 105, 107, 179
  A. cristatum 119, 169
  A. dasystachyum see *Elymus lanceolatus* ssp. *lanceolatus*
  A. repens see *Elymus repens*
  A. smithii see *Pascopyrum smithii*
  A. spicatum see *Pseudoroegneria spicata*
Agrostis 26, 43, 83, 86, 143, 175, 177–179
  A. canina 179
  A. capillaris 48, 85, 88, 103, 112, 123, 125, 145, 175–176, 179, 204, 207, 218
  A. gigantea 176
  A. megellanica 134
  A. stolonifera 37, 94, 105, 176, 178–179, 221
  A. tenius see *Agrostis capillaris*
  A. vinealis 179

Aira praecox 87
Albizia procera 166
Alchemilla monticola 220
  A. subcrenata 220
  A. wichurae 220
Allium polyrrhizum 167
Alloeochete 56
Allolepis 84
Alloteropsis 59
Alopecurus 49, 83, 177
  A. geniculatus 85
  A. pratensis 96–97, 176, 218
Alternanthera 59
  A. pratensis 45
Alysicarpus vaginalis 177
Amaranthaceae 59, 73
Amaranthus 59
Ambrosia psilostachya 200, 225
Ammophila arenaria 44, 95, 113
  A. brevigulata 113
Amorpha canescens 112, 180, 193, 225
Amphicarpum amphicarpon see *A. purshii*
  A. muhlenbergianum 49, 83
  A. purshii 49, 83
Amphipogon 27, 50
Andropogon 23, 27, 105, 165, 174
  A. bladhii see *Bothriochloa bladhii*
  A. brevifolius 94
  A. gayanus 194
  A. gerardii 14, 35, 42, 46, 49, 64, 66, 95, 101–103, 106, 108–109, 112, 116–117, 139–140, 173, 177, 180–180, 185, 193, 202, 208–209, 225, 241–242, Plate 2, Plate 11
  A. glomeratus 37
  A. halli 106, 117
  A. hypogynus 159

  A. ischaemum see *Bothriochloa ischaemum*
  A. lateralis 171
  A. semiberbis 93
  A. virginicus 103, 138
Androsace tapete 168–169
Antennaria neglecta 208
Anomochlooideae 25–26, 28, 30, 59
Anomochloa 31
  A. marantoidea 26
Anthoxanthum 83
  A. odoratum 45, 94, 103, 105, 175–176, 220
Apera spica-venti 88
Apiaceaea 76
Arabis hirsuta 120
Arachnis glabrata 177
Arenaria musciformis 168–169
Argyroxiphium 174
Aristida 21, 28, 50, 52, 56–57, 105, 116, 163, 166
  A. adscensionis 204
  A. bipartita 93, 197
  A. congesta 166
  A. dichotoma 28
  A. divaricata 170, 174
  A. hamulosa 170
  A. latifolia 169
  A. longiseta 28, 170
  A. oligantha 28, 119
  A. purpurea 106, 174, 223
Aristidoideae 25–28, 44, 55–57, 58
Arrhenatherum 43
  A. elatius 39–40, 42, 96–97, 106, 111, 117, 175–176, 218, 221
  A. elatius var. *bulbosum* 43
Artemisia 183
  A. filifolia 117
  A. frigida 169

Artemisia (cont.)
  A. ludoviciana 225
  A. pauciflora 167
  A. soongarica 183
  A. stracheyi 175, 183
  A. tridentata 117
  A. wellbyi 175
Arundinella hirta 183
Arundinaria 26
Arundinoideae 25–27, 31, 44, 50, 55–57, 77, 104
Arundo 27, 44
  A. donax 10, 14, 28, 75
  A. madagascariensis 166
Asclepias 225
  A. meadii 221
  A. syriaca 139
  A. tuberosa 139, Plate 11
  A. viridis 205
Aster ericoides see *Symphyotrichum ericoides*
Asteraceae 73, 76
Asthenatherum 56
Astragalus crassicarpus 47
  A. melilotoides 168
Astrebla 28, 216, Plate 8
  A. lappacea 28
Atriplex 59
Austrodanthonia tenuior 88
Austrostipa scraba 88
Avena 21, 49
  A. barbata 9
  A. fatua 9, 12, 52, 87, 100, 119
  A. sativa 36–37, 100
  A. sterilis 104
Axonopus 179
  A. anceps 163
  A. affinis 171
  A. canescens 163
  A. compressus 171
  A. purpusii 159, 165
  A. selloanus 163

Baccharis 174
Balsas teosinte 107
Bambusa 23, 26
  B. forbesii 44
Bambusoideae 25, 26, 28–29, 31, 44, 50, 55, 77, 104
Baptisia leucantha 101
Bassia 59
Beckmannia 43
Berberis heterophylla 168
Blysnus sinocompressus 175
Borassus flabellifer 166
Bothriochloa 105
  B. bladhii 166, 221

  B. barbinodis 170
  B. decipiens 166
  B. insculpta 93
  B. ischaemum 168, 183
  B. laguroides 161, 171
  B. macra 94
  B. saccharoides 177
Bouteloua 105, 179
  B. curtipendula 49, 65, 101–103, 106, 170, 173, 180, 223
  B. dactyloides 14, 28, 37, 49–50, 106, 117, 173–174, 192, 208–209, 225
  B. eriopoda 88, 116, 170, 173, 202
  B. gracilis 14, 42, 45, 48, 65, 103, 106, 117, 170, 173–174, 180, 192, 202, 208–209
  B. hirsuta 106, 170
  B. hirsuta var. *pectinata* 56
  B. pectinata see *B. hirsuta* var. *pectinata*
  B. radicosa 65
  B. rigidiseta 223
  B. rothrockii 170, 174
  B. scorpioides 170
Brachiaria 61
  B. mutica 165
  B. serrata 166
Brachyachne convergens 169
Brachyelytrum 31
Brachypodium distachyon 46
  B. phoenicoides 46
  B. pinnatum 143, 176
Brassica nigra 119
Brassicaceae 73
Brickellia eupatorioides var. *eupatorioides* 47
Briza 27
  B. media 52, 108
  B. subaristata 171
Bromus 12, 26, 44, 49, 75, 82, 153, 174, 179
  B. catharticus 96–97
  B. commutatus 208
  B. diandrus 9
  B. erectus 46, 65
  B. hordeaceus 115
  B. inermis 43, 48, 53, 101, 138, 183, 221, 240
  B. japonicus 43
  B. mollis 9, 119, 199, 221
  B. pictus 120
  B. polyanthus 207
  B. rigidus 119
  B. rubens 9
  B. secalinus 88
  B. sterilis 45, 88

  B. tectorum 43, 48, 87, 89–90, 104, 124, 194, 208, 216
  B. unioloides 38
Brysonima crassifolia 163
Buchloë 63, 84, 105, 179
  B. dactyloides see *Bouteloua dactyloides*
Buchlomimus 84
Bulbostylis 163
Burkea africana 150

Calamagrostis 49, 174–175, 191
  C. epigeios 168, 183
  C. lanceolata 94
  C. porterii ssp. *insperata* Plate 2
Calderonella 28
Calligonum mongolicum 168
Calylophus serrulatus 180
Calyptochloa 83
Caragana microphylla 168
Carduus acanthoides 9
Careya 166
Carex 22, 102, 177, 183, 225
  C. arenaria 95
  C. atrofusca 175
  C. bigelowii 175
  C. duriuscula 169
  C. moorcroftii 175, 183
  C. nivalis 175
  C. pensylvanica 112
  C. stenocarpa 175
  C. tetanica 112
Cassia 163
  C. chamaecrista 240
Casuarina junghuhniana 166
Catalepsis 52
Cenchrus 51
Centaurea 9
  C. diffusa 119
  C. maculosa 124, 140, 221
  C. solstitianis 140, 216, 225
Centotheca 28
  C. lappacea 28
Centothecoideae 25–28
Centrosema pubescens 14
Cerastium 113
Ceratoides compacta 183
  C. lateens 183
Chasmanthium 28
  C. latifolium 28
Chenopodiaceae 59, 73
Chionochloa 175
  C. antarctica 134, Plate 12
  C. rigida 193
Chloridoideae 25–28, 32, 44, 50, 55–57, 58–59, 63, 95, 179
Chloris 28, 61, 166

C. chloridea see Enteropogon
    chlorideus
C. gayana 14, 28, 177
C. ventricosa 88
C. virgata 166
Chromolaena odorata 10, 194
Chrysocoma ciliata 166
Chrysopogon fallax 166
Chrysothamnus 192
Chusquea 75, 174
Cirsium arvense 174
    C. vulgare 9
Cladoraphis spinosa 44
Claytonia lanceolata 207
Cleistochloa 83
Cleistogenes squarrosa 169
Clematis fremontii 180
Clerodendron serratum 166
Coelorachis selloana Plate 10
Colpodium 43
Colleguaya integerrima 169
Condaloa 174
Cortaderia 174
    C. selloana 56
Cottea 83
Conyza canadensis 89, 208
Crepis molis 220
Crinipes 27
Curatella americana 163
Cuscuta 120
Cyathea 174
Cyclostachya 84
Cymbopogon nardus 166
    C. plurinodis 166
Cynodon 63, 179
    C. dactylon 13, 14, 37–38, 79, 87, 94,
        177, 179, 202, Plate 10
    C. plectostachyus 75, 79
Cynosurus 179
    C. cristatus 124, 176, 179
Cyperaceae 22, 59
Cyperus 59
Cyperochloa 27

Dactylis 177–178
    D. glomerata 12, 42, 48, 65, 73,
        86–88, 96–97, 102, 106, 115,
        176–177
Dactyloctenium radulans 168
Dalea multiflora 225
    D. purpurea 101, 225, 240
Danthonia 28, 83, 174
    D. californica 221
    D. sericea 81, 93
    D. spicata 29
    D. unispicata 49
Danthonioideae 25–29, 44

Delphinium nelsonii 207
Dendrocalamus 23, 35, 52
    D. giganteus 75
Deschampsia 27, 43, 78, 83
    D. ceaspitosa 38–39, 88, 102
    D. flexuosa 123, 174
    D. klossii 134
Desmanthus illinoiensis 240
Desmodium 163
Desmostachya bipinnata 134
Dewalquea 30
Dianthus deltoides 121, 207
Diarrhena 26
Dichaeteria 27
Dichanthelium annulatum 134, 166
    D. clandestinum 83
    D. oligosanthes 102, 203, 208
    D. wilcoxianum 208
Dichanthium sericeum 88, 166
Digitaria 174
    D. californica 37, 170
    D. ciliaria 206
    D. decumbens 14
    D. eriantha 93, 166, 177
    D. ischaemum 85
    D. macroblephara 79
    D. sanguinalis 27
Diheteropogon filifolius 166
Dillenia ovata 166
Dinochloa 52
Distichlis 84
    D. scalarum 206
    D. spicata 85
    D. stricta 77
Dregeochloa 27
Dupontia fisheri 48

Ecdeiocolea spp. 30
Ecdeiocoleaceae 29–30
Echinacea angustifolia 180
    E. purpurea 204
Echinochloa 49
    E. crus-galli 27
Echinolaena 165
Echium plantagineum 115
Ehrharta 26, 43
Ehrhartoideae 25–26, 31, 44, 77, 104
Eleusine 52
    E. coracana 28
Elymus 14, 27, 78, 82, 105, 107
    E. canadensis 101–103, 173
    E. chinensis 169
    E. cinereus see Leymus cinereus
    E. lanceolatus ssp. lanceolatus 105, 173
    E. repens 78, 88, 93, 96–97, 140
Elyonurus 165
    E. barbiculmis 65, 170

Elytrigia repens 39, 94
Elytrophorus 27, 44
Enneapogon 83
Enteropogon chlorideus 49, 83
Entolasia leptostachya 88
Eragrostis 21, 28, 43, 45, 49, 52, 59, 61,
    63, 166
    E. benthamii 88
    E. curvula 177
    E. intermedia 170
    E. lehmanniana 204
    E. nutans 134
    E. oxylepis 50
    E. spinosa see Cladoraphis spinosa
    E. tef 28
    E. walteri 28
    E. xerophila 169
Eriachneae 25
Erigeron strigosus 115, 208, 225
Eriochloa sericea 41
Erioneuron pilosum 223
Eriosema 163
Eryngium yuccifolium 242
Escallonia 174
Espeletia 174
Eucalyptus 166, 192
    E. alba 166
Eugenia 192
Eupatorieae 174

Fabaceae 74, 76
Falopia japonica var. japonica 10
Festuca 21, 27, 43–44, 74, 82–84, 104,
    143, 174–175, 178–179
    F. altaica 221
    F. argentina 169
    F. arizonica 98
    F. arundinacea see Schedonorus
        phoenix
    F. contracta 134
    F. filiformis 179
    F. idahoensis 119, 221
    F. nigrescens see F. rubra ssp. fallax
    F. novae-zelandiae 175
    F. ovina 37, 44, 79, 87, 108, 113, 117,
        119, 145, 168, 175, 183
    F. pallescens 169
    F. pratensis see Schedononorus
        pratensis
    F. roemeri 221
    F. rubra 37–38, 85, 95, 103, 108,
        176–179, 200, 204, 218
    F. rubra ssp. fallax 179
    F. scabrella see F. altaica
    F. sulcata see F. valesiaca
    F. tenuifolia see F. filiformis
    F. valesiaca 167–168

Festuca (cont.)
 F. viviparoidea 43, 84
 F. vivipera see *F. viviparoidea*
Filifolium sibiricum 168
Fimbristylis 206
Flagellaria 29
Flagellariaceae 29
Flourensia cernua 116, 170

Galactia 163
Galium verum 123
 G. saxatile 175
Georgeantha spp. 30
Gigantochloa laevis 26
Glycine max (soybean) 171
Glyceria 44
 G. fluitans 85
 G. striata 94
Graminidites 30
 G. gramineoides 31
Grewia 166
Guaduella 26, 31
Gutierrezia sarothrae 170
Gynerium 84

Hakonechloa spp. 27
Hedychium gardnerianum 10
Helianthus annuus 124
 H. maximiliani 240
 H. rigida 112
Helichrysum 166
Helictotrichon pubescens 176
Hemerocallis minor 168
Hesperochloa kingii 85
Hesperostipa comata 48, 85, 103, 173
 H. curtiseta
 H. spartea 49, 103, 112, 173, 208–209
Heteropogon contortus 65, 93, 116, 134, 166, 183, 197, 206
 H. triticeus 166
Hieracium pilosella 117
Hilaria 49
 H. belangeri 37, 116, 170, 174
 H. jamesii 124, 126
 H. mutica 93, 116, 174
Holcus 177–178
 H. lanatus 52, 85, 95–97, 125, 176
 H. mollis 106, 176
Hordeum 23, 43, 49, 52, 76, 78, 82–83, 105
 H. marinum 9
 H. murinum 9
 H. pusillum 9, Plate 2
 H. secalinum, 176
 H. vulgare (barley) 21, 74, 78, 85
Hydrocharitaceae 59
Hydrilla verticillata 59

Hymenachne amplexicaulis 54
Hyparrenia filipendula 165
 H. rufa 165
Hypnum cupressiforme 175

Imperata cylindrica 10, 27, 166, 183, 197
Indigofera 163
Iris bungei 167
Iseilema 104
 I. membranaceum 169
 I. vaginiflorum 169
Isocoma tenuisecta 170

Joinvillea ascendems 30
 J. bryanii 30
Joinvilleaceae 25, 29–30
Jouvea 84
Juncaceae 22
Juncus 177
Juniperus 192–193
 J. deppeana 65

Knautia arvensis 123
Kobresia bellardii see *K. myosuroides*
 K. capillifolia 175
 K. humilis 175
 K. littledalei 175
 K. myosuroides 175
 K. pygmaea 175
 K. tibetica 175
Koeleria spp. 53, 183
 K. cristata see *Koeleria macrantha*
 K. gracilis 167
 K. laersseni 119
 K. macrantha 37, 47–48, 100–103, 106, 119, 169, 173, 208, 225
 K. pryamidata see *Koeleria macrantha*
Kuhnia glutinosa see *Brickellia eupatorioides* var. *eupatorioides*

Lantana camara 10
Labiatae 76
Larrea 174
 L. tridentata 116, 170
Lathyrus quinquervius 168
Leandra 174
Leersia 26, 83
 L. hexandra 26
 L. oryzoides 44, 81
Leontodon hispidus 218
Lepidium densiflorum 208
 L. virginicum 88
Leptaspis 26
Leptochloa filiformis 88
Leptocoryphium lanatum 163, 165
Lespedeza 14, 177
 L. capitata 120

Letagrostis 27
Leucaena leucocephela 10
Leymus arenarius 104
 L. chinensis 140, 168, 183
 L. cinereus 101
 L. triticoides 105
Liatris aspera 101
Lobelia 174
Lolium 27, 49, 74, 178–179
 L. multiflorum 85, 96–97, 176–177
 L. perenne 8–9, 12–13, 27, 41, 45, 48, 52, 72–73, 75, 81–82, 86, 88, 92, 96–98, 101, 108, 115, 121, 123, 145–146, 176–179
Lotus 75
 L. corniculatus 14, 70, 218
Lupinus perennis 120
Lycurus 49

Mabelia connatifila 30
Maytenus 174
Medicago 9, 12, 71
 M. ruthenica 168
 M. sativa 14, 68, 177
Melaleuca 192
Melica 43
 M. uniflora 45
Melinis minutiflora 165
Melocalamus 52
Melocanna 22, 52
Mertensia fusiformis 207
Merxmuellera rangei 28
Mesosetum 165
Micraira 22, 27, 43
Micraireae 25
Micrairoideae 27
Microlaena 26
Microstegium vimineum 37, 39, 49, 87
Mikania micrantha 11
Milium effusum 87
Mimosa 163
 M. microphylla 112, 180
 M. pigra 11
Minuartia verna 113
Miscanthidium teretifolium 44
Miscanthus 14
 M. floridulus 183
 M. sacchariflorus 183
Molinia 27, 43
 M. caerulea 27, 192
Moliniosis 27
Monachather 44
Monanthocloë 84
 M. littoralis 44
Monarda fistulosa 120
Monoporoites 30
Monoporopollenites 30

Muhlenbergia 83
  M. schreberi 88
Mulinum spinosum 120, 168
Myrtaceae 174

Nardus 27
  N. stricta 123, 175
Nassella cernua 9
  N. neesiana 171, Plate 10
  N. pulchra 9, 199
Nastus 49
Neeragrostis 84
Nematopoa 27
Neostapfia colusana 44
Neurachne 59, 61, 63
  N. minor 63
Neurolepis nobilis 44
Notodanthonia caespitosa 81

Ochlandra 22, 26, 50, 52
Oenothera 209
  O. macrocarpa 180
  O. serrulata see *Calylophus serrulatus*
  O. speciosa 225
Oligoneuron rigidum 180
Olyra 82
Opizia 84
Opuntia 174
  O. fulgida 116
  O. spinosior 116
Orcuttia 44–45
Ornis thoroldii 175
Oryzoideae see Ehrhartoideae
Oryza 26, 43
  O. sativa (rice) 21, 26, 53, 75–76, 109
Oryzopsis asperifolia 45
  O. hymenoides 88
Oxytopis microphylla 169

Panicoideae 25–27, 32, 50, 55, 58–59, 61, 76, 95, 104–105, 179
Panicum 21, 27, 43, 45, 49, 52, 59, 61, 63, 105
  P. agrostoides see *P. rigidulum*
  P. coloratum 14
  P. decipiens see *Steinchisma decipiens*
  P. decompositum 88
  P. effusum 88
  P. maximum 14, 27, 165
  P. miliaceum 27
  P. milioides see *Steinchisma hians*
  P. rigidulum 95
  P. simile 88
  P. spathellosum see *Steinchisma spathellosa*
  P. stenodes 159
  P. virgatum 14, 79–80, 87, 101–103, 105–106, 116–117, 173, 177, 180–181, 225, 236
Pappophorum 28, 83
Paractaenum 61
Paraneurachne 61
Pariana 26, 82
Pascopyrum smithii 48, 65, 89, 101, 105, 173, 192, 208, 225
Paspalidium distans 88
Paspalum 21, 52, 105, 165, 174, 179
  P. almum 165
  P. carinatum 163
  P. dilatatum 13, 41, 82, 171, 177
  P. floridanum 112
  P. hexastachyum 106
  P. notatum 13, 27, 171, 177
  P. pectinatum 165
  P. plicatulum 112, 165
  P. quadrifarium 106, 171
  P. urvillei 177
Pastinaca sativa 242
Pennisetum 21, 60, 83, 179
  P. clandestinum 14, 177
  P. glaucum (pearl millet) 27, 82, 109
  P. mezianum 206
  P. polystachyon 194
  P. purpureum 14, 27, 75, 177
Pentaschistis arabicus 28
  P. barbatus 28
  P. borussica 28
Pentzia globosa 166
Petalostemum multiflorum see *Dalea multiflora*
  P. purpureum see *Dalea purpurea*
Phalaris 49, 74, 78
  P. aquatica 74, 115
  P. arundinacea 14, 74, 94, 96–97, 106
  P. tuberosa 74, 135, 189
Pharoideae 25–26, 28, 30, 44, 59
Pharus 26, 31
Phaseolus 163
Phleum 43, 49, 177–179
  P. alpinum 123
  P. bertolonii 45, 123
  P. pratense 48, 65, 73, 82, 96–97, 106, 176, 179
Phragmites 27
  P. australis 27, 75
  P. communis 183
  P. mauritianum 75
Phyllanthus emblica 166
Phyllostachys nigra 42
  P. nidularia 82
Physalis pumila 203
Pinus 30
  P. caribaea 163
  P. cembroides 65
  P. merkusii 166
  P. pinaster 11
Piptochaetium 105
Piptophyllum 27
Plagiosetum 61
Plantago 176
  P. lanceolata 120, 218
Plectrachne 155, 216
Poa 21, 27, 32, 43, 49, 56, 75, 81–84, 86, 104, 174–175, 177–179, 241
  P. alpigena 84
  P. alpina 82, 84, 113, 175, 183
  P. annua 35, 82, 87, 123, 177
  P. artica 84
  P. bulbosa 81, 84
  P. cita 175
  P. compresa 241
  P. cookii 12, 134
  P. foliosa 134
  P. hiemata 116
  P. interior 207
  P. labillardieri 88
  P. lanuginose 169
  P. ligularis 169
  P. nemoralis 82, 179
  P. palustris 82
  P. pratensis 13, 22, 27, 48, 65, 81, 84–85, 87–88, 96–97, 102, 138, 140, 173, 176, 178–179, 183, 208, 216, 241–242
  P. rhizomata 37
  P. sinaicai 84
  P. sterilis 82
  P. trivialis 39, 85, 176–177, 179
Polystichum commune 207
Polygonum macrophyllum 175
  P. viviparum 175, 183
Pooideae 25–27, 31, 44, 50, 55, 76, 104, 179
Populus 192
Potamophila 26
Potentilla acaulis 169
  P. erecta 175
Poterium sanguisorba 218
Prepinus 30
Primula veris 120
Pringlea antiscorbutica 12
Pringleochloa 84
Prosopis 192
  P. glandulosa 116, 146
  P. juliflora var. *glandulosa* see *P. glandulosa*
  P. juliflora var. *velutina* see *P. velutina*
  P. velutina 116
Prunella grandiflora 107
  P. vulgaris 107

PLANT INDEX

Prunus virginiana 73
Psammochloa villosa 79
Pseudochaetochloa 84
Pseudoroegneria 14
    P. spicata 48, 105, 119, 124, 192, 221
Pseudoscleropodium purum 175
Pseudostachyum 52
Psidium cattleianum 11
    P. guayava 12
Psoralea 209
    P. floribunda see *Psoralidium tenuiflorum*
    P. tenuiflora see *Psoralidium tenuiflorum*
Psoralidium tenuiflorum 47, 180, 225
Pteridium aquilinum 73
Puccinellia 179
Puelia 26, 31
Puelioideae 25–26, 28, 30, 59
Pueraria montana var. *lobata* 11
    P. phaseoloides 14
Purshia 192
Puya 174

Quercus 192–193
    Q. arizonica 65
    Q. emoryi 65
    Q. hypoleucoides 65
    Q. marilandica 181
    Q. stellata 181

Rapanea 174
Ranunculus 176
Ratibida columnifera 112
    R. pinnata 242
Reaumuria songorica 168
Reederochloa 84
Restionaceae 29
Rhinan
    R. alectorolophus 120
    R. minor 120
Rhodiola algida 183
Rhus 193
Rhytidiadelphus squarrosus 175
Rosa 209
    R. bracteata 216
Roupala 174
Rudbeckia hirta 204
Ruellia 112
    R. humilis 203
Rumex acetosella 176
Rynchospora 163

Saccharum 27, 61, 76
    S. officinarum (sugar cane) 21, 58
    S. spontenum 113
Salsola passerine 168, 183
    S. tragus 174

Sanguisorba minor 120
Sapium sebiferum 194
Sartidia 28
Sasa 43, 45
Schedonnardus 105
    S. paniculatus 52
Schedonorus 179
    S. phoenix 13–14, 27, 41, 45, 74, 96–98, 101, 106, 123, 138, 151, 176–179, 204, 240
    S. pratensis 9, 45, 73, 176
Schinus polygamus 169
    S. terebinthifolius 11
Schismus 28
Schizachyrium 105, 174
    S. fragile 166
    S. scoparium 37, 41, 47–49, 65, 79, 87, 102–103, 112, 116–117, 120, 140, 151, 171, 173, 180–181, 185, 207–208, 225, 241–242
Schizostachyum 52
Scirpus 59
    S. yagara 183
Scleropogon 84
    S. brevifolius 116
Scrotochloa 52
Secale cereale 75, 78
Sehima nervosum 134, 166
Senecio 166, 174
    S. filaginoides 168
Seriphidium rhodanthum 183
    S. terrae-albae 183
Sesbania bispinosa 134
Sesleria 27
Setaria 43
    S. faberi 94
    S. flabellate 166
    S. glauca 85, 87
    S. incrassate 93
    S. verticillata 85, 88
    S. viridis 85, 87
Sida 169
    S. spinosa 206
Sieglingia 83
    S. decumbens 45
Silphium integrifolium 205
Sisyrinchium campestre 208
Sitanion 49, 107
    S. hystrix 85
Soderstromia 84
Sohnsia 84
Solanaceae 73
Solidago 242
    S. rigida see *Oligoneuron rigidum*
Sorghastrum 105
    S. nutans 14, 44, 49, 87, 95, 101, 103, 112, 116–117, 173, 177, 180–181, 185, 225, 242, Plate 11

Sorghum 21, 43, 46, 52, 60–61, 75, 166
    S. bicolor (sorghum) 27, 75–77, 109, 171
    S. halepense 13, 43, 109, 119
    S. intrans 91, 189, 194
    S. nitidum 166
Spartina 61, 83, 113
    S. alterniflora 107
    S. anglica 88, 107
    S. maritima 107, 113
    S. pectinata 106, 173
    S. x townsendii 107
Sphenopholis 105
Spinifex 52, 84
Sporobolus 52, 63, 83
    S. asper see *S. compositus*
    S. capensis 166
    S. compositus 49, 112, 180, 225
    S. cryptandrus 83, 85, 106, 117, 173
    S. cubensis 163
    S. flexuosus 116
    S. heterolepis 46, 112, 173, 209
    S. indicus 12, 42, 163
    S. pyramidalis 166
    S. rangi 43
    S. vaginiflorus 83
Steinchisma decipiens 59
    S. hians 59
    S. spathellosa 59
Stenotaphrum 179
Stipa 27, 50, 52, 83, 105, 161, 167–168, 174
    S. baicalensis 168
    S. breviflora 168
    S. bungeana 168, Plate 6
    S. capillata 167, 169
    S. caucasia ssp. *glareosa* 167–168
    S. charruana Plate 10
    S. comata see *Hesperostipa comata*
    S. gobica 167–168, Plate 6
    S. glareosa see *S. caucasia* ssp. *glareosa*
    S. grandis 140, 168
    S. humilis 169
    S. klemenzii 167–168
    S. krylovii 167–168, Plate 6
    S. lessingiana 167
    S. lettermani see *Achnatherum lettermani*
    S. neesiana see *Nassella neesiana*
    S. orientalis, Plate 6
    S. pulchra see *Nassella pulchra*
    S. pulcherrima 167
    S. purpurea 168–169
    S. purpureum 175, 183
    S. scrabra see *Austrostipa scraba*
    S. spartea see *Hesperostipa spartea*
    S. speciosa see *Achnatherum speciosum*

S. subsessiliflora 168, 175
S. zalesskii 167
Stipagrostis 28
　S. ciliata 28
　S. obtuse 28
　S. plumose 28
　S. uniplumis 28
Stoebe vulgaris 166
Streptochaeta spp. 26, 31
Stroptogyneae 25
Striga 120
Stylosanthes 12, 163
　S. guianensis 14, 177
　S. hamata 14
Styppeiochloa 27
Suaeda 59
Symphoricarpus 193
Symphyotrichum ericoides 102, 112, 180, 203, 208, 225
Symplocus 174
Swallenia alexandrae 93

Taeniatherum asperum see *T. caput-medusae*
　T. caput-medusae 194, 199
Talinum calycinum 180
Taraxacum officinale agg. 123
Tephrosia 163
Terminalia arostrata 169
　T. volucris 169
Tetradymia 192
Themeda triandra 14, 88, 93, 105, 111, 116, 165–166, 168, 183, 189–190, 193, 197, 202, 206
Thylascospermum caespitosum 168
Thymus drucei 117
　T. praecox 113
　T. pulegioides 207
Thysanolaena 28
Tibouchina 174

Tomanthera 120
Trachypogon 152, 165
　T. ligularis 165
　T. plumosus 163
　T. vestivus 163, 165
Tradescantia tharpii 180–181
Tragopogon dubius 225
　T. lamottei 89
Trevoa patagonica 169
Trifolium 12
　T. pretense 14, 70, 73, 142, 177
　T. repens 8–9, 12–14, 48, 70, 72, 81, 108, 121, 135, 142, 146, 176–178, 204, 225
　T. subterraneum 73, 115, 122–123, 142
Triglochin palustre 183
Trigonella ruthenica see *Medicago ruthenica*
Triodia 63, 155, 170, 216, Plate 9
　T. basedowii 170
Triplasis 83
　T. purpurea 83
Tripsacum 49, 107
Triraphis 63
Trisetum 53
　T. flavescens 96–97, 176
　T. spicatum 106
Tristachya 52, 165
　T. leucothrix 116
Tristania 192
Triticum 21, 49, 52, 76
　T. aestivum (wheat) 23, 74–76, 78, 107, 109, 119, 171
　T. dicoccum 23
　T. monococcum 23
　T. speltoides 78
　T. turgidum 107
　T. zhukovski 107

Ulmus 192
Uniola paniculata 113
Urochloa panicoides 166

Vallisneria spiralis 59
Vernonia 174
　V. baldwinii 200, 225
Veronica officinalis 207
Vicia 75
　V. sativa 14, 123
Viola 208
　V. nuttallii 207
　V. rafinesquii 115
Vitix negundo 168
Vochysia 158
　V. divergens 165
Vulpia bromoides 87, 115
　V. ciliata 100, 104
　V. fasciculata 91–92
　V. myuros 9
　V. octoflora 208

Xanthostemon 192

Zea 43, 61
　Z. diploperennis 107
　Z. perennis 107
　Z. luxuricans 107
　Z. mays (maize) 21, 27, 44, 49–50, 58, 60, 75, 77–78, 82, 107, 109, 171
Zenkeria 27
Zerna erecta 176
Zizania 26
　Z. aquatica 26
　Z. palustris 88
Ziziphus rugosa 166
　Z. sponosa 168
Zornia 163
Zoysia 61, 179
　Z. japonica 85
Zygochloa 84

Animal index

Forage animals such as cattle and sheep and general herbivores such as aphids are not indexed here as their effects on grassland are included more specifically in the subject index and under terms such as 'forage', 'grazing' and 'herbivory'.

Alcelaphus buselaphus (Bubal hartebeest) 195
Anabrus simplex (mormon cricket) 206
Antilocapra americana (pronghorn) 200
Antilope cervicapra (black buck) 196
Antistrophus silphii (cynipid gall wasp) 205
Apis cerana (Asiatic or Eastern Honey bee) 82
Aquila chrysaetos (golden eagle) 194
Axis porcinus (hog deer) 196

Bastocerus dichotomus (marsh deer) 196
Bathyergidae (molerats) 201–202
Bees 82
Bos bison (North American bison) 129, 173, 186, 195, 199–201, 214
Bubalus bubalis (wild water buffalo) 196

Caterpillars 75, 129, 206
Camelus bactrianus (Bactrian camel) 196
  C. dromedarius (dromedary camel) 196
Capra hircus (goat) 196
  C. ibex (ibex) 196
Catagonus wagneri (peccary) 196
Centrocercus urophasianus (sage grouse) 194
Cephalophus (duiker) 195
Cervus canadensis (American elk, wapiti) 196, 200
  C. duvauceli (swamp deer) 196
  C. elaphus (red deer) 196
Connochaetes taurinus (wildebeest) 165, 195

Cotesia marginiventris (parasitic wasp) 75
Crambus (sod webworms) 206
Ctenomyidae (tuco-tucos, curruros) 201–202
Cynomys (prairie dog) 195, 201–203
  C. ludovidianus (black-tailed prairie dog) 201

Damaliscus korrigum (topi) 165, 195
Dicerus bicornis (black rhinoceros) 195
Dicotyles tajacu (peccary) 196
Dipodomys spectabilis (banner-tailed kangaroo rat) 201–202

Equus burchelli (Burchell's zebra) 165, 195
  E. caballus (feral and domestic horse) 195–196, 200
  E. hemonius (wild ass) 196
  E. przewalksi (Mongolian wild horse) 196

Falco mexicanus (prairie falcon) 194
Formicidae (ants) 206–207

Gazella gazelle (chinkara) 196
  G. granti (Grant's gazelle) 195
  G. subgutturosa (goitered gazelle) 196
  G. thompsonii (Thompson's gazelle) 165
Geomyidae (pocket gophers) 201–202
Ground squirrels 173

Hemileuca olivia (range caterpillar) 206
Hemitragus jemlahicus (Himalayan thar) 196

Hippocamelus (guemal deer) 196
Hippopotamus amphibus (hippopotamus) 195, 197
Hypsiprymnodon moschatus (rat-kangaroo) 196

Irbisia (mirid grass bug) 206

Kobus (waterbuck) 195
  K. kobus (kob) 197

Labops (mirid grass bug) 206
Lagorchestes hirsutus (western hare-wallaby) 216
Lagomorpha (hares, rabbits, pikas) 201
Lama (camelid) 196
  L. glama (llama) 196
  L. guanicoe (guanaco) 169
Lasiorhinus krefftii (northern hairy-nosed wombat) 201
  L. latifrons (southern hairy-nosed wombat) 201
Leptopterna (mirid grass bug) 206
Lepus californicus (black-tailed jack rabbit) 194
Loxidonta africana (African elephant) 195
Lygaeus kalmii (small milkweed bug) 205
Odocoileus virginianus (white-tailed deer) 196, 200
Ozotocerus bezoarticus (pampas deer) 196

Macropus rufa (red kangaroo) 196
Macrotermes (termite) 112, 150, 204, 206
Mazama (brocket deer) 196
Melanargia galathea (marbled white butterfly) 78

Microtus longicaudus (longtailed vole) 203
   M. montanus (mountain vole) 78
   M. ochrogaster (prairie vole) 203
   M. pennsylvanicus (meadow vole) 204
Muridae (murid rodents) 201
Mustelidae (badgers) 201

Nasutitermes triodiae (termite) 206
Nemorhaedus goral (goral) 196
Nomadacris septemfasciata (red locust) 206

Ovis ammon (argali, mountain sheep) 196

Phyllophaga (white grubs) 206
Pocket gophers 173, 186, 201–202
Pontoscolex corethrurus (earthworm) 151

Potamochhoerus porcus (bush pig) 195
Procapra gutturosa (Mongolian gazelle) 196
   P. picticaudata (Tibetan gazelle) 196
Pseudois nayaur (bharal, Himalayan blue sheep) 196
Pudu (pudu deer) 196

Rhinocerus unicornis (Indian rhinoceros) 196
Rodentia (rodents including mice, voles, shrews, gophers, prairie dogs) 201
Rupicapra rupicapra (chamois) 196

Schistocerca gregaria (desert locust) 206
Schizaphis graminum (green bug aphid) 77
Spalacidae (molerats) 201–202
Spermophilus tridecemlineatus (ground squirrel) 202

Spodoptera (armyworms) 206
   S. exigua (beet armyworm) 75
Sylvicarpa grimmia (duiker) 195
Syncerus caffer (African buffalo) 165, 195

Tachyoryctes splendens (myrid molerat) 202
Tapirus (Tapir) 196
Taurotragus oryx (eland) 165, 195
Tayassu peccary (peccary) 196
Taxidea taxus (American badger) 90, 173, 186, 201–202
Termites (Isoptera) 112, 206
Tragelaphus scroptus (bushbuck) 195
Trinervitermes (termite) 206
Tumulitermes (termite) 206

Vicugna vicugna (camelid) 196
Vombatus ursinus (common wombat) 201

Subject index

Note: many words and terms are mentioned numerous times in the text, such as 'grassland', 'meadow', 'pasture', 'forage'. The index records mainly their first mention or definition. Names of fungi and bacteria are included here; plant and animal names are in separate indices.

Above-ground net primary production (ANPP) 131–137
Acaulospora 99
　A. longula 100
Acremonium see *Neotyphodium*
Adaptive management 169, 234–235
Agamospermy 83–84
Agricultural intensification 218
Agri-environment schemes (AES) 240–241
Agrostology 21
Alkaloids 74–75
Allelopathy 119
　Allelochemicals 73–78
Allopolyploidy 105
Alpha-diversity 185, 192
Alpine steppe 168–169, 175
Amenity grasslands 178–179
Amines 74–75
Amino acid uptake from soil 145
Anatomy 54–57
　Culms 54
　Leaves 54–57
　Roots 54
Anemophily (wind pollination) 32, 82
Animal unit (AU) 213
Animal unit month (AUM) 213
Annual grassland 85, 140–141, 145, 148, 199, 202, 227
Ants, Anthills 95, 207
Anthesis 82
Anthocyanins 76–78
Aphids and secondary plant compounds 74, 77, 98
Apomixis 83–84
Arbuscular mycorrhizal fungi–see mycorrhizae
Arthropod herbivory 205
Ascochyta (leaf blight) 97
　A. leptospora (leaf-spot fungus) 95

Azospirillum (a nitrogen fixing bacteria) 94, 142
Azotobacter (a nitrogen fixing bacteria) 142
Australian Collaborative Rangeland Information System (ACRIS) 233
Autopolyploidy 105
Auxin 80
Awn 36–37, 50–51

Beijerinckia (a nitrogen fixing bacteria) 142
Balansia (an endophytic fungus) 95
　B. henningsiana 95
Balansiopsis (an endophytic fungus) 95
Barley Yellow Dwarf Virus 204
Below-ground net primary productivity (BNPP) 132–137
Bews, John William 18–20
BEP clade 25–27, 29, 58
Bibury verges, UK 117, 123
Biodiversity in grasslands 15–16
Biofuels 13–15
Biogeographic origin of grasses 30
Bison, North American 129, 173, 186, 195, 199–201, 214
Blumeria graminis (powdery mildew) 97
Banching patterns
　Monopodial 37
　Sympodial 37
Blackland prairies 171
Breeding systems 82–85
　Apomixis 83
　Chasmogamy 83
　Cleistogamy 83
　Vivipary 83–84
Budbank 86, 192
Bulbs 43

C_3 grasses
　Compared with C_4 grasses–see entries under C_4 grasses
C_4 grasses 58–68
　Biochemistry of photosynthesis 60–63
　C_3/C_4 intermediates 63
　C_4 subtypes 28, 61–62, 65
　Differences compared to C_3 grasses 59
　Digestibility of C_4 grasses 68
　Distribution 65, 112–113
　Ecological ramifications 64–65
　Evolution 32–33, 58–60
　Mycorrhizal dependency 100–101
　Production compared with C_3 grasses 137
　Response to global climate change 65–68
　Use in amenity grasslands 178
　Use in semi-natural grasslands 177
　World without C_4 grasses scenario 188
Calothrix (a nitrogen fixing cyanobacteria) 142
Campos 158, 163–165, 171, 191, Plate 10
Campos de altitude 174
Carbon
　Carbon cycle 131–137
　Storage and sequestration in grasslands 14, 16–17, 95, 131, 239
Caryopsis–see Seed
Cattle 13, 74, 129, 158, 163, 165, 196, 201, 213–214, 216, 218, 220
　Grazing systems 213–215
　In pastoral systems 212
Cedar Creek, Minnesota, USA
　tallgrass-oak savannah 124, 127, 139–141

Cercosporium graminis (brown stripe) 96
Chasmogamy 83
Chinese Vegetation–habitat Classification System 182–183
Chromosome numbers in grasses 26–28, 104–107
Cladosporium phlei (Timothy eyespot) 97
Claviceps purpurea (ergot) 97
Cleistogamy 49, 83
Climate change 16, 95, 65–68, 115, 137
　Altered precipitation patterns 209–210
Climate diagrams Plate 3
Climax community concept 161, 222–223
Clonal growth 38–39, 79–80
Clostridium (a nitrogen fixing bacteria) 142
Commerical pastoralists 212
Communal rangelands 212
Community ecology 110–128
　Models of community structure 121–128
　Scaling issues 128
　Species interactions 115–121
　Succession 113–115
　Vegetation–environment relationships 110–113
Community-unit hypothesis 121
Competition 115, 118–119, 207, 237
　Affected by mowing or grazing 220
　Asymmetric 118
　Competitive effect and response 118
　Grime's CSR model 122–123
　Intraspecific 118
　Interspecific 118
　Symmetric 118
　Tilman's R* model 123–126
Conservation Reserve Program 239–240
Continuum hypothesis 121–122
Core-satellite species (CSS) hypothesis 126
Corms 43
Cool-season grasses–*see* C_3 grasses
Culms 36–37
　Cespitose habit 37
　Monopodial branching 37
　Sympodial branching 37
Cultivars 72, 105, 109
　In amenity grasslands 178
　In seeded pastures 177

Curtis Prairie 241–242, Plate 14
Cyanogenic glycosides 74–75
Cytokinin 80

Decomposition 149–153
　Decomposer organisms 149–150
　Decomposition rate 150–152
　Litter 152–153
Decreasers 223
Denitrification 146
Desert grassland 7, 169–170, 192
Desert steppe 168–169
Desertification 6, 169, 212
Digestibility of forage types 69
Dioecy 84–85
Dissolved organic nitrogen (DON) 144
Disturbance 90, 184–210
　Biotic versus abiotic 184
　Disturbance heterogeneity model (DHM) 185–186
　Drought 186, 207–210
　Endogenous versus exogenous 184
　Fire 115, 187–194
　Grazing 79, 111, 114, 185
　Herbivory 186, 194–207
　Intermediate disturbance hypothesis (IDH) 186–187
　Invasive exotic species 185
Disturbance regime 184
Diversity 137–140
　Biodiversity-ecosystem function hypothesis 137–138
　Declines with atmospheric nitrogen deposition 143–144
　Dynamic equilibrium model 137–138
　Effect of aridity on 168
　Mechanisms explaining effects on function 139
　Relationship to fire regime 193
　Shifting limitations hypothesis 137–138
　In restorations 236, 240
Domesday Book 8, 176
Drechslera dactyloides (leaf spot) 96
　D. siccans (brown blight) 96
Drought 139–140, 186, 207–210
　And pastoralism 212
　Influence on grassland communities 207
　North American drought of the 1930s 89, 207–209
Dustbowl 207–209

Earthworms 149–151
Ecoregions 6–7

Ecosystem engineers 202
Ecosystems 129–159
　Decomposition 149–152
　Energy and productivity 129–140
　Goods and services 12–13
　Grassland soils 153–159
　Nutrient cycling 141–149
　Function and restoration 237, 240
Ecotourism 17
Ecotypes 72, 81, 94, 102–104, 175,
　Use in restorations 237–238
El-Niño–Southern Oscillation 208
Endophytes 74, 95–98, 204
Energy flow 129–131
Entomophily (insect pollination) 82
Entrophospora 99
　E. infrequens 100
Environmental Stewardship scheme 240–241
Epichloë 95
　E. glyceriae (systemic fungus) 94
　E. typhina (choke) 97
Erosion 208
Erysphye graminis (powdery mildew) 94
Essential elements–*see* nutrient elements
EUNIS: the European Nature Information System 181–182
Evolution of grasses and grasslands 21–34, 94
　C_4 photosynthetic pathway 58–60
　Coevolution with grazers 31–34, 135, 198–200
　Hybridization 107–109
　Polyploidy 104–105
　Relationship to fire regimes 186, 188
　Spikelet evolution 51, 53, 82, 84
Exotic (non-native) plants 9–12, 115, 119, 140, 151, 165, 174, 227
　Relationship to fire regime 153, 194, 216
　Relationship to herbivory and grazing 171, 199
　Control by mowing 221

FACE (Free-air CO_2 enrichment) experiments 66
Facilitation 120–121, 237
Fertilizer 143, 146–147
　Effects on Park Grass Experiment 114
　Use in management 218–219
Fire 33, 93, 115, 133–135, 149, 152–153
　As management tool 215–218
　Effects on grasslands 191–194

Effects on invasive exotics 194, 216
Effects on production 135–136
Effects on soil 144–145, 149, 190–192
Litter affecting fire probability 153
Occurrence in grasslands 187–189
Slash-and-burn agriculture 5
Temperature, intensity and spread 189–191
Use by native peoples 188
Flavones 76–78
Flavanoides 76–78
Flooding, flooded grasslands 7, 159, 165
Fluctuations 115
Forage 1, 3
　Analysis 68
　Plants 13–15
　Quality 68–73
Forage and Grazing Terminology Committee 1, 3
Fossil history of grasses 30–31
Functional groups
　Annuals versus perennials in the soil seed bank 86
　CSR plant life histories 122–123
　Increaser and decreasers 224
　Population growth rates of annuals and perennials 93–94
　Raunkiaer life forms 123
　Regenerative strategies 90–91
Fungal relationships 94–102
　Diseases of grasses 94–96
　Endophytes 74, 95–98
　Mycorrhizae 99–102
Fragmentation 4–12

Genecology 102–109
　Ecotypes 72, 81, 102–104
　Genetic structure of grasslands 108–109
　Hybridization in grasses 107
　Metapopulations 104, 125–126
　Polyploidy 104–107
Genetic diversity 108, 237
Gigaspora 99
　G. gigantea 100
Global circulation models (GCMs) 66–68
Glomeromycota 99
Goats 196–197, 235, 211–212, 215, 235
Goose herbivory 200
Glomus 99
　G. aggregatum 100
　G. constrictum 100
　G. etunicatum 100
　G. fasciculatum 100

G. geosporum 100
G. mosseae 100
Glumes 36–37, 50–51
Grass-fire cycle 194
Grasshoppers 129, 136, 204–206
Grassland
　Biodiversity 15–16
　Carbon storage 15–16
　Definitions 1–2
　Disturbance 184–210
　Extent 2–4, Plate 1
　Fragmentation and loss 4–12, 174
　Goods and services 12–13
　Management and restoration 211–242
　Tourism 17–18
Grass Phylogeny Working Group (GPWG) 24
Grassland Reserve Program 239–240
Grazing 32, 79, 87, 111, 114, 147–*see* Herbivory
　As management tool 213–215
　Bison 145, 186, 200–201, 214
　Cattle 74, 163, 165, 171, 173, 176, 196, 212–214, 216, 218, 220
　Deterrence because of secondary compounds 77–78
　Effects on grassland 196–200
　Effects on production 135–137
　Effects on spatial patterning 111, 186, 200
　Goats 196–197, 235, 211–212, 215, 235
　Horses 34, 74, 195, 213
　Rabbits 95, 170
　Sheep 41, 111, 150, 169–170, 172, 175–176, 196–197, 213–216, 218, 220
Grazing lawns 186, 197, 200
Grazing systems 214
Grime's CSR model 122–123

Haying as management tool 87, 168, 176, 219–221
Heavy metal tolerant grasses 103
Helminthosporium 96
Hemiparasites 120
Herbage 3
Herbicide resistance 104, 221
Herbicide use as a management tool 221
Herbivores and herbivory 13, 74, 78, 186, 194–207, 237
　Coevolution with grasses 31–34
　Effects on ANPP and BNPP 135–136, 194, 199–200, 204
　Effects on diversity 198–199

Herbivore optimization hypothesis 199
　Interaction with invasive species 199
　Invertebrate herbivores 129, 204–207
　Large vertebrates 195–201
　Small vertebrates 129, 201–204
Hierarchical relationships 110, 128, 185, 210
Horses 34, 74, 195, 213
Hybridization in grasses 107, 175
　Ecological selection-gradient model 107
　Hybrid zones 107
　Tension zone model 107
Hydroxamic acids (DIMBOA, DIBOA) 74, 78
Hypsodonts and grass evolution 33–34

Increasers 222
Indigenous people and grasslands 5, 188, 211–213, 236
　Evolution of *Homo sapiens* in African savannah 5
Inflorescence 49
　Panicle 49, Plate 2
　Raceme 32, 49, Plate 2
　Spike 49, Plate 2
Inset herbivory–*see* entries under Invertebrate herbivory
Insect outbreaks 206
Integrated community concept 115
Interactions–*see* Species Interactions
Interpreting Indicators of Rangeland Health (IIRH) 228
Invertebrate herbivores 204–207
　Ants 206–207
　Aphids 74, 77, 204
　Bees 82
　Caterpillars 75, 129, 206
　Earthworms 149–151
　Effects on grasslands 33, 204–206
　Grasshoppers 129, 136, 204–206
　Nematodes 149, 204
　Termites 112, 150, 204
Invasibility 140
Invasive species 9–12, 119, 185
　Management 216, 220–221

Keystone species 13, 173, 200, 202
Köppen climate classification system 162–163
　Climate diagrams Plate 3
　Dry climates (desert, semi-desert & steppe, type B) 164, 166–170, Plates 3, 5, 6, 7, 8, & 9

Köppen climate classification system (*cont.*)
  Highland climates (montane grasslands, type H) 164, 174–175, Plates 3 & 12
  Moist continental mid-latitude (temperate, humid microthermal) climates (temperate grasslands, type D) 164, 171–172, Plate 3
  Moist subtropical mid-latitude (humid mesothermal, humid subtropical) climates (Pampas & Campos, type C) 164, 170–171, Plates 3, 10 & 11
  Tropical moist climates (savannahs, type A) 163–166, Plates 3 & 4
Konza Prairie, Kansas 100–102, 106, 115, 127, 133, 135, 140, 190, 193, 221, Plate 11, Plate 13
K-selection 122
Kranz anatomy 31, 55–57

Lambda (λ) 91–94
Landscape Function Analysis (LFA) 228, 233–234
Leaves 43–46
  Leaf sheath 44
  Ligule 36, 44–45
  Passive hypothesis, active hypothesis 45
  Phyllotaxy 43
Legumes 70
Lemma 36–37, 50–51
Leys and seeded pastures 176–178
Liebig's Law of the Minimum 114
Light-use efficiency (LUE) 59, 64
Lightning 188
Ligule 44–45
Litter 152–153
Llanos savannah grassland 163–165
Lodicules 36–37, 50–51
Long-term grassland studies 116–117

Macronutrients 70
Managed, semi-natural grasslands 175–179
  Amenity grasslands 178–179
  Leys and seeded pastures 176–178
  Secondary grasslands in the UK 176
Management 211–242
  Fertilizers 218 -219
  Fire 215–218
  Grazing 213–215
  Herbicides 221
  Mowing and haying 219–221
  Range assessment 221–235
  Restoration 235–242
  Techniques and goals 211–221
Mastigosporium album (leaf fleck) 96
  M. rubricosum (leaf fleck) 96
Matrix (transition, projection) models 91–94
Meadow 1, 3
  Meadow grasses 176
  Historical uses 219
  Conservation value 220
Meristematic growth 36, 43
Metacommunities 125–127
Metapopulations 104, 125–126
  Core-satellite species (CSS) hypothesis 126
  Brown's niche-based resource-use model 126
Microdochium (snow mold) 97
Micronutrients 70
Midewin National tallgrass prairie 236–237
Mima-mounds 202
Mitchell (*Astrebla* spp.) grasslands 169–170, 216, Plate 8
Mixed prairie 85, 130, 133–134, 136, 150–151, 172–173, 209
Models
  BOTSWA model 150
  CENTURY model 137, 150
  ELM model 137, 150
  Global Circulation Models 66–68
  Grime's CSR model 122–123
  Hurley Pasture Model 137
  Matrix (transition, projection) models 91–94
  Milchunas–Sala–Lauenroth (MSL) model 198
  Neutral models 127–128
  PHOENIX model 137
  Tilman's R* model 123–125
Montane grassland 7, 174–175, Plate 12
Mowing as management tool 219–221
Mycorrhizae (vesicular arbuscular) 70, 80, 99–102, 207
Myriogenospora (an endophytic fungus) 95
Mutualisms 121, 237
  Endophytes 74, 95–98
  Mycorrhizae 99–102
  Nitrogen fixing bacteria 121, 142–143

Nachusa Grasslands, Illinois, USA 236
Neotyphodium (an endophytic fungus) 95, 97
  N. ceonophialum 204
Neutral models 127–128
  Stochastic niche theory 127
Niche 126–127, 237
Nitrogen 41, 64, 70–73, 133–134, 138, 152
  Biological nitrogen fixation 121, 142–143, 191
  Denitrification 146
  Fertilizer amounts and effects 143, 218
  Leaky pipe / hole in the pipe model 146
  Mineralization 137, 144
  Nitrogen cycle 141–146
  Nitrogen deposition 143–144
  Nitrogen pools 141–142
  Secondary compounds 74–75
  Soil concentrations 71–73
  Tissue levels 71–73
  Uptake as amino acids 145
  Volatilization and losses 145–146
Nitrogen use efficiency 59, 63
Nomadic pastoralism 211
North American Great Plains 112–113, 170–174
Nostoc (a nitrogen fixing cyanobacteria) 142
Nutrient cycling 141–149
  Nitrogen cycle 141–146
  Phosphorus cycle 146–149
Nutrient elements 70–73
  Critical concentration 72
  Plant:soil concentration ratios
  Tissue nutrient levels

Outcrossing advantage 84
Overgrazing 169–170, 174, 184, 208, 213
Ovule 36–37, 50–51

PACCAD clade 25–29
Palea 36–37, 50–51
Pampas 147, 155, 157–158, 160–161, 170–171, 185, 198
Páramos 174, 191, Plate 3
Parasitism 119–120
  Hemiparasites 120
  Holoparasites 119
Park Grass Experiment, UK 94, 114–116, 124–125, 146, 218
Pastoralism 169, 211–213, 235
  Commercial pastoralists 212
  Tragedy of the commons 212–213
Pasture, pastureland 1, 3
Patch dynamics 186
Permanent pasture 3
Phenolics (i.e., anthocyanins, flavones, flavanoides) 76–78

SUBJECT INDEX

Phenology 81–82, 192, 209
Phosphorus 138
  Cycle 146–149
  Fertilizer 146–147, 218
  Inputs and cycling 146–148
  Labile inorganic phosphorus 147
  Mycorrhizal relationships 100–102
  Plant uptake 70, 148
  Returns and losses 148–149
  Soil concentrations 71–73
  Tissue levels 71–73
Photosynthesis
  C_3 pathway 58–60
  C_4 pathway 58–68
  C_4 subtypes 28, 61–62, 65
  C_3/C_4 intermediates 63
  CO_2 starvation hypothesis 60
  Distribution and evolution of pathways 31, 58–60, 112–113
  Ecological ramifications 64–65
  Global change effects on 65–68
  Light effects on 63–64
  Photorespiration 60
  Temperature effects on 63–64
Phytoliths 33, 78
Phytometer 35–36, 42
Pilot Analysis of Global Ecosystems (PAGE) Classification 2
Pistil 36–37, 50–51
Poaceae
  Artificial versus phylogenetic classifications 22–24
  Comparison with rushes and sedges 22
  Classification of subfamilies 22–24
  Earliest fossil 31
  Family characteristics 21–22
  Fossil history 29–31
  Subfamily characteristics 24–29
Pocket gophers 173, 186, 195, 201–202
Pollen
  Dispersal 82
  Fossil 30–31
Pollination and pollinators 12, 32, 82–84, 118, 121
Polyploidy 104–107
Population Ecology 81–109
  Fungal relationships 94–102
  Genecology 102–109
  Population dynamics 91–94
  Reproduction and population dynamics 81–94
Potassium 70, 138, 192, 206
  Fertilizer 218–219
  Soil concentrations 71–73
  Tissue levels 71–73
Prairie 3, *see* mixed prairie, short-grass prairie, tallgrass prairie

Prairie dogs 185–186, 202–203
Prairie restoration–*see* Restoration
Prescribed burning 216–218
Production, productivity
  Above-ground net primary production (ANPP) 131–137
  Below-ground net primary production (BNPP) 132–137
  Cool-season (C_3) versus warm-season (C_4) grasses 132, 137
  Diversity relationships 137–140
  Primary producers, primary productivity 129, 131
  Secondary producers 129
  Temporal variability 133, 135
Protein content 68–69
Puccinia coronata f. sp. *holci* (crown rust) 95–96
  P. graminis (black stem rust) 94, 96
  P. striliformis (yellow rust) 96
  P. recondite (brown rust) 96

r-selection 84, 122
Rachilla 36–37, 50–51
Ramet 35
  Longevity 38
Range assessment 221–235
  Range condition concept 222–225
  Rangeland health 227–235
  State-and-transition models 223–227
Rangeland 3
  Health 227
  Trend 223
Raunkiaer's life forms 192
Recruitment curves 91–92
Red: Far-red light ratio 80, 119
Regional grassland classifications 179–183
  Chinese Vegetation–habitat Classification System 182–183
  EUNIS: the European Nature Information System 181–182
  US National Vegetation Classification System 179–181
Reproduction
  Monocarpic plants 43
  Semelparous plants 43
  Vegetative prolifery 43
  Vegetative reproduction 36–43
Resource ratio hypothesis (Tilman's R* model) 115, 123–125
Restoration 235–242
  Agri-environment schemes (AES) 240–241
  Approaches 236–238
  Conservation Reserve Program 239–240

Curtis Prairie 241–242
Defining restoration 235–236
Environmental Stewardship scheme 240–241
Grassland Reserve Program 239–240
Grass roots volunteer groups 238–239
Issues and decisions 238–241
Relationship to ecological concepts 237
Society for Ecological Restoration International 239
Rhizobium (nitrogen fixing bacteria) 121, 142–143
Rhizomes 37, 42
Rhynchosporium secalis (leaf blotch) 96
  R. orthosporum (leaf blotch) 96
Roots 46–49
  Adventitious roots 46
  Distribution and turnover 47–49
  Root hairs 47, 71
  Seminal roots 46, 53
Rubisco 59, 61–64, 66, 141
Rusmularia holci-lanati (leaf spot) 97

Savannahs 3, 7, 32, 93–94, 111, 106, 133, 137, 149, 150, 152, 154–156, 163–166, 191, 193, 195, 197, 202, 206, 216, Plate 3, Plate 4
Satellite imagery 2, 228
  SPOT satellite image Plate 13
Scaling issues 128
  Relationship to effects of disturbance 185
  Species-pool hypothesis 128
Sclerocystis 99
  S. rubiformis 100
  S. sinuosa 100
Scutellospora 99
  S. pellucida 100
Secondary compounds 73–78
  Hydroxamic acids (DIMBOA, DIBOA) 74, 78
  Nitrogen compounds (e.g., alkaloids, amines, cyanogenic glycosides) 74–75
  Phenolics (i.e., anthocyanins, flavones, flavanoides) 76–78
  Terpenoids 75
Secondary grasslands 175
  In the UK 176
Secondary production 136
Seedlings 51-54
  Establishment 88–91
  Growth stages 53

Seed
  Caryopsis 51–53
  Disarticulation 52
  Dormany 85–88
  Germination 87–88
  Seedbanks 85–87, 195
Self-incompatibility mechanism in grasses 82
Semi-natural grasslands 8, 175–178
Serengeti grassland 18, 79, 111–112, 128, 163, 165, 195, 197–198, 206, Plate 4
Sheep 41, 74, 111, 135, 150, 169–170, 172, 175–176, 196–197, 213–215
Shortgrass prairie steppe 130, 149, 173–174, 192, 198, Plate 5
Silage 154, 219
Silicon 44, 78–79, 204
Society for Ecological Restoration International 239
Society for Range Management 222
Sod 3
Soils of grasslands
  Acrisols 154, 156–157, 159
  Azonal soils 153
  Antrosols 154–155
  Arenosols 154–155, 159
  Chernozems 149, 154, 157–158
  Ferralsols 154, 156–157
  Intrazonal soils 154
  Kastanozems 154, 157–158
  Lixisols 154, 156–157
  Nitisols 154, 156–157
  Phaeozems 154, 157–158
  Planosols 154, 158–159
  Regosols 154–156
  Solonetz 154, 157, 159
  Typical grassland soil 153
  Vertisols 154–155, 159
  Zonal soils 153
Soil organic matter (SOM) 129, 144, 151–152
Soil water, moisture 133, 135–136
Southeast Asian grasslands 166
Spatial heterogeneity 153, 197, 200
Species interactions
  Allelopathy 119
  Competition 115, 118–119
  Facilitation 120–121
  Mutualism 121
  Parasitism 119–120
Species-pool hypothesis 128
Spermospora lolli (leaf spot) 97
Spikelet 36–37, 49–51
  Primitive and advanced characters 51
Spinifex grassland 3, 155, 170, 216, Plate 9

Sporisorium amphiliphis (floral-smut fungus) 94
Stamens 36–37, 50–51
Standing crop 131
State-and-transition models 199, 223–227
  Ball-and-cup analogy 226
  Thresholds 226
  Transitions 226–227
Steppe 3, 137, 157–158, 167–169, 175, 182–183, 192, 198, Plates 3, 5, 6, & 7
Stochastic niche theory 127
Stocking density 213
Stolon 37
Styles 36–37
Subsidence pastoralism 211
Subsystems
  Decomposer 129
  Plant 129
  Herbivore 129
Succession 95, 113–115, 237
  Primary 113–114
  Secondary 114–115
  And the Range Condition Concept 222–223
Sustainable Rangeland Roundtable (SRR) 227–228
Sward 3

Tallgrass prairie 6–7, 66, 79, 95, 101- 102, 106, 108, 115–117, 127, 130–131, 133, 135–137, 140–144, 150–151, 154, 171–173, 180, 185–187, 190–193, 195, 200, 202–204, 216
Termites, termite mounds 112, 150, 152, 165, 204
Terpenoids 75, 192
Tillers/tillering 35
  Tiller appearance rate (TAR) 39–40
  Effects of environment on 41
  Survivorship 42
Tolypothrix (a nitrogen fixing cyanobacteria) 142
Tragedy of the Commons 212–213
Transhumant pastoralism 211
Transient maxima hypothesis 135–136, 153
Tree invasion of grassland 192–194
True prairie 65, 172–173, 209, and see tallgrass prairie
Turfgrass 178–179
Typhula incarnate (snow mold) 97

US National Vegetation Classification System 179–181
Urocystis (smut fungi) 97

Ustilago 97
  U. bullata (smut) 89
  U. cynodontis (smut) 94

Veld 3, 157, 163, 165–166, 213, 215–216
Vegetation classification
  Alliances 161
  Associations 161
  Existing natural versus potential natural vegetation 160–161
  Formations 161
  Natural versus cultural vegetation 160
  Pilot Analysis of Global Ecosystems (PAGE) Classification 2
Vernalization 81
Vertebrate herbivores 195–204
  American badger 186, 202
  Banner-tail kangaroo rat 202
  Effects on grassland 196–200
  Fossorial rodents (pocket gophers) 201–202
  Large vertebrates 195–201
  Milchunas–Sala–Lauenroth (MSL) model 198
  North American bison 129, 173, 186, 199–201, 214
  Pocket gophers 173, 186, 201–202
  Prairie dogs 185–186, 202–203
  Prairie voles 203
  Small vertebrates 201–204
  State-and-transition model 199
  Voles 203–204
Vivipary 83–84

Wahlund effect 108
Warm-season grasses *see* C_4 photosynthesis
Water use efficiency (WUE) 59, 63
Weaver, John Earnest 19–20
Western Australian Rangeland Monitoring System (WARMS) 233–234
World grasslands 160–183
  Crosswalk comparing grassland classifications 164
  Dry climates (desert, semi-desert & steppe) 164, 166–170, Plates 3, 5, 6, 7, 8, & 9
  Highland climates (montane grasslands) 164, 174–175, Plates 3 & 12
  Köppen climate classification system 162–163
  Managed, semi-natural grasslands 175–179

Moist continental mid-latitude (temperate, humid microthermal) climates 164, 171–172, Plate 3
Moist subtropical mid-latitude (humid mesothermal) climates (Pampas & Campos) 164, 170–171, Plates 3, 10 & 11

North American Great Plains 172–174
Regional grassland classifications 179–183
Tropical moist climates (savannahs) 163–166, Plates 3 & 4
Ways of describing vegetation 160–162

World Reference Base for Soil Resources 153–154

Xanthomonas oryzae (bacterial leaf blight of rice) 77

Zonobiomes 161

Printed in the USA/Agawam, MA
August 25, 2021